International Conference of Emerging Challenges in Engineering and Current Technology (ICECTIII 2021)

IOP Conference Series: Materials Science and Engineering
Volume 1176

Kuala Dungun, Malaysia
30-31 March 2021

Editors:

Nurbaiti Wahid
Fatimah Nur Mohd Redzwan
Norzila Mohd
Norhanifah Abdul Rahman
Nurul Huda Kamarulzaman
Nurul Hidayah Md. Noh
Sarah Yusoff

ISBN: 978-1-7138-4636-9
ISSN: 1757-8981

Printed from e-media with permission by:

Curran Associates, Inc.
57 Morehouse Lane
Red Hook, NY 12571

Some format issues inherent in the e-media version may also appear in this print version.

Institute of Physics
Dirac House, Temple Back
Bristol BS1 6BE UK

Phone: 44 1 17 929 7481
Fax: 44 1 17 920 0979

techtracking@iop.org

Additional copies of this publication are available from:

Curran Associates, Inc.
57 Morehouse Lane
Red Hook, NY 12571 USA
Phone: 845-758-0400
Fax: 845-758-2633
Email: curran@proceedings.com
Web: www.proceedings.com

TABLE OF CONTENTS

IOP Conf. Series: Materials Science and Engineering 1176 (2021) 011001 doi:10.1088/1757-899X/1176/1/011001

International Conference of Emerging Challenges in Engineering and Current Technology (ICECT III)
30 – 31 March 2021

PREFACE

ICECT III 2021 is organised by Universiti Teknologi MARA, Terengganu Branch to provide an opportunity to research scholars, delegates, academicians, and students to engage in dialogues on a wide range of topics in engineering challenges and current technology. Formerly launched in 2016 as the International Symposium of Engineering Technologies (ISET), the conference name was then changed to the International Conference on Engineering and Computing Technology (ICECT II) in 2018. Now the conference is rebranded as the International Conference of Emerging Challenges in Engineering and Current Technology (ICECT III) to ensure it is relevant to the latest emerging challenges in engineering and current technology. The main aims of ICECT III 2021 are to promote the emerging development of the current technology in the engineering field to encourage engineering applications on the current state-of-the-art in the various engineering fields. The conference provides a mechanism to communicate findings, solutions or innovations among scholars and scientists as well as motivate scholars, scientists, and engineers in related fields to provide to its body of knowledge. The conference focused on original research work in a multiple range of areas, including Chemistry and Chemical Engineering, Electrical and Electronics Engineering, Materials and Mechanical Engineering, Advanced Computing Technology, Applied Sciences and Mathematics. Also, it serves as a conduit for channelling advanced engineering research and applications to the regional industry. All submissions are subjected to rigorous review before an acceptance decision is made. The papers published in these proceedings will contribute to a better understanding of the related problem, the current solution and the future innovation as well as will serve as a reference for researchers and engineers in the field of electrical, mechanical, chemical, advanced computing technology and applied mathematics. The online live streaming presentation is used as a method of presentation within the two-day conference.

List of Conference Tracks:

- Future Innovation and Solution in Chemistry and Chemical Engineering
- Future Innovation and Solution in Electrical and Electronics Engineering
- Future Innovation and Solution in Material, Structure and Mechanical Engineering
- Future Innovation and Solution in Computer Science and Information Technology
- Future Innovation and Solution in Mathematics and Statistics
- Other Recent Trends in Science and Technology

Organizing Committee of Conference:

General Chair
Sharifah Fauziah Syed Draman (Assoc. Prof)

Conference Chair
Wan Dorishah Wan Abdul Manan (Assoc. Prof)

Secretary
Siti Aishah Che Kar

Editors of ICECT III 2021:

Chief Editor
Nurbaiti Wahid (Dr.)

Editors
Fatimah Nur Mohd Redzwan
Norzila Mohd
Norhanifah Abdul Rahman
Nurul Huda Kamarulzaman (Dr.)
Nurul Hidayah Md. Noh
Sarah Yusoff

IOP Conf. Series: Materials Science and Engineering 1176 (2021) 011002 doi:10.1088/1757-899X/1176/1/011002

Peer review declaration

All conference organisers/editors are required to declare details about their peer review. Therefore, please provide the following information:

- **Type of peer review: Single-blind**

  Single-blind peer review - reviewers' identities are kept hidden from authors.

- **Conference submission management system:** OpenConf Peer Review and Conference Management System

- **Number of submissions received:** 78 manuscripts.

- **Number of submissions sent for review:** 76 manuscripts.

- **Number of submissions accepted:** 51 manuscripts.

- **Acceptance Rate (Number of Submissions Accepted / Number of Submissions Received X 100):** 65%

- **Average number of reviews per paper:** 3 reviewers.

- **Total number of reviewers involved:** 108 reviewers.

- **Any additional info on review process:**

  Manuscript score with similarity check less than 30% after screening with Turnitin plagiarism software will be sent for a single-blind review process. The manuscripts were evaluated based on the technical criteria, quality criteria, and presentation criteria. The main aspect when assessing the manuscripts is quality and clarity of the content, originality, contribution to the related field, organization and readability as well as the quality of English. The reviewers were asked to fill in the peer review form and review questions from OpenConf Peer Review and Conference Management System. There are six reviewer's decisions on the review process including (i) must accept, (ii) clear accept, (iii) marginal tend to accept, (iv) marginal tend to reject, (v) probable reject, and (vi) reject with specifying their confidence (low, moderate, high) regarding qualifications to review the allocated paper. After editors received the review manuscripts, the authors were asked to revise and resubmit their manuscript according to the reviewer's comments.

- **Contact person for queries:**
  Dr. Nurbaiti Wahid
  School of Electrical Engineering, College of Engineering,
  Universiti Teknologi MARA, Terengganu, MALAYSIA
  nurbaiti@uitm.edu.my

IOP Conf. Series: Materials Science and Engineering 1176 (2021) 012001 doi:10.1088/1757-899X/1176/1/012001

Coal Bottom Ash as a Potential Adsorbent for CO_2 Capture

N H A Rani[1*], N F Mohamad[1], M Onn[2], M J Jalil[3] and N Muda[4]

[1]Faculty of Chemical Engineering, Universiti Teknologi MARA, Pasir Gudang Campus, Jalan Purnama, Bandar Seri Alam, 81750 Masai, Johor, Malaysia

[2]Faculty of Applied Science, Universiti Teknologi MARA, Pasir Gudang Campus, Jalan Purnama, Bandar Seri Alam, 81750 Masai, Johor, Malaysia

[3]Faculty of Chemical Engineering, Universiti Teknologi MARA, Penang Campus, 13500 Pematang Pauh, Pulau Pinang, Malaysia

[4]Low Carbon Power Generation Research Group, TNB Research Sdn Bhd, No.1 Lorong Ayer Itam, Kawasan Institusi Penyelidikan,43000, Kajang, Selangor, Malaysia

*Corresponding author: hidayurani@uitm.edu.my

Abstract. Coal bottom ash (CBA) is the solid residue produced from coal-fired power plant. It is classified as hazardous material, which can cause adverse effect to the environment and public health. In view of the situation, exploring and discovering possible utilization of these coal residues into value-added products has to be performed. Hence, the aim of this paper is to investigate the potential of raw CBA as an adsorbent for carbon dioxide (CO_2) gas adsorption. The physicochemical characteristics of CBA were determined using ultimate and proximate analysis, nitrogen adsorption-desorption isotherm, scanning electron microscopy (SEM), and elemental diffraction x-ray (EDX) analysis. CO_2 adsorption study was conducted using simulated flue gas by pressure swing adsorption process in a fixed-bed reactor to evaluate the suitability of CBA as an adsorbent material. High carbon content in CBA makes it's become a good precursor for production of carbonaceous adsorbent. Trace elements like calcium, silica and alumina, indicating that the material had a good adsorption capacity. The result indicated that CBA was able to adsorb CO_2 with the adsorption capacity of 0.000276 mol CO_2/g CBA. As a conclusion, CBA has the potential to be an adsorbent for its application in gas phase especially in CO_2 adsorption.

1. Introduction

Presently and in more years to come, the power generation in Malaysia will continue to rely heavily on gas and coal as the source of fuel. In 2016, gas usage has declined due to the to the retirement of several gas plants. The consumption of gas is also expected to further decrease in 2019 to 32% as coal usage increases and make up 64 % of the overall mix and overtake gas as dominant fuel [1]. This have resulted in the increase of carbon dioxide (CO_2) concentration in the atmosphere that causes a significant environmental problem such as global warming [2] Therefore, the national and international regulation for power plant regarding CO_2 emission became more stringent. Carbon Capture and Storage (CCS) is considered as one of the most suitable method to reduce the GHG emission to an acceptable range in order to avoid any permanent and irreversible damage of climate change [3]. Based on the report from the International Energy Agency (IEA), approximately 19% of energy related CO_2 emissions can be mitigated by CCS method by 2050 [4]. One of the most established technologies CO_2 capture is through sorbent adsorption. However, the commercial adsorbent is quite expensive. Therefore, the alternative ways to reduce the cost is utilizing a solid waste residue.

Coal fired-power plant not only as a major contributor to CO_2 emission problem but also further increased by coal ash (fly ash and bottom ash) generated in these plants as a solid waste residue which creates not only a solid waste disposal problem but also a dusting problem in the atmosphere and in the surrounding land area [5]. In 2010, 29 million of coal ash are produced where 15-20 % is coal bottom ash (CBA) and 75-80% is coal fly ash (CFA) [6]. Due to this, huge amount of coal ash is expected to form and thus, leads to storage and disposal issues as it requires large space. It is also feared that with the presence of heavy metals, coal ash will cause ground and surface water contaminations that can pose a threat to the environment [7]. Therefore, recycling and reuse of the coal ash can help mitigate the issues that may be encountered by the industry and at the same time improve the quality of the environment.

At present, the reutilization of coal bottom ash (CBA) as adsorbent materials have gained favorable interest in diverse research field due to its nature as low cost waste material and abundant availability which can easily be obtained from coal combustion processes. Moreover, the chemical composition of CBA which was reported to consist a high content of silica and alumina, contributed to an excellent adsorbent property [8]. However, CBA has a limited application and previous study found only 21% of the waste was recycled in the industries [9]. For example, a study done by Ishaq et.al., (2014) used coal ash adsorbent to remove xylenol orange from aqueous solution and revealed that the coal ash removed about 80% of xylenol orange from the aqueous solution within 40 min [10]. This is due to the availability of many active sites on the coal ash surface to react with dye molecules and give high adsorption capacity. Meanwhile, CBA that was activated at high temperature was able to adsorb acid red 114 until 91.2% at 80 min contact time [11]. However, most of the previous work on the utilization of CBA focuses on aqueous phase adsorption for instance dyes removal in textile industry wastewater and limited study was found on the usage of CBA as natural adsorbent in the gas phase.

The present study aims to explore the potential of CBA as adsorbent in a gas phase application for the adsorption of CO_2. The percentage of reduction of CO_2 in a pressure swing adsorption was determined for gas phase adsorption study. The properties of the CBA were analyzed using ultimate and proximate analysis, nitrogen adsorption-desorption isotherm based on Brunauer-Emmett-Teller (BET), scanning electron microscopy (SEM) and elemental diffraction x-ray (EDX) analysis for a better understanding of this industrial waste, prior to reuse. The results of this research are presented herein.

2. Materials and Methods

2.1. Sample Preparation

Coal bottom ash (CBA) was obtained from TNB Janamanjung Kapar (sub-bituminous coal). The CBA was dried in the oven at 110°C for 24 h to remove moisture. The dried CBA was sieved using sieve shaker machine (Thermo Fisher Scientific) to get 1-2 mm particle size.

2.2 Characterization

CBA was characterized by different techniques to determine structural properties of porous carbon. Proximate analysis was performed using thermogravimetric analyzer, TGA/DSC (Mettler Toledo, United States) according to the ASTM D7582-10 and the results are shown in terms of moisture, volatile content, fixed carbon and ash contents. Elemental analysis was carried out using CHNS-O Analyzer (FlashEA,1112 Series) to determine carbon (C), hydrogen (H), nitrogen (N) and oxygen (O) (by difference) of the sample. The pore architecture particularly specific surface area, pore diameter, total pore volume and pore size distribution were determined by nitrogen adsorption-desorption isotherm operating at 77K using gas sorption analyzer (AUTOSORB-iQ/MP-XR). Meanwhile, the microstructure, organic and inorganic compounds and weight percentage of sample was characterized by the scanning electron microscopy (SEM) and elemental diffraction x-ray (EDX) analyses.

2.3 Adsorption Study

For these experiments, 16 % CO_2 and 84 % N_2 gas mixture were used and was carried out in a pressure swing adsorption column (figure 1). In each experiment, 290 g of CBA was used in the adsorption column. The gas flow rate was 2 L/min under 4 bar pressure was passed until the saturation was reached, i.e., until the point at which the outlet CO_2 concentration equals that of the inlet stream, in order to assess the maximum dynamic adsorption capacity of adsorbent. The effluent gas from the outlet of the column was monitored and recorded at an online portable gas analyzer (MRU Model OPTIMA 7, Germany).

Figure 1. Vacuum pressure swing adsorption (VPSA) column

3. Results and Discussion

3.1. Characterization of Coal Bottom Ash

3.1.1. Proximate and Ultimate Analysis. The ultimate and proximate analyses were summarized in table 1. For ultimate analysis, it showed that CBA consists of carbon content is 18.10%, whereas 10.5% represents its fixed carbon. The element of carbon in CBA can be increase by physical separation. However, steam activation is not suitable for enrichment in carbon because the ash content in the sample is too high (82.7%) and it will lead to more ash content since the degree of burn-off is achieved. According to Rubio et al., (2007), chemical demineralization process is capable in reducing the ash content and simultaneously increases the carbon content [12]. With high carbon content in CBA, this makes the material a good precursor for production of carbonaceous adsorbent.

Table 1. Proximate and ultimate analysis for CBA

Sample	Proximate analysis (wt%)				Bulk density (kg/m^3)	Ultimate analysis (wt%)			
	Moisture	Volatile	Fixed carbon	Ash		C	H	N	O*
CBA	0.3	6.5	10.50	82.7	1020	18.10	1.36	0.29	80.25

3.1.2. BET Surface Area. The CBA was characterized by N_2 adsorption/desorption, BET surface and pore size distribution analysis. Table 2 shows the comparison of CBA in present study and previous researches in terms of surface area, pore volume and pore size. In this study, the surface area of CBA is 20.80 m^2/g and the pore size is 35.69Å, which falls under mesoporous group. Mesoporous particle mainly contributes around 2-50 nm (20-500Å). Meanwhile, for pore volume, CBA consists of 0.019 cm^3/g. A study conducted by Herman et al., (2016) obtained the surface area is 58.01 m^2/g and pore volume is 0.040 cm^3/g. This is due to the particle size, which plays an important role to enhance the CBA's surface area. From the point of view of adsorption process, low surface area of granular CBA is unfavorable, as it indicates lower adsorbate uptake during the adsorption process [13]. Thus, to enhance the absorptivity of the CBA material, crushing, grinding and sieving steps are crucial in order to obtain smaller particle size as well as improving surface area [14]. As a conclusion, with high pore volume and high surface area, it is expected that more molecules will be penetrated and retained in the pores, which can contribute to an increase in adsorption capacity. Therefore, a large surface area is favorable for an efficient CO_2 adsorption [15,16].

Table 2. Comparison of BET analysis

Sample	BET surface area (m^2/g)	Size particle (mm)	Pore volume (cm^3/g)	Pore size (Å)	Reference
CBA (Manjung)	20.80	1-2	0.019	35.69	Present study
CBA (Perak)	58.01	0.25	0.04	30.10	Herman et al., 2016 [8]
CBA (Korea)	1.41-8.47	1.19-4.75	-	-	Gorme et al., 2010 [9]

3.1.3. SEM and EDX Analysis. Figure 2 represent the SEM images at 1500 times magnification of CBA morphology and surface texture. It shows that the CBA consists of mixture of coarse and fine grains particles with no particular pattern. With this an irregular structure of the CBA's surface, indicates the presence of cavities and pores on the exterior side of the CBA [17]. Besides, Bumanis et al, (2013) stated that the bottom ash that has rough, gritty and angular structure is owed to the metallic compounds that deposit onto the surface [18]. After adsorption process, the pores probability filled with CO_2. Meanwhile, table 3 shows the EDX analysis of CBA, which presented the wt% of different elements in the CBA sample. The major constituent of the CBA are silica, SiO_2 (18.29 wt%) followed by alumina, Al_2O_3 (8.63 wt%) and iron, Fe_2O_3 (5.81 wt%). Other constituents included calcium, CaO (9.95 wt%), magnesium, MgO (1.28 wt%), potassium, K_2O (0.93 wt%) and sodium, Na_2O (2.29 wt%). Asokbunyarat et al, (2015) reported that high alumina and silica compositions in bottom ash make the material a very good adsorbent because both of elements are the best support matrix in order to capture CO_2 [8]. When the molecule of silica and CO_2 comes closer, it can promote the initial chemical interaction between of them and the reaction will be occurred only on the channel surfaces, forming unidentate, bidentate, and bridged carbonates [19]. Meanwhile, a high content of CaO is accountable for the alkaline characteristics and this help to increase CO_2 adsorption capacity via carbonation reaction which is denoted by Equation (1) [20, 21].

$$CaO\ (s) + CO_2\ (g) \longleftrightarrow CaCO_3 \qquad (1)$$

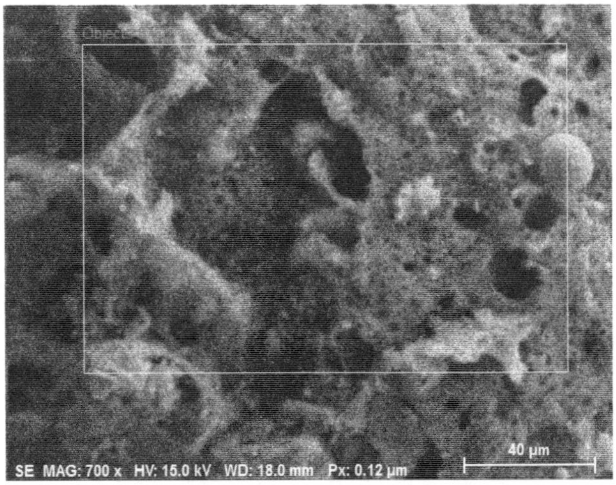

Figure 2. SEM of the CBA sample

Table 3. Composition of element in CBA from EDX analysis

Element	Wt%
Si	18.29
Al	8.63
Fe	5.81
Ca	9.95
Mg	1.28
K	0.93
Na	2.29

3.2. CO_2 Adsoprtion Study

Figure 3 exhibit the adsorption curve for raw coal bottom ash. C/Co represents the concentration of CO_2 adsorb by the adsorbent, and the Co represent the initial concentration of CO_2. It can be seen that the adsorption of CO_2 on raw bottom ash is inadequate. Initially, the CO_2 concentration in the exit gas is around 5% and within 100 seconds it reached to the inlet gas concentration 16% (Co). At 100 seconds the maximum value of C/Co is 80%. The adsorption capacity of the sample until saturation point was obtained at 0.000176 mol CO_2/g or 7.76 mg CO_2/g CBA which was much lower than the typically 0.001 mol/g for Norit activated carbon as reported by Hosseini et al., (2015) [22]. Although the adsorption capacity is low, CBA is able to adsorb CO_2 probably due to its physicochemical properties such as BET surface area. CO_2 molecules diffused into pores of CBA and were adsorbed onto the surface of pores of CBA by weak forces of intermolecular cohesion (physisorption) rather than a chemical interaction [23]. In addition, calcium oxide (CaO) found in CBA sample may also assist in the adsorption of CO_2. However, treatment of the ash samples with amines, acids or steam can significantly improve the adsorption capacity of coal ash because amine impregnation was found to help increase the selectivity of CBA towards non-polar molecules. For example, a study conducted by researchers in [24], where the fly ash was activated with steam and impregnated with amine gave adsorption capacity of 68.6 mg CO_2/g and Yahia et al., (2015) prepared activated carbon from fly ash using chemical activation achieve an adsorption capacity of 26 mg CO_2/g [24,25]. For those reasons, it is expected that by converting CBA to adsorbent i.e., activated carbon, the adsorption capacity of CO_2 will significantly increase.

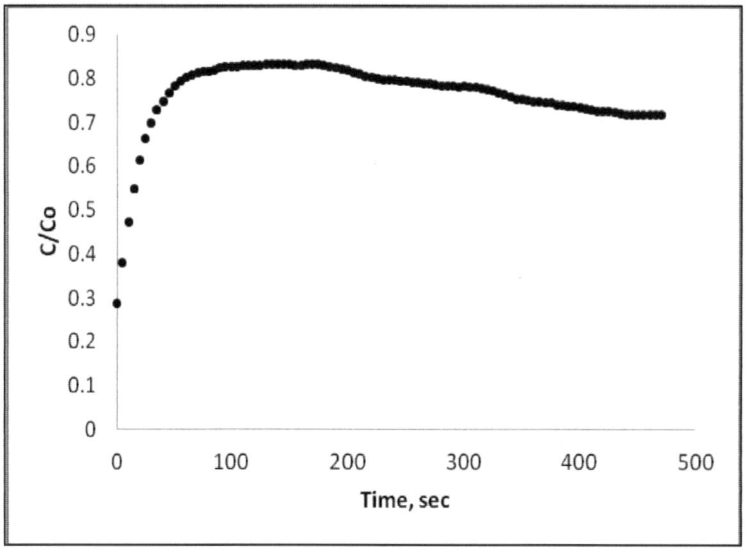

Figure 3. CO_2 adsorption breakthrough curve

4. Conclusion

Coal bottom ash is a waste product from the coal power plant and based on the characterization studies, it is evident that CBA has good potential to be used as adsorbent for CO_2 capture. With high carbon content confirms that CBA can be an effective adsorbent since its can make the material a good precursor for production of carbonaceous adsorbent. In addition, these elements such as Si, Ca and Al confirm its potential to be used as helper to enhance CO_2 adsorption capacity. Through adsorption testing, CBA is able to adsorb CO_2 with the adsorption capacity of 0.000176 mol CO_2/g CBA or 7.76 mg/g CBA. Even though the adsorption capacity is relatively low, the transformation of CBA into value-added products such as zeolites, activated carbon and mesoporous silica will increase the adsorption capacity. Therefore, the physiochemical properties of the CBA have been well-explained in order to have better understanding of these products prior to their utilization. As a conclusion, CBA has a potential to be an adsorbent for its application for CO_2 capture.

Acknowledgement

The authors gratefully acknowledgment the financial support by Universiti Teknologi MARA Johor under grant Bestari 600-UiTMCJ (PJIA.5/2). In addition, we thank the Management of TNB Research Sdn Bhd for providing the coal bottom ash samples for this research work.

References

[1] Peninsular Malaysia Electricity Supply Industry Outlook 2013 Suruhanjaya Tenaga (Energy Commission) http://www.st.gov.my
[2] Shariff A M, Murshid G, Keong L K, Bustam M A, Ahmad F 2014 Solubility of carbon dioxide in aqueous solutions of piperazine (PZ) at elevated pressure *Adv. Mat. Res.* **917** 144-150.
[3] Parshetti G K, Shamik C, Balasubramanian R 2015 Biomass derived low-cost microporous adsorbents for efficient CO_2 capture *Fuel* **148** 246-254.
[4] Energy and Climate Change 2015 World Energy Outlook Special Report International Energy Agency (EIA).
[5] Rashidi N A, Yusup S 2016 Overview on the Potential of Coal-Based Bottom Ash as Low-Cost Adsorbents *ACS Sustain. Chem. Eng.* **4** 1870-1884.

[6] Ramzi N I R, Shahidan S, Maarof M Z, Ali N 2016 Physical and Chemical Properties of Coal Bottom Ash (CBA) from Tanjung Bin Power Plant *IOP Conf Ser. Mater. Sci. Eng.* **160** 1-10.

[7] Abubakar A U, Baharudin K S 2012 Potential use of Malaysian thermal power plants coal bottom ash in construction. *Int. J. Sustain. Constr. Eng.* **3** 25-37.

[8] Asokbunyarat V, Van Hullebusch E D, Lens P N L, Annachhatre A P 2015 Coal bottom ash as sorbing material for Fe (II), Cu (II), Mn (II), and Zn (II) removal from aqueous solutions. *Water, Air, Soil Pollut.* **226** 1-17.

[9] Mohamed M, Yusup S, Bustam M A, Azmi N 2017 Effect of coal bottom ash and binder addition into CaO-based sorbent on CO_2 capture performance *Chem. Eng. Trans.* **56** 325-330.

[10] Mohammad I, Khalid S, Imtiaz A, Sirraj S 2014 Coal Ash as a Low Cost Adsorbent for the Removal of Xylenol Orange from Aqueous Solution *Iran. J. Chem. Chem. Eng.* **33** 53-58.

[11] Kuntaria, Aprilitab NH, Suherman 2017 Utilization of Coal Bottom Ash a Low-Cost Adsorbent for the Removal Acid Red 114 Dye *Eksakta: Jurnal Imu-Ilmu MIPA* **17** 11-19.

[12] Rubio B N, Izquierdo M T, Mayoral M C, Bona M T, Andres J M 2007 Unburnt carbon from coal fly ashes as a precursor of activated carbon for nitric oxide removal *J. Hazard. Mater.* **143** 561-566.

[13] Ding X Li Y, Guo L, Liao Q, Zhu X and Wang H 2019 Non-aqueous energy-efficient absorbents for CO_2 capture based on porous silica nanospheres impregnated with amine *Energy* **171** 109–119.

[14] Thaarrini J, Venkatasubramani R 2015 Feasibility studies on compressive strength of ground coal ash geopolymer mortar *Period. Polytech. Civ. Eng.* **59** 373–379.

[15] Herman A P, Yusup S, Shahbaz M, Patrick D O 2016 Bottom ash characterization and its catalytic potential in biomass gasification *Procedia Engineering* **148** 432-436.

[16] Choi W, Park J, Kim C and Choi M 2020 Structural effects of amine polymers on stability and energy efficiency of adsorbents in post-combustion CO2 capture *Chem. Eng. J.* **408** 127289.

[17] Gorme J B, Maniquiz M C, Kim S S, Son Y G, Kim Y T, Kim L H 2010 Characterization of bottom ash as an adsorbent of lead from aqueous solutions. *Env. Eng. Res.* **15** 207-213.

[18] Bumanis G, Bajare, D, Korjakins A 2013 The economic and environmental benefits from incorporation of coal bottom ash in concrete. *4th Int. Scient. Conf. Civil Eng.* 142-152.

[19] Yong X, Tse J S, Chen J 2018 Mechanism of Chemical Reactions between SiO_2 and CO_2 under mantle conditions *ACS Earth Space Chem.* **2(6)** 548-555.

[20] Zhang L, Zhang B, Yang Z, Guo M 2015 The role of water on the performance of calcium oxide-based sorbents for CO_2 capture: a review. *Energy Tech.* **3** 10-19.

[21] Khan Z, Yusup S, Ahmad M M, Rashidi N A 2014 Integrated catalytic adsorption (ICA) steam gasification system for enhanced hydrogen production using palm kernel shell *Int. J. Hydrog. Energy* **39** 3286-3293.

[22] Hosseini S, Marahel E, Bayesti I, Abbasi A, Abdullah L C, Choong T S Y 2015 CO_2 adsorption on modified carbon coated monolith: effect of surface modification by using alkaline solutions. *Appl. Surf. Sci.* **324** 569-575.

[23] Hidayu A R 2017 *Chapter 6-The preparation and application of activated carbon for gas adsorption* Nova Science Publisher p 165-184.

[24] Maroto-Valer M, Lu Z, Zhang Y, Tang Z 2008 Sorbents for CO_2 capture from high carbon fly ashes. *J. Waste Manag.* **28** 2320-2328.

[25] Alhamed Y A, Rather S U, El-Shazly A H, Zaman S F, Daous M A, Al-Zahrani A A 2015 Preparation of activated carbon from fly ash and its application for CO_2 capture. *Korean J. Chem. Eng.* **32** 723-730.

Effect of different red palm oil volume on characteristics and degradation of polylactic acid/zinc oxide film

S F A Karim[1*], J Jai[1], N A Saiful[1], K H K Hamid[1] and N Kamarudin[1]

[1]Department of Chemical Engineering, Universiti Teknologi MARA (UiTM) 40450 Shah Alam, Selangor, Malaysia

*Corresponding author : sitifatma9633@uitm.edu.my

Abstract. PLA/ZnO combination caused agglomeration, especially at a high concentration of ZnO. This condition improved by adding a plasticizer to smoothen the incorporation of ZnO into PLA. The work's objective was to study the changes in polylactic acid/zinc oxide (PLA/ZnO) film's properties incorporated with red palm oil (RPO) that acted as a plasticizer. The films were prepared using the solvent casting method. RPO's impact as a plasticizer was determined on the film's thickness, moisture content, functional group, mechanical properties, and solubility. More RPO increased the thickness because of the formation of more free volume in the polymer matrix. Otherwise, the moisture content has shown a different trend, which decreased with more RPO content. Fourier transform infrared spectroscopy proved that interaction occurred between each material in this film by sharpening the peak representing the ester group, C=O stretching bonds, 1365 cm^{-1} (C-H bending), and 1216 cm^{-1} (C-O stretching). The highest tensile strength obtained by PLA/ZnO/0.25RPO, showing the lowest performance in elongation at break. ZnO increased the solubility of PLA film, but the presence of RPO proportionally reduced the solubility. These results counsel that the PLA/ZnO/RPO films could be applied as an alternative to manufacture film packaging that needs high tensile strength and low elongation at break. Further study should be conducted for a range of one until 25 weight % of RPO over 100% PLA to determine the optimum formulation.

1. Introduction

Nowadays, polylactic acid (PLA) has become one of the recommended materials in film packaging. PLA is one of the raw materials to produce film packaging because PLA is renewable and expected to improve film packaging's biodegradability performance. Unique, most attractive biodegradable polymers, biocompatibility, and low price are the advantages of PLA [1]. However, pure PLA has limited properties such as low flexibility, brittle, rigid, poor thermal properties, poor crystallization behavior, and low water and gas barrier properties [2,3]. Another material is needed to enhance PLA film's weaknesses, such as the addition of filler, additives, or plasticizers. A slow biodegradation rate happened to PLA film because of the hydrophobicity and lack of reactive side-chain group in PLA film [4–7]. Combining PLA with other materials causes changes in the packaging materials' physical and mechanical properties after being mixed with additives. It could reduce the physicomechanical properties of composite materials [8].

Zinc Oxide (ZnO) was added into PLA films as a filler that improved the film's thermal stability, and it is safe for human cells. Producing significant dispersion in PLA ternary systems enhanced polymer interaction and increased the flexibility and mechanical properties [2]. However, the high interfacial tension between PLA and ZnO caused poor interfacial adhesion and led to ZnO's agglomeration on the PLA matrix [3]. The flexibility and extensibility of PLA film or other biopolymers could be improved by adding plasticizers [9,10]. Plasticizer overcomes the films' brittleness caused by extensive intermolecular forces [11]. Glycerol, sorbitol, and polyethylene glycol were the most commonly used plasticizer in the film [12]. Palm oil is an uncommon plasticizer that falls under a bio plasticizer used to modify the film's properties [13]. The triglycerides and unsaturated fatty acid in palm oil enhanced its ability to be used as plasticizers [14]. Based on the information obtained, no published paper has considered RPO's impact on PLA/ZnO films' properties. Therefore, the solvent casting method was used in this paper to enhance PLA/ZnO film with RPO's embedment. The changes in physical properties, chemical properties, and mechanical properties were determined. Besides, the solubility study was conducted to show the degradation ability of the film produced. The results provide an overview of replacing synthetic plasticizers with RPO.

2. Research Methodology

2.1. Materials

Polylactic Acid (PLA), (pallet from China), zinc oxide (ZnO), (Bendosen, Norway), chloroform (R&M chemical, Malaysia), and red palm oil (RPO), (Iko, Malaysia).

2.2. Methodology

2.2.1. Film Preparation. The film was prepared using a solvent casting method [15]. PLA/ZnO/RPO was mixed with mechanical stirring of 600 rpm at 80 $°C$ for 30 min with various RPO amounts, as shown in table 1. RPO was added as a plasticizer. For the first 15 min, 2g of PLA were mixed with chloroform and 5% (w/w) of ZnO. For the second 15 min, RPO was added to complete the solution. The film solution was poured over (80 mm x 15 mm) Petri dishes placed on leveled surfaces. The film solution was dried at room temperature until the film was produced. Finally, the films were separated from the Petri dishes for analysis and evaluation. The formulation of the films is tabulated in table 1.

Table 1. Formulation of the films

Sample name	PLA (g)	ZnO, (w/w)%	RPO, mL
PLA	2.00	0	0
PLA/ZnO	2.00	5	0
PLA/ZnO/0.25RPO	2.00	5	0.25
PLA/ZnO/0.50RPO	2.00	5	0.50
PLA/ZnO/0.75RPO	2.00	5	0.75
PLA/ZnO/1.00RPO	2.00	5	1.00

2.2.2. Film characterization. The thickness was measured using a digital micrometer at five various points to the nearest 0.001 mm, and the average value was calculated. The film's moisture content was then measured by a simple drying method in an oven at 60 °C using a universal oven [16]. The initial weight (M_i) and final weight (M_f) were recorded. The final weight was taken after a constant weight was obtained. The moisture content (%) was measured based on the weight loss ratio over the initial weight. Fourier transform infrared spectroscopy (FTIR), brand Mettler Toledo was used to monitor the functional properties changes at 4 cm^{-1} resolutions and 4000 - 600 cm^{-1} wavenumber [17]. The performance of the film is shown by the mechanical properties and solubility in water. The setup condition for mechanical properties using the Universal Testing Machine was 50 mm initial span, 50 mm/min crosshead speed, 2.5 kN load, and at room temperature condition [18]. The degradation was measured in measuring the solubility of the film in water [19]. A 40 x 20 mm sample was dried in an oven at 60 °C until a constant weight was achieved, called initial weight (M_o). The sample was then immersed in 50 mL distilled water, 24 hours, at room temperature, in a sealed beaker to prevent water and dust evaporation. The sample was taken out and dried to a constant weight at 60 °C in a universal oven, and the final weight, g (M_2), was measured. The film's solubility was the ratio percentage of the difference between M_o and M_2 divided by M_o.

3. Results and Discussion

3.1. Film thickness and moisture content

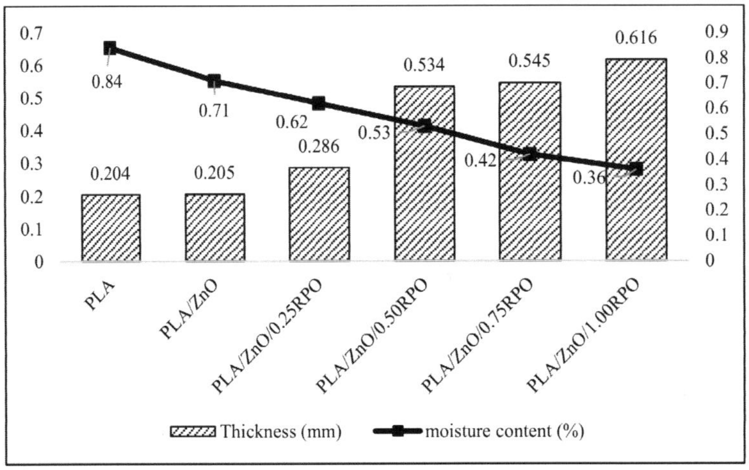

Figure 1. The thickness and moisture content of the film

The thickness of the film for packaging was critical since it affected the tensile strength and elongation at break [20]. Figure 1 shows the average thickness and moisture content for all films. It shows an increasing trend of thickness during a decreasing trend for moisture content. The highest thickness obtained was PLA/ZnO/1.00RPO, 0.616 mm, while the lowest, 0.204 mm, representing pure PLA film. Thickness increment only by 0.001 mm experienced by PLA film with ZnO's addition showing the insignificant function of ZnO in terms of film's thickness. Yasir likewise found that the expansion of ZnO did not altogether influence the thickness of the PLA film [21]. However, with the addition of RPO, the thickness increased proportionally. RPO's plasticizer character formed more hydrogen bonding and increased the film's solid content at once enhanced the thickness [22]. The secondary exist shows the percentage of moisture content. Low moisture content is needed to prevent the oxidation at lipid and water molecules' presence inside the film packaging [16]. PLA film has 0.84% moisture content in the film. ZnO's addition reduced the moisture content percentage to 0.71% because of the gap between PLA being replaced by ZnO. Therefore, the chances of moisture being trapped in the film become lower. The percentage of moisture content decreased with the addition of more RPO. RPO suggests the hydrophobic nature improved, which affects the ability of the film to retain water. Makhtar supported this condition that thermoplastic material has the highest moisture content without crude palm oil and palm olein oil [23].

3.2. Fourier transform infrared spectroscopy

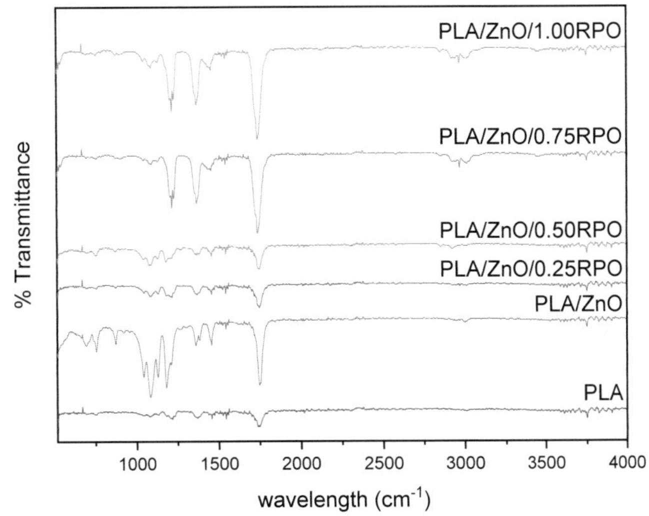

Figure 2. FTIR wavelength for the films

FTIR spectra were used to monitor the structural changes and functional groups of film samples at a molecular level through a detailed analysis [24]. Figure 2 displays the % transmittance for all film obtained from FTIR wavelength range between 515 to 4000 cm^{-1}. Four moderate peaks occurred at 1735 cm^{-1} for PLA film, indicating the ester group (C=O stretching bonds), 1456 and 1366 cm^{-1} for C-H bending, and 1217 - 1020 cm^{-1} regions represents ester group, the C-O stretching bonds. Almost similar peak region obtained by the previous researcher on PLA film, 1749 cm^{-1} designated to C=O of stretching of the ester group, 1450 and 1370 cm^{-1} for symmetric and asymmetric C-H bending deformation, and 1267-1045 cm^{-1} for C-O stretching [25]. The addition of ZnO had sharpened and shifted the 1735 cm^{-1} to 1748 cm^{-1}. The C-H bending of the blending becomes more vital with the occurrence of sharper and wider peaks of 1456 cm^{-1} and 1358 cm^{-1}. The peak is broader for the C-O stretching bond and decreased to a range of 1181 cm^{-1} to 1043 cm^{-1}. Kim mentioned that those changes occurred because ZnO increased the nucleation density of PLA [25]. Meanwhile, new peaks observed in PLA/ZnO, the peak of C-H bending are present at a peak range of 871 cm^{-1} to 755 cm^{-1} representing ZnO. These changes suggesting that PLA and ZnO sharing the same functional group, and interaction occurred at this functional group. The addition of 0.20 and 0.50 RPO into PLA/ZnO reduced all peaks observed. However, the peaks 1747 cm^{-1} (C=O stretching) sharpen tremendously for 0.75 and 1.00 RPO, proving linkage between PLA/ZnO with triglyceride ester in RPO [14]. Peaks 1365 cm^{-1} (C-H bending) experienced the same peak improvement due to RPO consisting of an equal ratio of saturated and unsaturated fatty acids that increased C-H bending peak. The same goes to peak 1216 cm^{-1} (C-O stretching) also experienced a sharper peak with 0.75 and 1.00 RPO where the bands near 1200 cm^{-1} referring to the CC(=O)-O functional group [26]. Plus, a new peak appeared at 2970 cm^{-1} that expressed the weak motion of C=C in *cis* position. It occurred in considering the unsaturated esters in palm oil. Similar results were obtained where the C=C motion was shown by the band at 3010 cm^{-1} [26].

3.3. Mechanical properties

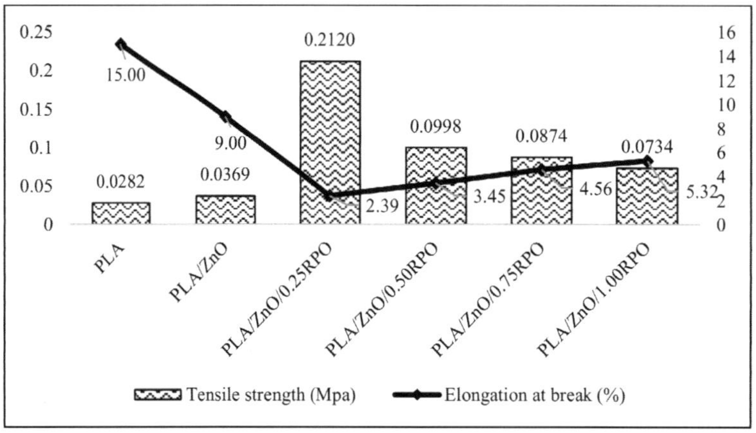

Figure 3. Mechanical properties performance of the films

Tensile strength and elongation at break are essential to film packaging because high tensile strength shows the ability of the film to sustain the tension given. In contrast, high elongation gave high ductility at once more likely to deform and not to break. Figure 3 shows the films' mechanical properties performance based on tensile strength (TS) and elongation at break (EAB). It shows the highest TS, 0.212 MPa obtained by PLA/ZnO/0.25RPO film, while the lowest TS, 0.0282 MPa represented by PLA film. Otherwise, the PLA/ZnO/0.25RPO film experienced the lowest EAB, 2.39%, and the PLA film has the highest EAB, 15%. ZnO increased the TS and decreased the EAB. Marra found ZnO's addition at 5% increased the TS and reduced the EAB at machine direction [27]. ZnO solidness caused positive collaboration among PLA and ZnO through hydrogen holding [3]. Contradict results obtained by Nasab [28] mentioning the addition of ZnO form 1 – 5 % reduced the TS by 16%, and ZnO stiffen the PLA film. A small increment of TS of PLA after the addition of ZnO being supported by FTIR results shows the interaction between PLA and ZnO occurred that fall on the same functional group with minor changes. RPO acted as a plasticizer to overcome the films' brittleness and fill the polymer chains' space. The presence of RPO shows higher TS and lower EAB if compared to PLA and PLA/ZnO film. However, more RPO decreased the TS and increased EAB. Jusoh [29] found similar results in which the decrement of TS occurred with more crude palm oil added into low-density polyethylene film by 9.34% with the addition of 5 w/w% crude palm oil. The presence of RPO increased the free volume network that causes an increment in separation between the polymer chain (EAB increased) and decrement in opposing forces of attraction cause the weakening of the interaction between the polymer chains (TS decreased) [16,29]. The results have been supported by FTIR finding that sharper peak formed at three points as explained in FTIR sections.

3.4. Degradation percentage

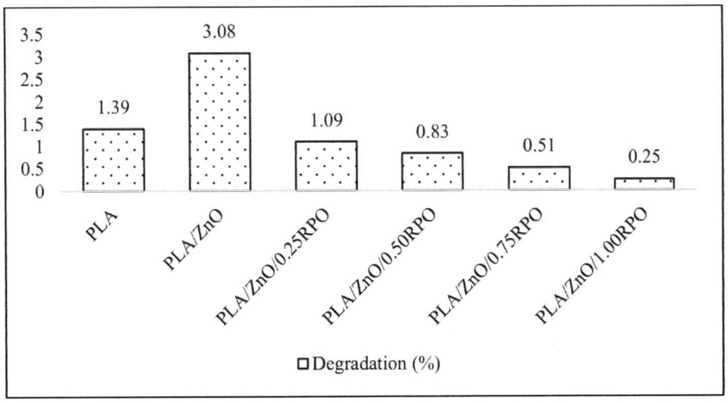

Figure 4. Degradation percentage in terms of percentage of solubility for the films

In this paper, the degradation percentage was measured based on the solubility of the film. Solubility is one of the indicators for the water-resistance of composite materials [30]. Potential food packaging applications may require minor water solubility to enhance product quality and improve moisture barrier properties and water resistance. Numerous investigations have exhibited the signs of diminishing the solubility of the film in water [29]. Figure 4 shows the level of solubility of the films that speak to the film's degradability. PLA/ZnO films have the highest degradation percentage, 3.08% shows the best film degraded in water. Degradation is high because the water molecules can quickly enter the void between PLA and ZnO [25]. A similar result obtained but using a different degradation method indicates ZnO as a reactant in the degradation reaction [31]. The films' higher degradation rate is probably because of the faster disruption and diffusion of molecular interactions [32]. The PLA/ZnO/1.00RPO has the lowest, 0.25% degradation percentage by cause of the RPO. More RPO reduced the degradation percentage (solubility) because of the hydrophobic character of RPO. There got 70% of tocotrienols in vitamin E that presence in palm oil. The tocotrienols are having hydrophobic properties [33]. Long-chain non-polar hydrocarbon chains like ester, fatty alcohols, and fatty acid also contributed to the hydrophobic character [34]. Hydrophobic compounds will decrease the film solubility, while hydrophilic compounds will increase it [16].

4. Conclusion

Plasticizers in PLA/ZnO films are needed to enhance the film's application. This paper successfully embedded RPO as a plasticizer into the PLA/ZnO film using a solvent casting method. RPO increased the thickness due to the formation of more free volumes in the polymer network. The hydrophobic character of RPO caused a reduction in moisture content present in PLA/ZnO film. PLA/ZnO/0.25RPO has the highest TS and lowest EAB, and at the same time, the TS increment was around 82% compared to PLA/ZnO film proving that RPO is an essential agent in PLA/ZnO film. Degradation by hydrolysis was performed, but RPO restricted the film's solubility because of RPO's hydrophobic character. Therefore, RPO can act as a plasticizer for PLA/ZnO film packaging. Based on the highest tensile strength, the recommended formulation was PLA/ZnO/0.25RPO. Further study should be conducted for additional RPO range between 1% to 25% over total amount of PLA to find the best film with the highest tensile strength and high elongation at break.

Acknowledgements
The authors acknowledge the Ministry of Higher Education Malaysia for supporting this work through two grants, FRGS/1/2018/TK02/UITM/02/9 and RAGS/1/2015/TK02/UITM/03/1.

REFERENCES

[1] Wang Y Y, Yu H Y, Yang L, Abdalkarim S Y H and Chen W L 2019 Enhancing long-term biodegradability and UV-shielding performances of transparent polylactic acid nanocomposite films by adding cellulose nanocrystal-zinc oxide hybrids *Int. J. Biol. Macromol.* **141** 893–905.

[2] Tang Z, Fan F, Chu Z, Fan C and Qin Y 2020 Barrier properties and characterizations of poly(lactic Acid)/ZnO nanocomposites *Molecules* **25** 1–12.

[3] Zhang R, Lan W, Ji T, Sameen D E, Ahmed S, Qin W and Liu Y 2021 Development of polylactic acid/ZnO composite membranes prepared by ultrasonication and electrospinning for food packaging *Lwt* **135** 1–9.

[4] Suryanegara L, Nakagaito A N and Yano H 2009 The effect of crystallization of PLA on the thermal and mechanical properties of microfibrillated cellulose-reinforced PLA composites *Compos. Sci. Technol.* **69** 1187–92.

[5] Taib R M, Ramarad S, Arifin Z, Ishak M and Todo M 2010 Properties of Kenaf Fiber / Polylactic Acid Biocomposites Plasticized with Polyethylene Glycol.

[6] Faludi G, Dora G, Renner K, Móczó J and Pukánszky B 2013 Biocomposite from polylactic acid and lignocellulosic fibers: Structure-property correlations *Carbohydr. Polym.* **92** 1767–75.

[7] Tudorachi N and Lipsa R 2011 Poly (vinyl alcohol)-g-Lactic Acid Copolymers and Films with Silver Nanoparticles *Journal of Applied Polymer Science* **122(2)** 1109 - 1120.

[8] Cooksey K 2007 Contaminants Effectiveness of antimicrobial food packaging materials Food Addit. Contam. **22(10)** 980-7.

[9] Ozkoc G and Kemaloglu S 2009 Morphology, Biodegradability, Mechanical and Thermal Properties of Nanocomposite Films Based on PLA and Plasticized PLA *J. Appl. Polym. Sci.* **114(4)** 2481-2487.

[10] Ljungberg N and Wesslén B 2002 The effects of plasticizers on the dynamic mechanical and thermal properties of poly(lactic acid) *J. Appl. Polym. Sci.* **86** 1227–34.

[11] Wang X, Sun X, Liu H, Li M and Ma Z 2011 Barrier and mechanical properties of carrot puree films *Food Bioprod. Process.* **89** 149–56.

[12] Nguyen Vu H P and Lumdubwong N 2016 Starch behaviors and mechanical properties of starch blend films with different plasticizers *Carbohydr. Polym.* **154** 112–20.

[13] Kamarudin S H, Jusoh E R, Abdullah L C, Ismail M H S, Aung M M and Ratnam C T 2019 Thermal and Dynamics Mechanical Analysis of Polypropylene Blown Films with Crude Palm Oil as Plasticizer *Indones. J. Chem.* **19** 545.

[14] Fatma S, Karim A, Aliah N, Mohammad A, Adawiyah R, Aziz A and Ibrahim K 2020 The Effect of Red Palm Oil Concentration towards Characteristics and Biodegradability of PLA-Starch Film **20** 1240–5.

[15] Heydari-Majd M, Ghanbarzadeh B, Shahidi-Noghabi M, Najafi M A, Adun P and Ostadrahimid A 2019 Kinetic release study of zinc from polylactic acid based nanocomposite into food simulants *Polym. Test.* **76** 254–60.

[16] Bilal Ahmad Ashwar, Asima Shah, Adil Gani, Umar Shah, Asir Gani, Idrees Ahmed Wani, Sajad Mohd Wani F A M 2014 Rice starch active packaging films loaded with antioxidants—development and characterization *Starch/Staerke* **67** 294–302.

[17] Orsuwan A, Shankar S, Wang L F, Sothornvit R and Rhim J W 2016 Preparation of antimicrobial agar/banana powder blend films reinforced with silver nanoparticles *Food Hydrocoll.* **60** 476–85.

[18] Wang X, Sun X, Liu H, Li M and Ma Z 2010 Food and Bioproducts Processing Barrier and mechanical properties of carrot puree films *Food Bioprod. Process.* **89** 149–56.

[19] Romero-Bastida C A, Bello-Pérez L A, García M A, Martino M N, Solorza-Feria J and Zaritzky N E 2005 Physicochemical and microstructural characterization of films prepared by thermal and cold gelatinization from non-conventional sources of starches *Carbohydr. Polym.* **60** 235–44.

[20] Datta D and Halder G 2018 Enhancing degradability of plastic waste by dispersing starch into

low density polyethylene matrix *Process Saf. Environ. Prot.* **114** 143–52.
[21] Arfat Y A, Ahmed J, Al Hazza A, Jacob H and Joseph A 2017 Comparative effects of untreated and 3-methacryloxypropyltrimethoxysilane treated ZnO nanoparticle reinforcement on properties of polylactide-based nanocomposite films *Int. J. Biol. Macromol.* **101** 1041–50.
[22] Karim S F A, Hamzah N A N, Aziz R A A and Ibrahim U K 2020 The effect of plasticizers towards the characteristics of methylcellulose film packaging *IOP Conf. Ser. Mater. Sci. Eng.* **845.**
[23] Mohd Makhtar N S, Muhd Rodhi M N, Musa M, Ku Hamid K H, Shuhada N, Makhtar M, Najwa M, Rodhi M, Musa M, Halim K, Hamid K, Mohd Makhtar N S, Muhd Rodhi M N, Musa M and Ku Hamid K H 2013 Thermal behavior of tacca leontopetaloides starch-based biopolymer *Int. J. Polym. Sci.* **2013.**
[24] Norfarahin A H, Sanny M, Sulaiman R and Nur Hanani Z A 2018 Fish gelatin films incorporated with different oils: Effect of thickness on physical and mechanical properties *Int. Food Res. J.* **25** 2503–10.
[25] Kim I, Viswanathan K, Kasi G, Sadeghi K, Thanakkasaranee S and Seo J 2019 Poly(lactic acid)/ZnO bionanocomposite films with positively charged ZnO as potential antimicrobial food packaging materials *Polymers (Basel).* **11.**
[26] Guerrero A, Anguebes F, Castelán M, Morales V, Química F De, Autónoma U, No C, Av E, Juárez C B and Carmen C 2014 Fourier Transform Infrared-Attenuated Total Reflectance (Ftir-Atr) Spectroscopy and Chemometric Techniques for the Determination of Adulteration in Petrodiesel/ Biodiesel Blends *Quim. Nova* **37** 392–7.
[27] Marra A, Silvestre C, Duraccio D and Cimmino S 2016 Polylactic acid/zinc oxide biocomposite films for food packaging application *Int. J. Biol. Macromol.* **88** 254–62.
[28] Nasab M S, Tabari M and Azizi M H 2018 Morphological and mechanical properties of Poly (lactic Acid) / zinc oxide nanocomposite films *Nanomedicine Res. J.* **3** 96–101.
[29] Jusoh E R, Ismail M H S, Abdullah L C, Yunus R and Rahman W A W A 2013 Effect of crude palm oil as plasticiser on the mechanical and morphology properties of low density polyethylene blown film *Int. J. Mater. Eng. Innov.* **4** 302–13.
[30] Wu Z, Huang X, Li Y C, Xiao H and Wang X 2018 Novel chitosan films with laponite immobilized Ag nanoparticles for active food packaging *Carbohydr. Polym.* **199** 210–8
[31] Anžlovar A, Kržan A and Žagar E 2018 Degradation of PLA/ZnO and PHBV/ZnO composites prepared by melt processing *Arab. J. Chem.* **11** 343–52.
[32] Kaya M, Khadem S, Cakmak Y S, Mujtaba M, Ilk S, Akyuz L, Salaberria A M, Labidi J, Abdulqadir A H and Deligöz E 2018 Antioxidative and antimicrobial edible chitosan films blended with stem, leaf and seed extracts of Pistacia terebinthus for active food packaging *RSC Adv.* **8** 3941-3950.
[33] Fatehi P, Baba A S, Eh suk V R and Misran M 2020 Preparation and characterization of palm oil in water microemulsion for application in the food industry *Br. Food J.* **122** 3077–88.
[34] Farooq A, Shafaghat H, Jae J, Jung S C and Park Y K 2019 Enhanced stability of bio-oil and diesel fuel emulsion using Span 80 and Tween 60 emulsifiers *J. Environ. Manage.* **231** 694–700.

The Separation of Oily Water Mixture Using Membrane

N S I Chik[1], N Z K Shaari[1*] and S A Shamsudin[1]

[1]Faculty of Chemical Engineering, Universiti Teknologi MARA, 40450 Shah Alam, Selangor, Malaysia.

*Corresponding author: norinzamiah@uitm.edu.my

Abstract. The increment of oil discharge in the water system gives a great impact to the environments. Membrane separation has been widely used to treat the oily water mixture due to its environmentally process, low operating cost and has better efficiency. Polysulfone (PSF) is widely used in the membrane fabrication due to its good physico-chemical, mechanical and hydraulic stability, but its hydrophobicity limits its application in the separation of oily mixture due to the fouling. The blend of PSF with poly (vinyl alcohol) (PVA) could compensate this limitation where the covalent bonding between PSF/PVA could form a protective layer to prevent the deposition of organic foulants on the membrane surface. Furthermore, PVA has been studied intensively for membrane applications because of its good chemical stability, film-forming ability and high hydrophilicity. To ensure the compatibility and high stability of the resultant membrane, the cross linking of the polymer blend (PSF/PVA) with tetraethylorthosilicate as silica nano precursor through sol-gel reaction was conducted in this study. Polyethylene glycol 400 was added as an additive to improve the properties of the integral membrane. The concentration of tetraethylorthosilicate was varied at 0wt.%, 0.5wt.%, and 1wt.% while the concentrations of other polymers are remained constant. Performance testing on the fabricated membrane through pure water flux and rejection of oil particles were conducted. The results exhibited that higher concentration of tetraethylorthosilicate led to higher water flux and higher rejection of oil particles respectively for the fabricated membrane. The crosslinked membranes achieved 100% of oil rejection after 45 minutes filtration time, meanwhile membrane from pure polymer blend only achieved 60.2% of oil rejection. This finding shows the potential of the fabricated membrane to be further explored in the area of oily water mixture treatment process.

1. Introduction

Population growth and rapid industrial developments such as petrochemical, metallurgy, coal tar, and metal-finishing initiating an increment of oil discharge in the water system which leads to water pollution. Oily wastewater that is not treated properly will give negative impact to the environment like affecting the clean water resources, endangering aquatic ecosystem, endangering human health, causing atmospheric pollution, and affecting crop production [1]. Due to different properties, oil is immiscible in water. Oil water treatment can be classified into six aspects, which includes flotation, coagulation, biological treatment, membrane separation technology, combined technology, and advanced oxidation process.

Membrane separation is one of the most popular processes and has been proven successful in separating oil from water. The membrane acts as semi-permeable layer between the two phases of oil and water and the membrane selectively filter from the two phases [2]. In addition, membrane technology shows more advantages compared to traditional methods such as better efficiency, relatively low operating costs, chemical stability, and environmentally friendly process. Different membrane separation process required different pore sizes and driving forces. The filtration process such as ultrafiltration (UF), microfiltration (MF), reverse osmosis (RO), and nanofiltration (NF) are depending on the membrane pore sizes.

The materials used to produce membranes include polymer, ceramic, and inorganic compounds. In many applications, polymer membranes have been widely used compared to ceramic based membrane because of their high efficiency in separating mixtures. The blended polymer will form the interaction between hydrophobic side chain combination with intramolecular and intermolecular hydrogen bonds and this interaction improved the membrane's filtration [3].

The example of polymer materials that are widely used for membrane preparation include polysulfone (PSF), polyethersulfone (PES), polypropylene (PP), cellulose acetate (CA), polyacrylonitrile (PAN), polyvinylidene fluoride (PVDF), and polyethylene (PE) [4]. One of the problems that commonly occurred in the membrane separation is fouling where it reduces the performance of membrane. Polysulfone is widely used due to its good physico-chemical, mechanical and hydraulic stability. Besides high tendency of fouling, the hydrophobic nature of polysulfone slows down the diffusion of non-solvent and retards coagulation during the phase inversion process, which will subsequently form a dense skin layer. Therefore, it causes a flux declining situation. Blending it with hydrophilic polymers could lessen these phenomenon [5]. Poly (vinyl alcohol) (PVA) has been studied intensively for membrane applications because of its good chemical stability, film-forming ability and high hydrophilicity. However, due to PVA is extremely hydrophilic material and polysulfone is a hydrophobic material, the cross linking of the polymer blend (Polysulfone/PVA) with silica nano precursor (tetraethylorthosilicate) was conducted in this study to ensure high stability for the resultant membrane. It has been proven that the combination of organic and inorganic polymers yields membrane with better mechanical and thermal properties due to the introduction of hydrogen and covalent bonds in the polymer matrix during the cross-linking reaction [6][7]. Furthermore, polyethylene glycol (PEG) was added in the formulation to overcome and reduce fouling by increasing pore size, developing a large pore structure, and enhancing antibacterial properties. The fabricated membrane was evaluated in term of rejection performance of oil particles from oily water mixture by incorporating different loadings of tetraethyl orthosilicate.

2. Methodology

2.1. Chemicals

Polyvinyl alcohol (PVA) with hydrolysis degree of 87-90% (MW: 85,000 – 124,000), polysulfone (PSF) with molecular weight (Mn) 22,000, and polyethylene glycol (molecular weight: 400 g/mol) used as additives. Tetraethylorthosilicate with 99% purity was used as a crosslinking agent and hydrochloric acid with 37% purity as the catalyst. All the chemicals above were obtained from Sigma Aldrich, Malaysia. In addition, 1-methyl-2- pyrrolidone (NMP) (purity 99%) and dimethyl sulfoxide (DMSO) used as solvents were obtained from Merck, Malaysia.

2.2. Membrane preparation

The integral membranes were prepared in 3 types of variations by using phase inversion method. Table 1 showed the formulations of each membrane (M1, M2, and M3).

Table 1. Integral membranes formulations.

Membrane	PSF Solution			PVA Solution		
	PSF (g)	PEG 400 (g)	NMP (g)	PVA (g)	TEOS (g)	DMSO (g)
M1	13	5	82	1	0	99
M2	13	5	82	1	0.5	99
M3	13	5	82	1	1	99

2.2.1 Preparation of PSF Solution. The PSF solution was prepared by dissolved 13 g of polysulfone pellet and 5 g of polyethylene glycol (PEG400) into 1-methyl-2-pyrrolidone (NMP) to prepare 13 wt.% of polysulfone. The mixture was heated and stirred continuously with 300 rpm at 80°C for 6 hours until it became homogeneous [8][9].

2.2.2 Preparation of PVA Solution. The polyvinyl alcohol (PVA) solution was prepared by adding 1 g of PVA in 99 g of dimethyl sulfoxide (DMSO) which was heated and stirred with 400 rpm at 80°C for 2 hours. Through a sol-gel method, 0.5 g of tetraethylorthosilicate (TEOS) was added as the crosslinker followed by 1 ml of HCl with 37wt.% as the catalyst. The mixture was stirred with 400 rpm for another 6 hours at 40°C. The same method was repeated for different TEOS concentrations; 0 wt.% and 1 wt.% [3][8].

2.2.3 Preparation of PSF/PVA Blend Membrane. The PSF / PVA solution was prepared by adding the 0.5 g of PVA solution into 100 g of PSF solution. The mixture was heated at 80°C and stirred for 1 hour with 750 rpm. Then, the mixture was left to cool at room temperature, 25°C [3].

2.2.4 Preparation of Integral Membrane. The blend membrane solution was poured onto a glass plate and spread using Baker Film Applicator at thickness 100 µm. Next, it was immersed in a coagulation bath for 24 hours for solidification. Then, the integral membrane was dried at room temperature for 24 hours. After that, the dried integral membrane was heat-cured at 45°C in an oven for 1 hour [3][10].

2.3 Oil Water Emulsion Preparation. Emulsifiers are the surfactants used to develop high surface pressure at emulsion surface. The oil water emulsion with 1000 mg/L oil concentration was prepared by mixing 1 L of distilled water with 30 mg of cetrimonium bromide (CTAB) surfactant and stirred for 5 minutes. Next, 1 g of vegetable oil was added into the distilled water for 10 minutes to ensure complete mixing of the mixture. The emulsion was stirred at 1250 rpm and heated at 30°C [11].

2.4 Permeation Performance of The Membranes. All the experiments were conducted at room temperature (~25°C). The permeation experiment of an ultrafiltration membrane was characterized by two different types of feed which are the pure water and oil water emulsion. The permeate flux was calculated by using Equation 1 and 2. The flux of pure water and oil water emulsion were measured by using Dead End filtration equipment, which was connected to a gas cylinder of compressed nitrogen that applied pressure to the feed. The filtration tests were carried out at 8 bar for 45 minutes, with 15 minutes interval each. The flux was measured gravimetrically with a digital electronic balance by weighing the permeate solution.

$$J_p = \frac{Q_p}{A\Delta t} \quad (1)$$

$$Q_p = \frac{W_p}{\rho} \quad (2)$$

Where J_p is the permeate flux (L/m^{2}h), Q_p the volume of permeate (L), A the effective membrane area (m^2), Δt is the sampling time (h), W_p the weight of permeate (g), and ρ the density of permeate (g/cm^3) [12][13].

2.5 Oil Removal Efficiency

The same method in 2.4 was used while using oil water emulsion as feed. The oil concentration was 1000 mg/L. High oil concentration in feed increased the oil adsorption and caused great resistance for permeating water [14]. Equation 3 was used to calculate the percentage oil rejection.

$$R(\%) = \left(1 - \frac{C_p}{C_p}\right) \times 100 \quad (3)$$

Where Cp is the concentration of oil in the permeate (mg/L) and CF is the concentration of oil in the feed (mg/L). The oil concentrations in permeate and feed were determined using a UV–vis spectrophotometer. The oil-in-water emulsion was initially calibrated for different known oil concentrations in terms of absorbance recorded at a wavelength of 235 nm which is the wavelength at which maximum absorption occurs [15]. The figure 1 showed the schematic diagram of the experimental set-up.

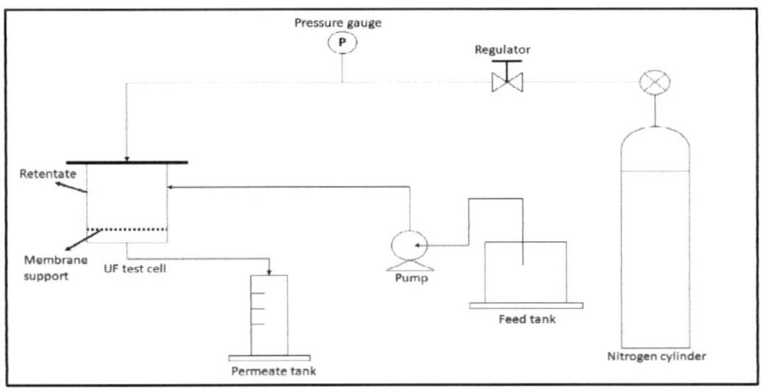

Figure 1. The schematic diagram of the experimental set-up [14]

3. Results and Discussions

3.1 Permeation Performance of The Membrane

Figure 2 shows the result of pure water flux for different type of membrane for filtration of 45 minutes. Based on Figure 2, the pure water flux of all membranes decreases as time increase. The increase in TEOS concentration resulted in higher water flux as observed from the figure where M3 recorded the highest water flux, M1 recorded the lowest water flux of water over time. The water flux related is with the size of pores and hydrophilicity of the membrane surface [12]. The flux declines as time increase during membrane filtration process is attributed to the concentration polarization on the membrane's surface. M3 had higher pure water flux because of enhanced hydrophilicity of the membrane with the presence of silica from TEOS [16]. Similar results found by [17] where hydrophilicity and permeability of the membrane increase with the increase in TEOS content from 0wt.% to 2wt.% [18].

Figure 3 illustrates oil water emulsion flux against time. Similar results were recorded with pure water flux where the fluxes of oil water emulsion also decrease with time increase. However, the oil water flux decreases with the increase in TEOS concentration in membrane. M1 recorded the highest oil water flux, followed by M2 and M3. The reduction of oil water flux on M3 was due to the crosslinking reaction occurred which had resulted in greater combining forces between SiO_2 and PVA and the formation structure of the membrane possibly more compact [19]. Therefore, it allowed more oil particles to deposit on the surface of the membrane thus, creates the concentration polarization phenomenon. As a result, M3 had the lowest flux of oil in water compared to M2 and M1. The use of 1000mg/L of oil concentration also contributes towards this situation because the concentration of oil in the feeds directly affect the oil adsorption (fouling). Fouling is mainly caused by adsorption of oil on the membrane structure, which modifies the critical surface tension and the wettability, as well as the effectiveness of pore diameter, resulting in reducing the membrane permeability [20].

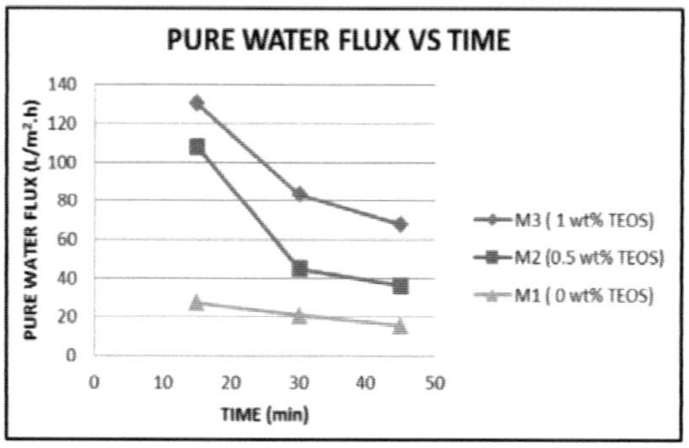

Figure 2. Pure water flux for various membranes

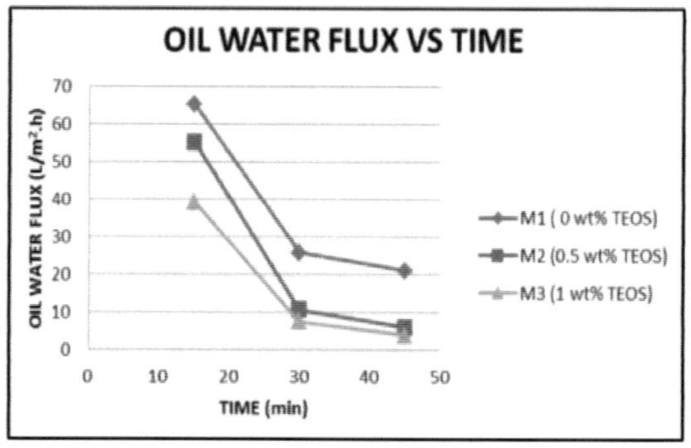

Figure 3. Permeate flux from oily water mixture for various membranes

3.2 Oil Removal Efficiency

Figure 4 shows the percent rejection of oil for each membrane. As time increase, the percentage of oil rejection increase. After 45 minutes, the M3 and M2 achieved maximum rejection of 100%. In contrast M1 only achieved 60.2% of rejection. This situation could be explained through the dense structure of M2 and M3 as a result of crosslinking blend of PSF/PVA with TEOS, where pore size of the membrane is smaller than the diameter of the oil molecule. Different pattern for M1 could be related to situation where longer retention time of oily mixture in the filtration rig leads to gravity settling of water and oil particles, where oily mixture started to demulsify so that the separation of the oily water mixture finally occurs [21][22].

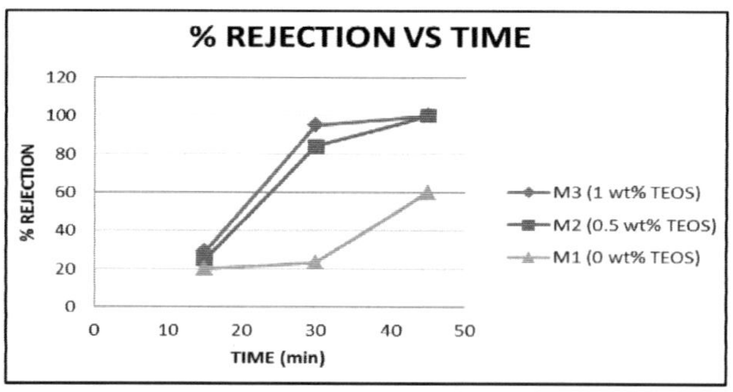

Figure 4. Percentage of oil rejection against time

Apart from observation on the performance of the fabricated membrane, the visual observation on the membrane solution was also conducted. The opposite characteristic of polysulfone and polyvinyl alcohol will create a different compatibility degree with solvent such as N-methyl 2-pyrrolidone, which will be resulted in inhomogeneous blend and results to low structural stability of the integral membrane [23]. Therefore, a good selection of solvents is important during the blending. Low structural stability is due to significant differences in surface tension between these two materials that results in the segregation of the two layers of the integral membrane under swollen condition. Based on the obtained result, a homogeneous blend was obtained from good selection of solvents and through the sol-gel method during crosslinking process respectively, which ensures the dispersion of the organic and inorganic components are at the molecular level in the membranes so that the resultant membrane solution is homogeneous [24].

4. Conclusion

Removing oil particles from oily water mixture by using the fabricated integral membrane was successful, where membrane M3 with 1 wt.% of TEOS had achieved the highest rejection of oil particles followed by M2 and M1. The presence of TEOS in the membrane had enhanced hydrophilicity of the membrane, that was reflected through the increase in pure water flux. The homogeneous blend of the membrane solution depicts the good selection of the solvents for both hydrophilic and hydrophobic polymers respectively, and as a result of crosslinking process. The finding shows the potential of the fabricated membrane to be further explored in the area of oily water mixture treatment process. Further study will be conducted by investigating on the effect of pressure during the separation process and the influence of different types of surfactants.

Acknowledgement

The authors also would like to thank Universiti Teknologi MARA (UiTM) for sponsoring the research through grant Bestari Perdana with file number 600-IRMI/DANA 5/3/BESTARI (P) (009/2018).

References

[1] Li Y, Mei H, and Fang H 2013 *Arab. J. Chem.* **10** 1913–1922.
[2] Gebrehiwet A G 2018 *J. Membr. Sci. Tech.* **8** (2) 1–16.
[3] Nurul A S, Norin Z K S, and Norazah A R 2016 *J. Eng. Sci. Tech* p 36–49.
[4] Nurul N A, Hatijah B, Zawati H, Muhamad Z Y, and Goh P S 2013 *J. Teknol.* **65** (4) 47–51.
[5] Aruna P, Saranya R, Arthanareeswaran G, Ismail A F, and Matsuura T 2014 *Desalination* **344** 280–288.
[6] Huang Y, Du J, Zhang Y, Lawless D, and Feng X 2016 *J. Membr. Sci. Tech.* **514** 229–240.
[7] Norin Z K S, Norazah A R, and Ramlah M T 2015 *J. Adv. Res.* **8** 16–38.
[8] Norin Z K S, Nurul A S, and Norazah 2019 *J. Environ. Chem. Eng.* **7** 102845.
[9] Ilya B, Vladimir V, Dmitry M, Anna O, Alexey & Vladimir V 2019 *Fibers* **7** 10.
[10] Gil J F, Amit C, and Xianshe F 2007 *J. Membr. Sci.***303** 54–63.
[11] Xia X, Hongxing Z, Guiying X, Yebang T, Jian Z, and Xin L 2013 *Colloids and Surfaces A: Physicochem. Eng. Aspects* **418** 60–67.
[12] Abdul L A, Yusuf N M, and Ooi B S 2012 *Desalination* **287** 35–40.
[13] Toraj M and Ehsan S 2009 *Desalination* **243** (1) 1–7.
[14] Ahmed F A, and Samah M A 2013 *J. Chem. Pet. Eng.* **14** (4) 53–70.
[15] Barnali C, Aloke K G, and Mihir K P 2008 *J. Membr. Sci.* **325** (1) 427–437.
[16] Hai P X, Yan H Y, Wan Z L, Xi Y, and Ya J G 2015 RSC Advances **5** 13733–13742.
[17] Kumari A, Choudhury A and Sarkhel G 2012 *Prog. Rubber Plast. Recycl. Technol.* **28**(3) 111-126.
[18] Archana K, Arup C, and Gautam S 2012 *Prog. Rubber Plast. Recycl. Technol.* **28** (3) 111–126.
[19] Haikuan Y, Jie R, Liang C, and Lian Shen 2013 *J. Appl. Polym. Sci.* **130** (6) 4066–4074.
[20] Sama A 2011 *J. Eng.* **17** (2) 252–264.
[21] Alaaeddin E, Shams A, Yahui Z, and John S 2020 *Separ. Sci. Tech.* 1–15.
[22] Chen L Y 2007 *Sep. Purif. Technol.* **54** 388–395.
[23] Filimon A, Albu R M, and Avram E 2015 *Int. J. Miner. Metall. Mater.* **9** 776–779.
[24] Cong H, Radosz M, Towler B F, and Shen Y 2007 *Sep. Purif. Technol.***55** 281–291.

IOP Conf. Series: Materials Science and Engineering 1176 (2021) 012004 doi:10.1088/1757-899X/1176/1/012004

Characterisation of microfibrils cellulose isolated from oil palm frond using high-intensity ultrasonication

M M Yashim[1,2], M Mohammad[1*], N Asim[1], A Fudholi[1] and N H Abd Kadir[3]

[1]Solar Energy Research Institute, Universiti Kebangsaan Malaysia, 43600 Bangi, Selangor, Malaysia

[2]Faculty of Chemical Engineering, Universiti Teknologi MARA, 23200 Dungun, Terengganu, Malaysia

[3]Faculty of Science and Marine Environment, Universiti Malaysia Terengganu, 21030 Kuala Nerus, Terengganu, Malaysia

*Corresponding author: masita@ukm.edu.my

Abstract. This study highlighted the utilization of agricultural byproducts as an alternative fiber resource to be used as one of the materials for reinforcement biocomposite. Cellulose was extracted from oil palm frond using the alkaline treatment and successfully isolated into microfibril via the combination of carboxymethylation pre-treatment ultrasonication to get highly crystalline and good thermal stable microfibers. 4% NaOH was used during alkaline treatment and followed by oxidative bleaching with 30% H_2O_2. Once the extracted cellulose is chemically pre-treated with monochloroacetic acid, it was subjected to 30 minutes ultrasonication treatment to reduce its size. The detailed comparative analysis using SEM, FTIR and TGA was conducted in this work revealed some breakages of intramolecular hydrogen bonds and glycosidic bonds that occurred during the alkaline and bleaching treatment of oil palm biomass. The SEM images showed significant morphology of rigid, organized and highly ordered cellulose fibrils changed into aggregated fibril bundles of microfibrils after ultrasonication. The results from the infrared spectrums revealed that the mild alkaline treatments and oxidative bleaching were able to remove a large fraction of lignin and hemicelluloses to leave a clean cellulose sample. The isolated microfibrils cellulose exhibit good thermal stability as almost 50% of its initial mass remains at a temperature of 300 °C. These findings demonstrate that oil palm fronds can be utilized for biocomposite reinforcement applications.

1. Introduction

Malaysia is located in the tropical region, where it receives a fair amount of sunshine and an abundance of rainfall throughout the year. This climate puts the country in an excellent position to promotes agricultural activities as its major commodity. Around 4.06 million hectares of land have been used for agriculture to contribute up to 56.4 % of Malaysia's national gross domestic products (GDP) [1]. However, activities from the agricultural sector create high amounts of biomass, particularly from palm oil plantations. It was reported that Malaysia produces approximately 168 million tons of biomass each year [2] at which 30 million tons of biomass produced are contributed from Malaysia oil palm industry, in the form of empty fruit branches (EFB), oil palm trunks (OPT) and oil palm fronds (OPF) [3]. The concern of environmental issues has encouraged efforts toward increasing the efficient utilization of raw resources and reduce waste productions. One of the approaches in minimizing the waste from this industry is by converting it into useful products such as fibers with thermal stability characteristics to be utilized as reinforced filler for biocomposite.

The detrimental environmental and ecological impacts of synthetic filler materials in polymer composite have forced researchers to pursue renewable biomass materials like cellulose for the replacement. Cellulose is a high molecular weight linear polymer consisting of a repeating D-glucose unit binded together between C1 and C4 through glycosidic bonds [4]. It is widely employed as starting materials to produce micro sizing cellulose, microcrystalline and microfibrillated cellulose. Despite various approaches to convert oil palm waste into wealth, OPF has not yet obtained an adequate industrial application for the past years. Studies have shown that OPF contains 84% holocellulose, out of which 44–58% is cellulose [5–7]. Cellulose constituent in the agricultural biomass together with the complex structure is responsible for the high thermal strength of the processed fiber produced. Recent studies have shown that cellulose becomes a desirable filler for biocomposite due to its thermally stable characteristic [8,9] and relatively strong (high crystallinity index).

The characteristic of extracted cellulose muchly depends on the cellulosic source and the treatment process during which the isolation is carried out, as reported by Chen *et al.* [10]. Multi-sized aggregate bundles of elementary microfibrillated cellulose (MFC) require intensive mechanical refinement of purified cellulose pulp to break the cell walls of fibers and release cellulose fibrils. Mechanical refinements such as refining or high-pressure homogenization [11], microfluidization [12], grinding [13] and twin screw extrusion [14] are often associated with high energy consumption. Therefore, the application of the hydrodynamic forces due to the high pressure associated with the collapse of cavitation bubbles during the ultrasonication treatment is now of great interest. A study has shown successful isolation of nanofibrils cellulose from wood using high-intensity ultrasonication with alteration in fiber properties [15]. To assist the degradation of the cellulosic fiber, oxidative or hydrolytic methods were suggested to swell the fiber and break down the cell wall at significantly lower mechanical forces. In the past few years, 2,2,6,6-tetramethylpiperidin-1-oxyl (TEMPO)-mediated oxidation was used to assist the ultrasonication isolation method as the presence of carboxylate group effectively break the inter-fibril hydrogen bonding [16]. Until recently, the combination of carboxymethylation and ultrasonication is still not widely reported.

The isolation methods cellulose fiber creates a distinct distribution of amorphous and crystallinity value of the fiber and hence influences the character of the cellulose fibers differently. So, this work aims to extract the cellulose from the oil palm frond using simple alkalization treatment and to isolate the oil palm frond cellulose microfibrils by using the combination of carboxymethylation treatment and ultrasonication method (chemo-mechanical method). The physicochemical characteristics of the cellulose fiber extracted after ultrasonication treatment were studied. The reuse of this agricultural biomass for a potential value-added product will become one of the additional revenues as well as develop the diversification of the agricultural industry.

2. Methodology

2.1. Raw Materials

Samples of the oil palm fronds used for the cellulose extraction were obtained from an oil palm plantation in Terengganu, Malaysia. The raw samples were washed for disinfection, cut into small pieces, and dried at a temperature of 60–85 °C for 48 h. The dried samples were ground into a powder and sieved until specific sizes were reached. Commercially available microcrystalline cellulose, NaOH, H_2O_2, sulphuric acid, monochloroacetic acid, isopropanol, acetic acid, methanol, and ethanol were from Sigma-Aldrich and used as received. All chemicals were of analytical reagent grade (suitable for laboratory use).

2.2. Extraction of cellulose

The oil palm fronds (OPF) powder was directly treated with 4 wt.% NaOH at 90–100 ºC for 2 hours. The obtained black slurry from alkaline treatment was filtered and washed solid residue with distilled water several times until the pH of the filtrate is neutral. The residue was then dried in the oven at 55 ºC to constant weight for 24 hours. At this stage, oil palm frond cellulose was obtained. The OPF cellulose was bleached with 30% H_2O_2 at 90 ºC for an hour. The bleaching step was repeated three times until the cellulose turns white. The bleached OPF cellulose was rewashed using distilled water and filtered using the Buchner filter system. After filtration, the bleached OPF cellulose was dried in the oven at 55 ºC.

2.3. Carboxymethylation treatment

Five grams of bleached cellulose powder was alkalized in 30 % (w/v) NaOH at 30°C for 60 min with the presence of isopropanol as a solvent. The carboxymethylation process is followed by the addition of 6 g of monochloroacetic acid at 55°C for under constant stirring conditions for 180 mins. The slurry was filtrated, and the solid residue was soaked in methanol overnight to remove impurities. The suspension is neutralized with 90% acetic acid. The solid residue was washed with ethanol, and the obtained carboxymethylcellulose was dried using a freeze dryer for 48 hours at -82 °C temperature.

2.4. Ultrasonication

Carboxymethylated cellulose powder was soaked in distilled water (0.1 % w/v) before the suspension was subjected to ultrasonic treatment for 30 min using an ultrasonic processor at 20 kHz and output power of 130 W. The final suspension of fibrillated cellulose was collected and freeze-dried for characterisation.

2.5. Characterization

2.5.1. Fourier Transform Infrared (FTIR). Infrared spectroscopy of extracted cellulose from oil palm fronds was carried out using Bruker Tensor 27. Bands were recorded in the region from 4000 cm^{-1} to 600 cm^{-1}. The spectrum was compared with commercial crystalline cellulose for analysis.

2.5.2. Scanning Electron Microscopy (SEM). The morphological studies of extracted cellulose from OPF and ultrasonicated CMC were carried out using a scanning electron microscope. The samples were first coated with gold/palladium by using a sputter coater device from Bal-Tec (Multi Coating System MED20), which is a good sample preparation technique. Then, SEM (JEOL-JSM-6010LA) was applied at an acceleration voltage of 15 kV for the characterisation of the samples.

2.5.3. Thermogravimetric analysis/Differential thermal analysis (TGA/DT). Thermal behaviour of raw OPF, cellulose extracted and carboxymethylated cellulose (CMC) was investigated using simultaneous thermal analyser NETZSCH STA 449F3. The samples were heated from ambient temperature up to 900 °C with a heating rate of 10 ºC/min^{-1} under a nitrogen gas atmospheric condition.

2.5.4. X-ray Diffraction. The crystallinity phase of the extracted cellulose samples was evaluated through an X-ray diffractometer, (Bruker AXS-D8 Advance). The diffraction patterns were recorded using Cu radiation at 40 kV and 40 mA with a wavelength (λ) of 0.15406 nm.

3. Results and Discussion

3.1. Infrared spectroscopy

The FTIR spectra of the raw oil palm frond and the extracted cellulose were measured to investigate changes in the chemical structures occurring during the alkalinization and oxidative bleaching treatment. From the FTIR spectra shown in figure 1, the spectrum of the extracted cellulose is closely similar to commercial crystalline cellulose indicates the effectiveness of the extraction method. The peaks in the regions 3200 and 3300 cm^{-1} correspond to the stretching vibration of the hydroxy groups (O-H), the main functional groups for cellulose. The intensity of the peak reduced in the extracted cellulose sample compared to raw OPF indicates the reduction in the O-H concentration as the alkaline treatment reduced the hydrogen bonding in cellulosic hydroxyl groups [17]. The absorption band at 2897 cm^{-1}, associated with the stretching vibration of the C-H bond of cellulose, appears in all the extracted samples but intensified in the extracted cellulose sample.

The peaks between 1650 and 1630 cm^{-1} were attributed to C-H and O-H stretching in the lignin and hemicellulose, which has shown a significantly low intensity in extracted cellulose samples when comparing to raw OPF. The mercerization (alkaline treatment) has removed hemicelluloses and partially removed the lignin during this process [18]. Glycosidic bonds in the hemicelluloses are hydrolyzed during mercerization treatment and promote the ester bonds' cleavage between the hydroxyl groups of lignin and the carboxyl groups of the hemicelluloses. The low intensity of the peak within this region, attributed to the lignin aromatic ring, was also recorded due to the oxidative bleaching process. The bleaching process was responsible for oxidizing the ether bonds between the hydroxyl groups of lignin and the carbohydrates [7]. The peaks, which are also typical for cellulose compounds present in the bands at 1160 and 896 cm^{-1}, were also recorded in the extracted cellulose from the oil palm frond. The peaks were associated with the C–O and C–H stretching of glycosidic linkages.

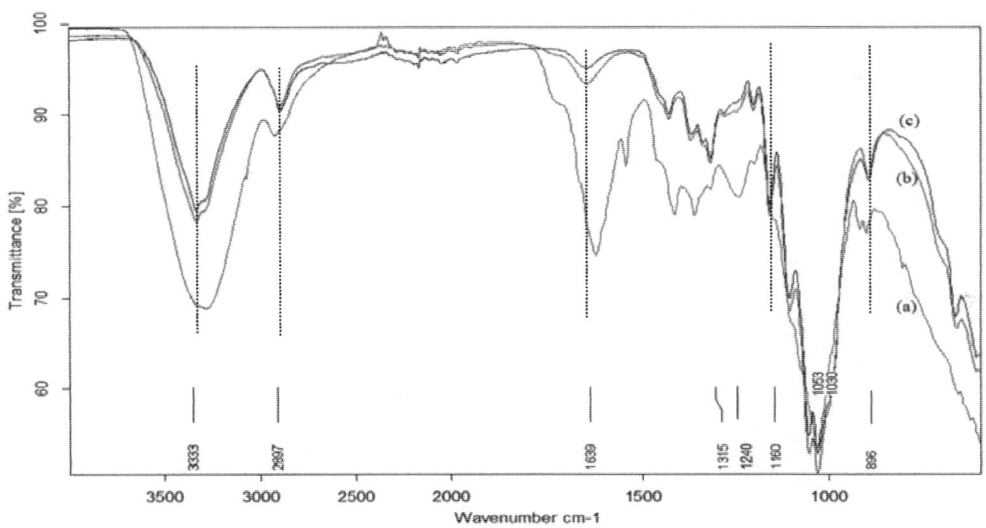

Figure 1. Comparison of FTIR spectra of; a) raw oil palm frond (OPF), b) extracted cellulose from the oil palm frond, and c) commercial microcrystalline cellulose.

3.2. Morphology Observations

It is observed that the colour of the raw oil palm frond fibers changed from brown to white after treatment, as shown in figure 2. Significant colour changes were due to the removal of lignin and other extractives during mercerization and bleaching treatments, resulting in a lighter colour and smoother surface of the fibers, respectively. The same physical observation was recorded during cellulose extraction from OPT using potassium hydroxide and hydrogen peroxide bleaching process [19]. The presence of phenolic substances such as lignin and extractive are responsible for the wood's dark colour. The removal of these substances has caused a notable colour change in the extracted cellulose sample [20].

Figure 2. Comparison of the appearance of a) raw OPF, and b) extracted cellulose.

The individual micro-sized fiber bundles were captured in SEM images of the surface structure of the extracted cellulose sample, as shown in figure 3(a), which is caused by the solubilization of hemicellulose and the disruption of lignin. The irregular and rougher fiber cells with many fibril bundles attached were observed in the SEM images of cellulose from the alkali-treated bamboo [20]. The combination of carboxymethylation and ultrasonication treatment has caused the surface morphology changes from a fibrous cellulose structure to a finely laminated structure, as can be seen in figure 3(b). The lateral size of cellulose fibrils isolated using ultrasonication was recorded within 2.5–5 µm in width. Therefore, the interconnected network microfibers obtained in this study could provide a great reinforcing capability for biocomposite applications.

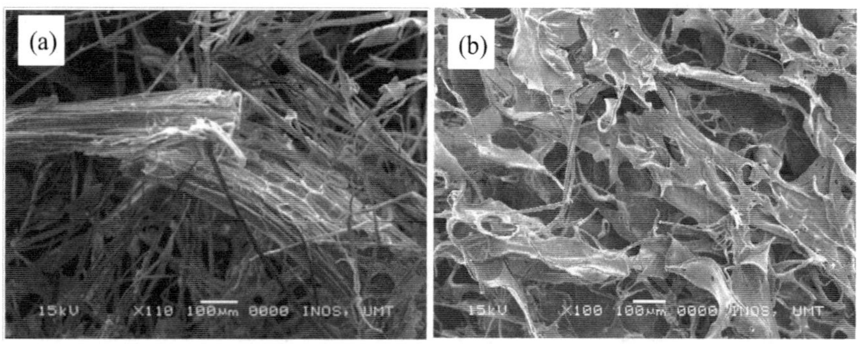

Figure 3. Comparison of the SEM image of; a) extracted cellulose from OPF and b) ultrasonicated carboxymethyl cellulose (CMC) under 100× magnification.

3.3. Thermogravimetry analysis (TGA) and differential thermal analysis (DTA)

Figure 4 shows the thermogram of TGA of all three different types of samples; raw oil palm frond, extracted cellulose and ultrasonicated carboxymethyl cellulose (CMC) as compared to microcrystalline cellulose (commercial). The thermograms of all the samples exhibited two distinctive mass loss stages: 1) temperature range below 150 °C and 2) temperature between 160 and 600 °C. The small drop of mass loss that occurs at a temperature lower than 100 °C was attributed to the water evaporations in all cellulosic samples. The second stage of weight loss occurs in the raw OPF sample at a lower temperature (165°C) than the extracted cellulose sample (275 °C) signify the mass loss due to decomposition of lignin and hemicellulose [21]. The occurrence can be deduced from the DTG image of all the samples, as shown in figure 5. The mass loss at a temperature ranges from 180 to 300 °C represents the degradation of lignin and hemicellulose in the raw OPF sample. Two small stages of mass loss occurring in raw OPF samples do not appear in the DTG image of the extracted cellulose image. The observation could indicate a successful removal of hemicelluloses and lignin during alkalinization and bleaching procedures. The major mass loss at a temperature range of 300 to 350 °C represents cellulose degradation in the OPF-cellulose sample.

The effects of ultrasonication treatment on the thermal stability of carboxymethylated cellulose were also conducted using thermogravimetry analysis. The thermogram of the CMC sample in figure 4(d) has shown two distinct weight losses are observed from the results. The first decomposition occurs in the temperature range of 30 and 130 °C, where the weight loss is about 14.5 %. The initial weight was attributed to the loss of moisture in the sample. A similar observation has been reported by Moussa *et al.* [22] for the carboxymethyl cellulose extracted from almond shells, almond stems, and fig stems. The second decomposition temperature started at 245 °C showing a rapid weight loss up to 300 °C due to the loss of COO^- from the polysaccharide [23]. This second stage weight loss in ultrasonicated CMC shifted to a comparatively lower temperature than that of cellulose confirms a relatively lower thermal stability of ultrasonicated CMC within the temperature range. Though, its thermal stability is still good. The thermogram shifted a little bit to the left due to the presence of carboxymethyl functional group catalyse the degradation of carboxymethyl cellulose [24]. The conversion of extracted cellulose from OPF changes both the molecular structure and bonding energy which subsequently affects the thermal behavior of CMC differently. Alkaline treatment during the production of CMC increases the amorphous structure of CMC [25], which degrades at a lower temperature than that of crystalline structure. At 300°C, the residue mass of CMC was about 50 %, and decomposition slowly continues as temperature increases. Complete decomposition of the carboxymethylated cellulose (CMC) did not take place within the temperature range, as the residue mass remained about 24 % even when the temperature reached 900 °C. The same observations were reported by Han *et al.*, as the residue mass of CMC remains about 60 % when the temperature reached 600 °C [26].

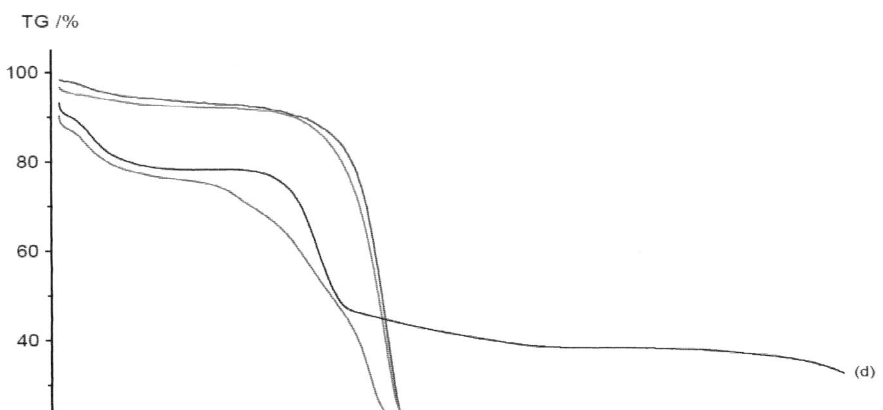

Figure 4. Thermogram image of the (a) raw oil palm frond, (b) extracted cellulose, (c) commercial microcrystalline cellulose and (d) ultrasonicated carboxymethyl cellulose (CMC).

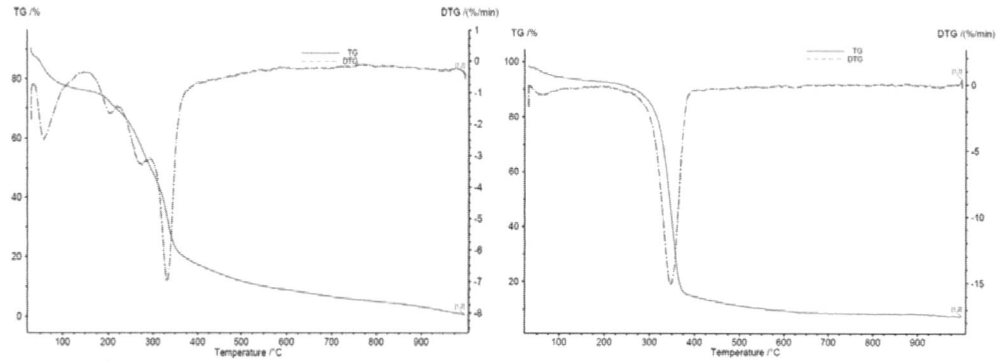

Figure 5. DTG curve of raw OPF (left) and extracted cellulose from OPF (right).

3.4. X-ray Diffraction

Crystallinity is another essential property of cellulosic materials as it may affect their mechanical properties as biocomposite reinforcement materials [27]. Figure 6 shows the x-ray diffraction spectra of extracted cellulose, CMC and ultrasonicated carboxymethylated cellulose to study the change of crystallinity upon chemical and physical treatments. The diffractogram of extracted cellulose from the oil palm frond has shown a major peak at $2\theta = 22.5°$, which is believed to represent the typical cellulose (I) [28] and indices the 002 crystal phase of cellulose. This type of cellulose is naturally found in the plant. The cellulose extracted from OPF was highly amorphous, as indicated by a relatively low peak intensity in the diffractogram, which implies a more disordered structure. As the extracted cellulose is converted into carboxymethyl cellulose, the major peak of the diffractogram shifted to $2\theta = 20°$ due to the broadening or cleavage of hydrogen bonds by carboxymethyl substitution at the hydroxyl groups of cellulose [22].

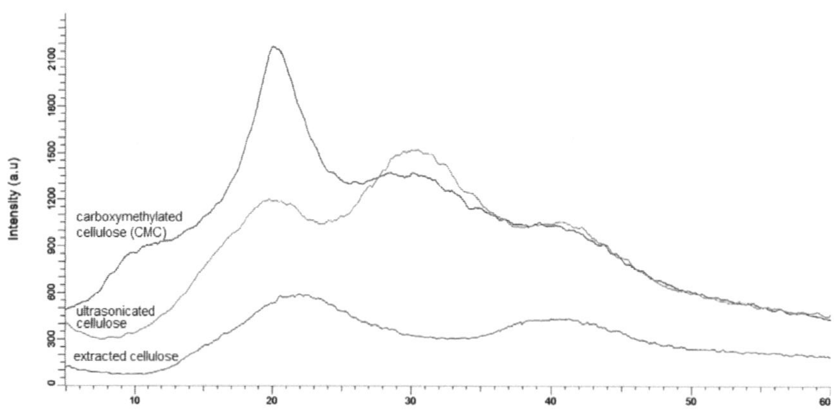

Figure 6. X-ray diffraction spectra of extracted cellulose from OPF, carboxymethylated cellulose (CMC), and ultrasonicated carboxymethylated cellulose.

During the production of carboxymethyl cellulose, alkalinization and etherification destroy the crystalline form of cellulose, and some hydrogen bonds are opened for the formation of cellulose ether [29]. The swelling of the cellulose in an alkaline solution causes a stretching to the neighboring crystalline cellulose molecules and tends to distort them. As the CMC is subjected to high-intensity ultrasonication, the crystallinity reduces due to the breakdown of the hydrogen bonds of the cellulose microfibrils. This reveals that the ultrasonication treatment caused damage to the crystalline domain of the microfibrils. The same observation was reported by Lu *et al.* [30] in their study related to the production of cellulose nanocrystal from bamboo pulp via ultrasonication-assisted $FeCl_3$-catalyzed hydrolysis process.

4. Conclusions

High-quality cellulose was successfully extracted from oil palm frond using simple alkalinization pre-treatment followed by oxidative bleaching technique using hydrogen peroxide. FTIR measurements of the fibers revealed that alkaline and oxidative treatment effectively remove the hemicellulose and lignin. Cellulose microfibrils were isolated from oil palm frond using the combination of carboxymethylation and high-intensity ultrasonication, with smaller apparent width than that isolated untreated pulp. This concludes the presence of the charged ether group reduces the effect of hydrogen bonding, thereby easing fibrillation. Carboxymethylated cellulose fiber also exhibited enhanced thermal properties, with residual mass remains 50% with thermal degradation temperatures higher than 300°C. The crystallinity of the ultrasonicated cellulose fibers also higher than their original extracted cellulose. Micro-sized cellulose extracted from oil palm frond using relatively cheap chemo-mechanical process has the potential to be used as reinforcement materials for biocomposite to increase its mechanical properties.

Acknowledgments

Funding from the Fundamental Research Grant Scheme (FRGS/1/2019/TK10/UKM/02/1) and Dana Modal Insan (MI-2019-018) by the Malaysian Ministry of Education as well as the Ministry of Higher Education of Malaysia is gratefully acknowledged for the completion of our original work.

References

[1] Ahmad T A T 2015 Dimensions of Food and Livelihood Security of Agricultural Trade: The Case of Malaysia *Emerging Economies* **chapter 5** p 113–32.

[2] Ozturk M, Saba N, Altay V, Iqbal R, Hakeem K R, Jawaid M and Ibrahim F H 2017 Biomass and bioenergy: An overview of the development potential in Turkey and Malaysia *Renew. Sustain. Energy Rev.* **79** 1285–302.

[3] Onoja E, Chandren S, Abdul Razak F I, Mahat N A and Wahab R A 2019 Oil Palm (Elaeis guineensis) Biomass in Malaysia: The Present and Future Prospects *Waste and Biomass Valorization* **10** 2099–117.

[4] Klemm D, Philipp B, Heinze T, Heinze U and Wagenknecht W 1998 *Comprehensive Cellulose Chemistry* **chapter 4** pp197–207.

[5] Karp S G, Woiciechowski A L, Soccol V T and Soccol C R 2013 Pretreatment strategies for delignification of sugarcane bagasse: A Review *Brazilian Arch. Biol. Technol* **56** 679–89.

[6] Devi L U, Bhagawan S S and Thomas S 1997 Mechanical properties of pineapple leaf fiber-reinforced polyester composites *J. Appl. Polym. Sci* **64** 1739–48.

[7] Rasila S, Rasli A M, Ahmad I, Lazim A M and Hamzah A 2017 Extraction And Characterization Of Cellulose From Agricultural Residue - Oil Palm Fronds *Malaysian J. Anal. Sci.* **21** 1065–73.

[8] Formela K, Hejna A, Piszczyk Ł, Saeb M R and Colom X 2016 Processing and structure–property relationships of natural rubber/wheat bran biocomposites *Cellulose* **23** 3157–75.

[9] Nasution H, Harahap H, Al Fath M T and Afandy Y 2018 Physical properties of sago starch biocomposite filled with Nanocrystalline Cellulose (NCC) from rattan biomass: The effect of filler loading and co-plasticizer addition *IOP Conf. Ser. Mater. Sci. Eng* **309** 012033.

[10] Chen S, Schueneman G, Pipes R B, Youngblood J and Moon R J 2014 Effects of crystal orientation on cellulose nanocrystals-cellulose acetate nanocomposite fibers prepared by dry spinning *Biomacromolecules* **15** 3827–35.

[11] Zhang J, Song H, Lin L, Zhuang J, Pang C and Liu S 2012 Microfibrillated cellulose from bamboo pulp and its properties *Biomass and Bioenergy* **39** 78–83.
[12] Wang W, Mozuch M D, Sabo R C, Kersten P, Zhu J Y and Jin Y 2015 Production of cellulose nanofibrils from bleached eucalyptus fibers by hyperthermostable endoglucanase treatment and subsequent microfluidization *Cellulose* **22** 351–61.
[13] Afra E, Yousefi H, Hadilam M M and Nishino T 2013 Comparative effect of mechanical beating and nanofibrillation of cellulose on paper properties made from bagasse and softwood pulps *Carbohydr. Polym.* **97** 725–30.
[14] Ho T T T, Abe K, Zimmermann T and Yano H 2015 Nanofibrillation of pulp fibers by twin-screw extrusion *Cellulose* **22** 421–33.
[15] Chen W, Yu H, Liu Y, Chen P, Zhang M and Hai Y 2011 Individualization of cellulose nanofibers from wood using high-intensity ultrasonication combined with chemical pretreatments *Carbohydr. Polym.* **83** 1804–11.
[16] Wu J, Du X, Yin Z, Xu S, Xu S and Zhang Y 2019 Preparation and characterization of cellulose nanofibrils from coconut coir fibers and their reinforcements in biodegradable composite films *Carbohydr. Polym.* **211** 49–56.
[17] Kumneadklang S, O-Thong S and Larpkiattaworn S 2019 Characterization of cellulose fiber isolated from oil palm frond biomass *Mater. Today Proc.* **17** 1995–2001.
[18] Johar N, Ahmad I and Dufresne A 2012 Extraction, preparation and characterization of cellulose fibres and nanocrystals from rice husk *Ind. Crops Prod.* **37** 93–99.
[19] Mazlita Y, Lee H V. and Hamid S B A 2016 Preparation of cellulose nanocrystals bio-polymer from agro-industrial wastes: Separation and characterization *Polym. Polym. Compos.* **24** 719–28.
[20] Xie J, Hse C Y, De Hoop C F, Hu T, Qi J and Shupe T F 2016 Isolation and characterization of cellulose nanofibers from bamboo using microwave liquefaction combined with chemical treatment and ultrasonication *Carbohydr. Polym.* **151** 725–34.
[21] Hivechi A, Bahrami S H, Arami M and Karimi A 2015 Ultrasonic mediated production of carboxymethyl cellulose: Optimization of conditions using response surface methodology *Carbohydr. Polym.* **134** 278–84.
[22] Moussa I, Khiari R, Moussa A, Belgacem M N and Mhenni M F 2019 Preparation and Characterization of Carboxymethyl Cellulose with a High Degree of Substitution from Agricultural Wastes *Fibers Polym.* **20** 933–43.
[23] Rani M S A, Rudhziah S, Ahmad A and Mohamed N S 2014 Biopolymer electrolyte based on derivatives of cellulose from kenaf bast fiber *Polymers (Basel)* **6** 2371–85.
[24] Onyianta A J, Dorris M and Williams R L 2018 Aqueous morpholine pre-treatment in cellulose nanofibril (CNF) production: comparison with carboxymethylation and TEMPO oxidisation pre-treatment methods *Cellulose* **25** 1047–64.
[25] El-Sakhawy M, Tohamy H A S, Salama A and Kamel S 2019 Thermal properties of carboxymethyl cellulose acetate butyrate *Cellul. Chem. Technol.* **53** 667–75.
[26] Han F, Xiong D, Wang Q, Shao B and Chen M 2013 Thermal properties of carboxymethylcellulose and methyl methacrylate graft copolymers *J. Macromol. Sci. Part B Phys.* **52** 1242–9.
[27] Van Hai L, Son H N and Seo Y B 2015 Physical and bio-composite properties of nanocrystalline cellulose from wood, cotton linters, cattail, and red algae *Cellulose* **22** 1789–98.
[28] Hastuti N, Kanomata K and Kitaoka T 2018 Hydrochloric Acid Hydrolysis of Pulps from Oil Palm Empty Fruit Bunches to Produce Cellulose Nanocrystals *J. Polym. Environ.* **26** 3698–09.
[29] Wang X, Liu J, Gong Z, Huang C, He S, Yu L, Gan L and Long M 2019 Influence of degree of substitution of carboxymethyl cellulose on high performance silicon anode in lithium-ion batteries *Electrochemistry* **87** 94–9.

[30] Lu Q, Tang L, Lin F, Wang S, Chen Y, Chen X and Huang B 2014 Preparation and characterization of cellulose nanocrystals via ultrasonication-assisted FeCl3-catalyzed hydrolysis *Cellulose* **21** 3497–506.

Mechanical properties of bio-epoxy resins and synthetic epoxy resins blends

A Derahman[1*] and Z Z Abidin[2]

[1]Department of Chemical Engineering, Faculty of Engineering, UiTM Bukit Besi,23200 Terengganu;

[2] Department of Chemical and Environmental Engineering, Faculty of Engineering, University Putra Malaysia (UPM) 43400 Seri Kembangan, Selangor

*Corresponding author: aishahderahman@uitm.edu.my

Abstract. This paper studies the mechanical properties of bio-epoxy resins blended with synthetic epoxy resins (epoxamite). The bio-epoxy resins were derived from Jatropha methyl esters through epoxidation method. They were formulated with epoxamite and hardener at different compositions and then cured at different temperature and time settings. The cured blends were subjected to tensile and flexural tests using Instron machine. Tensile and flexural strength of the mixtures were compared with the 100% epoxamite in order to evaluate the suitability of bio-epoxy resins as an alternative to synthetic epoxy resins with respect to mechanical properties. Tensile strength of 100% epoxamite is 38.32 MPa and flexural strength is 63.32 MPa. The mixtures of bio-resins and epoxamite demonstrated very low mechanical strengths compared to the 100% epoxamite. Therefore, they are not suitable to be used as an alternative to synthetic epoxy resins in industrial applications. However, they may find other usage due to high reactivity of the bio-epoxy resins.

1. Introduction

In recent years, the formulation of polymers from bio-based materials and sustainable resources has gained attention due to polymers such as epoxy resins are used extensively in engineering and composite materials. However, these resins are synthesized using raw materials derived from non-renewable resources such as petroleum. Various issues associated with these resources in the context of depletion of the reserves and environmental problems have urged the need of an alternative or replacement [1,2].

Bio-epoxy resins derived from plant materials have been synthesized and their properties have been studied to evaluate the suitability for various engineering applications including paints, coating, adhesives, and biomedicine [3]. However, they have a tendency to form aggregates during processing and demonstrates a low resistance to moisture absorption which reduces the compatibility between matrix-fibre compatibility resulted in poor performance of composite material. Alkali treatment was introduced to enhance the compatibility of the composite materials by modifying the structure and the surface properties of the composite materials. Good adhesiveness, rigidity, specific resistance, dimensional stability, chemical resistance and high fluidity prior to curing which allows for an easy processing are some of the properties of epoxy ring that enabled them to be used in mechanical applications [4,5,6]. Matrix components reinforced with fibres formed a composite material. The selection of the material's composition depends on the desired composite material properties. It is important that the combined material's properties have improved compared to matrix or fibre materials alone [7]

Abundant researches have been conducted using synthetic epoxy resins in the formulation of composite materials [8,9,10]. However, with regards to the previous mentioned depletion of non-renewable resources and environmental issues, this research used epoxy resins prepared from fatty acid methyl esters (FAMEs) derived from Jatropha oil which is an interesting natural resource alternative as FAMEs are biodegradable, non-toxic, non-depletable, non-volatile, low cost and available in abundant [11,12]. The utilization of renewable resources is growing rapidly worldwide, and becoming a popular choice for alternatives. This includes Malaysia which is looking forward to achieve 50% utilization of renewable resources such as palm oil by 2025 [13] since palm oil is a major crop in Malaysia. However, it is still regarded as major source of vegetable oil all over the world rather than a fuel. Therefore, utilizing Jatropha oil will not become any issue since it is inedible and toxic for both human and animal consumption [1].

Few studies and experiments have been conducted utilizing epoxidized Jatropha oil in various applications including coating, adhesives, tribology, and polymers [14,15,16]. However, research about FAME or biodiesel in production of bio-epoxy resins is scarce. FAMEs or biodiesels have been used mainly in the automobile industry. The availability of its structure to undergo for chemical modifications has reopened new dimension to oleochemical synthesis, which primarily transforms the double bonds into oxirane ring that produces the bio-epoxy resins. The mechanical properties of bio-epoxy resins which are often evaluated are the tensile and flexural strength. Tensile strength is the ability of a material to resist a force that tends to pull it apart, while flexural strength means the maximum stress in a material just before it yields in a bending test.

This project is relevant to this aim in the exploration of finding alternative and improve awareness on the importance of renewable resources in this country. Previous studies have been done to synthesize epoxy bio-resins from Jatropha oil and evaluate its physical and chemical properties. However, in this study, bio-epoxy resins are synthesized from Jatropha methyl esters (JME). There is limited research on their mechanical properties. Therefore, this research is essential to evaluate their mechanical properties in order to observe their suitability as an alternative to synthetic epoxy resins.

2. Materials and methods

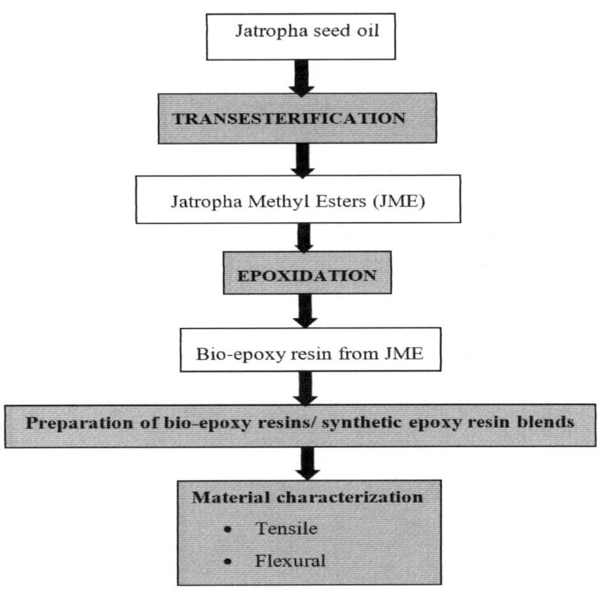

Figure 1. Process flow diagram

The process flow diagram for the production of bio-epoxy resins from Jatropha oil is shown in figure 1. Jatropha seed oil used in this work was purchased from BATC Development Berhad (Bionas) without further purification. Production of Jatropha methyl ester was prepared through transesterification, conducted in a 1000 ml tree-necked flask completed with mechanical stirrer, reflux condenser, and thermometer under atmospheric pressure. Bio-epoxy resins was synthesized using epoxidation method with peroxyacetic acid generated in-situ [17,18].

For the polymerization process, commercial synthetic epoxy resins (epoxamite) and hardener were used. For sample preparation, homogenous mixtures of bio-epoxy resins from this study and epoxamite were prepared in different compositions [19]. A mechanical stirrer was used to mix the composition and stirred about 10 min continuously. Then, the mixture was poured into steel mould plates (230 × 230 × 10 mm) and treated according to the curing conditions set based on the composition's percentage. Those resins blends were prepared to investigate the properties of the blends and also to evaluate their mechanical properties. The thickness of sample was 5–6 mm to ensure the compatibility with tensile and flexural machine for analysis. After curing was completed, all samples were cut using a vertical cutting machine (Metabo Bas 260 Swift, Germany) equipped with a three-roller guide; therefore, the saw blade is guided optimally and renders high precision cutting results for subsequent analysis.

2.1 Mechanical test

The mechanical properties of sample which are tensile and flexural strength were conducted using universal testing machine.

2.1.1 Tensile. Tensile tests were carried out following ASTM D3039-08 using universal test machine, Instron model 3382 with a load cell of 100 kN and a crosshead rate of 8 mm/min. This simple but reliable machine is originated from North America, and is ideal for tension and/or compression applications. Cured synthetic epoxy resins and bio-epoxy resins were cuts in the dimensions of 250 × 25 × 5 mm before being tested using Instron machine. Four specimens of each sample type were used in each test. The dimensions of samples were programmed into the machine's computer and then, the sample was positioned vertically and the grip was tightened at the top before being lowered down slowly. Subsequently, the lower grip was tightened firmly to prevent any slippage. The value of crosshead was entered and load was applied to the sample. The testing of the sample was prolonged until the sample broke. The procedure was repeated for other samples as well. The values of tensile strength and modulus were obtained and analysed accordingly.

2.1.2 Flexural. With regards to flexural tests, a universal test machine, Instron model 30 K with a load cell of 500 N was used as suggested by ASTM D790-10. A span:depth ratio of 16:1 were used with specimen dimensions 220 × 20 × 5 mm. The sample was mounted on two supports and loaded with crosshead speed of 8 mm/min. The test was prolonged until the sample broke. The procedure was repeated for other samples as well, and the values of flexural strength were obtained and analysed accordingly.

3. Results and discussion

In order to investigate the mechanical properties of bio-epoxy resins/synthetic epoxy resins (epoxamite) blends, five samples were prepared in different blend compositions as shown in figure 2, with curing temperature and time for each composition shown in table 1. Increasing the percentage of synthetic epoxy resins will require the increase of curing temperature and time. Bio-epoxy resins are expected to have lower performance compared to synthetic epoxy resins. The polymerization of bio-epoxy resins with hardener is slower, hence requiring higher temperature and longer curing time [19].

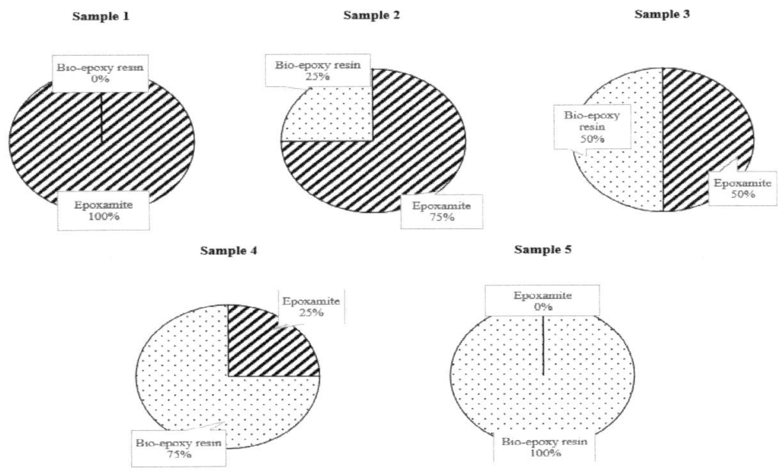

Figure 2. Percentage compositions five samples of bio-epoxy resins/ epoxamite blends

The mechanical strength of epoxamite blended with bio-epoxy resins was investigated by the measurement of tensile and flexural strength. Data for Samples 4 (75–25% bio-epoxy resins/epoxamite) and 5 (100% bio-epoxy resins) were unable to be obtained because the blends failed to crosslink and polymerization attempt resulted in high viscosity of liquid, even after curing. This could be due to the non-reacted hardener with the epoxides in the composition [19]. In order to undergo the tensile and flexural tests, the sample must be in solid form. Hence, Samples 4 and 5 were unable to be tested. As demonstrated in table 1, only data for Samples 1, 2, and 3 were obtained and discussed in further sections. [20] stated that tensile strength is taken as the stress value at the maximum of the nominal stress–strain curve. The ductility is expressed as percentage of total elongation at fracture. The average values for tensile strength and tensile modulus measurements are reported in table 2. The value obtained is in MPa.

Table 1. Curing temperature and time of bio-epoxy resins and synthetic resins blends

Samples	Curing conditions
Sample 1	24hr RT + 3hr (60 °C) + 3hr (80 °C)
Sample 2	24hr RT + 3hr (60 °C) + 3hr (80 °C)
Sample 3	24hr RT + 4hr (80 °C) + 4hr (120 °C)
Sample 4	24hr RT + 4hr (80 °C) + 8hr (120 °C)
Sample 5	24hr RT + 4hr (80 °C) + 12hr (120 °C)

3.1 Tensile

Tensile strength and modulus of blends for Samples 1, 2, and 3 are shown in table 2. Clearly, at 100% epoxamite composition (Sample 1), it had the highest values for tensile strength and tensile modulus with 38.32±3.16 MPa and 751.00±96.45 MPa, respectively. These values correlated with the strain-stress curve depicted in figure 3. An object or medium under stress becomes deformed and the quantity that describes this deformation is called strain. Strain is given as a fractional change in either length of the sample. Hence, it is a dimensionless number. The relationship of stress and strain is proportional as can be seen in the figure below; the greater the stress, the higher the strain.

Table 2. Tensile strength and tensile modulus

Sample	Bio-epoxy resins/epoxamite (%)	Tensile strength (MPa)	Tensile modulus (MPa)
1	0/100	38.32±3.16	751.00±96.45
2	25/75	2.42±1.2	6.00±1.5
3	50/50	0.49±0.24	0.75±0.34

From figure 3, it can be seen that Sample 1 has the lowest strain of 6% which indicated that the molecules in the cured resins are high in the interatomic bonding that is due to the formation of three-dimensional network and the strong chains making the material hard and tough [19]. Thus, it led to the difficulty of the sample to absorb stress which resulting a high value of tensile strength and tensile modulus. These results also explained the physical state of Sample 1 which was rigid and hard to bend by hand. These findings also conformed with the previous study which mentioned that rigid structure or the highly crosslinking density contributed to the high mechanical strength of [21]. Sample 2 (25–75% bio-epoxy resins/epoxamite) had tensile strength and tensile modulus of 2.42±1.2 MPa and 6.00±1.5 MPa, respectively, and the failure strain of Sample 2 was at 80% as depicted in figure 4.

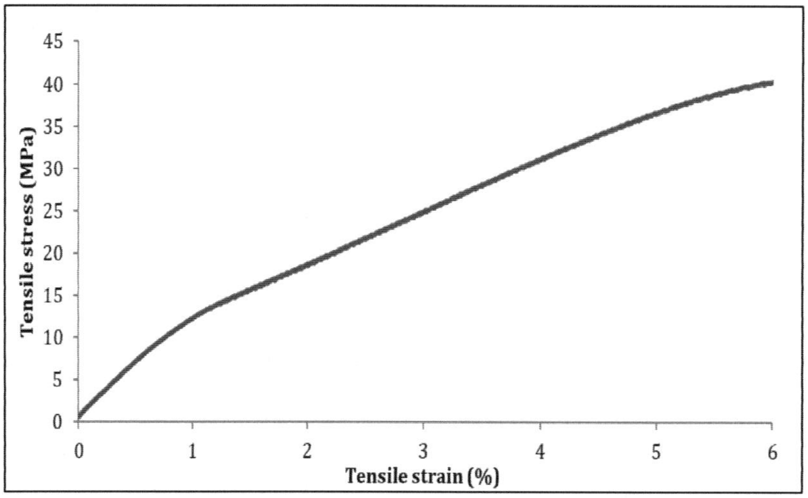

Figure 3. Tensile stress-strain curve of Sample 1(0–100% bio-epoxy resins/epoxamite)

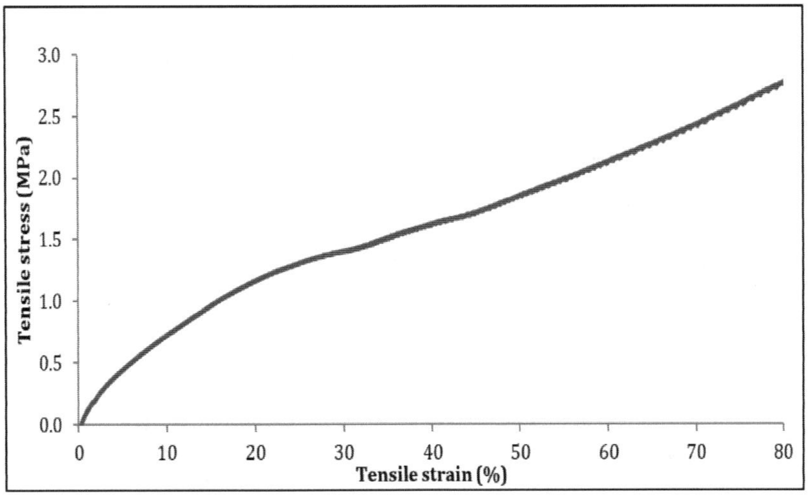

Figure 4. Tensile stress-strain curve of Sample 2 (25-75% bio-epoxy resins/Epoxamite)

Sample 3 (50–50% bio-epoxy resins/epoxamite) had the lowest tensile strength and tensile modules with the values of 0.49±0.24 MPa and 0.75±0.34MPa, respectively. The lowest values of tensile strength and tensile modulus of Sample 3 can be related to the stress-strain curve (figure 5) which illustrated the strain of 150%. High failure of strain in Sample 3 showed that the molecules allowed to withstand more stress when subjected to stress. These results also explained the physical state of Sample 3 which was more ductile compared to Sample 2.

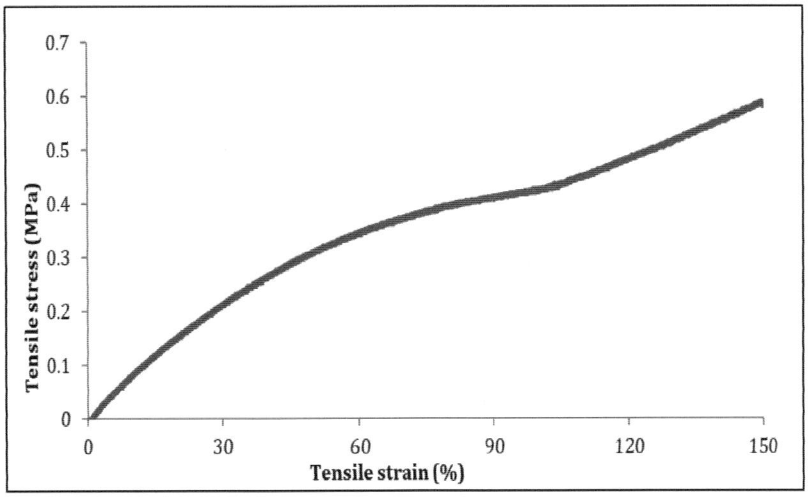

Figure 5. Tensile stress-strain curve of Sample 3(50-50% bio-epoxy resins/epoxamite)

Therefore, it can be concluded that the addition of bio-epoxy resins in the blend of the cured resins led to a low value of tensile strength and tensile modulus, which resulted into a high value amount of failure strain. Thus, it can be seen that the tensile strength and tensile modulus decreased as the amount of epoxamite composition decreased. A similar trend was observed in the studies of epoxidized hemp oils/jute fibre composites [4,7,22].

3.2 Flexural

Conducting flexural test generally helps to determine the degree of brittleness of the material. The brittle materials tend to break abruptly without any specific warning. However, symptoms of breakage are usually visible before the fracture point [23]. Flexural properties of biopolymers composites are strongly affected by the quality of the interface between the materials, including the static adhesion strength and the interfacial stiffness [24]. Flexural strength and flex modulus for tested samples are summarized in table 3. As can be seen, Sample 1 with 100% epoxamite had the highest value of flexural and flex modulus which were 63.32±7.52 MPa and 2260.82±86.32 MPa, respectively. The addition of bio-epoxy resins reduced values of flexural strength and flex modulus tremendously.

Table 3. Flexural strength and flex modulus

Sample	Bio-epoxy resins/epoxamite (%)	Flexural strength (MPa)	Flex Modulus (MPa)
1	0/100	63.32±7.52	2260.82±86.32
2	25/75	0.63±0.2	11.90+3.12
3	50/50	0.055±0.013	1.20±0.54

Graph of flexural stress-strain can be used to evaluate the changes in the properties of pure synthetic epoxy with addition of bio-epoxy resins. Figure 6 illustrated the graph of flexural stress-strain pure synthetic epoxy. At the highest flexural strength, the strain value was very low, 6%. Flexural stress and modulus were clearly reduced with the increment of bio-epoxy resins composition from 25% to 50%. Figure 7 depicted that as the composition of bio-epoxy resins was increased to 25%, the flexure stress and modulus of the sample reduced to 0.63±0.2 MPa and 11.90+3.12 MPa, respectively. It was observed that increasing the amount of bio-epoxy resins contributed to the higher formation of epoxy groups which leads to the brittleness of the materials through the reaction with the hydroxyl groups. As the sample easily bends and breaks during the tests, it indicated the weak linkages between the hardener and the bio-epoxy resins [19].

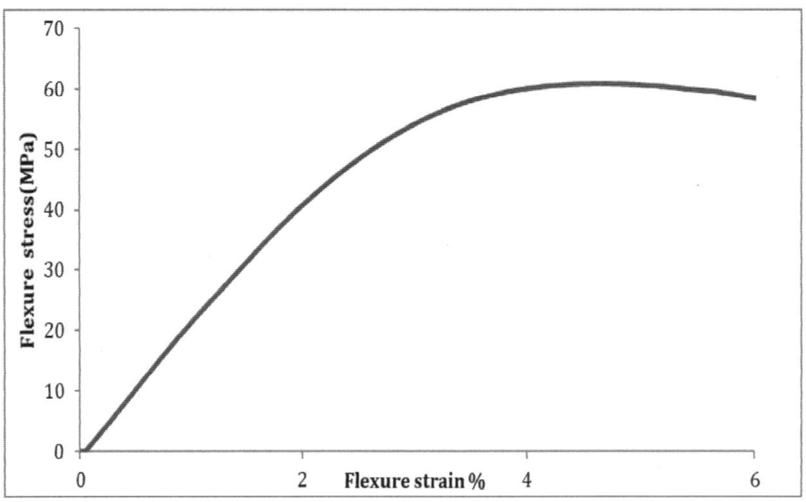

Figure 6. Flexural stress-strain curve of Sample 1 (0–100% bio-epoxy resins /Epoxamite)

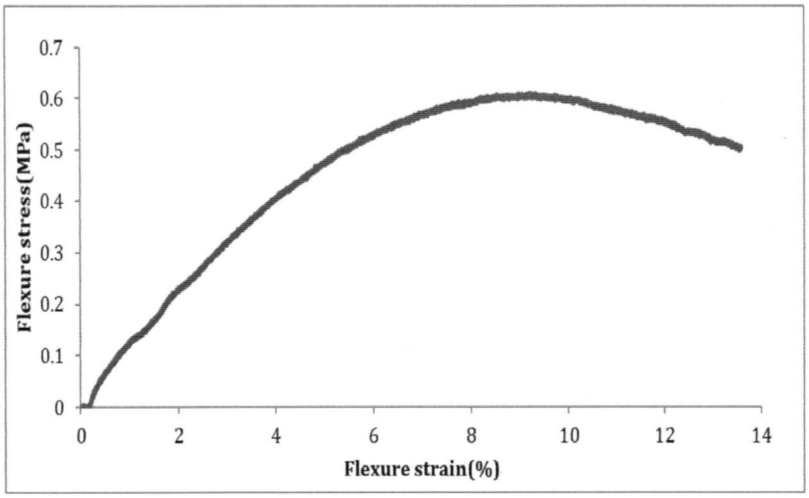

Figure 7. Flexural stress-strain curve of Sample 2 (25–75% bio-epoxy resins/epoxamite)

There is limited literature that discussed about bio-epoxy resins and epoxamite resins blends. In general, it is reported that the tensile strength and flexural strength of synthetic epoxy resins (epoxamite) are higher compared to the mixture of bio-epoxy resins and epoxamite blend [23]. This effect was due to the fact that the bio-epoxy resins is formed from monoglycerides which may contain unsaturated double bonds in the molecular structure. The absence of the double bonds in some of the transesterified methyl esters was the reason for the mechanical properties (tensile and flexural) of the bio-epoxy resins. A research conducted by [25] reported a similar trend, and stated that higher content of bio-epoxy resins decreases crosslinking network of epoxy by partially unreacted epoxidized methyl esters and fatty acid chains. In addition, the internal oxirane rings of bio-epoxy resins are less reactive than petroleum-based epoxies. Therefore, the mechanical performances of bio-based resins are low compared to a petroleum-based epoxy resins.

4. Conclusion

Synthetic epoxy resins and bio-epoxy resins blends were successfully prepared, and were subjected to mechanical tests. Based on the obtained results, the modulus of elasticity of blend materials is significantly lower than the 100% synthetic epoxy resins. Blends with higher percentage of bio-epoxy resins in the composition had lower mechanical because excess amount of epoxy groups leads to the formation of the ether groups and the copolymerization, making the material weak and easy to be compressed. The performance of bio-epoxy resins in the aspect of mechanical strength were still incomparable to the synthetic epoxy resins, hence are unsuitable to be used as an alternative for engineering purposes. However, since the content of the oxirane ring in the bio-epoxy resins makes it highly reactive, bio-epoxy resins synthesized from Jatropha methyl esters may find applications as biolubricants or reactive diluents. All in all, findings of this research are very important to provide another pathway for the production of high-quality materials using safe and economical techniques.

The following recommendations are suggested for future works:

- Evaluation of other properties of synthetic epoxy resins and bio epoxy resins blends for other applications such as lubricants, reactive diluents plasticizers, toughening agents and etc. These properties are included dynamic mechanical analysis (DMA), differential scanning calorimetry (DSC), drying time, scratch hardness, chemical resistance, rheology analysis and etc.
- The study of curing kinetics because time and temperature are essential to produce bio-epoxy resins-based polymers for specific purposes

ACKNOWLEDGEMENTS
The authors would like to thank the universities, UiTM and UPM, for all contributions that have made it possible to carry out this research.

REFERENCES
[1] Hamzah N H C, Khairuddin N, Siddique B M and Hassan M A 2020 *Processes* **87** 1.
[2] Tobib H M, Rostam H, Mossa M A A, Aziz Hairuddin A and Noor M M 2019 *IOP Conf. Ser.:Mat. Sci. Eng.* **469** 1.
[3] Kumar S 2017 *Polym.Plast. Technol. Eng.* **11.**
[4] Fernandes F C, Kirwan K, Lehane D and Coles S R 2017 *Eur. Polym. J.* **89** 449.
[5] Goncalves J, Adalberto D, Campos T, Oliveira G D J and Elze J R 2014 *Mater. Res.* **174** 878.
[6] Kumar S, Mohanty S and Nayak S K 2020 *J. Macromol. Sci. Part A: Pure d Appl. Chem.* **579** 654.
[7] Loong M L and Cree D 2018 *J. Polym Environ.* **261** 224 .
[8] Jiang L, Walczyk D, Mcintyre G, Bucinell R and Li B 2019 *J. Cleaner Prod.* **207** 123.
[9] Joel R D, Arumugaprabu V, Rajasekar E and Santhosh G 2018 *Mater. Today: Proceedings* **52** 6815.
[10] Ronald M and Mwasiagi J I 2018 *Fibres Text East. Eur.* **26** 87 .
[11] Gairola S P, Tyagi Y K, Gangil B and Sharma A 2020 *Mater. Today: Proceedings* 1.
[12] Hasnan N H A, Yaakob N, Abu Kassim M N and Mohd Noh U A 2020 *Indonesia J. Chem*, **204** 858.
[13] Zakaria S U, Basri S, Kamarudin S K and Majid N A A 2019 *IOP Conf. Ser.: Earth Environ. Sci.* **268** 1.
[14] Carbonell-Verdu A, Bernardi L, Garcia-Garcia D, Sanchez-Nacher L and Balart R 2015. *Eur. Polym. J.* **63** 1.
[15] Hafiezal M R M, Khalina A, Zurina Z A, Azaman M D M and Hanafee Z M 2019 *J.Compos. Sci.* **31** 6 .
[16] Latif F E A, Abidin Z Z, Cardona F, Biak D R A, Abdan K, Tahir P M, and Ern L K 2020 *Processes* **8** 1.
[17] Derahman A, Abidin Z Z, Cardona F, Biak D R A, Tahir P M, Abdan K and Liew K E 2019 *Braz. J. Chem. Eng.* **362** 959.
[18] Tegene D and Yadessa M 2017 *Am. J. Appl. Chem.* **51** 1.
[19] Agarwal K K and Agarwal G 2019 *Technol. Innovation Mech. Eng.* **6.**
[20] Narducci R, Chailan J F, Fahs A, Pasquini L, Di Vona M L and Knauth P 2016 *J. Polym. Sci., Part B: Polym. Phys.* **5412** 1180.
[21] Qi M, Xu Y J, Rao W H, Luo X, Chen L and Wang Y Z 2018 *RSC Adv.* **847** 26948.
[22] Manthey N W, Cardona F, Francucci G and Aravinthan T 2013 *J. Reinf. Plast. Compos.* **3219** 1444.
23] Jenarthanan M P, Marappan K,] and Giridharan R 2019. *Pigment Resins Technol.* **483** 243.
[24] Al-Namie I, Ibrahim A A and Hassan F M 2011 *Iraqi J. For Mech. Mater. Eng.* **113** 486.
[25] Mustapha R, Rahmat A R, Abdul Majid R and Mustapha S N H 2019 *Polym.-Plast.Technol Mat.* **58** 1311.

IOP Conf. Series: Materials Science and Engineering 1176 (2021) 012006 doi:10.1088/1757-899X/1176/1/012006

Toxicity Study of *Ananas Comosus* and *Centella Asiatica* as Active Ingredients in Anti-inflammatory Lotion

N S A Rahman[1*], N Saadon[1], F M Tap[1], N A Yusof[2]

[1]Faculty of Chemical Engineering, Universiti Teknologi MARA Cawangan Terengganu, Kampus Bukit Besi, 23200 Dungun Terengganu

[2]Academy of Language Studies, Universiti Teknologi MARA Cawangan Terengganu, Kampus Bukit Besi, 23200 Dungun Terengganu

*Corresponding author: nursyukriah@uitm.edu.my

Abstract. Inflammation, a common disease faced by people, could lead to chronic inflammatory diseases, and worst, cancer. As a result, many inflammatory patients use nonsteroidal anti-inflammatory drugs (NSAIDs) as to relieve their inflammation, which are available in the market. However, these medicines do not cure the disease but relieve it temporarily, and the patients are exposed to their dangerous long-term effects. Since Malaysia is rich with fruits and herbs that are rich in therapeutic properties, more natural remedies from the local fruits and herbs should be discovered. Based on the previous research done, the combination of *Ananas comosus* and *Centella asiatica* could have the potential in relieving inflammation. Therefore, this preliminary research was aimed in determining the toxicity in combining the two main active ingredients using *in vitro* MTT assay against HaCat Cell lines. The findings proved that at the highest concentration (10000 µg/ml), no cytotoxic effect has been recorded. The findings from this research would encourage more natural remedies to be offered to the inflammation patients, who eventually could avoid themselves from depending solely on the harmful medicines.

1. Introduction

Inflammation is said to be experienced by everyone [1]. It is a response from our immune system to protect our body from infection, injury or disease. Our body's white blood cells release chemicals into the blood or affected tissues in order to protect our body from foreign substances. As a result of the release, the blood flows to the injured or infected area increases, and redness and warmth may occur [2]. Inflammation has two types, which are acute and chronic inflammation and the later can have long-term and whole-body effects, which could lead to diseases like heart attack, stroke and cancer [3]. Many acute inflammatory patients use nonsteroidal anti-inflammatory drugs (NSAIDs) as to relieve their inflammation, like rashes. NSAIDs are also found to inhibit the protein target such as COX-1 [4].

The NSAIDs are chemical-based drugs often used as the inhibitors for both acute and chronic inflammation [5]. They are chemically synthesized and designed to block the production of inflammatory mediators. There are many NSAIDs available in the market, such as aspirin and ibuprofen, but these medicines do not cure the disease but relieve it temporarily, and the patients are exposed to their dangerous long-term effects, such as stomach ulcers, kidney damage and increase the risk of heart attack or stroke [6]. Based on a previous study, NSAIDs were used to relieve the pain rather than to cure it [7]. Besides that, these drugs are also associated with several complications. The long-term used of NSAIDs often leads to a negative impact on the liver and renal system. The most common side effect is gastric bleeding [8]. The long-term use of steroid, either alone or combined with NSAIDs, will disturb the immune system of our body rigorously [9]. Thus, the negative complications of the usage give a clear signal for more discoveries of natural remedies, specifically in the treatment of inflammation.

Abundant of bioactive compounds can be found in food such as fruits and herbs. Researchers have high interest in phytochemicals because of their roles in the prevention of degenerative diseases such as cancer [10]. Furthermore, these compounds also have many active substances that can control the activities of a wide range of enzymes and cell receptors [11]. Bioactive compound, such as polyphenol, is also a natural antioxidant found in food [12]. It is capable of scavenging radicals such as hydroxyl (OH') and peroxyl (ROO') [13]. Protein in plant, such as bromelain in *Ananas comosus*, has attracted high attention in therapeutic application because of its non-toxic properties. The use of bromelain as a hydrolysing agent in releasing the antioxidant and antimicrobial peptides had also been reported [14].

Treatment with a single anti-inflammatory drug often causes multiple adverse effects such as hepatoxicity, gastrointestinal haemorrhage, meningitis and asthma [15]. Therefore, a novel strategy combining two or more potential compounds that exhibit inhibitory effects against inflammation activity has been proposed to overcome the problem. Moreover, this approach has previously been shown to enhance the thermal stability of the combination [16]. Inflammation is categorised as a complex disease, thus it needs a multi target to reduce the inflammation. The designing of multi-targeted drug combination is crucial in the development of a sustainable treatment on inflammation. Drug combination is a method of combining two or more active pharmaceutical ingredients into a single dosage form and this is a promising approach in interfering a complex disorder [17]. For example, a drug combination was used for treating complex diseases such as cancer [18] and HIV [19]. Combinations of natural products, such as alkaloids, have been discovered as new potential compounds that can treat sleeping sickness (trypananosomiasis) [20]. Previous study claimed that the combination compound also provides synergistic effect on inhibition of inflammation [21].

Since Malaysia is rich with fruits and herbs which are rich in therapeutic properties, more natural remedies should be discovered from the local fruits and herbs. Based on the previous research done, the combination of *Ananas comosus* and *Centella asiatica* could have the potential in relieving inflammation since studies have been conducted on them separately and each was found to be an anti-inflammatory agent. The leaf extract of *Ananas comosus* shows significant lesser inflammation effect as compared to that in the carrageenan control group. At 48 hours post injection of carrageenan, a significant decrease in paw volume was observed with 500 mg/kg b.w. (0.78 ± 0.13 ml) of *Ananas comosus* leaf extract as compared to that in the carrageenan control group (1.3 ± 0.18 ml) [15]. Meanwhile, *Centella asiatica* exhibited promising effect on inflammatory responses by macrophages in an atopic dermatitis (AD) mouse model. The histopathological analysis showed that *Centella asiatica* inhibited hyperkeratosis, proliferation of mast cells and infiltration of inflammatory cells [22].

Extracts of plant components can be achieved using different processes and using different medium such as alcohol, glycerine, oil and vinegar [23] and some of them use very low or no heat to produce extracts. To avoid damages of a plant's components, glycerine is used as it requires no heating and it dissolves the soluble components of the plant. As a result, the active ingredients which are soluble to glycerine will be recovered in the extract. In a skincare product, an active ingredient is the chemical in the skincare that fulfils the function or purpose of the product [24] and in this research, *Ananas comosus* and *Centella asiatica* are the active ingredients since they are anti-inflammatory agents.

Hence, based on the needs to discover more natural remedies, this research was conducted to determine the toxicity in combining the two main active ingredients, which are *Ananas comosus* and *Centella asiatica*, before they are used in an anti-inflammatory lotion formulation. Ensuring the safety of the product is part and partial of the manufacturer's responsibility on the consumers [25].

2. Research method

2.1. Materials
The emulsifier derived from pure castor and glycerin were purchased from Gobiotics, while sweet almond oil, virgin olive oil, virgin grape seed oil, rice bran oil, jojoba oil, and coconut oil were purchased from an organic oil shop. $MgCl_2$ and glycerine were purchased from a chemical supplier. The rose hydrosol, distilled water, ananas glycerite, *Centella asiatica* glycerite were prepared by the formulator. Preservative ECO or Geogard ECT which consists of Benzyl Alcohol, Salicylic Acid, Glycerin and Sorbic Acid was obtained from Aromantic Limited. Phosphate buffer saline, (3-(4,5-dimethyl-thiazol-2y) 2,5-diphenyl-tetrazolium bromide (MTT), and acetic acid were purchased from Merck Sdn. Bhd., Malaysia. Modified eagle's medium (MEM contains amino acids, vitamins, inorganic salts, dextrose and phenol red), fetal bovine serum, platelet derived growth factor, phosphate buffer saline (PBS), trypsin-EDTA (0.25% trypsin, EDTA 4Na), and dimethyl sulfoxide (DMSO) were purchased from GIBCO®, USA.

2.2. Preparation of Extract
Fresh *Ananas comosus* was peeled and *Centella asiatica* leaves were cleaned using distilled water. The plant materials were allowed to dry and later, chopped in a thin and small size. The *Ananas comosus*, *Centella asiatica*, glycerine and preservative ECO were weighted. The 100 g of *Ananas comosus* and *Centella asiatica* infusions were prepared separately. All ingredients were mixed in an airtight and sanitized glass container. The container was placed away from the direct light and agitated for 10 days and on the 11th day of infusion, the mixtures were strained using a coffee filter. The strained liquid or the extract was then transferred into a final container and stored at 2°C.

2.3 Cell cultures
Human Skin epidermal keratinocytes (HaCaT) were purchased from Elabscience and were cultured according to the protocol provided by the depositor. HaCaT cell lines were grown in DMEM supplemented with 10% FBS and 10% FCS with additional 1% PS at 37°C under humidified atmosphere of 5% CO_2.

2.4 MTT assay
The cytotoxicity study on the combined extract of Ananas comosus and Centella asiatica was examined by using 3-[4, 5-dimethyl-thiazol-2-yl]-2, 5-dipehenyl tetrazoliumbromide (MTT) assay [26].

2.5 Toxicity test
The cells were seeded at a density of $2x10^5$ cells/well in a 96-well plate, excluding the first row, and incubated in humidified atmosphere for 24 hours prior to the treatment. The test samples were prepared by dissolving the combined extract in DMEM to give the final concentrations of the crude extracts at 10000, 5000, 2500, 1250, 625, 312.5, 156.25, 78.125, 39.06, and 0 µg/ml using serial dilution calculation. The medium was replaced after 24 hours with 200 µL of complete DMEM and a serial dilution of the combined plant extract. The cells with and without the serial dilution of combined extract were incubated again in humidified atmosphere for 24 hours. After the incubation, the cells were washed using PBSA, which was removed later. 20 µL of freshly prepared MTT solution (5 mg/ml dissolved in PBS) was added into each well and the cells were incubated again at 37°C for 5 hours. The MTT solution was then removed and replaced with 200 µL DMSO for 15 minutes to allow the dissolution of the purple MTT formazan crystal. The absorbance was measured at 570 nm and samples were run in 6 replicates. Untreated cells were used as the control.

2.6 Statistical analysis
All data were expressed in mean±SEM. Statistical analysis was performed using Graphpad prism with one way ANOVA and Tukey test. Significant differences were considered as $p<0.05$.

3. Result and discussion

The effect of different concentrations of the combined extract on cell viability was measured by using MTT assay. MTT assay is a useful *in vitro* model for testing cytotoxicity because it can show the ability of the compound to stimulate the proliferation of the human dermal keratinocytes (HaCaT). HaCaT is one of the cells that will be affected by toxic substances also known to be involved in a various of immune responses and inflammatory of the skin. Therefore, HaCaT has been selected as it is the most appropriate target cells for evaluating sensitivity of the skin to toxicants. Basically, toxicants will affect basic functions of the cells and the toxicity can be determined by assessing cellular damage. Thus, the effect of the combined extract on the HaCaT proliferation was determined and the results were tabulated in figure 1.

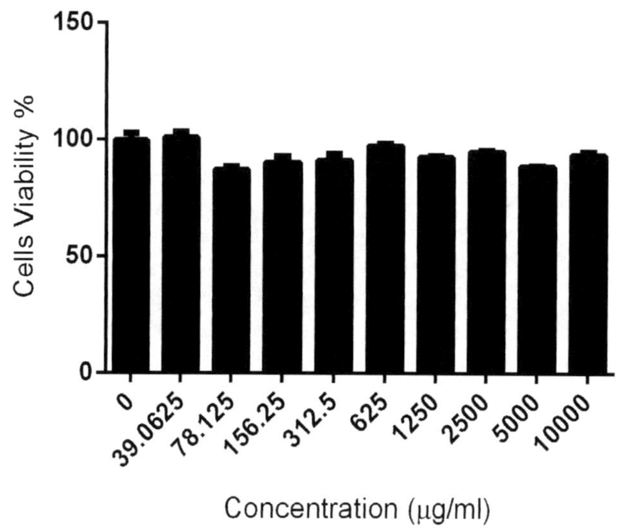

Figure 1. The Effects of the combined extract of *Ananas comosus* and *Centella asiatica* on HaCaT cell proliferation during treatment for 24 hours.

In overall, all of the extracts give percentage of cell viability higher than 80% with the highest percentage of cell viability was shown at the concentration of 39.0625 μg/ml while the lowest cell viability was recorded at 78.125 μg/ml. However, no significant difference ($p>0.05$) has been observed among all the concentrations and control (concentration = 0 μg/ml) which indicates no cytotoxic effect was observed even at the highest concentration.

In vitro MTT assay was performed in order to determine the toxicity effect of active ingredients towards the basic cellular functions and generally based on the enzymatic conversion of MTT in the mitochondria. Thus, the cytotoxicity effect of the plant extracts at high concentration can be related to the alterations of the basic functions of mitochondria which might cause the apoptosis and necrosis occurred, thus resulted in the death of cells [27,28]. Surprisingly, at the highest concentration (10000 μg/ml), no cytotoxic effect has been recorded which showed that the combination extract does not alter the basic functions of mitochondria thus possesses a wide margin safety level upon exposure of high dose.

This current finding is in accordance with the findings from previous studies, where they found that the extract of *Ananas comosus* and *Centella asiatica* exhibited low cytotoxicity effect against human ovarian SKOV-3 and human dermal keratinocyte respectively[29,30]. Combined crude extracts only contain natural ingredients in various amounts which may explain its low cytotoxicity effect against HaCaT. Furthermore, the use of herbal in the combination is believed to produce synergistic therapeutic activity but with a lesser side effect. This recent finding suggests that the combined extract of *Ananas comosus* and *Centella asiatica* is safe on human dermal keratinocytes where all tested concentrations showed above 80% on cell viability. The high percentage of cell viability showed that the combined extract could enhance the proliferation of HaCaT cell lines thus initiate the release of repair and inflammatory mediators in response to Interleukin-1β and TNFα.

With the said results, the ethnopharmacological approach in selecting this combination for study is useful for current knowledge contribution and this arises the needs for future exploitation on its potential anti-inflammatory activity.

4. Conclusion

As a conclusion, the finding from this preliminary study gives a secure assurance that *Ananas comosus* and *Centella asiatica* are a suitable combination as active ingredients in an anti-inflammatory lotion formulation. The finding provides alternatives in providing a remedy for inflammation and could encourage more natural remedies to be offered to the inflammation patients, who eventually could avoid themselves from depending solely on the harmful medicines. Further research should be conducted on the formulated lotion, which has the combination of *Ananas comosus* and *Centella asiatica* as its active ingredients, to determine its effectiveness in reducing inflammation.

References

[1] Singh R, Mishra M K and Aggarwal H 2017 Inflammation, Immunity, and Cancer *Mediators Inflamm.* **2017.**

[2] Levine B, Mizushima N, Virgin H W and Manuscript A 2011 Autophagy in immunity and inflammation *Nature* **469** 323–35.

[3] Furman D, Campisi J, Verdin E, Carrera-Bastos P, Targ S, Franceschi C, Ferrucci L, Gilroy D W, Fasano A, Miller G W, Miller A H, Mantovani A, Weyand C M, Barzilai N, Goronzy J J, Rando T A, Effros R B, Lucia A, Kleinstreuer N and Slavich G M 2019 Chronic inflammation in the etiology of disease across the life span *Nat. Med.* **25** 1822–32.

[4] Varga Z, Sabzwari S R A and Vargova V 2017 Cardiovascular Risk of Nonsteroidal Anti-Inflammatory Drugs: An Under-Recognized Public Health Issue ed A Muacevic and J R Adler *Cureus* **9** e1144.

[5] Rodrigues E B, Farah M E, Bottós J M and Aggio F B 2016 Nonsteroidal anti-inflammatory drugs in the treatment of retinal diseases *Retinal Pharmacotherapeutics* **55** p 212–20

[6] Davis A and Robson J 2016 The dangers of NSAIDs: look both ways *Br. J. Gen. Pract.* **66** 172–3.

[7] Nalamachu S 2013 An overview of pain management: the clinical efficacy and value of treatment. *Am. J. Manag. Care* **19** s261-6.

[8] Goldstein J L and Cryer B 2015 Gastrointestinal injury associated with NSAID use: a case study and review of risk factors and preventative strategies *Drug. Healthc. Patient Saf.* **7** 31–41.

[9] Bancos S, Bernard M P, Topham D J and Phipps R P 2009 Ibuprofen and other widely used non-steroidal anti-inflammatory drugs inhibit antibody production in human cells *Cell. Immunol.* **258** 18–28.

[10] Singh S, Sharma B, Kanwar S S and Kumar A 2016 Lead Phytochemicals for Anticancer Drug Development *Front. Plant Sci.* **7** 1667.

[11] Medini F, Bourgou S, Lalancette K, Snoussi M, Mkadmini K, Coté I, Abdelly C, Legault J and Ksouri R 2015 Phytochemical analysis, antioxidant, anti-inflammatory, and anticancer activities of the halophyte Limonium densiflorum extracts on human cell lines and murine macrophages *South African J. Bot.* **99** 158–64.

[12] Masisi K, Beta T and Moghadasian M H 2016 Antioxidant properties of diverse cereal grains: A review on in vitro and in vivo studies *Food Chem.* **196** 90–7.

[13] Nimse S B and Pal D 2015 Free radicals, natural antioxidants, and their reaction mechanisms *Rsc Adv.* **5** 27986–8006.

[14] Salampessy J, Phillips M, Seneweera S and Kailasapathy K 2010 Release of antimicrobial

peptides through bromelain hydrolysis of leatherjacket (Meuchenia sp.) insoluble proteins *Food Chem.* **120** 556–60.

[15] Kargutkar S and Brijesh S 2018 Anti-inflammatory evaluation and characterization of leaf extract of Ananas comosus *Inflammopharmacology* **26** 469–77.

[16] Talib S Z and Hasyim K K 2008 Stability of bromelain-polyphenol complex in pineapple juice *Chem. Eng.* **49** 27–38.

[17] Majid A 2011 Thermal stability of free bromelain and bromelain-polyphenol complex in pineapple juice *Int. Food Research J.* **18** 1051–60.

[18] Wu G-S, Lu J-J, Guo J-J, Huang M-Q, Gan L, Chen X-P and Wang Y-T 2013 Synergistic anti-cancer activity of the combination of dihydroartemisinin and doxorubicin in breast cancer cells *Pharmacol. Reports* **65** 453–9.

[19] Wildum S, Zimmermann H and Lischka P 2015 In Vitro Drug Combination Studies of Letermovir (AIC246, MK-8228) with Approved Anti-Human Cytomegalovirus (HCMV) and Anti-HIV Compounds in Inhibition of HCMV and HIV Replication *Antimicrob Agents Chemother.* **59** 3140–8.

[20] Sommerfelt R M 2014 *Molecular mechanisms of inflammation-a central role for cytosolic phospholipase A2*, PhD Thesis, NTNU-Trondheim.

[21] Mohamed Tap F, Abd Majid F A, Ismail H F, Wong T S, Shameli K, Miyake M and Ahmad Khairudin N B 2018 In Silico and In Vitro Study of the Bromelain-Phytochemical Complex Inhibition of Phospholipase A2 (Pla2) *Molecules* **23** 73.

[22] Ju Ho P, Jun Sung J, Ki Cheon K and Jin Tae H 2018 Anti-inflammatory effect of Centella asiatica phytosome in a mouse model of phthalic anhydride-induced atopic dermatitis *Phytomedicine* **43** 110–9.

[23] Kumadoh D and Kwakye K 2017 Dosage Forms Of Herbal Medicinal Products And Their Stability Considerations-An Overview *JCR* **4(4)** 1-8.

[24] Zappelli C, Barbulova A, Apone F and Colucci G 2016 Effective active ingredients obtained through biotechnology *Cosmetics* **3** 39.

[25] Ross G 2006 A Perspective on the Safety of Cosmetic Products: A Position Paper of The American Council on Science and Health *Int. J. Toxicol.* **25** 269–77.

[26] Khazaei S, Esa N M, Ramachandran V, Hamid R A, Pandurangan A K, Etemad A and Ismail P 2017 In vitro Antiproliferative and Apoptosis Inducing Effect of Allium atroviolaceum Bulb Extract on Breast, Cervical, and Liver Cancer Cells *Front. Pharmacol.* **8(5)**.

[27] Bakkali F, Averbeck S, Averbeck D and Idaomar M 2008 Biological effects of essential oils – A review *Food Chem. Toxicol.* **46** 446–75.

[28] Rafiquzzaman R 2010 Cytotoxic Effects of Quercus Infectoria Extracts towards Cervical (Hela) and Ovarian (Caov-3) Cancer Cell Lines *Heal. Environ. J.* **1.**

[29] Abdul Halim N A, Ramasamy S, Tan B C, Khalid N and Yaacob J S 2018 In vitro shoot regeneration and analysis of biochemical, antioxidant and anticancer properties of Ananas comosus var. MD2 *Malaysian J. Fundam. Appl. Sci.* **14** 263–8.

[30] Azis H A, Taher M, Ahmed A S, Sulaiman W M A W, Susanti D, Chowdhury S R and Zakaria Z A 2017 In vitro and In vivo wound healing studies of methanolic fraction of Centella asiatica extract *South African J. Bot.* **108** 163–74.

IOP Conf. Series: Materials Science and Engineering 1176 (2021) 012007 doi:10.1088/1757-899X/1176/1/012007

Effectiveness of *Annona Squamosa* and *Annona Muricata* Seed Extracts as Ingredients in Bio-pesticides Spray

Z Irwan[1]*, W F W Kamarudin[1], U A S A Korish[2], N A Z[2], A S Rusli[2] and S Sallehuddin[2]

[1]Faculty of Applied Sciences, Universiti Teknologi MARA (Terengganu), 23200, Bukit Besi, Terengganu

[2]Faculty of Chemical Engineering, Universiti Teknologi MARA (Terengganu) 23200, Bukit Besi, Terengganu

*Corresponding author: zilda213@uitm.edu.my

Abstract. Pesticides released chlorofluorocarbon (CFC) that can harm to the environment and human health. Therefore, it is a need for development of an alternative methods in pest management due to negative effects of pesticides for human health and the environment. The objective of this study was to evaluate the effectiveness of *Annona Squamosa* and *Annona Muricata* seed extracts as ingredients in bio-pesticides spray on killing crickets (*Gryllidae*) and mealworm (*Tenebriomolitor*).The seeds of *Annona Squamosa* and *Annona Muricata* were dried, powdered and extracted in soxhlet's apparatus using *n*-hexane as a solvent at 60°C. The extracted oil was placed in the rotary evaporator for purification process. As a result, mixture of extracts of *Annona muricata* and *Annona squamosa* induced high mortality rate to mealworm and cricket compared to extracts of *Annona squamosa* and extract of *Annona muricata* at all concentrations tested. Therefore, the extracts of these *Annona muricata* and *Annona squamosa* may be used as one of the alternatives as bio-pesticides that offer cheaper and environmentally friendly method.

1. Introduction

Pesticides are substance, that is being used as a plant growth regulator intended for preventing, destroying, or mitigating any pest. However, synthetic pesticides that has been widely used may cause harmful health effects to humans such as cancer, shorts and long-term injury to the lung damage, nervous system, reproductive dysfunction of the hormone and immune systems [1]. Despite, pesticides can affect environment by contaminating water, soil, turf, and other vegetation. Moreover, pesticides can be toxic to a host of other organisms including birds, fish, beneficial insects, and non-target plants in addition to killing insects or weeds [2]. The common chemical ingredients that used in pesticides are Bendiocarb, Bifenthrin and Boric Acid. Due to its toxicity for human health and its impact on environment, it is become an urgent need to search for more environmental and toxicologically safe natural pesticides. Natural pesticides are one of the cheaper and safer alternatives in pest management. Bio-pesticides derived from natural materials such as plants, bacteria, animals, and certain minerals are naturally occurring substances that can control pests by non-toxic mechanisms [3]. It is usually less toxic than conventional pesticides which by contrast, produced by synthetic materials that directly kill or inactivate the pest.

Plant extracts were likely the earliest agricultural biocontrol, as history records that nicotine was used to control plum beetles as early as the 17th century. Experiments involving biological controls for insect pests in agriculture date back as far as 1835, when white-muscadine fungus (Beauveria bassiana) has been demonstrated as one of causes for an infectious disease in silkworm [4]. In most cases, the active ingredients of biopesticides are formulated in the same way as the synthetic pesticides and most convenient for farmers to use the same equipment for application [5]. Recently, several plants such as *Annona Muricata* and *Annona Squamosa* have been reported to have insecticidal properties and therefore can be used as bio-pesticide to control pests.

Annona Muricata is an evergreen blooming tree that is mostly distributed in tropical and subtropical regions of the world. It is a member of family *Annonaceae* known as soursop, graviola and guanabana. The fruit is locally referred as *durian belanda*. Research showed that more than hundreds of bioactive compounds such as alkaloids, acetogenins, and phenols were found in *Annona Muricata* plant. These bioactive compounds in vitro study have been identified as antioxidant, anti-inflammatory, anti-microbial, anti-cancer, insecticidal and larvicidal while in vivo studies found that the presence of these bioactive compounds as anti-stress, anxiolytic, immunomodulatory, anti-inflammatory, antidepressant, wound healing, antimalarial, hypoglycemic, hepato-protective, gastro protective, anticancer and anti-tumoral activities [6]. Furthermore, phytochemical studies reveal that most of annonaceous acetogenins usually been found in leaves, barks, seeds, roots and fruits of *Annona Muricata* which this annonaceous acetogenins are the major constituents of *Annona Muricata* [6]. Active chemical compounds in the form of acetogenin found in leaves and seeds of soursop can be used to control the trips attack, suppress the intensity of the attack of green ladybugs and reduce the population [7]. Acetogenin also has a properties as an insecticidal (anti-insect-worm), antifeeding or lowering appetite properties [7]. Investigation on an effectual level of *Annona Muricata* (soursop) extracts on mosquito vectors namely, *Aedes aegypti*, *Anopheles stephensi* and *Culex quinquefasciatus has also been done*. The study proved the potential of soursop extract as larvicidal agent in larvae of *A. aegypti.* with modifications in biochemical constituents[8].

Annona squamosa, commonly known as Custard Apple, is a native of West Indies and is cultivated throughout India, mainly for its edible fruit. The plant contains aporphine alkaloids, carvone, linalool, limonene, squamosin and quercetin [9]. Acetogenins, another a characteristic group of compounds isolated from *Annona Squamosa* seeds have been suggested to act as potential anti-neoplastic agents and are also the principal insecticidal constituents of *Annona* seed extract [10]. Custard apple extracts have shown promise for pest control against a range of insect pests. Laboratory and field tests showed that extracts from custard apple kernels were effective against crop pests like caterpillar. The result showed that soursop leaf extract is effective as a biopesticide material to control leaf caterpillar pest for delima rose apples. [11]. Insecticidal activity of the seed extracts of custard apple, *Annona Squamosa* in petroleum spirit, ethyl acetate, acetone and methanol against on larvae and adult beetles was also studied. It is reported that the seeds of *Annona Squamosa* were reported to have insecticidal and abortifacient properties. The crude oils from seeds of *Annona Squamosa* at 2.5 percent concentrations significantly reduced the leaf damage caused by larvae and adult beetles [10].

Recent research has been focusing on extraction of mixture of *Annona Squamosa* and *Annona Muricata* to increase its performance as bio-pesticides. Research by [13] found that the seed extracts of mixture of *Annona Squamosa* and *Annona Muricata* can be used as one of the alternatives in controlling mosquito that offer a new a cheaper and environmentally friendly method. Meanwhile, research by [12] proved the potential utilization of secondary metabolites in soursop (*Annona Muricata*) and sugar apple (*Annona Squamosa*) as plant-based pesticides to control pests and diseases of plants. Due to its (extraction of mixture of *Annona Squamosa* and *Annona Muricata*) potential as natural bio-pesticides, therefore, the objective of this research was to determine the effectiveness of mixture of Annona *Squamosa* and *Annona Muricata* extract in killing crickets (*Gryllidae*) and mealworm (*Tenebriomolitor*). The extraction using soxhlet extractor method is used with *n*-hexane as solvent.

2. Research Method

2.1. Materials/Instruments

The materials and instruments that are used to prepare the bio-pesticide are *Annona Squamosa* and *Annona Muricata* seed, 250ml flask of Soxhlet apparatus for extraction, rotary evaporator, the n-hexane as a solvent for the extraction of seed.

2.2. Method

After collection of seeds, both *Annona Squamosa* and *Annona Muricata* seeds were removed manually from these fruits, washed in water, and dried under the fan and was crushed into fine powder by using blender. The seeds were then further dried for 24 hours inside the laboratory oven to remove moisture content. A total of 500 grams of dried *Annona Squamosa* and *Annona Muricata* seeds were obtained, respectively.

2.3. Extraction Bio-Pesticide

After the drying process, 30 grams of each sample were weighed and placed inside the Soxhlet extraction thimble. 250ml of n-hexane was taken in the round bottom flask. The temperature was maintained at 60°C for boiling 4 hours. After 4 hours hexane with the extracted oil was taken out from the flask; and the seed sample was allowed to be cooled before further process. The sample then is placed into the rotary evaporator for purification process; to remove the solvent and the extract was separated out.

2.4. Preparation and Application of Bio-Pesticide

Bio-pesticide spray was categorized into three types as tabulated in table 1; *Annona Squamosa* with water based, *Annona Muricata* with water based and the mixture of both *Annona Squamosa* and *Annona Muricata*. These three types of sprays were prepared in different ratio of oil and base.

Table 1. Percent composition of *A. Squamosa*, *A. Muricata* and mixture of *A.Squamosa* and *A.Muricata*

Percent of *Annona Suqamosa* (%)	**Percent of *Annona Muricata* (%)**	**Percent of mixture *annona squamosa* and *Annona muricata* (%)**
90	90	90
45	45	45
15	15	15

2.5. Testing Against Insects

The different ratio of sprays was tested to two types of insects which are mealworms and crickets as shown in figure 1. The prepared sprays were tested against two types of insects: mealworms and cricket. The insects were placed in the dish containing food and drink and the insects were sprayed in various proportion and composition of the extract. Observation on the mortalities of the insects were done day by day. Mortality is a major change observed in the toxicity test. Animal mortality data were required to determine the efficacy of the extract. Observation on mealworms mortality was done within 1-3 days. Test of mealworms mortality was performed on 3 treatment groups with 3 replications. Each treatment was given 10 mealworms. The dead mealworms were calculated to find out the average of mealworms mortality. The data were obtained by determining the average number of mortality of insects within 3 days to know if there is a mortality difference of each treatment and find out the best concentration for custard apple, soursop, and mixture of custard apple and sour soup.

Figure 1. (A) Mealworm (*Tenebriomolitor)*; (B) Cricket (*Gryllidae*); (C) Spray of extracts of *Annona Squamosa*, *Annona Muricata* and mixture of *Annona Squamosa* and *Annona Muricata*

3. Results and Discussion

3.1. Effectiveness of Annona Squamosa, Annona Muricata and mixture of Annona Squamosa, Annona Muricata extracts in controlling mealworms. From the figure 2, the graph showed how many mealworms killed by three different types of sprays at (A) 15%, (B) 45% and (C) 15% composition. Types of extract or spray chosen was extract of *Annona Squamosa*, extract of *Annona Muricata* and mixture of extract (*Annona Squamosa* and *Annona Muricata*). Based on the results, number of mealworms killed after 1-3 days were proportional to the concentrations of the extracts: 15%, 45 % and 90% respectively. When 90% of aqueous extracts of *Annona Squamosa*, *Annona Muricata* and mixture of extract (*Annona Squamosa* and *Annona Muricata*) were used, mortalities of mealworms increased compared to 45% and 15 % concentration. For the three concentrations 15%, 45% and 90%, the mortalities of mealworms of *Annona Squamosa* after 3 days were respectively 1 ± 1, 1 ± 1 and 5 ± 1, and 2 ± 1, 2 ± 1 and 6 ± 1 for the aqueous extracts of *Annona Muricata*, respectively and for the mixture of *Annona Squamosa* and *Annona Muricata*, the mortalities of mealworms after 3 days were respectively 6 ± 1, 7 ± 1 and 8 ± 1. For all extracts and at all concentrations tested, the mixture of *Annona Squamosa* and *Annona Muricata* has better performance compare to others. Therefore, this research support research conducted by [13]. The extracts of *Annona Squamosa* and *Annona Muricata* mixtures showed higher insecticidal properties compared to extracts of *Annona Muricata* and extracts of *Annona Squamosa*. Furthermore, research by [12] stated the potential of mixture soursop (*Annona Muricata*) and sugar apple (*Annona Squamosa*) as a plant-based pesticides in controlling pests and diseases of plants. Phytochemical screening found the secondary metabolite products in the seed extracts of Soursop (*Annona Muricata*) and sugar apple (*Annona Squamosa*) that is: alkaloids, flavonoids, polyphenols, terpenoids, coumarin, steroids, lactones, anthraquinone, tannins, glycosides, cyanadin, and saponins.

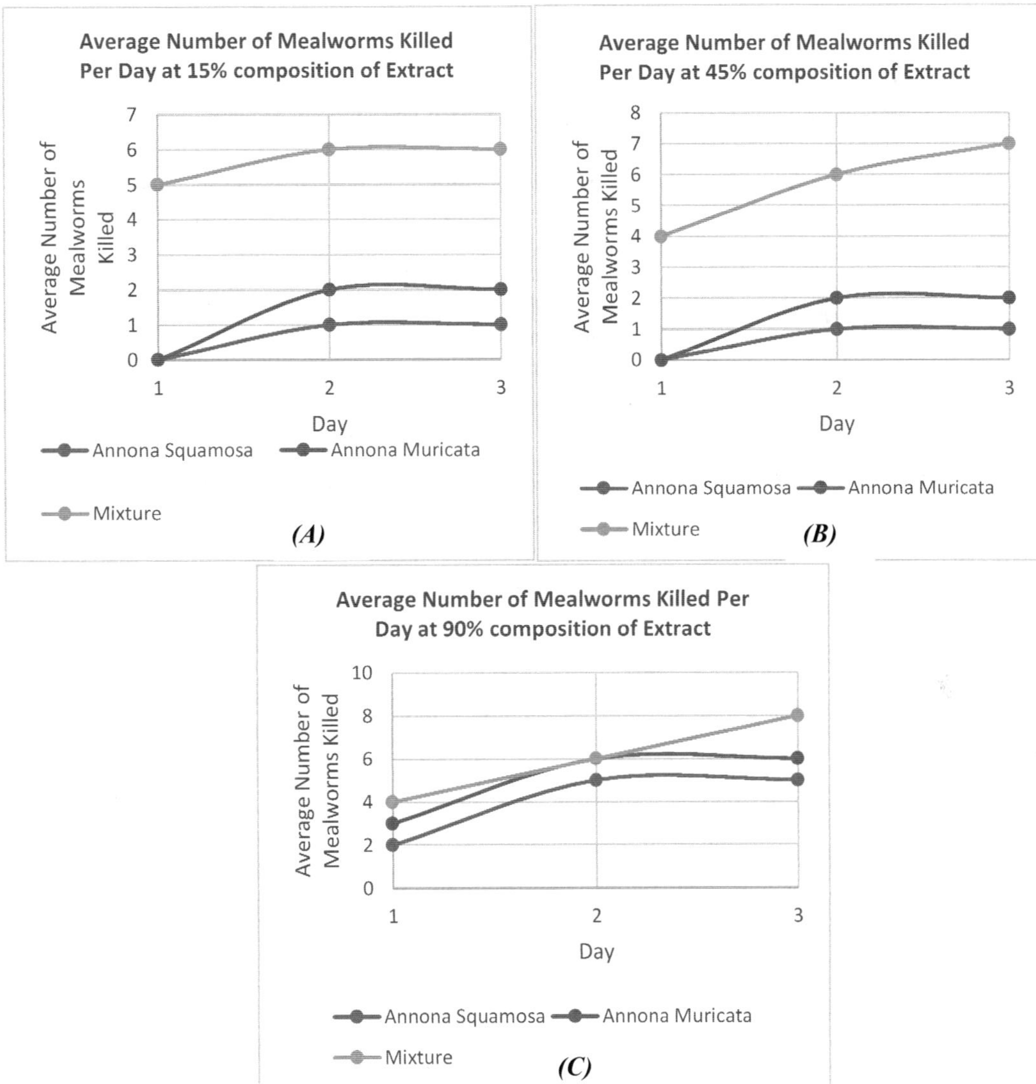

Figure 2. Average number of mealworms killed per day at different concentration of extract (A) 15% composition; (B) 45% composition; (C) 90 % composition.

3.2. Effectiveness of Annona Squamosa, Annona Muricata and mixture of Annona Squamosa, Annona Muricata extracts in controlling crickets.

Figure 3 explains how many crickets killed by three different types of sprays; extract of *Annona Squamosa*, extract of *Annona Muricata* and mixture of extract (*Annona Squamosa* and *Annona Muricata*) extract at different concentration; 15%, 45% and 90%. For three concentrations 15%, 45% and 90%, the mortalities of crickets of *Annona Squamosa* after 3 days were respectively 0, 3 ± 1 and 3 ± 1, and 2 ± 1, 4 ± 1 and 4 ± 1 for the aqueous extracts of *Annona Muricata*, respectively and for the mixture of *Annona Squamosa* and *Annona Muricata*, the mortalities of mealworms after 3 days were respectively 4 ± 1, 5 ± 1 and 6 ± 1. From the results, *Annona Squamosa* has the least performance rather than *Annona Muricata* and mixture of *Annona Squamosa* and *Annona Muricata*. The similar result was observed by the analysis using mealworms in figure 2 stated that mixture of *Annona Squamosa* and Annona *Muricata* showed the highest efficacy in killing mealworms. Based on the percent composition of the extract, the percent composition at 90% showed the highest number of crickets killed. Therefore, the optimum condition to kill the crickets is at 90% concentration using mixture of *Annona Squamosa* and *Annona Muricata*. According to [13], chemical identifications showed that these extracts contain alkaloids and flavonoids compounds that probably confer their biological insecticidal properties.

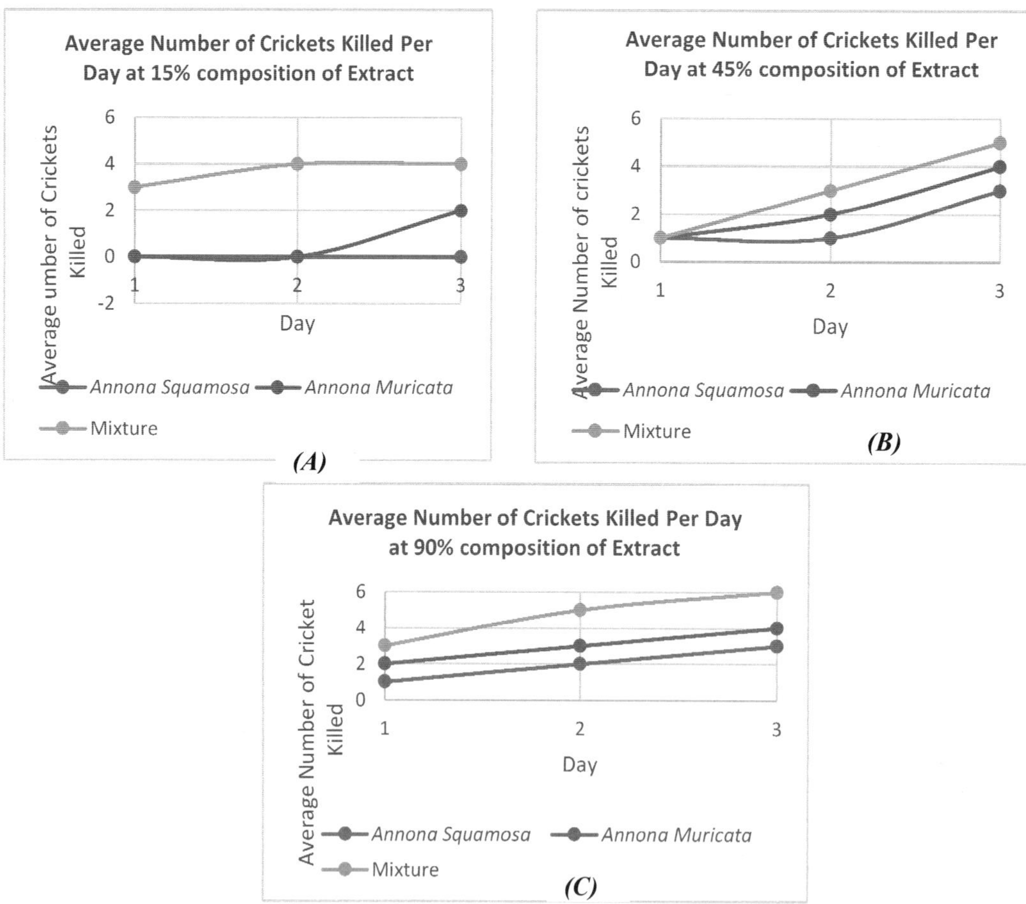

Figure 3. Average number of crickets killed per day at different concentration of extract (A) 15% composition; (B) 45% composition; (C) 90 % composition.

4. Conclusion

The overall findings of the study indicated the effectiveness of soursop seed extract (*Annona Muricata*) and custard apple (*Annona Squamosa*) seed extract on mealworms and cricket's mortality. It is concluded that the extracts of *Annona Squamosa* and *Annona Muricata* both showed insecticidal properties though at varying levels of composition. Extracts of Annona muricata displayed higher insecticidal potential compared to extracts of *Annona Squamosa* but among them, mixture of *Annona Squamosa* and *Annona Muricata* showed the best performance on insecticidal potential. The most effective concentration of soursop and custard apple extract (*Annona Squamosa* and *Annona Muricata*) in killing leaf mealworms and crickets is the extract of 90% concentration. Therefore, if used properly at an efficient dose, these extracts of *Annona Squamosa* and *Annona Muricata* were revealed to be a low cost and powerful environmentally friendly pesticides.

Some recommendations that can be applied to improve this research are by testing the other *Annona* family fruit because all *Annona* fruit contain acetogenin in their seed. Based on the experiment, we used water based with high water composition where the next experiment can be improved such as instead of using high water composition, a high oil extracted composition more efficient. Next, the research can be made by investigating the sustainability of the extract since the bio-pesticide is made up from organic matter that is easily rotten. Besides, the best preservative also needed to be identified to keep the bio-pesticide longer.

References

[1] Tolera S T 2020 Systematic Review on Adverse Effect of Pesticide on Top Ten Importers of African Countries *J. Appl. Sci. Environ. Manag.* **24(9)** 1607–1616.

[2] Meftaul M I, Venkateswarlu K, Dharmarajan R, Annamalai P and Megharaj M 2020 Pesticides in the urban environment: A potential threat that knocks at the door *Sci. Total Environ.* **711** 134612.

[3] Kamarulzaman N H, Mazlan N, Rajendran S D and Mohayidin M G 2012 Role of biopesticides in developing a sustainable vegetable industry in Malaysia *Int. J. Green Econ.* **6(3)** 243–259.

[4] Pavela R 2016 History, presence and perspective of using plant extracts as commercial botanical insecticides and farm products for protection against insects - A review *Plant Prot. Sci.* **52(4)** 229-241.

[5] Tijjani A, Kutawa B A, Muhammad A, Gambo A and Musa H 2017 Biopesticides for pests control: A review *J. Biopestic. Agric.* **3(1)** 6–13.

[6] Moghadamtousi S Z, Fadaeinasab M, Nikzad S, Mohan G, Ali H M, and Kadir H A 2015 Annona muricata (Annonaceae): A review of its traditional uses, isolated acetogenins and biological activities *Int. J. Mol. Sci.* **16(7)** 15625–15658.

[7] Wibawa I P A H 2019 Perbandingan Efektifitas Ekstrak Legundi (Vitex Trifolia L.), Sirsak (Annona muricata L.) dan Damar (Agathis borneensis Warb.) Terhadap Mortalitas Larva Artona flavipuncta Hama Pada Tanaman Kecombrang (Etlingera elatior (Jack) R.M.Sm.) *Agro Bali Agric. J.* **2(1)** 21–27.

[8] Parthiban E, Arokiyaraj C and Ramanibai R 2020 Annona muricata: An alternate mosquito control agent with special reference to inhibition of detoxifying enzymes in Aedes aegypti *Ecotoxicol. Environ. Saf.* **189** 110050.

[9] Begum N 2013 Calotropis procera and Annona squamosa: Potential Alternatives to Chemical Pesticides *Br. J. Appl. Sci. Technol.* **3(2)** 254–267.

[10] Khalequzzaman M and Sultana S 2006 Insecticidal Activity of *Annona squamosa* l Seed Extracts against the Red Flour Beetle *Tribolium castaneum* (Herbst) *J. Bio-Science* **14** 107–112.

[11] Amalia A V and Yusa M H 2018 Control pest of leaf caterpillars (Plutella xylostella) in delima rose apples using soursop leaf extract (Annona muricata) *J. Pendidik. IPA Indones.* **7(1)** 1–8.

[12] Tando E 2018 Review : Potensi Senyawa Metabolit Sekunder dalam Sirsak (Annona Murricata) dan Srikaya (Annona squamosa) sebagai Pestisida Nabati untuk Pengendalian Hama dan Penyakit pada Tanaman *J. Biotropika* **6(1)** 21–27.

[13] Ravaomanarivo L H R, Razafindraleva H A, Raharimalala F N, Rasoahantaveloniaina B, Ravelonandro P H and Mavingui P 2014 Efficacy of seed extracts of Annona squamosa and Annona muricata (Annonaceae) for the control of Aedes albopictus and Culex quinquefasciatus (Culicidae) *Asian Pac. J. Trop. Biomed.* **4(10)** 798–806.

IOP Conf. Series: Materials Science and Engineering 1176 (2021) 012008 doi:10.1088/1757-899X/1176/1/012008

Effects of Disposed-water from Barapukuria Coal Mine, Dinajpur, Bangladesh on Agriculture and Aquaculture

M T Rahman[1,2*], A H Khan[4], B M Negash[1,2], K Sultana[4], M A Hasan A Askary[5], F Rubbi[3], L Das[3]

[1]Department of Petroleum Engineering,
[2]Shale Gas Research Group (SGRG),
[3]Department of Mechanical Engineering,
Universiti Teknologi PETRONAS, 31750 Tronoh, Perak, Malaysia
[4]Department of Petroleum and Mining Engineering
[5]Department of Mechanical Engineering,
Chittagong University of Engineering and Technology, Chittagong, Bangladesh

*Corresponding author : tauhidur.pme@gmail.com

Abstract. The environment is a prime concern in the mining industry because of its adverse impact on the solid, liquid, and gaseous surroundings. These three parts of the environment are directly connected to agricultural production. As the Barapukuria coal mine is located in the agricultural zone, its effects can be crucial. Previously, the impacts of the subsidence of the Barapukuria coal mining area were studied. So far, there is no significant study regarding its effects on agriculture. This research has focused on the impacts of the disposed water from the Barapukuria coal mine on the nearer aquaculture and agricultural fields. Therefore, the BOD, COD, DO, TDS, TSS, SS, residual chloride, hardness, alkalinity, carbon, sodium, calcium, magnesium, iron, potassium, sulphate, etc., being present in the disposed-water were analyzed in this study. The values of BOD, COD, DO, pH, TSS, TDS, and TS were lied in the best optimum range to cultivate crops and fisheries. It is also found that the disposed-water of the Barapukuria coal mine is tolerable for agriculture.

1. Introduction

In this present world, the population is increasing in a booming way. Besides, the balance of the environment is degrading galloping with it [1-2]. Therefore, this degradation of the environment directly jumps to the human being through some specific fields. In most cases, this field is the agricultural sector [3-4]. In agriculture, the effect of a little change in the environment is spectacularly large. Now, the fact is, environmental degradation is a very elaborate sphere. It might be affected by many causes. Some might be human-caused, and some might be natural. Human-caused degradations are generated by industrialization, especially by an unprocessed human, chemical, industrial, and nuclear waste. However, all of these degradations can be processed by applying engineering recycling and reproduction systems through governmental regulations. But another major cause of environmental pollution, with which the people of the world are struggling, is mining extraction. The fact is that we cannot but approve of the terrible impact of mining extraction. In many areas around the world, the extraction process has been running out for years. Therefore, their massive effects on the environment have also been occurring in the same way [5]. These effects cause tons of agricultural production loss and harm human beings' lives and all the living beings related to that ecology.

Production of natural resources through mining processes draw negative impacts on ecological and social sectors [6]. From the beginning of a mining operation, it produces dust and radiation, which directly impacts the surrounding environment. Some studies claimed that long term mining operations are responsible for the disappearance of river, lake and forest [7]. However, Hota and Behra showed that coal mining activities positively and negatively impacted rural households [8]. A continuous mining operation disposed of million gallons of wastewater to the nearby locality, which degrade the agricultural field's soil and the water of the aquaculture [9]. Several studies also discovered that groundwater quality is also facing degradation due to some harmful metallic substances with more than tolerable limits.

Mining significantly coal mining is increasing year by year, and the highest impact is implemented in the agricultural system [11]. The situation is terrific, no doubt. Hence, it is a crying need to take immediate measures to reduce this effect. First, it is necessary to calculate how much pollution can be caused by each mining field and its actual impact on the nearer locality. In Bangladesh, a vast population of around 160 million demands a considerable volume of energy. So, the energy use growth rate is increasing day by day [12]. Hence, the energy demand is a crucial issue in this country that has no way to be neglected. In this circumstance, the running of coal mining is a crying need here. However, the effects on the environment caused by the run of this coal mine should also be considered. In Bangladesh, the Barapukuria coalfield is the only field where coal mining is running out. This field is the only running internal source of the coal energy generation of this country. Therefore, this coalfield needs to take responsibility to match up the whole countries' energy demand. However, the impact on the environment caused by this field is also a considerable matter. Therefore, at first, it should be found out how much influence is generated on the environment by this mining field. In this research, the data related to aquaculture and agriculture (water) at the Barapukuria Mining field's nearer locality has been collected. Thoroughly analyzing these data, it is decided whether running this mining field in this location is environmentally distressful or not.

2. Geographic location, methods and data collection

2.1. Geographic Location

Barapukuria coalfield is the only mining field in Bangladesh on which production is going on. The field is located at Parbatipur, an Upazila of Dinajpur District in the Rangpur Division, Bangladesh. Parbatipur is located at 25°39′12″N and 88°54′56″E. It has 53,146 households and a total area of 395.1 km^2. Saidpur upazila on the north, Phulbari upazila on the south, Badarganj upazila on the east and Chirir Bandar upazila on the west. Figure 1 represents the geographic location of the studied area.

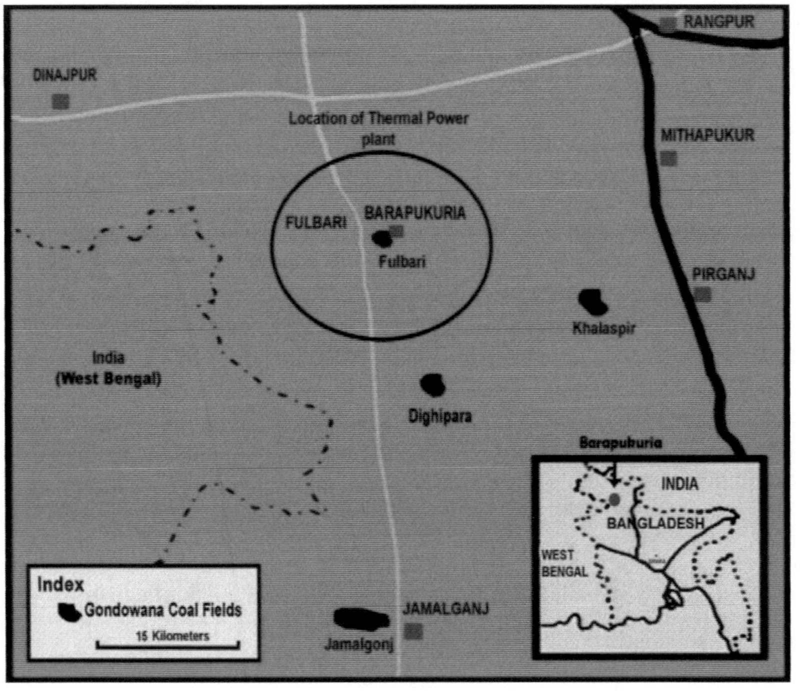

Figure 1. Location of the studied area [13].

To analyze the effects of the Barapukuria coal mine on the environment, the following analyses are conducted.

2.2. Biological Oxygen Demand (BOD)

The Biological Oxygen Demand (BOD) is mainly correlated to enough oxygen in water and soil sample to make the biological living being alive in that sample. The amount of oxygen in the sample must remain in an optimum condition. The greater or lesser value of oxygen than the optimum may cause a respiratory problem to the biological beings. Hence, BOD's value must be remaining in the optimum standard value, which is up to 50 ppm [14].

2.3. Chemical Oxygen Demand (COD)

The Chemical Oxygen Demand (COD) indicates the necessary chemical composition in the water, air, and soil samples. These chemicals are mainly organic and inorganic materials that lie in the sample, which is required to preserve the environment and pH of water and soil balanced to live the living beings. The optimum level of this COD value is up to 200 ppm [15].

2.4. Dissolved Oxygen (DO)

The DO is the measure of dissolved oxygen in any sample. It is the minimum amount of oxygen particle that is needed to dissolve in water. The optimum value of DO lies from 4.50 to 8.00 ppm [16]. This measure is essential to make the water sample used for any purpose.

2.5. Total Solid (TS)

TS means total solid particles lie in a water sample. This solid particle consists of two types of particles. The first one is Total Dissolved Solids (TDS) particles in ppm, and the second one is Total Suspended Solid (TSS) particles in ppm. The accumulated solid particles are called TS and are measured in ppm. The optimum amount of TS that needs to lie in the water sample is 2100 ppm in total [17].

2.6. Total Dissolved Solid (TDS)

Total Dissolved Solids (TDS) are the solid particles in colloidal form and mixed in the water in all ratios. They cannot be separated from the water sample through filtering; boiling is needed to separate the solution's particles. The presence of these types of particles in the optimum amount is significant to make the water drinkable. It is also very substantial to make the water usable for farming and doing fisheries activity. Exceed the optimum level of these particles is not desirable.

2.7. Suspended Solid (SS)

The presence of suspended solid is termed SS. This term is also significant to make the water drinkable. After analyzing the data mentioned earlier, some additional compound analyses have also been performed that are specially related to agriculture and aquaculture farming. All these tests were performed in the laboratory of BCSIR, Dhaka, and Laboratory of Department of Environment, Bogra (DOB). The results of these tests that have been found out in the laboratory are given in table 1, table 2, and table 3.

3. Data analysis and discussion

3.1. Deviation determination

Following the provided data, table 1, table 2 and table 3 indicate the deviation of the essential compounds badly needed to agriculture and aquaculture comparing to the standard value.

Table 1. Obtained parameter of 1207 Coal Face 1, open off cut water

Chemical Parameters	Laboratory test value (mg/L)	Standard Value	Comments
P^H	7.48 (no unit)	6.5-8.5 (WHO,1984)	Within the tolerable limit of agriculture and livestock
Na	20.00	200 (WHO,1984)	
K	4.55	12 (BWPCB,1976)	
Ca	19.40	200 (WHO,1984)	
Mg	7.34	150 (WHO,1984)	
Al	<0.05	0.2 (WHO,1993)	
Fe	0.32	50 (WHO,1984)	
Cl^-	1.63	250 (WHO,1993)	
CO_3	No detectable	-	
SO_4	12.00	500 (WHO,1984)	
AS	<0.005	0.01 (WHO,1984)	

Table 2. Obtained parameter of 1214 Coal Face 2, open off cut water

Chemical Parameters	Laboratory test value (mg/L)	Standard Value	Comments
P^H	7.34 (no unit)	6.5-8.5 (WHO,1984)	Within the tolerable limit of agriculture and livestock
Na	28.60	200 (WHO,1984)	
K	8.32	12 (BWPCB,1976)	
Ca	45.40	200 (WHO,1984)	
Mg	14.30	150 (WHO,1984)	
Al	0.95	0.2 (WHO,1993)	
Fe	7.46	50 (WHO,1984)	
Cl^-	1.93	250 (WHO,1993)	
CO_3	No detectable	-	
SO_4	68.00	500 (WHO,1984)	
AS	<0.005	0.01 (WHO,1984)	

Table 3. Outside Boundary Drain Water (obtained from BCSIR)

Name of the Test	Obtained value (mg/l)	Standard Value (mg/l)
TS	--	2100
DO	7.78	4.50-8.00
BOD	2.53	50
Oil & Grease	6.88	-
TDS	----	2100
TSS	156	150

3.2. Impacts determination

In a word, the impacts of the coal mining operation can be generalized in some specific categories. Each of all types has a remarkable influence on human life. Significantly, the farming sector is greatly influenced by this operation, whatever it is about agriculture or aquaculture or even cattle and poultry farming. Nevertheless, some other impacts of coal mining operations can also be put into consideration. One and the most crucial of them is land subsidence.

Considering all the impacts mentioned above, here is some discussion about all the significant effects of the Barapukuria coal mine on the nearer locality, farming, and others' human economy. This whole discussion has been organized in some categories below.

3.2.1. Impacts on agricultural farming. To say about agricultural farming, Dinajpur (the region of Barapukuria coal minefield) can be specialized in paddy, potato, mustard, and in the fourth the jute production. Among them, the paddy can be given the most important as it is the main crop of that region. Hence, if there is any coal mine impact, the paddy field will face the most significant effect. However, in the first portion of this discussion, the impact of every property of soil that can affect paddy production because of the coal mining operation will be discussed elaborately.

For that, we ought to know about the essential properties of soil needed for paddy production at first. The value of the P^H of the water used for irrigation purposes is the basic need for paddy cultivation. The electric conductivity (EC) got in the irrigated water is another basic need. About the inorganic materials needed for paddy cultivation, Na, K, Ca, Mg, Al, Fe, Cl^-, CO_3, As, SO_4 can be said as some of the essential materials in the irrigated water. Moreover, the presence of turbidity, Total Dissolved Solid (TDS), Total Suspended Solid (TSS), Biological Oxygen Demand (BOD), Chemical Oxygen Demand (COD) in the irrigated water are also some vital part of paddy cultivation. In some steps, the discussions with all these properties are given below.

Electric conductivity (EC):

When an excess number of salts gather and accumulate at the root area of any crop, crop production is adversely affected. Therefore, the salinity of the water, which will be used for irrigation, should be determined. The most popular indirect method for calculating the amount of salt content present in the irrigation water is measuring its EC value. A higher EC value indicates the presence of greater value of salt contents in the water. Generally, the EC values are expressed using deci Siemens per meter (dS/m) unit [18].

However, the water used for irrigation has a long-range of total salinity levels. The irrigation water comes from snow-fed rivers, shows a total salinity level within 0.5 to 0.6 dS/m. This value usually becomes from less than 1 dS/m to higher than 12 or 15 dS/m for groundwater. For seawater, the value is about 50 dS/m as it is highly saline. If the salinity of irrigation water is higher, then the possibility of production reduction is also higher.

The data, tested in the Department of Environment, Bogra (DOB), found that the EC in the water, extracted from the Barapukuria coal minefield, is 390 μS/cm before the treatment. But this water is not allowed directly to go to the irrigation field without treatment. After the treatment, the EC is found 370 μS/cm, which indicates the optimum rate of salinity in the water extracted from the coalfield. Hence, this water does not possess any crucial impact on the yield of paddy or other crops in the nearer locality.

P^{H}:

P^{H} is the Hydrogen ions' density present in any sample whatever it might be a water or soil sample. Hydrogen ion density is a crucial parameter to detect the acid and base condition of that sample. The detection of acid or base is mainly significant because it shows whether the water or soil sample's behaviour to the living beings relies on that soil or water. If the acidic condition gets higher than the moderate, it will not be possible to live any living being in that soil or water because of decay. On the other hand, a primary environment is also very harmful because of its extreme reducing activity. Hence, a moderate rate of both acidic and basic properties is desirable in any soil or water sample. As the rate of this acidic and basic condition is measured by pH, the pH value must be remaining in a desirable condition. The desired value of pH for drinking water is 6.5 to 8.5 (WHO,1984). That means this rate show both the optimum availability of acid and base in the water and soil sample. If the pH is 7.00 that means, there is neither acid nor base in the sample. When the value increases from 7.00 to 8.00 or 9.00 or more, it refers to the increasing rate of basic properties.

On the other hand, the decreasing value from 7.00 (6.00 or 5.00 or less) shows the presence of acidic materials at an increasing rate. Nothing of these is acceptable for any kind of living being as well as cultivation or farming. Therefore, the presence of an optimum value of pH is a crying need.

Though most of the plants can withstand a wide range of pH values, they cannot tolerate the soil's higher acidic environment. The acidic environments make the soil mineral dissolved and increase the metal ion's concentrations to a toxic level. On the other side, the higher alkaline environments decrease the soil minerals solubility and create deficiencies of necessary nutrients. A little deficiency in the required nutrients inhibits the growth of the plants.

However, the goal of managing pH is not just achieving value. The aim is to maintain an optimum pH environment where there are enough nutrients and no toxic materials for the plants. These favourable environments are achieved when the pH value remains between 5.8 and 6.5. So far, few plants need an exceptional acidity environment [19]. The pH of water extracted from the Barapukuria coalfield is about 7.34 (1214 coal face) (BCSIR) and 7.48 (1207 coal face) (BCSIR). This value of pH is acceptable in the sense that as the bacteria like actinomycetes need 7-7.5 pH to develop oxygen in the yield field, this value of pH will come to a generous side of the overall paddy and other cultivation of the nearer locality.

Sodium (Na):

The chlorophyll production slightly deviates from the change of the sodium ion. The change in water efficiency also gets a slight difference in the shift of sodium ion. However, it can be said that 17 mg/g of sodium ion is the optimum condition for any paddy field. The results tested by BCSIR also showed that the amount of sodium in the water extracting from the 1207 face coalfield and the 1214 coalfield of Barapukuria are 20.00 and 28.60, respectively. Those are very close to the optimum condition. Hence, the water will not adversely impact the decline of the paddy production of the nearer community of the Barapukuria coal minefield.

Potassium (K):

Sometimes, some common nutritional disorders are observed in lowland rice production. However, potassium fertilizers help to omit these types of issues, such as rice bronzing, etc. Moreover, it increases the resistance capability of the plants to some pests and diseases. K deficiency usually appears between the beginning of internode elongation and seed heading. Developing symptoms include stunted plants with thin stems and a dark green colour, necrotic spots at leaf tips progressing to the leaf base, including drying of leaf edges, sterile flowers, and root damage. K plays essential roles in protein synthesis, enzyme activation, photosynthesis, stomatal movement, osmoregulation, energy transfer, cation-anion balance phloem transport, and stress resistance [20].

Barapukuria coal mine did not make any lack or excess of potassium contentment in its extracted water, which went to the nearer paddy field as the source of irrigation. The test occupied by BCSIR showed that potassium contentment in the extracted water from the Barapukuria coal mine was about 4.55 mg/l at coal face 1207 and 8.32 mg/l at coal face 1214, respectively. However, the highest limit for the optimized value is about 12 mg/l [21].

Calcium (Ca):

Calcium is a source of immobile nutrition for any plant whatever it is a rice plant (*Oryza Sativa*) or the other else. It patronizes the growing activity of a plant. The xylem sap of a plant is only supplied by calcium, and this sap is the base of plane widening. As the source of plant mobilization, calcium mobilizes the older tissue of the plants. As a cause of that, the cell wall of the plant gets hardened and healthy. Another business of calcium that also bears tremendous significance is its transpiration activities. When the transpiration gets reduced, the instant calcium starts to reduce the tissues' rapid growth to make a balance with the reduction of transpiration [22].

Calcium is the weapon of soil to reduce salinity. The removal of carbohydrates is mobilized in the presence of calcium. All the cell acids of a plant get nutrients from calcium. In total, calcium plays a vital role in balancing the chemical properties of soil, reducing soil salinity, and providing plants' proper nitrification.

Barapukuria coal mine field extracts the water without making any deficiency of calcium content lies in it. BCSIR tests that the calcium contentment in 1207 and 1214 coal face in the Barapukuria coal mine extracted water contains 19.40 mg/l and 45.40 mg/l, respectively, while the standard limit of calcium presence is below 200 mg/l [23].

Magnesium (Mg):

The principal deed of magnesium is to alive the growth cycle of a plant. This process is managed by specific stages like several enzyme activities in plant, assimilation, protein synthesis, balancing the cellular pH, and the cation-anion balance activation [24]. Each step is directly depended on the availability of magnesium. The magnesium in a plant is generally sourced by the irrigated water which is applied on. Hence, the proper amount of magnesium in the irrigated water is a must to regulate all the important deeds covered with magnesium.

According to BCSIR, the presence of magnesium in the extracted water of the two faces, in specific coal face 1207 and coal face 1214 at Barapukuria coal minefield is 7.34 and 14.30 mg/l, respectively, where the highest limit of the presence of magnesium in any water sample is 150 mg/l [23]. Hence, magnesium is supplied to irrigation water without an excessing limitation.

Sulfate (SO_4^{2-}):

Sulfate is one of the most important constituents for straw-like plants like rice, jute, etc., which are some prime plants in the nearer locality of Barapukuria coal minefield. The prime deed of Sulphur or SO_4^{2-} ion increases the chlorophyll production and protein synthesis [25]. Also, the plant structures and functions are depended on it. There is less chance to get enough Sulphur from the irrigation water. Most of the Sulphur comes to the plants through the implementation of sulfate fertilizer. Nevertheless, the higher the amount of Sulphur remains in the irrigated water, the lower the amount of fertilizer is needed.

BCSIR shows that the amount of Sulphur was approximately 12.00 and 68.00 mg/l in the water of 1207 and 1214 coal face, respectively. The optimum value certified by WHO, 1984 is 500 mg/l in this case. Therefore, there is no more than the usual amount of Sulphur in the coalfield's water; instead, it is optimum to apply an economical amount of Sulphur fertilizer.

Arsenic (As):

Arsenic comes with irrigation water to the soil and degrades the grains production. Some reports suggested that for groundwater irrigation, the As-soil concentration is higher during the winter season than in the summer [26-27]. However, the addition of As through the irrigation water to soil also increases As concentration in the grains [28].

Drinking As containing water (>0.05 mg/L) for a long time may cause unexpected death sometimes, as it enters the human body through different foods containing it. For example, the remaining in the rice creates considerable health hazards to its consumers. Continuous consumption of As-infected rice causes a series of health issues such as cancer in the liver, bladder, lung, skin, kidney, and some skin diseases. The consumption of As contaminated rice straw by cattle increases the As contamination in meat and milk, which also a medium for entering the human body [29].

However, the amount of as found by BCSIR in the 1207 and 1214 coal face extracted water is below 0.005 mg/l, which is an impressive sign for the irrigation water to the nearer paddy field.

Chloride (Cl^-)

Balancing the presence of Chloride ion is a precondition for planting. It generates both positive and negative effects on the plants simultaneously. Photosynthesis, the osmotic adjustment, occurs in the presence of an appropriate amount of chloride. On the contrary, the plant is affected by chlorine toxicity when chlorine's presence exceeds the plant's tolerable limit, and yield reduction occurs significantly.

As chloride is bearing negative charges, it does not get attached to the soil particles and moves with the water flow. Therefore, it is significant to check the irrigation water's chloride concentration before entering grain fields. Generally, low to medium saline water contains 100-300 mg/L chlorine anions [30].

Table 4 and table 5 are showing the susceptibility range and effects of chloride-containing water on different crops. These tables summarized that the chloride concentration of irrigation water less than 150 mg/L is safer for most crops. The Chloride ion presence in the extracted water of Barapukuria coal mine 1.63 and 1.93 mg/L in the coal face of no1207 and 1214, respectively. This value is significantly within the optimum limit such that it neither exceeds the tolerable limit of the plant nor touch the toxicity level. Instead, it meets ups at least amount, which is essential for photosynthesis and osmotic activities.

Table 4. Effects of chloride (presents in irrigation water at different levels) on crops [31].

Chloride (ppm)	Effects on crops
<70	Safer
70-140	Only sensitive plants exhibit injury.
141-350	Mid-range tolerable plants exhibit injury.
>350	Exhibit severe issues.

Table 5. Susceptibility levels for various crops to foliar injury from chloride-containing water [31].

Cl concentration	Range	Foliar injury for crops
<175 (mg/L)	Low	Apricot, plum, tomato
175- 350 (mg/L)	Medium	Pepper, potato, corn
351- 700 (mg/L)	High	Alfalfa, barley, sorghum
>700 (mg/L)	Excessive	Sugarbeet, sunflower

3.2.2. Impacts on fish farming. In fish farming, some extensively important required components should be in the water with a definite range. The more or less of that amount of the necessary element might cause severe degradation of fishes' birth and growth. Even in many cases, the early termination of life might occur. To get rid of this, the fast and foremost task is to ensure the essential components to triggerfish farming and shrink the possibility of adverse products.

The first deed will know about the essential elements for fishing that must lie at the farming ponds' desired rate to combat that. The crucial parameter is pH for fish farming which shows the acidic and basic properties of the water. Some other equally essential parameters are the amount of As, SO_4^{2-}, Ca, and Mg presence in the water. Additionally, COD (Chemical Oxygen Demand), BOD (Biological

IOP Conf. Series: Materials Science and Engineering 1176 (2021) 012008 doi:10.1088/1757-899X/1176/1/012008

Oxygen Demand), and suspended and dissolved solids in the water are also important. As these refer to the total fish-farming conditions of a pond processed for aquaculture. A brief discussion is drifted below concerning all the essential parameters of fish-farming:

pH:

Fish lives in water and depends on water for breathing, growing, feeding, excreting wastes, and reproduction. Hence, understanding the physiochemical qualities of water is crucial for successfully operated aquaculture.

Table 6 represents the effects of pH on fishes. It summarizes that both the significantly lower and higher pH values are harmful to aquatic organisms. Lower pH levels help to release the metals from the sediments. These metals can affect the fish's metabolism processes. A higher pH value helps to denature the cellular membranes and harms fishes. Moreover, it converts most of the ammonium atoms to ammonia, and this toxic ammonia can kill the fishes.

The above discussion emphasizes the importance of pH for aquaculture. Most of the fishes cannot survive for a long time in the water whose pH is above 11 or below 4. The suitable pH range for fishes is between 6.5 and 9. The pH of the water exerted by the Barapukuria coal mine was within a range of 7.34 to 7.48. These values indicate that this water is safer and suitable for fish farming.

Table 6. Effects of pH on the fishes living warm water pond [32].

pH Range	Effects on Fish
4	Acidic death point
4-5	No reproduction
4-6.5	Slow growth
6.5-9	Suitable ranges for fish reproduction
9-10	Slow growth
>11	Alkaline death point

Calcium and Magnesium (Ca and Mg):

Nutrients such as magnesium, calcium are essential for the excellent growth rate of fishes. Fishes can get nutrients from their food or the surrounding water. However, the presence of nutrients should be optimum because much Ca and Mg increases the hardness of the water.

For freshwater, in most cases, the alkalinity and hardness remain within the following range < 5 mg/L to > 150 mg/L. BCSIR reported that the amount of Mg and Ca presented in the 1207 and 1214 coal mine face is 7.34 and 14.30 mg/L, respectively. This value lies below 100 mg/L, and it should not put any adverse effect to change the natural content of magnesium and calcium lies in the pond water.

Biological Oxygen Demand (BOD), Dissolved Oxygen (DO), and Chemical Oxygen Demand (COD):

Biological oxygen demand (BOD) means the amount of oxygen used by a different microorganism. It is believed that when the oxygen concentration becomes less than 5 ppm, then most of the fishes cannot survive for a long time [33]. According to BCSIR, the BOD in the coal mine field extracted water is 2.53 mg/l which is less than 50 mg/l or 50 ppm, in a safer range for fish cultivation. Besides, the amount of dissolved oxygen (DO) according to BCSIR is 7.78 mg/l which is greater than the least desirable limit for fish culture, 5 mg/l. Hence, this extracted water did not provide any adverse condition in this case. The report of the test supplied by DOB (Department of Environment, Bogra) showed that the number of COD in the outside boundary drain water of the Barapukuria coal minefield is 54 mg/l after treatment. Compared with the optimum value, 200 mg/l, this value of COD is in a desirable range.

4. Conclusion

To meet the increasing energy demand, Bangladesh needs to focus on the production of underground natural resources. Due to the limited reserves, this country has to continue the coal production from its only running coalfield. However, environmental rules and regulations are getting strict. Moreover, some ecological regulatory organizations are raising questions on the ecological feasibility of Barapukuria coalfield. Therefore, this research determined its impacts on agriculture and aquaculture. According to the discussion, it is concluded that the effects of discharged water from the Barapukuria coal mine on agriculture and fish farming are negligible. The logical and scientific explanation in favour of this conclusion has been discussed thoroughly in this research paper. Almost all the parameters that refer to the pond's conditions or agriculture water were in a safer range. This water can fairly mix with the water for irrigation and aquaculture and work instead of creating any ecology degradation. The values of BOD, COD, DO, pH, TSS, TDS, and TS were lied in the best optimum range to cultivate crops and fisheries. Therefore, it can be said that the mining discharged water from Barapukuria is environmentally and ecologically feasible for agriculture and aquaculture and to lead on. However, this research only focused on the impacts of disposed water. Future research should be on determining the environmental impact on soil and air.

Nomenclature

As	Arsenic
BOD	Biological oxygen demand
Ca	Cancium
Cl^-	Chloride
COD	Chemical oxygen demand
DO	Dissolved oxygen
EC	Electrical conductivity
Mg	Magnesium
Na	Sodium
SO_4^{2-}	Sulfate
SS	Suspended solids
TS	Total solid
TDS	Total dissolved solids

References

[1] Wang Y 2004 Environmental degradation and environmental threats in China *Environ. Monit. Assess* **90** (1-3) 161-169.
[2] Kalipeni E 1992 Population growth and environmental degradation in Malawi *Afr. Insight* **22 (4)** 273.
[3] Rozelle S, Veeck G and Huang J 1997 The impact of environmental degradation on grain production in China 1975–1990 *Economic Geography* **73(1)** 44-66.
[4] Olanipekun I O, Olasehinde-Williams G O and Alao R O J S O T T E 2019 Agriculture and environmental degradation in Africa: The role of income **692** 60-67.
[5] Ripley E A and Redmann R E 1995 *Environmental effects of mining* CRC Press.
[6] Mishra N and Das N 2017 Coal mining and local environment: A study in Talcher coalfield of India *Air, Soil Water Res.* **10.**
[7] Deonandan K and. Dougherty M L 2016 *Mining in Latin America: Critical approaches to the new extraction* Routledge.
[8] Hota P and Behera B 2016 Opencast coal mining and sustainable local livelihoods in Odisha, India *Miner. Econ.* **29(1)** 1–13.
[9] Singh G 1988 Impact of coal mining on mine water quality *Int. J. mine water* **7(3)** 49–59.

[10] Senapaty A and Behera P 2012 Study of selected trace elements in coal and groundwater of the Ib-valley coalfield of Odisha, India and their possible impact on human health *Int. J. Earth Sci. Eng.* **5** 1241–1248.
[11] Baruah S and Dutta J 2009 Nanotechnology applications in pollution sensing and degradation in agriculture: a review *Environ. Chem. Lett.* **7(3)** 191-204.
[12] Islam S and Khan M Z R 2017 A Review of Energy Sector of Bangladesh *Energy Proc.* **110** 611-618.
[13] Howladar M F 2016 Environmental impacts of subsidence around the Barapukuria Coal Mining area in Bangladesh *Energy, Ecol. Environ* **1(6)** 370–385.
[14] "Biochemical oxygen demand," https://en.wikipedia.org/wiki/Biochemical_oxygen_demand (accessed December 9, 2019).
[15] "Chemical Oxygen Demand (COD)," https://realtechwater.com/parameters/chemical-oxygen-demand/ (accessed December 11, 2019).
[16] "Water Testing Total Dissolved Solids Drinking Water Quality," https://www.water-research.net/index.php/water-treatment/tools/total-dissolved-solids (accessed December 10, 2019).
[17] "U.S. Environmental Protection Agency," https://archive.epa.gov/water/archive/web/html/vms58.html (accessed December 11, 2019).
[18] Wiatrak P, Khalilian A, Mueller J and W. Henderson 2009 Applications of soil electrical conductivity in production agriculture **93** 16-17.
[19] "Soil pH Importance in Agriculture," https://www.asiafarming.com/soil-ph-importance/ (accessed December 25, 2019).
[20] Wang M, Zheng Q, Shen Q and Guo S 2013 The Critical Role of Potassium in Plant Stress Response *Int. J. Mol. Sci.*, **14(4).**
[21] Chowdhury M M A, Jahan M, Sakib M N and Rahman M A 2016 A Statistical Assessment of Groundwater Quality in Nilphamari Pourashava, Nilphamari District, Bangladesh *Proc. Of 3rd Int. Conf. on Advances in Civil Engine.* CUET.
[22] Aslam M, Mahmood I, Qureshi R, Nawaz S, Javaid A and Ahmad Z 2001 Nutritional Role of Calcium in Improving Rice Growth and Yield under Adverse Conditions *Int. J. Agric. Biol.* **3(3).**
[23] "World Health Organization," https://apps.who.int/iris/handle/10665/311318 (accessed December 29, 2019).
[24] "UNIERSITY OF MINNESOTA EXTENSION," https://extension.umn.edu/agriculture/nutrientmanagement/secondarymacronutrients/magnesium-for-crop-production-in-minnesota/ (accessed December 26, 2019).
[25] "EXPERT SYSTEM FOR PADDY," http://www.agritech.tnau.ac.in/expert_system/paddy/nutrientmanagement.html (accessed January 5, 2020).
[26] Dittmar J, Voegelin A, Roberts L C, Hug S J, Saha G C, Ali M A, Badruzzaman A B M and Kretzschmar R 2007 Spatial Distribution and Temporal Variability of Arsenic in Irrigated Rice Fields in Bangladesh. 2. Paddy Soil *Environ. Sci. Technol.* **41** (17) 5967-5972.
[27] Huq S M I, Joardar J C, Parvin S, Correll R and Naidu R 2006 Arsenic contamination in food-chain: transfer of arsenic into food materials through groundwater irrigation *J. Health Popul. Nutr.* **24(3)** 305-316.
[28] Akinbile C O and Haque A M M 2012 Arsenic contamination in irrigation water for rice production in Bangladesh: a review *Trends Appl. Sci. Res.* **7(5)** 331.
[29] Smith N M, Lee R, Heitkemper D T, Cafferky K D N, Haque A and Henderson A K 2006 Inorganic arsenic in cooked rice and vegetables from Bangladeshi households *Sci. Total Environ.* **370(2-3)** 294-301.
[30] "Chloride In Plants, Water And Soil- Essential Or Harmful," https://www.smart-fertilizer.com/articles/chloride (accessed August 5, 2019)

[31] "smart-fertilize.com," https://www.smart-fertilizer.com/articles/chloride/ (accessed December 28, 2019).
[32] "pH in Fish Farming," https://www.yokogawa.com/library/resources/application-notes/ph-in-fish-farming/ (accessed December 18, 2019).
[33] Doudoroff P, Anderson B G, Burdick G E, Galtsoff P S, Hart W B, Patrick R, Strong E R, Surber E W and Van Horn W M 1951 Bio-assay methods for the evaluation of acute toxicity of industrial wastes to fish *Ind. Wastewater* **23(11)** 1380-1397.

IOP Conf. Series: Materials Science and Engineering 1176 (2021) 012009 doi:10.1088/1757-899X/1176/1/012009

Removal of Reactive Orange 16 Dye using Casuarina equisetifolia seeds as Packing Media for Microbial Biofilm Formation

W H W Osman[1*], M M Tan[1], A A A Rashid[1], N Mohd[2] and R Mohammad[1]

[1]Faculty of Bioengineering and Technology, Universiti Malaysia Kelantan, Jeli Campus, 17600 Jeli, Kelantan, Malaysia

[2]Faculty of Chemical Engineering, Universiti Teknologi MARA Terengganu Campus, Bukit Besi Branch, 23200 Dungun, Terengganu, Malaysia

*Corresponding author: hasnidah.osman@umk.edu.my

Abstract. Dyes effluent mostly is toxic and mutagenic to living organism. Casuarina equisetifolia seeds are usually found in coastal and have potential as a packing media for microbial biofilm formation. The present study was designed to evaluated the performance of three laboratory scale reactors namely, sequencing batch biofilm reactor (CES-SBBR), a combination of adsorption and biological process, sequencing batch biofilm reactor (FC-SBR), and packed bed reactor (CES-PBR) on the removal of RO16 dye. The CES-SBBR and CES-PBR were packed with Casuarina equisetifolia seeds. The microorganism was collected from textile industry and undergo the preadaptation period with the 2 mg/L of RO16 solutions in CES-SBBR and FC-SBR. The biomass concentration initially reduced from 0.12 to 0.03 mg/L. Upon the addition of nutrients, the biomass concentration was increased to 0.22 mg/L. The performance study of three reactors was performed with the initial concentration of RO16 solution and hydraulic retention time (HRT) of 2 mg/L and 24 hr, respectively. The RO16 dye removal efficiencies were in range of 80.00-82.26%. The removal efficiencies were reduced to 70.00-71.28% when the initial concentration of RO16 dye increased to 4 mg/L. The removal efficiencies were continued to reduce with only 48.06-50.00% as the initial concentration of RO16 dye increased to 6 mg/L. The removal efficiencies were improved when the HRT was extended to 48 hr. The biomass concentration of CES-SBBR and FC-SBR increased up to 0.50 and 0.33 mg/L, respectively. The RO16 removal percentage for FC-SBR and CES-PBR were found to be lower compared to as for CES-SBBR. This study shows that the combination of adsorption and biological process enhanced the removal efficiency of RO16 dye.

1. Introduction
Dyes have extensive application in textile, paper, equipment, cosmetic, pharmaceutical, aquaculture, food and beverage, as well as plastic industries [1]. The invention of synthetic dyes solved the natural dyes problems which normally fade when exposed to sunlight and washing, however new issues arise from improper treatment or untreated dyes by dye-utilizing industries [2]. More than 280,000 tons of untreated synthetic dyes are expected to have been discharged into water bodies globally [3].

Synthetic dyes are made from complex organic substances to strengthen their capability to withstand degradation when contact with water and detergents even when under extreme heat [2]. Synthetic dye molecules are complex which contain auxochromes and chromophores which is water soluble bonding compound and colour giving compound respectively allowing them to have stable structure.

Casuarina equisetifolia also known as Rhu or Beach She-Oak and are commonly found in Southeast Asia which grows at coastal habitat such as sandy coasts and coastal forests. Casuarina equisetifolia is a plant which is cultivated for applications in paper manufacturing, firewood, furniture and building timber since its woods are extremely hard and dense [4]. However, Casuarina equisetifolia seeds are considered as biowaste and have potential as a packing media for bioreactor. It served as a packing media for the formation of microbes and helped to adsorb the pollutant on its surface. Benefits of using Casuarina equisetifolia seeds includes the easily accessible, provide a high surface area and most crucially its suitability for attachment and growth of microorganisms.

Biological dye removal methods were an appealing approach compared to conventional methods. Generally, removal of dyes by bacteria strains are more effective compared to fungi and algae [5]. Besides, biofilm process for dye removal is proved it's among the most effective methods. The steps in biofilm development start with the initial attachment of the bacteria to the surface, establishment to the biotic and abiotic surfaces, biofilm maturation and lastly detachment of cells from surface. Extracellular polymeric substances (EPS) help the biofilm community to adapt towards the changes of pH, concentration of toxic substances and temperature thus play important roles in biofilm growth and function as a part of protection mechanisms for the biofilm community. The benefits of packing media with biofilm formation are to increase the chance of extended usage of the absorbent media because microbial degradation helps extend the saturation point. Microorganisms that usually found in batik wastewater are Bacillus sp., Brevibacillus parabrevis and Alcaligenes faecalis [6][7]. Biofilm reactor configurations applied in dye wastewater removal are tricking filters, high-rate plastic media filters and membrane immobilized cell reactors.

Thus, the aim of this study is to monitor the performance of three laboratory scale reactors namely, sequencing batch biofilm reactor (CES-SBBR), a combination of adsorption and biological process, sequencing batch biofilm reactor (FC-SBR) and packed bed reactor (CES-PBR) on the RO16 dye removal. This study also conducted to determine the effect of two operational parameters which are initial concentration of dye and hydraulic retention time (HRT) on RO16 dye removal.

2. Materials and Methods

2.1. *Chemicals*
Reactive Orange 16 dye and glucose were purchased from Sigma-Aldrich Chemicals. Ammonium nitrate and potassium dihydrogen phosphate were purchased from Merck Chemicals. Hydrochloric acid, sodium hydroxide, magnesium chloride (Hexahydrate), and calcium chloride (Dihydrate) were purchased from R&M Chemicals. Iron (III) chloride-6-hydrate, manganese (II) chloride-4-hydrate, and sodium bicarbonate were purchased from HmbG Chemicals.

2.2. *Microorganisms*
The source of microorganisms in this study was collected from a textile industry located in Pantai Cahaya Bulan, Kelantan, Malaysia. The sample was culture in a reactor. Growth of microorganisms was monitored using mixed liquor suspended solids (MLSS). The MLSS were measured by filtering 10 mL of samples through the 0.45 μm Whatman® cellulose nitrate membrane filter. The samples were heated in oven at 105 °C for 1 hr and the weight of sample before and after the heating process was were determined [8].

2.3. *Preparation of Packing Media*
Casuarina equisetifolia seeds were collected from Pantai Batu Burok, Kuala Terengganu, Terengganu, Malaysia. *Casuarina equisetifolia* seeds were washed with distilled water and soaked for 24 hr. Next, the seeds were dried in oven at 70°C for 5 hr. The specific surface area and pore size distribution were determined by Brunauer, Emmett and Teller (BET).

2.4. Design of Laboratory Scale Reactors

Three laboratory scale reactors namely, CES-SBBR, FC-SBR and CES-PBR, with the working volume of 1.0 L were operated in this study (figure 1). The reactors were maintained at the condition as in table 1. The CES-SBBR, FC-SBR and CES-PBR were operated in aerobic condition in which oxygen was pumped into the reactor by a peristaltic pump. The CES-SBBR and CES-PBR were packed with the 80 g of *Casuarina equisetifolia* seeds as the packing media for microbial biofilm formation (figure 1). The CES-SBBR and FC-SBR were supplied with the nutrients for the microbial growth. The composition for the nutrient supplied were 300 mg/L glucose as carbon sources, ammonium nitrate as nitrogen sources and potassium dihydrogen phosphate as phosphorus sources with the ratio C:N:P of 100:5:1. During performance studies, the reactors were fed with the concentration of RO16 dye solutions ranged from 2 to 6 mg/L. In order to the determine the effect of the initial concentration of RO16 dye solution, the HRT was maintained at 24 hr. In order to determine the effect of HRT on the RO16 removal the HRT was varied from 24 to 48 hr while the initial concentration of RO16 dye solutions was maintained at 6.0 mg/L.

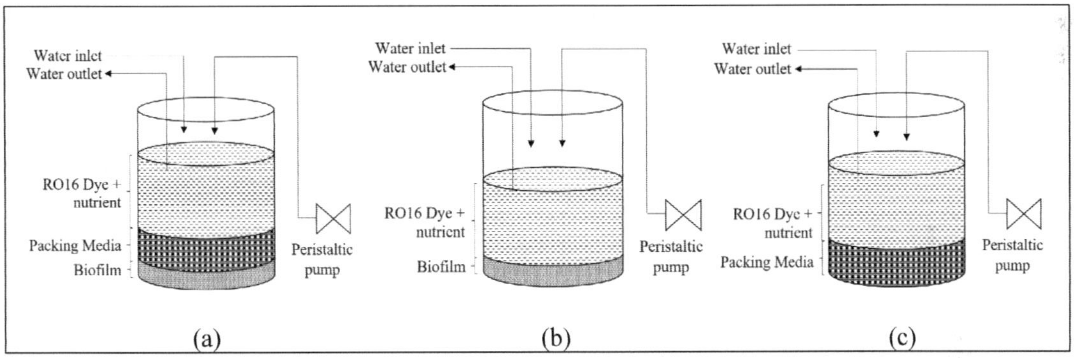

Figure 1. Schematic diagram for the laboratory scale reactor. (a) CES-SBBR; (b) FC-SBR and; (c) CES-PBR.

Table 1. Operation conditions of the reactors.

Parameter	Value
Working volume (L)	1.0
Temperature (°C)	25.0±1.0
pH	7.0±0.7
HRT (hr)	24-48

2.5. Determination of RO16 dye removal

The spectrophotometer (Thermo ScientificTM GENESYS 20) was used to analyse the UV-Visible spectra and was set to the wavelength at maximum absorbance (λmax) of RO16 dye [9]. The samples were filtered using a syringe filter with a pore size of 0.45um. The absorbance reading of before and after the removal of RO16 dye was determined using spectrophotometer at wavelength of 493 nm.

3. Results and Discussion

3.1. Characterization of CES Media

Through BET result the surface area of CES Media is 0.731 m^2/g (table 2).

Table 2. Specification of CES Media using BET.

BET	Value
Surface area	0.731 m^2/g
Average pore radius	4.78545e^{+01}Å
Total pore volume	1.749e^{-03} cc/g for pores smaller than 1285.1Å (radius) at P/Po = 0.99249

3.2. Preadaptation of Microorganisms

The preadaptation period is completed within 19 day. The purpose of the preadaptation period is to allow attachment of microbial biofilm formation on packing media. 2 mg/L of RO16 dye was added into the CES-SBBR and FC-SBR reactors to retain the microbe be able to withstand RO16 dye solutions. Initially, the MLSS value for CES-SBBR was 0.12 mg/L, before reduced to 0.05 mg/L on day 4 and 0.03 mg/L on day 7 (figure 2). The reduction in biomass concentration is due to insufficient nutrients in both reactors. The source of energy for the synthesis of new cellular material is carbon and inorganic nutrient for example nitrogen, phosphorus, sulphur, potassium, calcium, iron, sodium, magnesium, and chloride are required for microbial growth and reproduce [10]. On day 7, the nutrients were added to both reactors with C:N:P ratio of 100:5:1. On day 10, the MLSS reading slightly increased to 0.06 mg/L. This increment in MLSS reading could be explained that nutrients are required for the microbial growth. The biomass concentration in CES-SBBR and FC-SBR were continued to increase up to 0.22 and 0.18 mg/L, respectively on day 19. From the results obtained, the microbe grew faster in CES-SBBR compared to FC-SBR (figure 2 and figure 3). Both reactors were then supplied with higher concentration of RO16 dye in order to study their performance in the removing of RO16 dye.

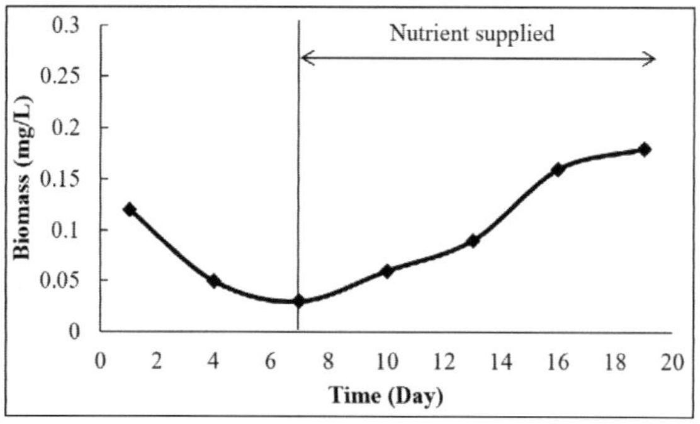

Figure 2. Biomass growth in CES-SBBR.

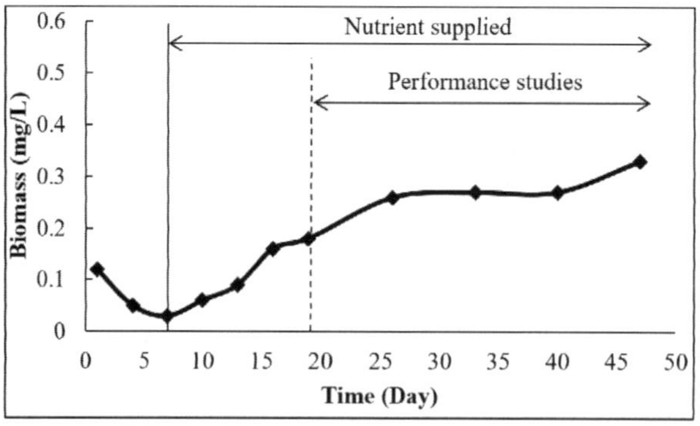

Figure 3. Biomass growth in FC-SBR.

3.3. Performance studies of Reactive Orange 16 dye removal process

During performance study, the effect of initial concentration of dye and HRT to the RO16 removal have been monitored. The performance of reactors was analyzed on the basis of the removal efficiency. The study was carried out for 32 days. The nutrients were supplied into the CES-SBBR and FC-SBR reactors. The concentration of biomass (mg/L) was also been monitored throughout the study.

3.3.1. CES-SBBR. Initially, the CES-SBBR reactor was operated at HRT of 24 hr and supplied with the initial concentration of RO16 dye of 2 mg/L. At this stage, the removal efficiency was in the range of 80.0-82.3% at HRT 24 hr. Next, the reactor was supplied with the 4 mg/L of RO16 dye, and the removal efficiency was in the range of 69.1-71.3%. The initial dye concentration was further increased to 6 mg/L and the removal percentage was reduced to the range of 48.1-50.0%. As shown in figure 4, the removal efficiencies of the CES-SBBR reactor were decreased when the initial dye concentration was increased. However, the removal efficiency was improved when the HRT was extended to 48 hr (figure 4). The removal efficiency was in range of 74.6-76.6%. Similar results were found in the previous study conducted by [11]. It is noted the dye removal efficiency was increased with HRT. For the microbial growth, the concentration of biomass was 0.29 mg/L at the first week of the study prior been increase to 0.36 mg/L at the second week of the study. The concentration of biomass was gradually increased to 0.48 mg/L at week 4 of the study. An even thickness of the biofilm effects the removal efficiency and simultaneously shorten the HRT, the CES Media is an ideal carrier since it has porosity and rough surface were improving attachment and growth of biofilm [12]. According to previous studies, the dye removal percentage increased with HRT which recorded 52.2±3.1% RO16 removal at HRT of 24 hr and increased to 60.6±3.9% at HRT of 48 hr and further improved to 68.1±3.4% at HRT of 72 hr [11].

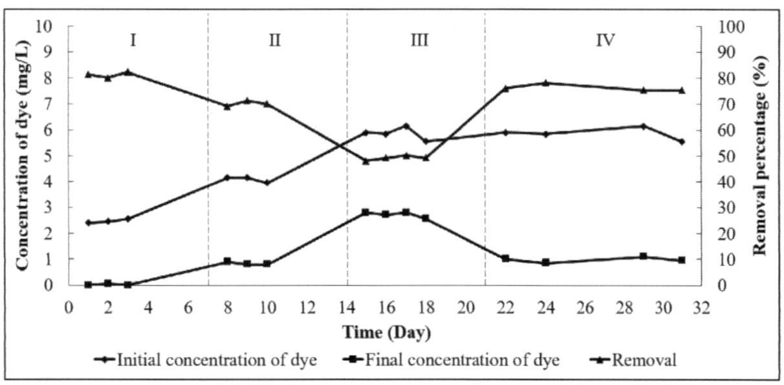

Figure 4. Performance study monitoring for CES-SBBR. Initial concentration (♦); Final concentration (■); and Removal Efficiency (▲).

3.3.2. FC-SBR. Initially, the FC-SBR reactor was operated at HRT of 24 hr and supplied with the initial concentration of RO16 dye of 2 mg/L. At this stage, the removal efficiency was in the range of 60.0-62.9% at HRT 24 hr. Next, the reactor was supplied with the 4 mg/L of RO16 dye, and the removal efficiency was in the range of 39.4-40.4%. The initial dye concentration was further increased to 6 mg/L and the removal percentage was reduced to the range of 34.4-42.6% at HRT of 24 hr and showed improvement during HRT of 48 hr with 52.5-55.9% dye was removed. As shown in figure 5, the removal efficiencies of the CES-SBBR reactor were decreased when the initial dye concentration was increased. However, the removal efficiency was improved when the HRT was extended to 48 hr (figure 5). The removal efficiency was in range of 74.6-76.6%. The removal of the azo dye is most likely to happen under microaerophilic condition which means under low oxygen concentrations condition since aerobic respiration dominates the use of NADH, thus preventing electron transfer from NADH to the azo bond [12]. Isolated microbe Bacillus sp able to degrade 100 mg/L of RO16 at HRT of 72 hr with 90% removal under microaerophilic condition [6]. Isolated microbe Proteus sp. under microaerophilic condition able to remove 150 mg/L of RO16 dye at HRT of 72 hr with 68.1±3.4% removal [11].

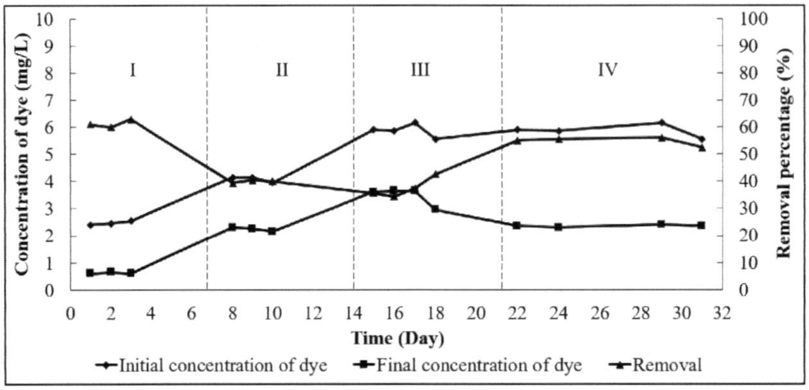

Figure 5. Performance study monitoring for FC-SBR. Initial concentration (♦); Final concentration (■); and Removal Efficiency (▲).

3.3.3. CES-PBR. Initially, the CES-PBR reactor was supplied with the initial concentration of RO16 dye of 2 mg/L and operated at HRT of 24 hr. At this stage, the removal percentage was in the range of 3.2-5.0% at HRT of 24 hr. When the reactor was supplied with 4 mg/L of RO16 dye with HRT of 24 hr, the removal percentage of dye was only in the range of 1.1-4.4%. When influent dye solution was increased to 6 mg/L, the removal percentage of dye recorded decreased to range from 0.0-2.2%. Although the HRT was increased to 48 hr, the removal percentage was only 0.78-2.99%. This result indicated that CES Media alone is inefficient in removing RO16 dye. Figure 6 shows overall performance of CES-PBR in removal of RO16 dye. At lower concentration of dye, the ratio of surface-active sites to the total dye molecules in the solution is high, therefore all the dye molecules can interact with the absorbent and be removed from the solution [13].

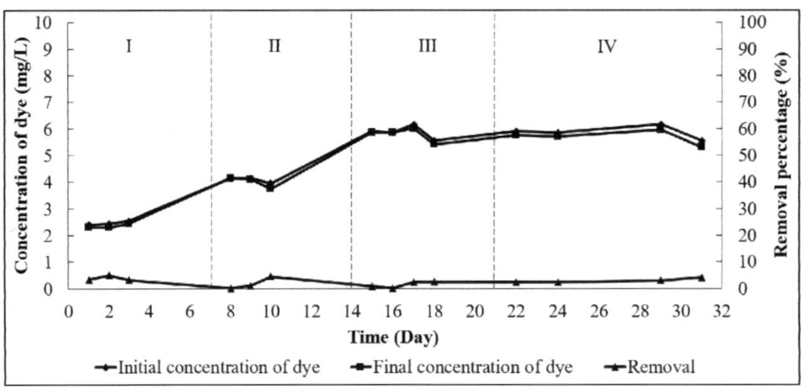

Figure 6. Performance study monitoring for CES-PBR. Initial concentration (♦); Final concentration (■); and Removal Efficiency (▲).

3.4. Effect of Initial Concentration of Dye

Based on the findings, CES-SBBR is better in terms of effectiveness for RO16 removal compared to both FC-SBR and CES-PBR (figure 7). The result indicates that the combination of microbial biofilm and CES Media in CES-SBBR improved the efficiency of RO16 removal which the maximum RO16 removal for HRT of 24 hr had reached 82.3%. While, CES-PBR is ineffective in removing RO16 dye which maximum RO16 removal for HRT of 24 hr was in range of 3.2-5.0%, whereas in FC-SBR maximum RO16 dye removal for HRT of 24 hr was in range of 60.0-62.9%.

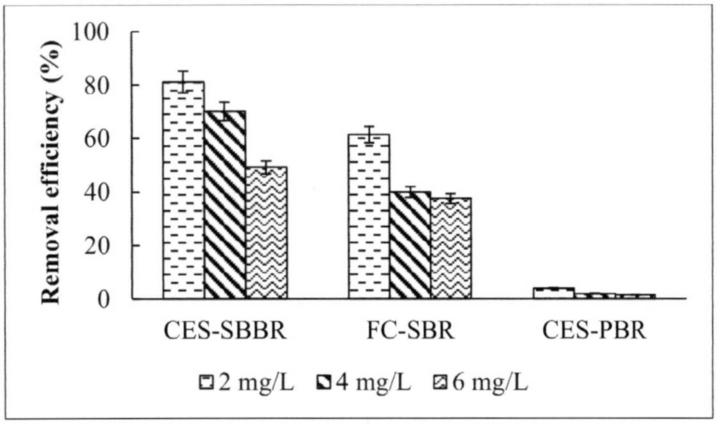

Figure 7. Effect of initial concentration of RO16 dye on removal efficiency of RO16 in CES-SBBR, FC-SBR and CES-PBR.

3.5. Effect of HRT

Removal percentage increased at HRT of 48 hr compared to HRT of 24 hr (figure 8). Removal percentage in CES-SBBR reached 78.13% at HRT of 48 hr but maximum removal in CES-SBBR reached only 50.00% at HRT of 24 hr. Same goes to FC-SBR, removal percentage reached 55.97% at HRT of 48 hr while highest removal only 42.62% at HRT of 24 hr. Also, for CES-PBR, the maximum removal percentage is 2.94% at HRT of 48hr compared to 1.37% at HRT 24 of hr. This indicate that a combined adsorption and biological process is better in terms of efficiency of RO16 dye removal compared to single biological process and single adsorption process.

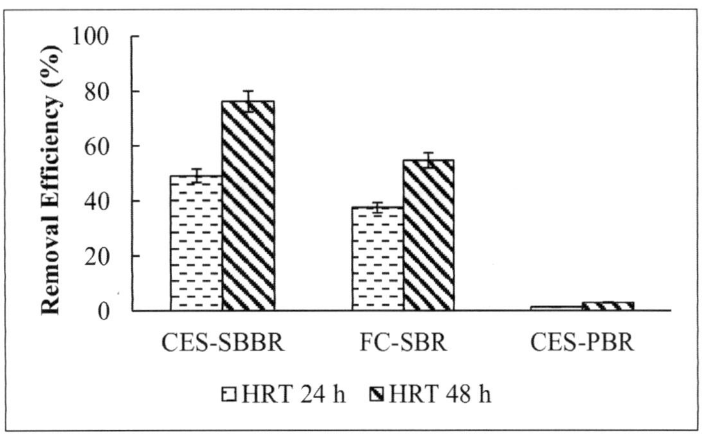

Figure 8. Effect of HRT on removal efficiency of RO16 in CES-SBBR, FC-SBR and CES-PBR.

3.6. Comparison with the Previous Studies

In this study, the performance study of RO16 dye removal using Casuarina equisetifolia seeds as packing media for microbial biofilm formation showed that the maximum removal efficiency was 81.13%. Table 3 shows the comparison on present study with previous studies. The continuous bioreactor packed with Casuarina seeds immobilized with mixed bacteria cultures reached 69.27% removal efficiency for 500 mg/L of RO16 dye [15]. The fungal culture (Pleurotus ostreatus and yeast Candida zeylanoides) immobilized on polyamide mesh carrier revealed that fungus culture could be applied on removal of RO16 dye [16]. Apart from microbial biofilm from mixed culture, there was a study that using isolated culture of Halomonas sp. immobilized in volcanic rocks in a continuous packed bed reactor which was able to remove 95% of 50 mg/L RO16 dye at HRT of 24 hr [12].

Table 3. Comparison with previous studies for RO16 removal.

Microbe	Experimental Condition	Initial concentration (mg/L)	HRT	Removal (%)	Citation
Mixed culture	Biofilm attached on CES media Aerobic, pH 7.3±0.4, 25±1°C	2	24 hr	81.13±1.13	Present study
		4	24 hr	70.22±1.07	
		6	24 hr	49.03±0.97	
		6	48 hr	76.75±1.38	
Halomonas sp.	*Halomonas sp.* Immobilized in volcanic rocks in continuous packed bed reactor, pH 10, 38°C	50	24 hr	95	[12]
Mixed culture	Cells immobilised on *Casuarina* seeds Aerobic, pH 7±0.3, 27±5°C	500	10 d	69.27±5	[15]
Mixed culture (*Pleurotus ostreatus* and yeast *Candida zeylanoides*)	Fungal culture immobilized on polyamide mesh carrier. Aerobic, pH 6.5, 28°C	150	11 d	87.5	[16]

4. Conclusion

From the findings of the performance study, it revealed that the combination of biological and adsorption processes in CES-SBBR is more efficient in removing dye compared to the single biological process (FC-SBR) and single adsorption process (CES-PBR). It is also noted that the removal efficiencies decreased as the initial concentration of dye increased. When 2 mg/L of RO16 was fed into the reactors, the maximum removal efficiency reached 82.26%. The maximum removal efficiency then reduced to 71.28% when initial concentration increased to 4 mg/L and experienced further reduction to 50.00% as initial concentration of dye increased to 6 mg/L. On the other hand, the removal efficiencies increased as the HRT extended from 24 to 48 hr. With initial concentration of 6 mg/L, the maximum removal efficiency reached up to 78.12% at HRT of 48 hr while only 50.00% at HRT of 24 hr. In conclusion, study on CES-SBBR showed that the combination of biological (biofilm) and adsorption (CES media) able to remove RO16 dye and CES Media was suitable for microbial biofilm formation. Further research on combination of biological and adsorption processes could be the future solution in waste water treatment plants especially for textile industries.

Acknowledgement

The authors acknowledge the financial support received in the form of grants from the Universiti Malaysia Kelantan via UMK Fundamental Research Grant, R/FUND/A13000/01405A/001/2020/00736.

References

[1] Al-Fawwaz A and Abdullah M *2016 Int. J. Environ. Sci. Dev.* **7(2)** 95–99.
[2] Katheresan V, Kansedo J and Lau SY 2018 *J. Environ. Chem. Eng.* **6(4)** 4676–4697.
[3] Abu Talha M, Goswami M, Giri B S, Sharma A, Rai B N and Singh R S 2018 *Bioresource Technol.* **252** 37–43.
[4] Krishnaraj S, Deshik Sandeep S, Poovarasan R, Magesh D, Ajay Sooria KV and Kiran R 2018 *Mater. Today: Proc.* **5(1)** 1800–1806.

[5] Vikrant K, Giri B S, Raza N, Roy K, Kim K H, Rai B N and Singh R S 2018 *Bioresource Technol.* **253** 355–367.
[6] Shean O S C and Rahim A A 2019 *ARPN J. Eng. Appl. Sci.* **14(9)** 1669–1674.
[7] Bharti V, Vikrant K, Goswami M, Tiwari H, Sonwani R K, Lee J, Tsang D C W, Kim K H, Saeed M, Kumar S, Rai B N, Giri B S and Singh RS 2019 *Environ. Res.* **171** 356–364.
[8] Escolà Casas M, Chhetri R K, Ooi G, Hansen K M S, Litty K, Christensson M, Kragelund C, Andersen H R and Bester K 2015 *Sci. Total Environ.* **530–531** 383–392.
[9] Vo T S, Vo T T B C, Suk J W and Kim K 2020 *Nano Convergence*, **7(1)** 4.
[10] Metcalf W and Eddy C 2004 *Wastewater Engineering: Treatment and Reuse, fourth ed.* McGraw Hill New York.
[11] Abbas N, Hussain S, Azeem F, Shahzad T, Bhatti SH, Imran M, Ahmad Z, Maqbool Z and Abid M 2016 *World J. Microbiol. Biotechnol.* **32(11)** 181.
[12] Montañez-Barragán B, Sanz-Martín J L, Gutiérrez-Macías P, Morato-Cerro A, Rodríguez-Vázquez R and Barragán-Huerta BE 2020 *Extremophiles.* **24(2)** 239–247.
[13] Shahryari Z, Goharrizi A S and Azadi M 2010 *Int. J. Water Resour. Environ. Eng.* **2(2)** 16 – 28.
[14] Anjaneya O, Shrishailnath SS, Guruprasad K, Nayak A S, Mashetty S B and Karegoudar T B 2013 *Int. Biodeterior. Biodegrad.* **79** 64–72.
[15] Bharti V, Shahi A, Geed S R, Kureel M K, Rai B N, Kumar S, Giri B S and Singh R S 2017 *Indian J. Biotechnol.* **16(2)** 216–221.
[16] Šlosarčíková P, Plachá D, Malachová K, Rybková Z and Novotný Č 2020 *Microbiologica.* **65** 629–638.

Optimization of AB Fertilizer for Plant Growth in Fertigation Using Central Composite Design

S F S Draman[1]*, S M Shukri[1], N S I Ishak[1], A S A Samad[1] and A F Pauzi[1]

[1]Faculty of Chemical Engineering, Universiti Teknologi MARA Terengganu Campus, Bukit Besi Branch, 23200 Dungun, Terengganu, Malaysia

*Corresponding author: sfauziah@uitm.edu.my

Abstract. A plant needs to consume sufficient essential nutrients for production speed. This study investigated the optimum condition for water spinach growth of wick technique fertigation. Fertigation is a fertilizer application process in which the drip system mixes fertilizer within the irrigation water. The fertilizer solution is uniformly distributed in the irrigation system. The chosen method is a hydroponic method and suitable in an urban area with no soil for planting. Fertigation is also an easy and cost-effective method. The water spinach height's optimum condition was determined through the central composite design (CCD) as the experiment (DOE) design. The critical factors selected for this study were the concentration of AB fertilizer and wick length. Obtaining an adequate AB fertilizer concentration is crucial because it will affect the growth of water spinach. The statistical analysis based on a CCD showed that 1.5 mL of AB fertilizer concentration (volume/volume) and 12 cm of wick length were the optimum conditions to achieve the best water spinach growth in 21 days. The CCD as a response surface method was proved to help obtain optimum water spinach plant growth conditions.

1. Introduction

In the late year of 2019, the COVID-19 pandemic developed quickly and widely across the globe had profound consequences for food safety and nutrition. The unfolding crisis has impacted food systems and, through multiple factors, disrupted people's access to food. In the aftermath of lockdowns caused by the global health crisis, people worldwide have experienced a massive disturbance to food supply chains and a major global economic recession [1]. The combination of consequences of COVID-19 itself through the long run, such as without a wide scale of concerted action, the corresponding mitigating initiatives, and the growing worldwide recession, could threaten food systems' functioning [2]. In Malaysia, the Ministry of Agriculture and Food Industry started the Urban Agriculture Program to help urban households reduce living costs. This program is seen to be a good program through the self-production of some of the foodstuffs needed [3].

Fertigation is widely used in industry agricultural purposes as dispenser units, making it more convenient and comfortable for general landscape projects [4]. A dosing device for the standard solutions where nutrients are inserted, and good irrigation is needed for fertigation. For an urban area, the hydroponic method is more suitable since no soil is necessary for planting. The soil-less production of plants growing using mineral fertilizer solutions. Nutrients are transmitted directly to the roots in this process [5]. Hydroponic vegetables are usually produced in closed systems techniques. Compared to field farming, this method ensures less water is required [6]. The hydroponics method was reported to successfully apply to cultivate vegetable crops and other plants [7]. Inorganic fertilizers are usually used in hydroponic systems, whereas organic fertilizers cannot be added. These chemical fertilizers are a significant component of hydroponics. There are two formulated functional solutions for fertilizer: solution A and solution B, representing any hydroponic formula. To expand, all plants require seventeen elements. These contain carbon (C), hydrogen (H), oxygen (O), nitrogen (N), phosphorus (P), potassium (K), calcium (Ca), magnesium (Mg), sulfur (S), copper (Cu), iron (Fe), sodium (Na), manganese (Mn), boron (B), molybdenum (Mo), and chlorine (Cl). Most of the materials used in the formulation of hydroponics AB nutrients are obtained through these natural sources of minerals in the soil [5,6]. It is expected, the plant growth will affect due to the controlled amount of AB fertilizers concentration. Therefore, obtaining the optimization of chemical fertilizer (AB fertilizers) concentrations is essential.

Irrigation is the direct consumption of liquid fertilizer to plant growth. Irrigation aims to specify ample water for plants to reduce the pressure that can decrease productivity. The amount of water and its frequency rely upon local climatic factors, crop and development phase, and soil-moisture-plant features [4]. There is an approach to examining crop measures such as colour or leaf angle transitions, but the result can become too late to prevent crop yield or quality from decreasing. The faster a plant grows, the more fertilizer and water it will require. Consequently, as watering is increased, so is leaching and nutrient loss [8]. Thus, obtaining the need for irrigation that will not involve analysis of evapotranspiration rates is crucial.

In this study, water spinach growth was explored. Water spinach is a semi-aquatic plant that flourishes with little or no supervision. Water spinach contains a fair amount of vitamins A, B6, and C, along with magnesium and fibre. The plant has been considered medicinal in Asia for hundreds of years for its purge effects. The plant also has been studied for its use in India to treat lead toxicity [9]. This study examined the optimum condition for water spinach growth of wick technique fertigation using Central Composite Design (CCD) as experimental (DOE) design. CCD is the typical design of response surface methodology (RSM) and widely used in various experiments. RSM collects mathematical and statistical techniques for designing experiments, building models, evaluating the effects of factors, and searching for optimum condition factors for desirable responses [10]. The optimization process of this methodology involves four steps. Firstly, the statistically designed combinations' responses are studied, followed by estimating the coefficients and fitting them in a mathematical model that best works the experimental conditions. Then, the response of the fitted model is predicted. CCD advantages are that it allows the experimental designer to know what effect the factors had on response. It also provides high-quality predictions of linear and quadratic interaction effects of parameters affecting the process [11]. In this study, two factors (AB fertilizer's concentration and wick length) were used for the optimum height of water spinach growth.

2. Methodology

The material used was AB fertilizer in liquid form, cocopeat, water spinach seed, aluminium foil 1.5-litre bottle, and water. The stock of AB fertilizer used is complimentary from Jabatan Pertanian Negeri Terengganu. The Nutri-pot method was constructed in this hydroponic study. First of all, the cocopeat has been washed and let it wet about 60% and covered with a wet towel. Then, sprout the seed using the egg board and cocopeat for three days. After three days, transfer them into a bottle that has been cut according to the experiment. Next, add AB fertilizer into the water at the bottom of the bottle and cover the bottom with aluminium foil to prevent algae from growing, as shown in figure 1. The experiment was conducted for 21 days to obtain the complete data of water spinach's length.

The CCD again using a coded factor level for the variable, as shown in table 1. Meanwhile, the high and low levels defined for two independent variables were shown in table 2. The manipulated variables were the concentration of AB fertilizer and wick length from the CCD, while the response variable was the water spinach plant's height. The total volume of water used is 500 mL for all 12 experiments. Design Expert 13 software was used to analyze all the obtained data. By using electrical conductivity (EC) meter concentration of AB fertilizer (volume/volume) for coded level -1, 0 and +1 is 1.0 mS/cm, 2.2 mS/cm and 3.8 mS/cm, respectively.

Figure 1. Example of Nutri-pot for the experiments

Table 1. The concentration of AB Fertilizer and Wick Length using Coded Factor Level in Central Composite Design.

Run	The concentration of AB Fertilizer (Volume/Volume)	Wick Length (cm)
1	-1	-1
2	1	-1
3	-1	1
4	1	1
5	0	0
6	0	0
7	0	0
8	0	0
9	0	1.414
10	1.414	0
11	-1.414	0
12	0	-1.414

Table 2. Experimental Factor and Level in CCD for Concentration of AB Fertilizer and wick length

Variables	**Level**		
	-1	0	1
Concentration of AB fertilizer (volume/volume)	0.5 mL A + 0.5 mL B	1.0 mL A + 1.0 mL B	1.5 mLA + 1.5 mL B
Wick length (cm)	6	9	12

3. Results and Discussion

Table 3 shows the water spinach height after 21 days of the experiment according to the run, respectively, with the concentration of AB fertilizer (volume/volume) and wick length (cm) as the factors.

Table 3. Height of Water spinach

Run	Factor 1 The concentration of AB Fertilizer (Volume/Volume)	Factor 2 Wick Length (cm)	Response Height of Water spinach (cm)
1	0.5 mL A + 0.5 mL B	6.0	3.1
2	1.5 mL A + 1.5 mL B	6.0	6.5
3	0.5 mL A + 0.5 mL B	12.0	5.5
4	1.5 mL A + 1.5 mL B	12.0	9
5	1.0 mL A + 1.0 mL B	9.0	11.3
6	1.0 mL A + 1.0 mL B	9.0	5.1
7	1.0 mL A + 1.0 mL B	9.0	11.5
8	1.0 mL A + 1.0 mL B	9.0	8.1
9	1.0 mL A + 1.0 mL B	13.2	6.0
10	1.7 mL A + 1.7 mL B	9.0	12.3
11	0.3 mL A + 0.3 mL B	9.0	9.6
12	1.0 mL A + 1.0 mL B	4.8	7.7

Based on the results obtained in table 3, statistical analyses were carried out using Desing Expert 13.0 software. The ANOVA results in table 4 show the reaction condition value for F-value, P-value, and R^2 value. These three components are related to each other by the performance of the experiment.

Table 4. ANOVA for Response Surface Quadratic Model

Source	Sum of Squares	df	Mean Square	F-value	p-value	
Model	44.73	5	8.95	1.10	0.4376	not significant
A-Concentration of AB. Fertilizer	14.36	1	14.36	1.77	0.2257	
B-Wick length	0.7787	1	0.7787	0.0957	0.7660	
AB	0.0025	1	0.0025	0.0003	0.9865	
A^2	0.0003	1	0.0003	0.0000	0.9955	
B^2	29.06	1	29.06	3.57	0.1007	
Residual	56.95	7	8.14			
Lack of Fit	24.39	3	8.13	0.9987	0.4794	not significant
Pure Error	32.56	4	8.14			
Cor Total	101.67	12				

R^2 = 0.4399, predicted R^2 = -1.2061, adjusted R^2 = 0.0398

F-value is a value on the F distribution that measured the distance between individual distributions. Many statistical tests generate F-value. This value can determine whether the test is statistically significant. It is also used to analyze variance (ANOVA) and calculated by dividing two mean squares. This calculation is to determine the ratio of explained variance to unexplained variance. The F-value must correspond to a p-value where if the F-value is high, the P-value will low, which means that when the F-value goes up, the P-value will go down.

As shown in table 4, the F-value for this model is 1.10. It implies the model is not significant relative to the noise. There is a 43.76% chance that an F-value this large could occur due to noise. Then, the lack of fit F-value of 1.00 implies the lack of fit is not significant relative to the pure error. Therefore, a 47.94% chance of a lack of fit F-value could occur because of noise. The non-significant lack of fit is good [12], and the most important is when the model is fitted.

A p-value is a probability measurement to observe a difference that may occur just by a random chance. Generally, the lower the p-value shows, the greater the statistical significance of the observed difference. The p-value can be used as pre-selected confidence levels for hypothesis testing. The p-value that has less than 0.0500 indicates the model term is significant [13]. If the p-value is more than 0.1000, it tells the model term is not significant. This model's p-value is 0.4376 causing the R-square value to be relatively high, which is 0.4399.

R-squared (R^2) is a statistical measurement explaining the percentage of the variance for a dependent variable response by the movement of an independent variable [14]. It shows how well the model fits the data or how much variation was explained by the model. R^2 explains the extent to which the variance of one variable will be presented to the second variable even though the correlation explains the strength of the relationship between independent and dependent variables. Other than that, the higher the value of R^2, means the more explanation of variation by the model. If this situation happened for a model, it is undoubtedly the best scenario in that model [15]. Next, if there is a low value of R^2, the model does not explain much about the data variation. As shown in table 5, the results of the value of R^2 is 0.4399. It means about 43.99% of the variation of the model was explained within the data. This value is a bit low, but the predictor variable still provides information about the response. Even short R^2 can have a unique contribution to this study. Next, the adjusted and predicted R^2 value is 0.0398 and -1.2061, respectively. Adj R-squared is a measure of the variation around the mean explained by the model, adjusted for the model's number of terms. The adjusted R-squared decreases as the number of terms in the model increase if those additional terms don't add value to the model. This statistical model showed that the positive value of R^2 adj is produced. It is an ordinary model to be chosen because there are no warning issues about the variation explained by the model.

Table 5. Model Summary Statistic

Source	Std. Dev.	R^2	Adjusted R^2	Predicted R^2	PRESS	
Linear	2.94	0.1489	-0.0213	-0.4677	149.22	
2FI	3.10	0.1489	-0.1348	-1.2541	229.18	
Quadratic	2.85	0.4399	0.0398	-1.2061	224.29	Suggested
Cubic	3.13	0.5172	-0.1588	-9.9064	1108.88	Aliased

Next, Pred R-squared measures the amount of variation in new data explained by the model. table 5 the R2 pred value is not within 0.2 to each other, meaning that the terms in this model are not significant, and the number of design points to the unstable prediction. However, the R2 pred value is lower than the R2 adj that shows the warning sign, but there is no negative impact on the analysis or the conclusions [15].

Figure 2 indicates the relationship of the result surface plots for the height of water spinach (cm) vs wick length (cm), the concentration of AB fertilizer (mL). The factors influencing the variables on response can be seen according to the figure 2. These graphs are used to comprehend the interaction of two reaction variables. The plot below illustrates how the concentration of AB fertilizers (mL), and wick length (cm), impact the height of water spinach (cm) [16].

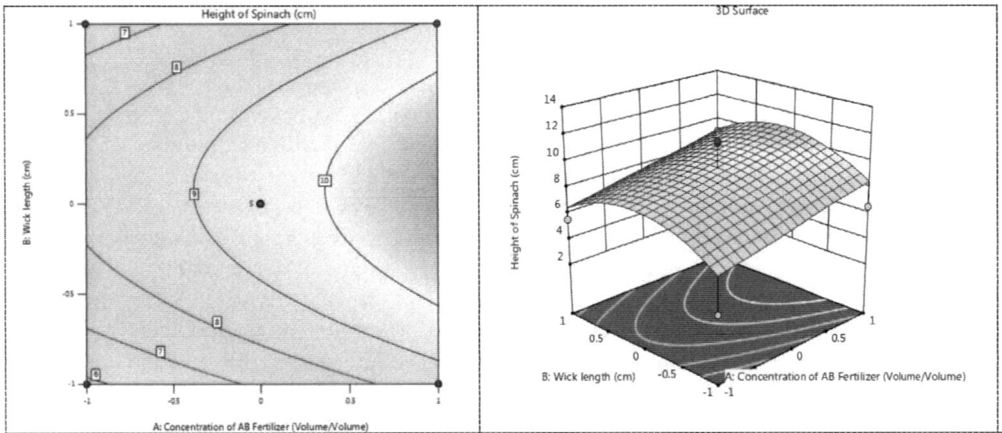

Figure 2. Response Contour Plot (2D) and Surface Plot (3D) Display Interactive Effect of The Variables on The Height of Water spinach

Based on the contour plot graph, the darker orange regions have a maximum water spinach height (cm). In this graph, a peak of wick length is ten, and showing the value of the concentration of AB fertilizers is at 0.45 at the contour ranges. This response indicates that the higher water spinach length depends on the wick length and AB fertilizers' concentration. Those plots display the influence of two variables at a time on the response.

Generally, the height of water spinach increases with the increase in AB fertilizer concentration. This result is in line with the outcome reported by previous researchers in canola cultivation [17]. It is because fertilizers are crucial for plant growth as a source of nutrients, and the consistency of liquid fertilizers is determined by their pH and nutrient content [18]. In this experiment, the result explains that the concentration of AB fertilizer and wick lengths is not always necessarily in a tremendous amount to obtain a higher value of water spinach's height. The average rate of wick length and AB fertilizer concentration admitted the height of water spinach to increase. The slope of the water spinach height enhanced by the average wick length rates was found to be under the average amount of AB fertilizer's concentration level than at a lower level. Thus, the water spinach plant is accelerated at the mean value of wick lengths and AB fertilizer concentration, resulting from the higher water spinach height. This result agrees with previous researchers' reported work in optimizing AB-mix fertilizer varieties of hydroponic lettuce (Lactuca sativa L.) [19].

4. Conclusion

The response surface method is useful for optimizing the concentration of AB fertilizers. Statistical analysis was developed as a valuable and efficient tool for creating the optimum AB fertilizer concentration for plant growth. The statistical analysis based on the central composite design showed that containing 1.5 mL of A fertilizer and 1.5 mL of B fertilizer concentration of AB in 500 mL water (volume/volume) and 12.0 cm wick length was the best condition to achieve the optimum water spinach growth. Further studies need to carry out to verify the optimum condition for the plant's growth.

References

[1] HLPE 2020 *Impacts of COVID-19 on food security and nutrition: developing effective policy responses to address the hunger and malnutrition pandemic* Rome

[2] Policy Brief: The Impact of COVID-19 on Food Security and Nutrition. (2020, June). Retrieved from https://www.un.org/sites/un2.un.org/files/sg_policy_brief_on_covid_impact_on_food_security

[3] *PENJANA: Lebih ramai disasar ceburi pertanian bandar* (June 16,2020) and *Program pertanian*

bandar perlu dikomersial (February 13, 2021) Retrieved from https://www.bharian.com.my/berita/nasional/

[4] Sivanappan R K 2015 *Principles of Drip/Micro Or Trickle Irrigation* in M. R. Goyal, Water and Fertigation Management in Micro Irrigation. New York: Apple Academic Press.

[5] Phibunwatthanawong T and Riddech N 2019 Liquid Organic Fertilizer Production for Growing Vegetables under Hydroponic Condition *Int. J. Recycl. Org. Waste Agric.* **8** 369-380.

[6] Jones Jr J 2012 *Hydroponic. In-Plant Nutrition and Soil Fertility Manual (2nd ed.)* 109-208.

[7] Malaysia Hydroponics AB Nutrients is suitable for the leafy vegetables and vegetative stage of fruiting plants. (n.d.). Retrieved from Malaysia Hydroponics: http://malaysiahydroponics.com/malaysia-hydroponics-powder-fertilizer-10l.html

[8] Sambo P, Nicoletto C, Giro A, Pii Y, Valentinuzzi F, Mimmo T, Lugli P, Orzes G, Mazetto F, Astolfi S, Terzano R and Cesco S 2019 Hydroponic Solutions for Soilless Production Systems: Issues and Opportunities in a Smart Agriculture Perspective Front *Plant Sci.* **10** 923.

[9] Davanjee S, Dua T, and Das S. Water spinach 2015 Ipomoea Aquatica (Convolvulaceae), Ameliorates Lead Toxicity by Inhibiting Oxidative Stress and Apoptosis. *PLoS ONE.* **10(11).**

[10] Syed Draman S F, Daik R El-Sheikh S M and Latif F A 2013 Doping optimization of polypyrrole with toluenesulfonic acid using BoxBehnken design *AIP Conf. Proc.* **1571(2013)** 782–788.

[11] Sathavornvichit N, Bookkamana P, and Plubin B 2006 Using Central Composite Design for Process Optimization. https://www.weibull.com/hotwire/issue146/hottopics146.htm

[12] Ivana S and Stefano di G *24th. Eur. Symp. on Comput.-Aided Process Eng.* **(3).**

[13] Faye A, (June 15, 2016), What is the relationship between R-squared and p-value in a regression from https://www.researchgate.net/post/What_is_the_relationship_between_R squared_and_p-value_in_a_regression

[14] Jason F, (November 18, 2020), R-Squared Definition, from https://www.investopedia.com/terms/r/r-squared.asp

[15] Corporate Finance Institute, (2015-2021), R-Squared, from https://corporatefinanceinstitute.com/resources/knowledge/other/r-squared/

[16] Bouzid Ait-A, Philippe P and Abdelkhalak El H 2020 *6 - Meta-Model Development Embedded Mechatronic Systems 2 (Second Edition)* ISTE Pages 157-187.

[17] Koocheki A, Mahallati M N, Moradi R and Mansooru H 2014 Optimizing Water, nitrogen and Crop Density In Canola Cultivation Using Response Surface Methodology And Central Composite Design *Soil Sci. Plant Nutr.* **60(2)** 286-298.

[18] Sunaryo Y, Purnomo D, Darini M T and Cahyani V R 2018 Effects of goat manure liquid fertilizer combined with AB-MIX on foliage vegetable growth in hydroponic *IOP Conf. Ser. Earth Environ. Sci.* **129(1).**

[19] Rosnina and Sarah Mauliza 2020 Optimization of AB-mix Fertilizer on Varieties of Hydroponic Lettuce (Lactuca sativa L.) *Journal of Tropical Horticulture* **3(2)** 86-91.

Elucidation of morphology developed of Ca_xCe_y/ZrO_2 solid catalyst for the production of lactic acid from glycerol conversion

N Razali[1*], A Z Abdullah[2]

[1]Faculty of Chemical Engineering, Universiti Teknologi MARA Cawangan Terengganu, Kampus Bukit Besi, 23200 Dungun, Terengganu, Malaysia,

[2]School of Chemical Engineering, Universiti Sains Malaysia, Kampus Kejuruteraan, 34200 Nibong Tebal, Pulau Pinang, Malaysia,

*Corresponding author : norainirazali@uitm.edu.my

Abstract. Monometallic and bimetallic noble metal supported catalysts have been tested for the catalytic glycerol oxidation to lactic acid. Modification of zirconia, (ZrO_2) using CaO, which is an alkaline earth metal as the base source and Ce as the metal source to give, 30wt%Ca_xCe_y/ZrO_2 was investigated. The catalysts were prepared using co-precipitation method at different of Ca_x to Ce_y ratios (x:y = 1.0:0.5 – 1.0:3.0) and subsequently calcined at 800 °C. Subsequently, the glycerol reaction to lactic acid was carried out at a 250 °C for 2 h in a base free condition using pure glycerol. The interaction between the metallic sites and CeO_2 support showed was characterized to elucidate the developed morphology of catalyst. The textural and structural properties of the catalysts were studied using Field emission scanning electron microscope (FESEM) and Transmission electron microscope (TEM). It was found that the morphology of the catalyst well developed and contribute to the significant reaction activity through the reaction mechanism. The spent catalyst also could produce the lactic acid and the morphology is still in a good condition showing capability of cyclability.

1. Introduction

Glycerol is a very important industrial feedstock for conversion into several high value chemicals. Their applications are found in food, cosmetic, drug and tobacco industries. Chemical products formed from glycerol valorization are such as lactic acid, polyglycerol, 1,2 propanediol, hydrocarbons, glyceric acid, dihydroxyacetone, hydrogen and olefins [1–5]. Recently, lactic acid has emerged as one of the promising products. A number of literatures have reported several effective means to produce significant yield of lactic acid such as through hydrothermal reaction, hydrogenolysis, selective oxidation and fermentation processes [3,6,7]. However, drawbacks of fermentation methods are such as low productivity, low efficiency, high cost of enzymes, limited space–time yield, poor scalability and the need for complicated purification methods. Thus, a greener and more economical production of lactic acid using catalytic conversion has been extensively explored.

Lactic acid is an important high-value chemical that is typically used as a raw material for the production of medicines, cosmetics, fragrances, and polylactic acid, which is a carbon-neutral bioplastic [8]. Oxidation reaction is a high temperature reaction, requiring the catalyst with a high thermal stability. CeO_2 catalyst has attracted much attention due to its good oxygen storage capacity, thermal resistance and redox promotion property [9–11]. The chemical basicity properties of the catalyst was contributed by CaO, as the important modification according to Marinkovic et al. [12]. It was evident that the strong basicity of the catalyst could enhance catalytic dehydration and benzylic acid rearrangement of the dihydroxyacetone intermediates to lactic acid. In addition, Ca leaching could be minimized by coupling with La_2O_3 or CeO_2 as an appropriate support [13]. Atsushi et. al. [14] found that the combination of the two metal oxides which have both Brønsted acid and Lewis acid sites accelerated rates of lactic acid formation. ZrO_2 could be a suitable solid support due to its surface catalytic sites properties such as high concentration of hydroxyl groups and coordinative unsaturated Lewis acidic-basic $Zr4^+$ O^{2-} pairs. These features were believed to potentially enhance the catalytic ability of the transition metal active sites [15]. Thus, this potential created the interest to explore the suitability of a Ca_xCe_y/ZrO_2 mixed oxide catalyst in this oxidative reaction. The catalyst was hypothesized to be effective for this reaction on the basis of its basicity and oxygen storage capacity roles to selectively convert glycerol to lactic acid.

Bimetallic Ni-Co catalysts supported on CeO_2 were developed by Zhenchen et. al. [16] and found that the introducing Co into the supported catalysts was important to prevent the aggregation of Ni into large particles. A comprehensive study on the incorporation of Ca_xCe_y onto ZrO_2 support is still needed for enhancing stronger the metal-support interaction which can affect the catalytic activity. In this study, the structure of Ca_xCe_y/ZrO_2 catalysts in the oxidation reaction of glycerol is reported. The morphology of metal-support developed were analysed towards lactic acid production, which might give the fundamental insight into the catalytic mechanism of the catalyst. In-depth characterization provided new understanding on the nature of active sites structure of the resulting Ca_xCe_y/ZrO_2 catalysts, while their activity and selectivity in the oxidation reaction of glycerol to lactic acid were demonstrated. The performance of the catalysts was specifically studied at different Ca to Ce ratios as an attempt to correlate the activity data with the structural properties of these solids. The catalytic properties and performance of the catalyst were to be correlated, to generate significant conclusions on their behaviours in the glycerol selective conversion to lactic acid.

2. Materials and Method

2.1 Materials

Glycerols (98 % purity), calcium nitrate hexahydrate, $Ca(NO_3)_2.6H_2O$, cerium nitrate hexahydrate, ($Ce(NO_3)_2.6H_2O$), zirconyl chloride octahydrate ($ZrOCl_2.8H_2O$), ammonium hydroxide, NH_3OH (99 % solution in water), sodium hydroxide, NaOH (90%) and sodium bicarbonate, Na_2CO_3 (90%) were purchased from Fluka. Deionized water (DI) was used throughout the study for the preparation of the support and catalyst. All the chemicals were of reagent grade and were used as received without further purification.

2.2 Preparation of support material

Zirconia (ZrO_2) was prepared based on a work reported by Yang et al. [17]. An aqueous solution of zirconium chloride octahydrate ($ZrOCl_2.8H_2O$) was first prepared using DI water and stirred for 30 min. Then, the pH value of the solution was adjusted to pH 10-11 by dropwise addition of ammonium hydroxide solution (25 %). Stirring of the solution was then continued for 3 h. The resultant solution was then filtered with DI water to remove all the contaminants and then dried overnight at 100 °C. The sample was then calcined at 500 °C for 3 h and used as the catalyst support.

2.3 Preparation of mixed oxide catalysts

The support material was first dispersed in DI water and the required amounts of $Ca(NO_3)_2$ and $Ce(NO_3)_2$ solutions were added dropwise under stirring. Then, calculated amounts of 1M NaOH and Na_2CO_3 were also added dropwise to maintain the pH at 11-12. The temperature of the mixture was set at 80 °C for 5 h on a hotplate stirrer. After that, the suspension was filtered and the solid was washed with deionized water to eliminate all traces of base (pH test). Then, the catalysts were dried overnight at 80 °C followed by calcination at 500 °C for 4 h. The catalysts are represented as 30% wt/wtCa_xCe_y/ZrO_2 (x and y indicate the serial ratios of the active components in the samples).

2.4 Characterization of catalyst

The catalyst's surface morphology was studied by means of a field emission scanning electron microscope (FESEM Carl Zeiss SMT, Oberkochen, Germany). Transmission electron microscope (TEM, Eftem Libra 120) was used to observe the internal morphology of the catalysts. Catalyst sample of about 0.05 g was first dispersed in 3 ml of 100 % acetone. Then, the suspension was shaken vigorously and the suspended powder (light powder) was sucked out slowly with a micropipette and dropped on the grid for the analysis.

2.5 Catalyst activity

Batch hydrothermal glycerol reaction was carried out in a 50 mL stainless steel autoclave at a stirring speed of 250 rpm as shown in figure 1. The standard reaction was carried out under the following reaction conditions: 533 K, 2 bars of initial O_2 pressure, 99% aqueous glycerol, 20 wt. % of catalyst for 2 h of reaction. After being purged for 15 minutes, the reactor was heated to the desired reaction temperature of 250 °C and the O_2 pressure was then increased to about 10 bar. The liquid-phase products were analysed by means of a gas chromatograph with a SHODEX capillary column (50 m × 0.32 mm) and a flame ionization detector. The catalytic performance demonstrated by the catalysts was evaluated based on glycerol conversion as well as the yield and selectivity of lactic acid using high performance liquid chromatograph (HPLC). To confirm the presence of intermediate products in identifying the mechanism of the oxidation reaction, some liquid samples were also analysed with Gas chromatography–mass spectrometry (GCMS) equipped with a BPX5 capillary column (30 x 0.25 mm) at a heating rate 30° C/min and in the range of 40- 225 °C.

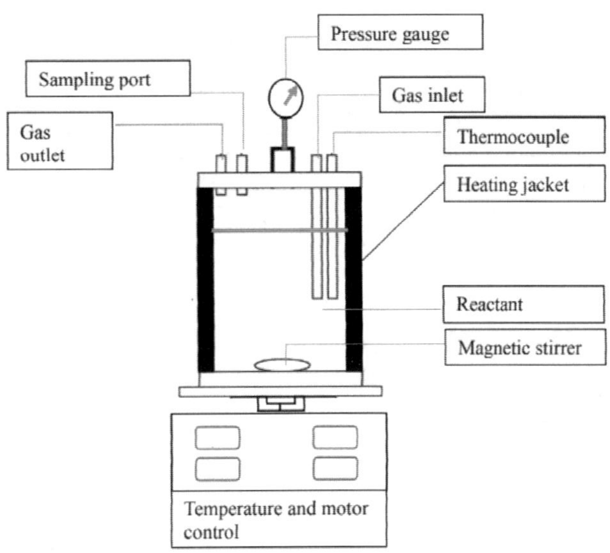

Figure 1. Schematic diagram of batch pressurized reactor

3.0 Results and Discussion

3.1 Morphology of mixed oxide catalyst by FESEM analysis

Clear changes in the morphologies of the catalyst after modification of the ZrO_2 support with different Ca_xCe_y ratios were observed by FESEM. Pure ZrO_2 support, (figure 2 (a)) appears to have a nubby structure, the surface of which was dense and compact. The morphology was apparently uniform and it contained quasi-spherical like microparticles with inner-aggregate pores particles as described by Majedi et al. [18]. As the ratio of Ca_xCe_y was further increased from 1.0:0.5 (figure 2 (b) to 1.0:2.0 figure 2(c), clearer but much smaller external pore structure was formed. Ca_1Ce_2/ZrO_2 (figure 2 (c)) exhibited much different morphological structure as compared to that of ZrO_2 support, showing high dispersion of metal on the ZrO_2 support and was consistent with that of platinum on Ce/ZrO_2 catalysts [19]. In addition, calcination process could also contribute to the formation of a hollow, porous, and fluffy products as the reactions that occurred would release a large amount of gas and heat.

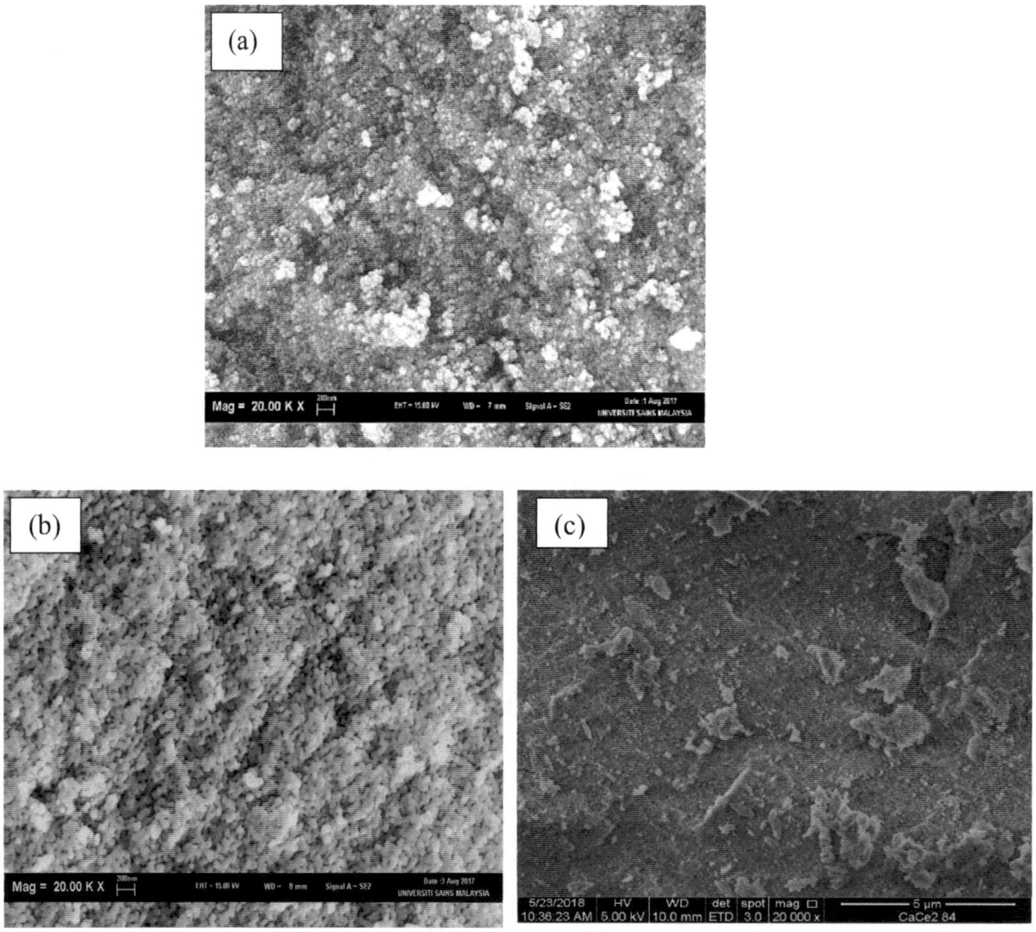

Figure 2. FESEM of images at a 20kX magnification of (a) ZrO_2, (b) Ca_1Ce_1/ZrO_2 and (c) Ca_1Ce_2/ZrO_2 catalysts

3.2 Morphology of mixed oxide catalyst by TEM analysis

The TEM image in figure. 3 (a-d) shows the differences morphology developed by addition of different amount of Ca and Ce on particles of the ZrO_2 support. The pure ZrO_2 support (figure 3 a) exhibited particles with diameters in the range of 15 to 30 nm. As the combined introduction of Ca and Ce, particles agglomeration was clearly observed (Fig. 3 (b, c)) onto the ZrO_2 surface suggesting that the metallic phases were fairly well dispersed, and this was consistent with a report by Wang et al. [20]. This agglomeration was probably due to the fact that the presence of Ca_xCe_y on the modified zirconia particles allowed their interaction and caused the particles to sinter, forming a new catalyst structure. As the Ca_xCe_y ratio was increased from 1.0:0.5 to 1.0:2.0, figure 3 (d), the formation of more organized and uniform spheres that were attached together was clearly observed with no visible difference in the porosity between the two types of catalyst. Thus, the presence in certain amount of Ca and Ce could play a significant role in transforming nonporous ZrO_2 support to porous catalyst materials and caused an increment in the particle size of the mesostructures support. The enlargement of the pores suggested that the dispersion of Ca_xCe_y onto the support, i.e. ZrO_2 most probably occurred on the external pores which resulted in a partial blockage of the internal pore structure.

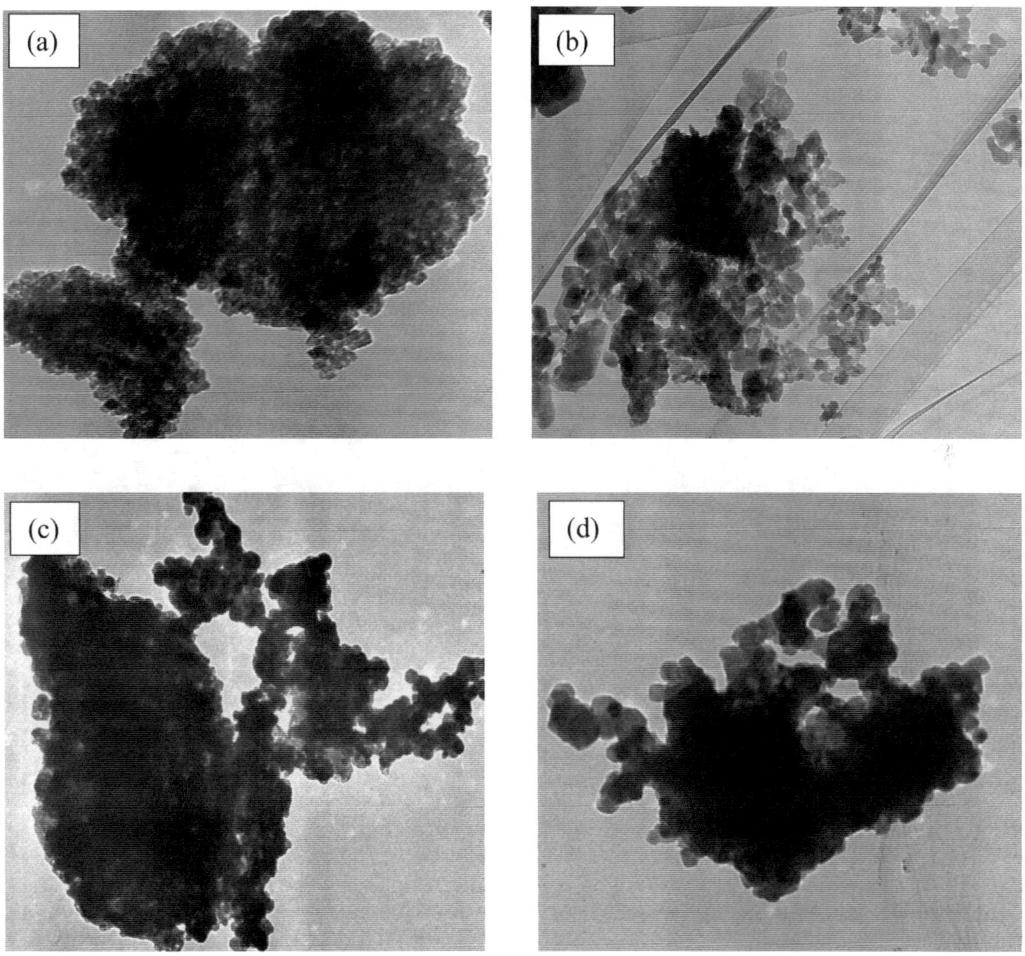

Figure 3. TEM images at a 25kX magnification of (a) pure ZrO_2 support (b) Ca_1Ce_1/ZrO_2 (c) Ca_2Ce_1/ZrO_2, and (d) Ca_1Ce_2/ZrO_2 catalysts.

3.3 Comparison fresh and spent catalyst by FESEM_EDX analysis

The morphology of $CaCe/ZrO_2$ catalyst fresh and spent in one cycle of oxidation reaction has been compared. The comparison also has been analyzed in term of glycerol conversion, lactic acid yield and selectivity percentage as well as tabulated in table 1. The corresponding EDX (table 1) analysis (figure 4) was shown in order to know quantitatively the differences in amount of active metal leached and the structure itself. It can be clearly observed in EDX analysis that the amount of Ca, Ce and O decreased throughout the cycle. The amount of Ca dropped from 5.29 to 1.42 wt % and the amount of Ce dropped from 12.86 to 4.64 wt%. The percentage of Zr less after spent in a one cycle about 7.45%, however, the oxide form is higher. The increment of oxide, O form maybe due to the high temperature of pressurized reaction, then some intermediate metal was transformed to the oxide state, thus oxide, O is increased about 13.6%. As for the ZrO_2, no significant reduction was observed for the spent catalysts compared to that of the fresh catalyst. These results indicate the interaction and cooperation of active metal with zirconia which was responsible for the stability of the catalyst.

Table 1. EDX results of the fresh and spent catalysts

	Weight percent (%)				Glycerol conversion (%)	Lactic acid (%)	
Catalyst	Ca	Ce	O	Zr		Yield	Selectivity
Fresh	5.29	12.86	27.35	54.50	94.1	42.8	45.5
Spent	1.42	4.64	31.07	50.44	96.1	37.3	38.9

Figure 4 (a) and (b) shows no significant differences on the structure before and after catalyst has been spent by observing the small external pore structure was not collapsed. This morphology was synchronized with the value of Zr from EDX analysis that there is no significant reduction on the weight percent of Zr detected. However, more agglomerate particle was not form for spent catalyst that showed that is the Ca and Ce agglomerate has been leached throughout the reaction.

Figure 4. SEM images of (a) fresh and (b) spent Ca_1Ce_2/ZrO_2 mixed oxide catalyst

3.4 Mechanism of lactic acid production

To explore the reaction pathways and mechanism for the glycerol conversion to lactic acid catalyzed by Ca_xCe_y/ZrO_2 in the presence of O_2, several control experiments were performed at different reaction time. The products were analyzed with GCMS and HPLC. It was demonstrated that glycerol oxidation is an environmentally clean method by using oxygen as oxidant and could be operated at mild reaction conditions (low temperature and pressure) [21]. Generally, the selective oxidation of glycerol tends to follow the base catalyse steps in which benzylic acid rearrangement to lactic acid is facilitated by four equivalents of base in the reaction [22,23]. From GCMS analysis results in table 2, all the intermediates were detected at small percentages indicating that the intermediate were converted to the desired final products i.e. lactic acid. According to Feng et. al..[24], 1,2-PDO could also be oxidized to lactic acid. Then, from HPLC analysis, lactic acid was detected as final product with the addition of low concentration of sulphuric acid.

Table 2. Products detected in glycerol oxidation reaction by catalysed by Ca_xCe_y/ZrO_2 catalyst

	Area (%)		
Elements/Reaction time	1 h	1.5 h	2.0 h
Glycerol	36.77	35.64	3.1
Glyceraldehyde	3.71	-	0.51
Pyruvaldehyde	1.16	5.29	-
Acetol	-	0.05	0.21
1,2 PDO	1.67	5.2	37.48

From both GCMS and HPLC results by using Ca_xCe_y/ZrO_2 as mixed oxide catalyst in oxidation reaction of glycerol to produce lactic acid, the reaction mechanisms seem to follow route 1 of oxidation mechanism as proposed by Ftuoni et al. [25] (figure 5).

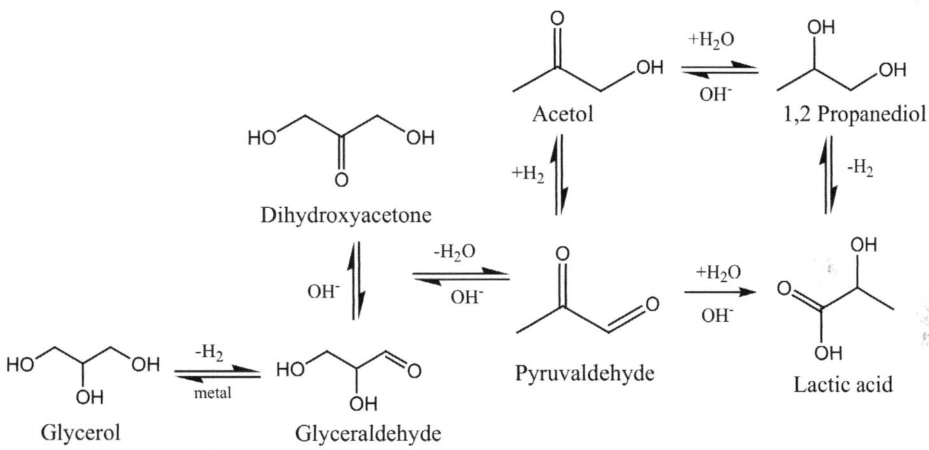

Figure 5. Selective oxidation mechanism of glycerol to lactic acid [25].

The glycerol oxidation mechanism over the $30wt\%Ca_1Ce_2/ZrO_2$ starts with kinetically relevant oxidative dehydrogenation of glycerol to glyceraldehyde and pyruvaldehyde intermediates. On the other hands, the presence of OH- from basic component i.e. CaO then assists in catalysing the dehydration and rearrangement of these intermediates. Lactic acid is then produced by further oxidation of intermediates over the metal component (Ce). The improvement of catalytic performances over mixed oxide $CaCe/ZrO_2$ is attributed to the enhancement of the formation of surface oxygen vacancies and facile Ce^{4+}/Ce^{3+} reduction as confirmed by XPS analysis. It has been identified as crucial points in contributing to the high catalytic activity. In order to assist the activation of the hydroxyl groups in glycerol, basic sites of the active metal plays important roles in increasing the rate of the reaction and enhances the selectivity to lactic acid in the selective oxidation reaction.

4. Conclusions

Solid mixed metal oxide catalysts, with Ca, as the base component in the form of $30wt\%Ca_1Ce_2/ZrO_2$ were prepared by precipitation route and then evaluated for the catalytic oxidation reaction of glycerol to lactic acid. The nature and the composition of active metal could significantly influence the catalyst structure and lactic acid production. The morphology formation of supported mixed oxide was proven by FESEM_EDX and TEM analysis. The pathway of mechanism has been evaluated by HPLC and

GCMS were successful. The higher ratio of Ce at a Ca:Ce composition of 1:2 and deposited onto ZrO_2 was found success. The activity of the catalysts was explained by the formation of the Ca-Ce-O mixed oxide phase by metal-support interactions showed by FESEM and TEM images. Even though the metal interaction has been leached in spent catalyst but still could produce lactic acid and maintain the structure itself and proven capability to be cycle. For next research, the ability to cycle the catalyst will be further study to further improve the cyclability in regeneration method.

Acknowledgements
This work was funded by the Ministry of Higher Education, Malaysia under the Transdisciplinary Research Grant Scheme (TRGS 6762001) and the Fundamental Research Grant Scheme (FRGS 6071366).

References
[1] Chinchilla L E, Olmos C M, Villa A, Carlsson A, Prati L and Chen X, 2015 *Catal. Today* **253** 178–89.
[2] Ayoub M, Abdullah AZ 2013 *Chem. Eng. J.* **225** 784–9
[3] Checa M, Marinas A, Marinas J M, Urbano F J 2015 *Appl. Catal. A. Gen.* **507** 4–43.
[4] Sullivan JA, Burnham S 2015 *Renew Energy* **78** 89–92
[5] Zakaria Z Y, Linnekoski J and Amin NAS 2012 *Chem. Eng. J* **207–208** 803–13.
[6] Crotti C and Farnetti E 2015 *J. Mol. Catal. A Chem.* **396** 353–9.
[7] Jin F and Enomoto H 2011 *Energy Environ. Sci.* 382–97.
[8] Naoki M, Natsumi M, Norihito H, Osamu S, Aritomo Y 2020 *Catal. Today* [In press].
[9] Zeng S, Zhang X, Fu X, Zhang L, Su H, Pan H 2013 *Appl. Catal. B Environ.* **136–137** 308–16.
[10] Kambolis A, Matralis H, Trovarelli A, Papadopoulou C 2010 *Appl. Catal. A Gen.* **377(1–2)** 16–26.
[11] MacIel C G, Silva T D F, Hirooka M I, Belgacem M N, Assaf J M 2012 *Fuel.* **97** 245–52.
[12] Marinkovic D M, Stankovic M V, Velickovic A V, Avramovic J M, Miladinovic M R, Stamenkovic O O 2016 *Renew Sustain Energy* **Rev 56** 1387–408.
[13] Kim M, DiMaggio C, Yan S, Salley SO, Ng K Y S 2011 *Green Chem.* **213(2)** 334.
[14] Atsushi T, Hiroshi G, Ryuji K S, Ted O 2019 *Appl. Catal. A Gen.* **570** 200-208.
[15] Kim Y-H, Hwang S-K, Kim JW, Lee Y-S. 2014 *Ind. Eng. Chem. Res.* **53(31)** 12548–52.
[16] Zhenchen T, Huatang C, Yehan T, Hero J H, Paolo P, 2020 *Appl. Catal. B. Environ.* **263** 118273.
[17] Yang G-Y, Ke Y-H, Ren H-F, Liu C-L, Yang R-Z, Dong W-S 2016 *Chem. Eng. J.* **283** 759–67.
[18] Majedi A, Davar F, Abbasi A 2018 *Environ. Chem. Eng.* **6** 701–9.
[19] Ozawa M, Takahashi-Morita M, Kobayashi K, Haneda M 2017 *Catal. Today.* **281** 482–9.
[20] Wang Q, Li Z, Zhao B, Li G, Zhou R 2011 *J. Mol. Catal. A. Chem.* **344(1–2)** 132–7.
[21] Crotti C, Farnetti E 2015 *J. Mol. Catal. A. Chem.* **396** 353–9.
[22] Lakshmanan P, Upare PP, Le N-T, Hwang YK, Hwang DW, Lee U-H, 2013 *Appl. Catal. A. Gen.* **468** 260–8.
[23] Purushothaman R K P, van Haveren J, van Es D S, Melian-Cabrera I, Meeldijk J D and Heeres H J 2014 *Appl. Catal. B. Environ.* **147** 92–100.
[24] Feng S, Xue W, Yin H, Meng M, WAng A and Liu S 2015 *Royal Soc. of Chem.* **5** 106918.
[25] Ftouni J, Villandier N, Auneau F, Besson M, Djakovitch L, Pinel C 2015 *Catal. Today* **257** 1–7.

The Design of IoT Dengue Alert Supervisions System with RFID Access

W M K W Mamat[1], M Kassim[1]*, S Shahbudin[1], N Ya'acob[1]

[1]Faculty of Electrical Engineering, Universiti Teknologi MARA, 40450 UiTM Shah Alam, Selangor, MALAYSIA

*Corresponding author: murizah@uitm.edu.my

Abstract. This research presents a development of a patient's Dengue Alert Supervisions (DAS) System. The system is designed adaptively to alert patients and can be monitored over the Microsoft Windows Interfacing Applications system. The purpose is to alert patients on dengue attacks if they are possibly infected with dengue. Problem identifies that patients are difficult and hard to go for regular medical check-ups especially to the hospital. They face issues like long queues and long waiting times at public hospitals and it takes time to diagnose the dengue attacks. The objective of this system is to design a handy system for dengue detectors and to accurately warn the patients so that, they are being treated properly. The elements designed in this system comprised of a temperature sensor and blood dengue kit which is combined with push buttons as a tool to determine the condition of blood cells. This system used Radio-Frequency identification (RFID) access card to detect personal details. The database is directed to Windows application s on the DAS system. Results present the value of the blood level of infected dengue and body temperature in an online graph. The data can automatically send to hospitals online and access by medical officers on current situations. The system is beneficial to society as it is an essay system that can provide the user with simple and timely feedback. It would also be a significant invention to medical officers in diagnosing patients with much easier and quicker collection data using the Internet of Things (IoT).

1. Introduction

Dengue fever virus is a common disease that happened all around the world these days. Estimates of the global incidence of dengue infections per year have been increased drastically in both tropical and subtropical countries [1]. This disease is transmitted when the mosquitoes that are called Aedes mosquitoes, which live in dirty places like unclean water or vases that carry the virus and attack humans[2]. Dengue is among the most important mosquito-borne diseases, with more than half of the world's population at risk of infection in dengue endemic countries. Environmental management, which includes any activities that involve environmental modification, environmental manipulation and changes to human behaviour have been used to mitigate the risk of dengue transmission[3] . The disease began to multiply in the number of cases from year to year and country to country. From 2009 to 2011, the disease began to infect humans in Brazil as the incidence were 205.5, 530.3, and 400.5 per 100,00 persons[4]. In Malaysia, the fatality rate has increased month per month due to the disease. Adulticiding using permethrin, deltamethrin, and malathion and larviciding with temephos and *Bacillus thuringiensis isrealensis(Bti)* are the main methods used to control the dengue vector in Malaysia. The scientist asserts that two types of species occur in Malaysia which are *aegypti* that found in areas that are developing

IOP Conf. Series: Materials Science and Engineering 1176 (2021) 012012 doi:10.1088/1757-899X/1176/1/012012

rapidly with less vegetation and *albopictus* that prefers conditions with more vegetation and more exophilic[5].

The dengue disease was firstly reported in 1902 and the first outbreak happened in 1962 at Georgetown, Penang, and followed by Kuala Lumpur in 1973 in Malaysia. Subsequently, the outbreak of the disease happened in both the urban area and rural area [6, 7]. Hence, the infected area where there is a breeding ground for mosquitoes within 100 meters to 200 meters or quarantine an area experiencing an outbreak of dengue has been studied. The transmission of the disease on specific also depends on the mosquito's growing period[8]. For every person infected with dengue, there are different levels for different ages. According to the past research, there were 128 subjects which involved extensive experiment on dengue disease including 80 males and 48 females from various sages who clinically suspected to have this disease and were referred for ultrasound scanning for both abdomen and thorax [9, 10]. Humans which have been infected with dengue disease will have common clinical manifestations like fever, severe headache, retro-ocular pain, pain in the muscles and joints, and purpuric spots on the body [11].

There are two general types of dengue fever, which can be considered as the most severe which are Dengue Hemorrhagic (DHF) and another one is Dengue Shock Syndrome (DSS) [12, 13]. According to research mosquitoes were strictly associated with standing water, with the belief that mosquitoes must have water to lay their eggs. They figured out by analyzing the vulnerability of water which is associated with health hazards. Dengue virus would be active in the patient's body for about 3-14 days or on average the symptoms are usually detected on the 4th to 6th day. The patients infected by dengue would experience sudden fever that would later be followed by a variety of signs and symptoms that are not specific. This is because the blood of the patients is not stable and low volume of platelet[14]. The sign that indicated the dengue fever are sudden high fever and persistent, red spots on the skin, severe pain in bones, joints, muscles, eyes, and head, bloody body, nose and mouth, loss f appetite, vomiting, stomachache, bitter taste on the tongue and weak and also feeling of listless. There are several ways for the infectious dengue fever virus to be transmitted, such as dengue fever germs carried by Aedes mosquitoes, the healthy person is bitten by mosquitoes, dengue fever patients with dengue germs in his body and Aedes mosquitoes bite dengue patients.

The advancement in modern technology devices has driven the users to incorporate the is electronic devices to perform their daily tasks and improve the basic operation of the device [15]. Therefore, this system is developed and implemented in Windows 10 Application, enabled the interactive control panel for monitoring the dengue system. Nowadays, the IoT has grown to become the most preferred system that utilizes the local area networking and global area networking to user details [16]. IoT can access out the range that localhost used. The application is suitable to use by using with Graphical User Interface (GUI) on the Windows 10 device. The system also provides an interface that allows users to explore every detail of this software [17]. The webpage localhost is developed by using the XAMPP software. XAMPP is an Apache distribution package that is small, portable, and contains the common web development technologies in its set. This is a very effective tool to develop and test applications in PHP and MySQL database [18]. The Windows 10 applications are developed by using the Microsoft Visual Studio Community and Arduino.

2. Methodology

This research comprises of three main phases which are the design phase, development phase, and testing phase of the hardware and software system for the Dengue Alert and Supervision System.

2.1 Flowchart

Figure 1 shows the project flowchart of the Dengue Alert Supervisions (DAS) System. Both hardware and software evaluations were made to analyze and characterize important parameters to be used in the system's design. The hardware used in this project were microcontroller Arduino Uno, USB 2.0 Cable Type A/B, Radio Frequency Identification (RFID) RCC 552 with RFID Card, Temperature Sensor LM 35, 3 components 6x6x8mm push button that linked with Dengue Kits. The software used in this project was Microsoft Visual Studio, Operation System Windows 10, and XAMPP Server by Online Database. Both hardware and software need configure to access the database and serial port. It will automatically be connected after the configuration database and serial port. The user connects the card to the RFID sensor and the value of integers as user ID display in Microsoft Visual Studio Applications. The user ID which has already been stored in the database can access the system else patients failed to access the system. New users need a new user ID which the patient needs to fill in their details into a new form register account. After the system allows access, the patient also can check their blood by using Dengue Kits. The result on Dengue Kits links with level blood needs to click push buttons. Each push buttons have a blood condition level. Temperature sensors detect human body temperature. Users need to click the push button while holding the temperature sensor. The value of both blood condition level and human body temperature load Microsoft Visual Studio Applications. Finally, both values were recorded and saved into the database.

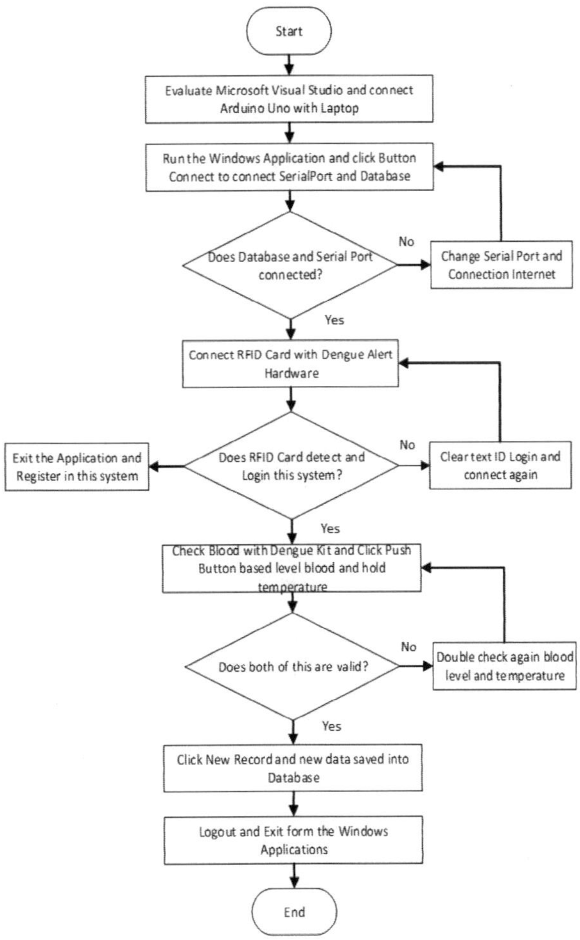

Figure 1. Flowchart of the DAS System

2.2 Block Diagram

This project used some chemical kits that are needed to complete this system. Dengue Kits is one of the required tests in the system to check the user's blood sample. It was combined with push buttons to indicate the level that the user can refer to have used. There are three levels, indicating that the Dengue kits provide three-level blood condition. Since dengue tests can be easily be checked using the human

IOP Conf. Series: Materials Science and Engineering 1176 (2021) 012012 doi:10.1088/1757-899X/1176/1/012012

body's temperature, there is a sensor that is specifically used to check the body's temperature. The sensor is LM35 which is expertly used to detect body temperature. Another more advanced system, RFID RC552 sensor used as well, and the user has to log in and provide their details which are saved in a secured database. All data is controlled by one microcontroller which controlled the system. For this project, Arduino Uno is used since it is more suitable and easier to configure in terms of the coding for the three mains sensors input. The codes in Arduino is read by Microsoft Visual Studio Community 2015 using Visual Basic Interface. The condition and output of the program will be displayed on the Visual Basic Interface.

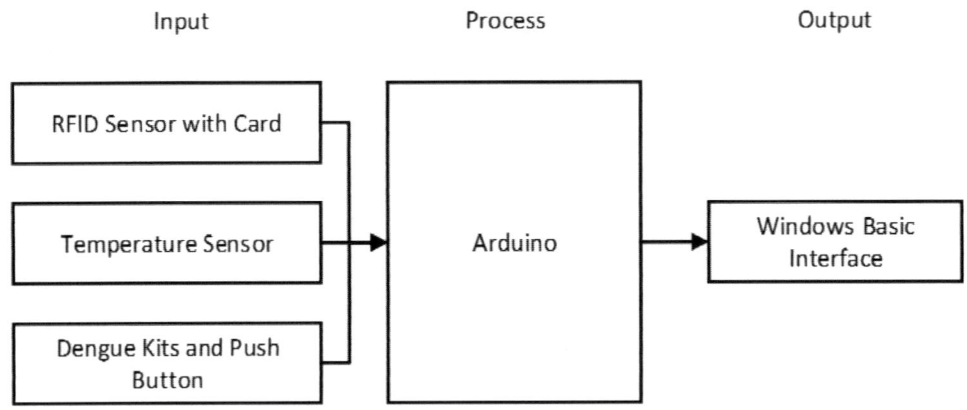

Figure 2. Block Diagram of the System

Figure 2 shows the functional block diagram of the system. Three of the components are the input of this system which is RFID Sensor with Card, Temperature Sensor, and Dengue Kits and Push Button. Data from the three components are captured and are processed by the Arduino Uno. Finally, the result is displayed on the Windows application. Most of the common Windows application used is Visual Basic Interface. Visual Basic is a program similar to another language that more to Windows Interface that is suitable for computers and linked with Arduino through serial Port.

3. Results and Analysis

Results show the successful functions of the developed system on windows application Interface for DAS System. Since the project is a system application, the application is called DAS System (DAS). Figure 3 shows the combined hardware and software DAS System with Dengue Kits. Figure 3 shows the result of the hardware and software that was being developed and produced. The hardware is connected to the device by Serial Port Cable and linked with Visual Studio. Before initiating the system, both hardware and software need to be connected. The devices require internet access Wi-Fi or Lan Cable. The Visual Studio Application was run with the connected hardware. Figure 4 shows the displayed Windows Application.

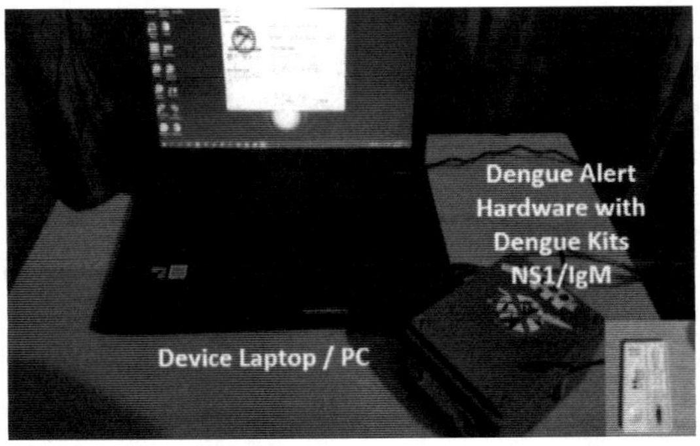

Figure 3. DAS System with Dengue Kit.

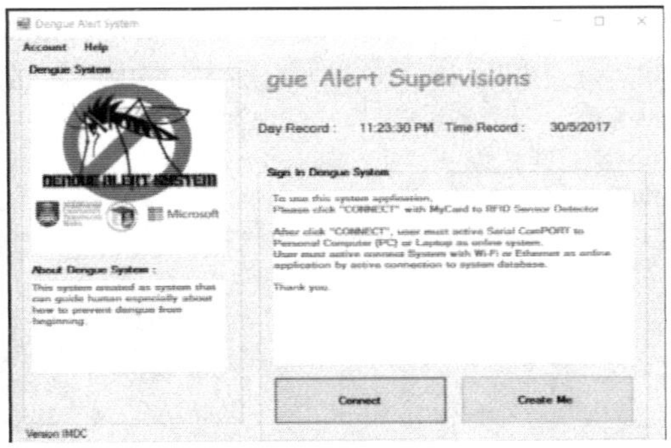

Figure 4. DAS System

The Windows applications is displayed when the Visual Studio is run. The patients that have already registered in this system would need to click the connect button, and If they have not, click the Create Me button. After connected then the next form is shown. Figure 5 shows how the patients were briefed about the configuration on this system and the connection between Arduino Uno and Visual Studio using Serial Port and Visual Studio with Database Online using MySQL server. There are two configurations before initiating this application. First, the selection process for serial port to find a suitable device, either PC or Laptop. It is automatically shown that one serial port can be used by the PC. Only one port depends on the PC or Laptop. Click until the Timer display ON. Next step, click "Connect Database". A network of this device is connected. Figure 6 shows the patient database.

There are two conditions after the device is being connected to this application and database. The success of database online already setup has shown a MessageBox that displayed "Connection Successful" else "Connection Lost" or "Reading the stream fail" is displayed. The RFID in the hardware has completed the setup process and it is ready to be used after the two configurations are completed. RFID card needs to be scanned by RFID RC552 on this hardware to login into this system. It is automatically displayed ID numbers depending on the registered RFID card. Login through this system and all the details that have been saved in the database can be displayed and filled into all textboxes and data table. After the data about patients displayed, the patients can do a checkup on the blood level and body temperature. Before starting, the patient needs to check the blood manually using the provided Dengue Kit. In the Dengue Kits, 3 levels need to be updated into this system. Level 1 shows one indicator and the patient needs to click push button number one. This means that the patient does not have dengue. Level 2 shows two indicators and the patient needs to click push button number two which means the patient has dengue for the first time and needs to be checked by a doctor. Level 3 shows three indicators and the patient needs to click push button number three. This level is critical and the patient is required to be quickly treated by a doctor.

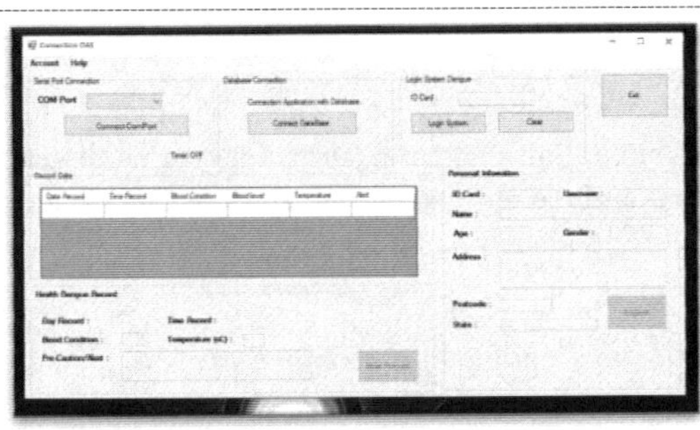

Figure 5. Configuration Serial Port.

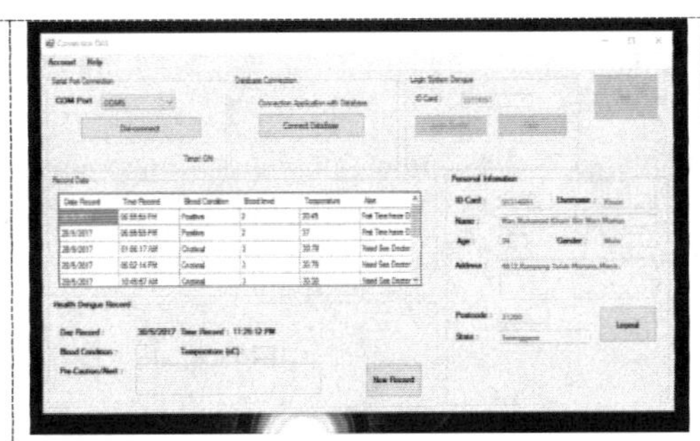

Figure 6. Data about patients using this system.

IOP Conf. Series: Materials Science and Engineering 1176 (2021) 012012 doi:10.1088/1757-899X/1176/1/012012

Patients have to click and hold the temperature push-button temperature to detect the value of body temperature. Both value temperature values and push-button value will be sent to the textbox as shown in figure 7. Patient click "New Record" and data table updated as similar with database updated. The patient can log out and exit for this application. All the data is recorded and saved into the online database.

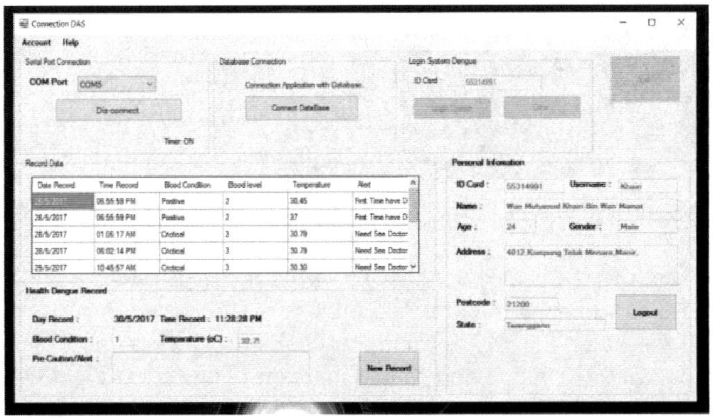

Figure 7. Value of Temperature Sensor detects and Blood Level.

Figure 8 shows an online database that has collected all data about the patients' body temperature by LM35 as human body temperature and saved into the database. Blood level and condition were checked by Dengue Kits and click the button on hardware will be displayed in Windows application. The database is updated securely in the online database. All data recorded into one form of Windows application s that displayed the result from the data collected.

←T→	No_Record	Date Record	Time Record	Blood Condition	Blood Level	Temperature	Alert	id_card
Edit Copy Delete	1	26/5/2017	06:55:59 PM	Positive	2	30.45	First Time have Dengue. Need See Doctor	55314991
Edit Copy Delete	2	28/5/2017	06:55:59 PM	Positive	2	37	First Time have Dengue. Need See Doctor	55314991
Edit Copy Delete	3	28/5/2017	01:06:17 AM	Cridical	3	30.79	Need See Doctor Quickly!!	55314991
Edit Copy Delete	4	28/5/2017	06:02:14 PM	Cridical	3	30.79	Need See Doctor quickly!!	55314991
Edit Copy Delete	5	29/5/2017	10:45:57 AM	Cridical	3	30.30	Need See Doctor quickly!!	55314991
Edit Copy Delete	6	30/5/2017	02:07:17 AM	Cridical	3	27.86	Need See Doctor quickly!!	55314991
Edit Copy Delete	7	30/5/2017	03:14:31 AM	Negative	1	31.77	Normal	55314991
Edit Copy Delete	8	30/5/2017	12:38:33 PM	Positive	2	31.28	First Time have Dengue. Need See Doctor	55314991
Edit Copy Delete	9	30/5/2017	12:57:28 PM	Negative	1	27.86	Normal	55314991
Edit Copy Delete	10	30/5/2017	11:28:47 PM	Negative	1	32.75	Normal	55314991
Edit Copy Delete	11	31/5/2017	09:18:49 AM	Negative	1	21.99	Normal	55314991
Edit Copy Delete	12	31/5/2017	09:32:33 AM	Negative	1	21.99	Normal	55314991

Figure 8. Online Database

Figure 9 shows the collected data from selected patients. In this form, there are two chart graphs, each display about body temperature, blood condition, and the date when the data was recorded. This data shows the balanced and manipulated data of blood and temperature.

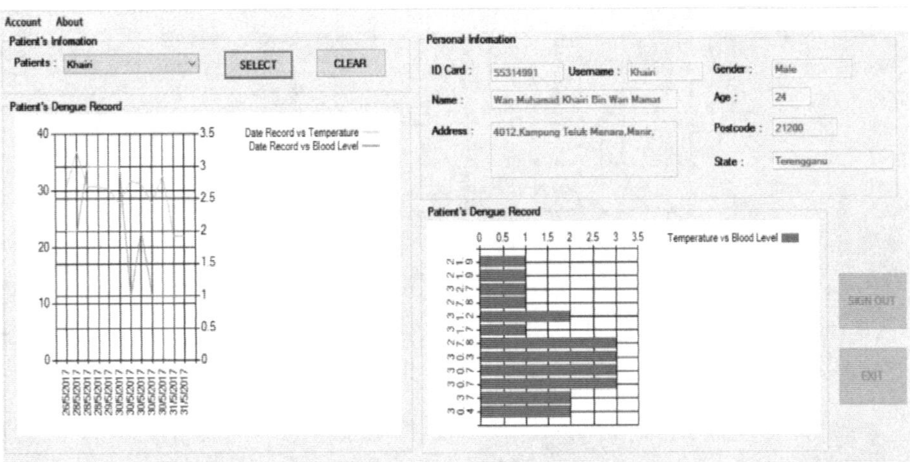

Figure 9. Data that was recorded.

Figure 10 shows the data for Blood Level and Body Temperature. The system checked the condition of the patient. Both blood level and temperature are displayed in a condition that is suitable to patients if the blood level and temperature are varying in condition. Figure 11 illustrated the actual condition of the patient. The patient is considered as infected with Dengue disease but no fever if the blood level is higher and the temperature is normal. If the blood level is normal and the temperature is higher, it indicates that the patients only have a high fever. The data is automatically tracked by a recorded database and it is submitted to the doctor if the condition is critical and high fever. The data can be sent to Government hospitals for them to act in treating the patient or quarantining the location of the patient's living location. The system used Arduino Uno as the main component for the system hardware and was used as the controller of the whole system. Arduino was connected to the Microsoft Visual Studio Community by the installation of some add-ons for configuration purposes to both hardware and software. Since this system consists of hardware and software applications, it is suitable to be developed by using the Windows Basic Interface. Microsoft Visual Studio has a wider range of programming language which can easily be used. As a Windows application, Visual Basic is easy to work with, by simply selecting some form or event that needs to be displayed as the user interface. The codes that used to connect to the database has use localhost and global network, as it is easier since Visual Basic as many language styles to configure it from VB.net until Visual Basic 2015. Microsoft Windows 10 is a suitable running application of Microsoft Platform Interfaces without crashing. Connecting Visual Basic to the Database is also convenient and a high uplink to both software and hardware. Since this project requires software, the hardware part also had to be completed for this system. Figure 12 shows the connection of all input to the one microcontroller Arduino Uno. The inputs that are required are RFID Sensor RC552, Temperature Sensor LM35, and Push-button. The combination of inputs has been fed to the controller that saved the data acquired into the database. All processes from input to output were controlled by one microcontroller. All data will be connected and linked up with the database and saved on each table in the database.

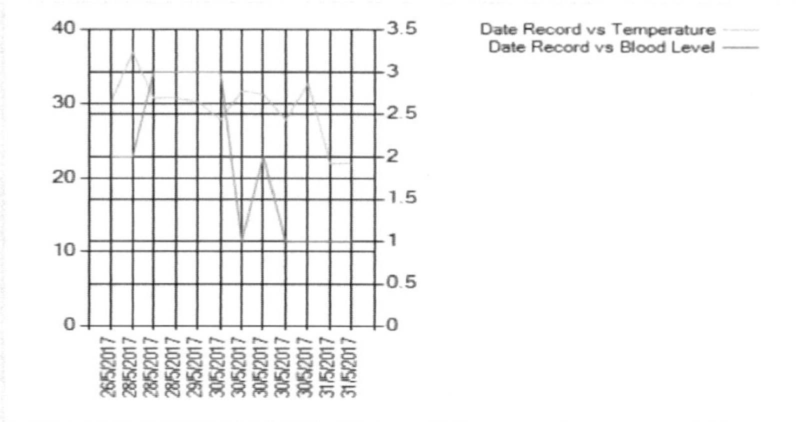

Figure 10. Graphs about Date Record, Blood Level, and Body Temperature.

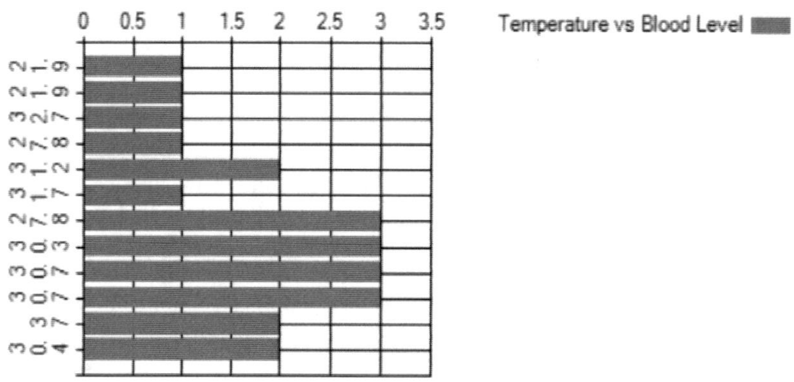

Figure 11. Data Temperature and Blood Level

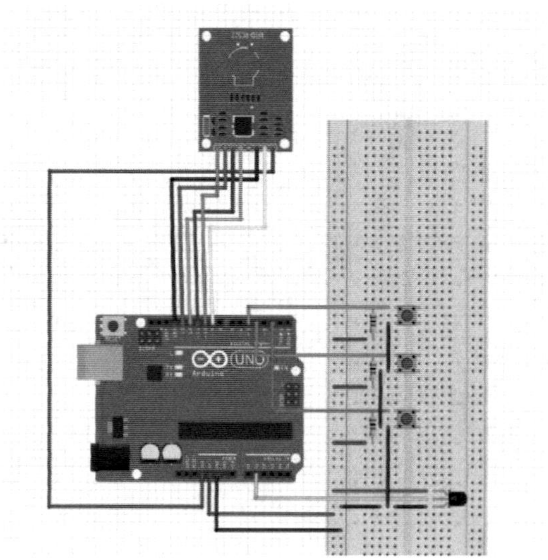

Figure 12. Connection all input sensors with Arduino Uno.

4. Conclusion

In a conclusion, this DAS System on Windows Platform using Arduino Uno that had been developed is a great alternative to the traditional time-consuming approach easy to use by anyone. The combination of chemical kits and electronic components has been designed in the system that able to explore very deep and more suitable ways of recognizing the condition of the blood. This system had utilized the Microsoft Platform like Windows 10 and Visual Basic Interfacing wand was designed to run on laptops or desktops. As the connection with the IoT, all the measured data has been recorded in the database provided, which is easy to store and upload. Furthermore, Windows 10 application like Visual Basic made it easier for the users to connect and control this application with high security without bugs and system crashes either in its hardware or software.

5. Acknowledgement

The author would like to thanks the Faculty of Electrical Engineering, UiTM Shah Alam for the support grants in publishing this paper.

References

[1] Arham A F et al 2021 Participatory: Stakeholder's Engagement Toward Dengue Control Techniques in Klang Valley, Malaysia *SAGE Open* **11(**1).

[2] Yahya W N.F W and Dom N C 2020 Profile of aedes breeding in Urban high-rise buildings: A of dengue outbreak in Malacca, Malaysia Planning Malaysia **18**(3) 251-260.

[3] Mahmud MAF, Abdul Mutalip MH, Lodz NA, *et al* 2019 Environmental management for dengue control: A systematic review protocol *BMJ Open* **9**(5).

[4] Ferraz F O, Bomfim M R, Totola A H, Ávila T V, Cisalpino D, Pessanha J E, da Glória de Souza D, Teixeira Júnior A L, Nogueira M L, Bruna-Romero O, Teixeira M M 2013 Evaluation of laboratory tests for dengue diagnosis in clinical specimens from consecutive patients with suspected dengue in Belo Horizonte, Brazil *J. Clin. Virol.* **58**(1) 41-46.

[5] Tan W, Liew J W K, Selvarajoo S, Lim X Y, Foo C J, Refai W F, Robson N, Othman S, Hadi H A, Mydin F H M, Malik T F A, Lau Y L, Vythilingam I 2020 Inapparent dengue in a community living among dengue-positive Aedes mosquitoes and in a hospital in Klang Valley, Malaysia. *Acta Tropica* **2014.**

[6] Adnan R A, Ramli M F, Othman H F, Asha'ri Z H, Ismail S N S, Samsudin S 2021 The Impact of Sociological and Environmental Factors for Dengue Infection in Kuala Lumpur, Malaysia. *Acta Tropica* **216.**

[7] Ahbirami R and Zuharah W.F 2020 School-based health education for dengue control in kelantan, malaysia: Impact on knowledge, attitude and practice *PLoS Neglected Tropical Diseases* **216.**

[8] Ali S, et al 2017 Environmental and Social Change Drive the Explosive Emergence of Zika Virus in the Americas **11(2)**.

[9] Chai X T et al 2020 Ultrasound findings of plasma leakage as imaging adjunct in clinical management of dengue fever without warning signs *Med. J. Malaysia* **75**(6) 635-641.

[10] Gabaglia C R Zika virus and diagnostics 2017 *Curr. Opin. Pediatr.* **29**(1) 107-113.

[11] Rothan H A et al 2014 Identification of natural antimicrobial agents to treat dengue infection: in vitro analysis of latarcin peptide activity against dengue virus *BMC Microbiol.* **14**(1) 140.

[12] Islam M R et al 2021 Knowledge attitude and practice of dengue syndrome management on the basis of national guideline by the physicians of dhaka city *J. Med.* **22**(1)18-21.

[13] Hashan M R et al 2021 Association of dengue disease severity and blood group: A systematic review and meta-analysis *Rev. Med. Virol.* **31**(1) 1-9.

[14] Saif K and Rahman S Z 2021 Oral bleeding in dengue fever - our perspective. *Bangladesh J. Med. Sci.* **20**(1) 11-16.

[15] Kassim M et al 2019 Dengue attack analysis system on mobile application *Int. Conf. Syst. Eng. and Tech. Proc.* **20**(1) 11-16.
[16] Razali M A A et al 2020 A ThingSpeak IoT on Real Time Room Condition Monitoring System *2020 Int. Conf. on Autom. Cont. and Intell.(Malaysia).*
[17] Thaisongkroh P and Rattanasiriwongwut M 2018 Development of a software management for capital investment decision using microsoft visual basic programming *Int. Conf. on Eng. Appl. Sci. and Tech. Exploring Innovative Solutions for Smart Society. (Phuket, Thailand).*
[18] Agustin F et al 2019 Comparative Analysis of Application Quality between Appserv and Xampp Webserver Using AHP Based on ISO/IEC 25010:2011. *Int. Conf. on Cyb. and IT Serv. Manage.*

Three-Phase Induction Motor Controller Design Based on Hysteresis Voltage Control

M S N Ismail @ Marzuki[1]*, A I M Arifin[1], W A K W Chek[1], A H Hussian[2], F R M Tazudin[3], and S A Zulkifli[4]

[1]Faculty of Electrical Engineering, Universiti Teknologi MARA, 23000 Dungun, Terengganu, Malaysia

[2]Faculty of Engineering Technology, University College TATI Campus Teluk, Kalung, 24000 Kemaman, Terengganu, Malaysia

[3]Jabatan Kerja Raya Ketengah, 23400 Dungun, Terengganu, Malaysia

[4]Faculty of Electrical and Electronic Engineering, University Tun Huseein Onn Malaysia, 86400 Parit Raja, Johor, Malaysia

*Corresponding author: saifulnajib@uitm.edu.my

Abstract. This paper describes a controller for Three-Phase Induction Motor. Induction motor is an electromechanical actuator widely used due to its reliability and relatively low maintenance cost. However, the control problem of an induction motor is complex due to nonlinearities, load torque perturbation, and parameter uncertainties. An element included in this study is voltage source control, which is to control the voltage fed from three-phase inverter to Three-Phase Induction Motor. Hysteresis Controller was proposed in this paper to minimize voltage error. Hysteresis controller is seen as an input–output phase lag corrector. The implementation of the designed hysteresis controller is performed in simulation using MATLAB/Simulink. The result proved that the proposed controller design can minimize the line voltage error of the induction motor, preventing the line voltage from being kept low hence avoiding excess current at low speeds.

1. Introduction

Induction motors, either single phase or three-phase are widely used in industry due to their reliability and low cost. However, three-phase induction is the most interesting and has been attracted the attention of many researchers due to their non-linearity [1]. Many controllers have been developed and can be classified into passive and adaptive power controls. The example for passive power control is hysteresis, relay, and sliding mode control. Examples for adaptive power control are PID, fuzzy, and P-resonant controller [2]–[5].

Basically, motor is driven by an inverter. Inverter is a device which supplies various frequencies of power supply to equipment [6]. Based on this function, motor revolution speed can be controlled causing reduction in energy consumption. One of the ways of how it functions is voltage source inverter (VSI) where it takes in a fixed voltage from a device, such as a dc power supply, and converts it to a variable frequencies AC supply [7].

Other papers have discussed the strategy of controller for induction machine. Yassine Kali et al. [8] have found a good performance in terms of current tracking by proposing a robust nonlinear current controller of a six-phase induction motor drive. The technique proposed is a combination of Discrete-time Sliding Mode Control and the Time Delay Estimation method for the inner current control loop of an Indirect Rotor Field-Oriented Control of a six-phase induction motor drive. The proposed technique provides high accuracy in estimation of uncertainties, perturbations, and unmeasurable rotor current.

The low saturation effect was resulted by using Fuzzy-Anti Windup PID controller for induction motor controller as discussed by Premkumar et al. [9]. The technique considered a fuzzy based proportional (kp), fuzzy base integral (ki), and fuzzy based derivative (kd). In simulation results, the authors found that the proposed technique yielded better performance in providing settling time, overshoot, undershoot, peak, and peak time compared to PI controller and Anti windup PID controller. However, this technique drawback on lower speed and is time consuming.

Hichem Hamla et al. [10] presented a modified direct torque control for induction motor drives based on variable hysteresis band current control with a considerable improvement of its characteristics. This technique was proposed to control the stator currents and the torque and magnitude of stator flux. In comparison to the conventional direct torque control, this technique has proven to be superior as it included one more parameter on the controller, provided better quality of stator current waveforms, and reduced torque ripples.

In this paper, hysteresis voltage control is proposed due to its ability to perform variable frequency switching, very fast response, and simple operation compared to other control methods. The main principle is to switch the motor "on" and "off" very rapidly. The AC output of the inverter can be controlled in both amplitude and frequency which are the requirements of induction motor at any desired point of operation [11]. The switching operation in inverter is to connect the power switches when line voltage is low and disconnect when it is high. In addition, Arduino will be used as a digital signal processing (DSP) [12]. All simulation works was done in MATLAB Simulink.

2. Research Methodology

2.1. The Proposed Controller (Hysteresis Controller)

Hysteresis controller derives the switching signals of the inverter power switches in a manner that reduces the voltage error. The switches are controlled so that they follow the references. The voltage ramping up and down between two limits is illustrated in figure 1. When the voltage through the induction motor exceeds the upper hysteresis limit, a negative voltage is applied by the inverter to the induction motor. This causes the voltage in the induction motor to decrease. Once the voltage reaches the lower hysteresis limit, a positive voltage is applied by the inverter to the induction motor and this causes the voltage to increase and the cycle repeats.

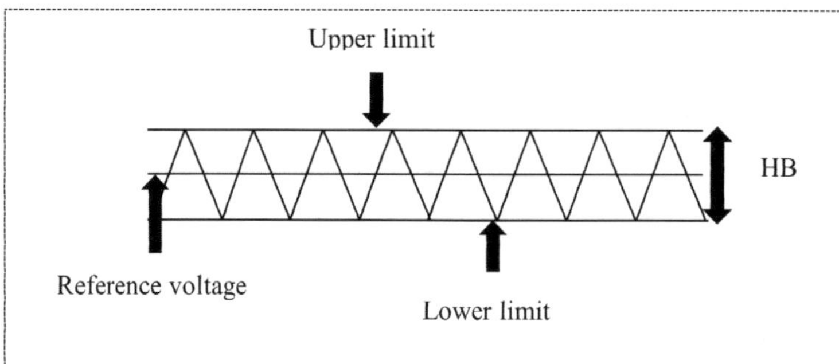

Figure 1. Hysteresis controller operation.

The voltage controllers of the three phases are designed to operate independently. Each voltage controller determines the switching signals to the inverter. The switching logic for phase A is formulated as below:

IOP Conf. Series: Materials Science and Engineering 1176 (2021) 012013 doi:10.1088/1757-899X/1176/1/012013

If Va < (V_a ref -HB) upper switch is OFF and lower switch is ON
If Va < (V_a ref +HB) upper switch is ON and lower switch is OFF

In the same action, the switching of phase B and C are derived. However, they happen in different phase shifts which are phase B in 120° shift and phase C in 240° shift.

2.2. The Proposed Block Diagram of Controller

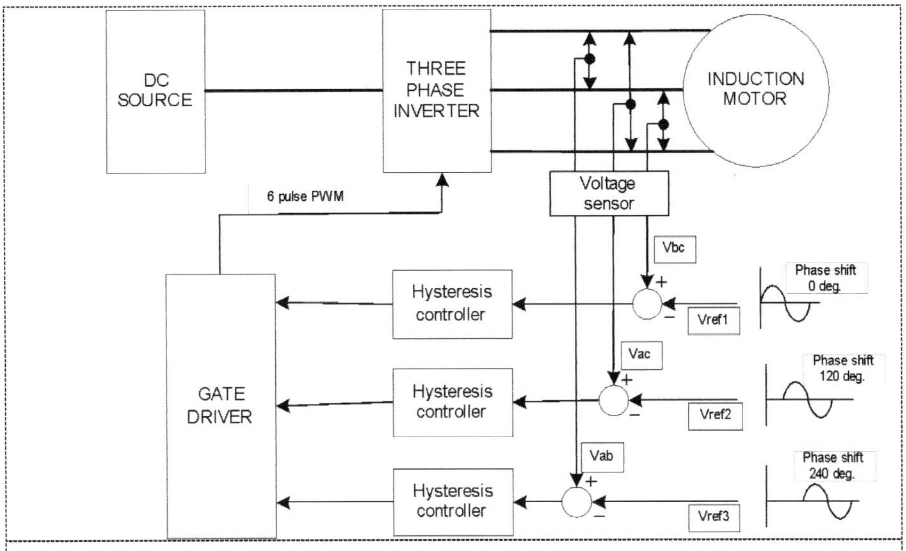

Figure 2. The proposed block diagram of controller.

Figure 2 describes the overall simulation block diagram. The main setup is the input as in DC source, three-phase inverter, and the Three-Phase Induction Motor. Arduino is employed as the voltage sensor which then sends the data to MATLAB Simulink for simulation. The controller setup is at the feedback of the system, starting from voltage reading from connection between three-phase inverter and Three-Phase Induction Motor. The reading will then be compared with reference voltage. The error from that comparison will be sent to Hysteresis Controller. The Hysteresis Controller will control the PWM (Pulse Width Modulation), increasing and decreasing the voltage based on references to the three-phase inverter.

2.3. Hysteresis Controller Design

Hysteresis controller is the main part in the controller design. The output of the hysteresis controller is in PWM signal. These PWM signals will be used by the gate control for each of the power switches in three-phase inverter. In the project, there are three hysteresis controllers that create six PWM. Three of them are the PWM of each phase, and the other three are the inverted signals of the first three PWMs. Figure 3 (a) and (b) show hysteresis controller block and how to get PWM and its inverse.

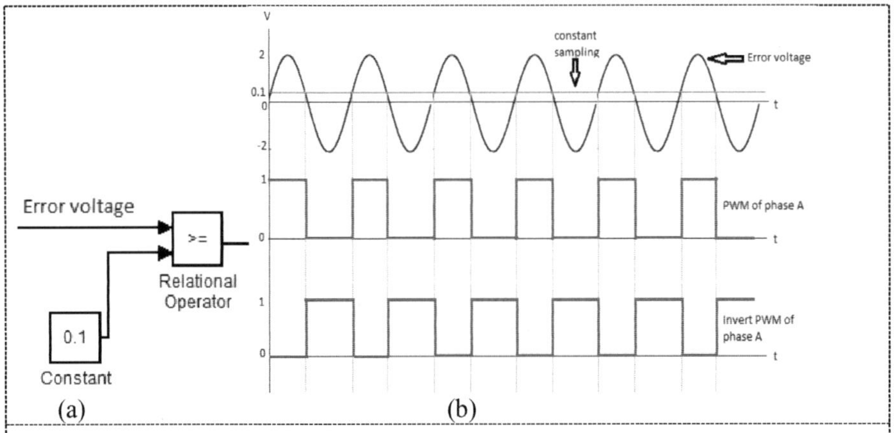

Figure 3. (a): Hysteresis controller block, (b): The PWM produced in phase A

The error voltage which is the input to the hysteresis controller is 2VAC. It was cut by constant 0.1 as a sampling signal. From this action, the PWM and its inverse of "Phase A" were created.

2.4. Hysteresis Controller Model

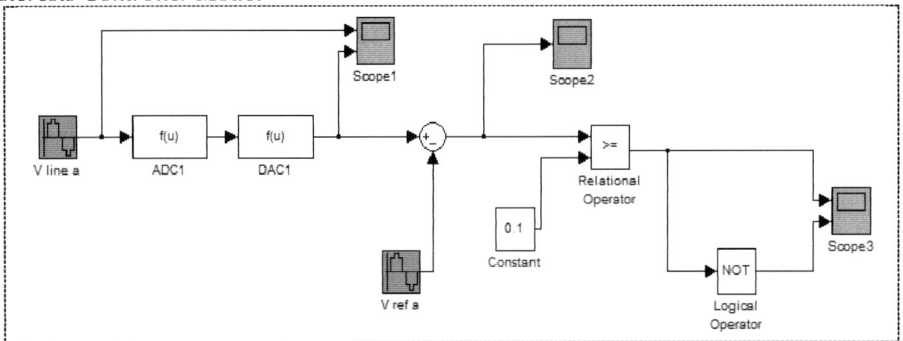

Figure 4. Hysteresis controller model for phase A.

Figure 4 above shows the controller for phase A in MATLAB Simulink environment. The same arrangement is found for phase B and C with difference phase shift. V line is the input voltage or feedback voltage from the line voltage inverter to induction motor. ADC is the analog to digital converter and DAC is the digital to analog converter. The MATLAB Simulink model created 6 pulse PWM signals, where each phase has 2 pulses. The first two signals are at 0° (phase A), then the next two signals are at 120° (phase B) and at 240° (phase C). The voltage line (V line) is set to 18Vac and the voltage reference (V ref) is set to 20Vac. This means that the error carried is 2Vac.

3. Results and Discussions

Discrete second order filter was used, with the cut off frequency at 100Hz and the damping factor zeta is 0.707 of lowpass filter. The result of the simulation is illustrated in figure 5.

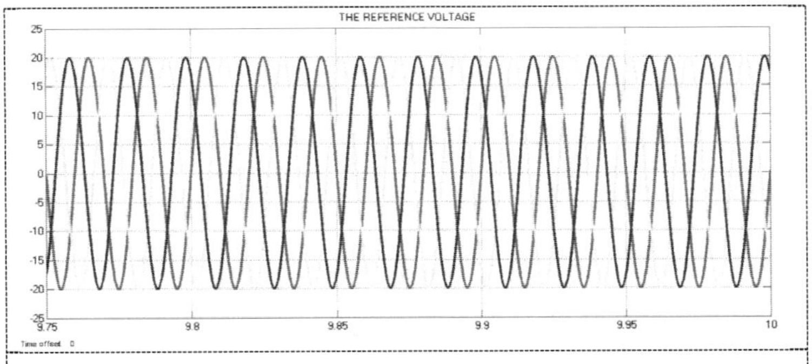

Figure 5. The three-phase component for reference voltage signal (voltage vs time).

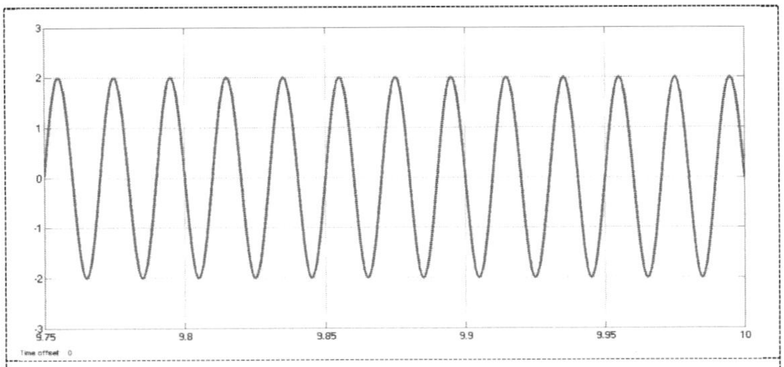

Figure 6. The voltage error after hysteresis controller (voltage vs time).

The waveform in figure 6 shows the error voltage as 2Vac. It comes from the comparison of 18Vac as the analog input with 20Vac as the reference voltage.

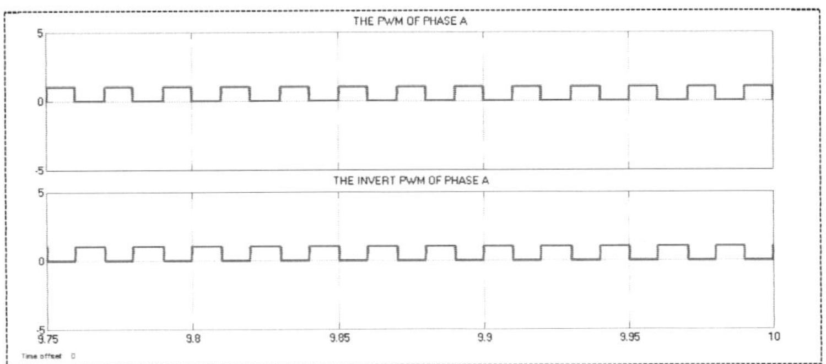

Figure 7. The PWM and invert PWM for phase A (voltage vs time).

Figure 7 depicts the PWM for Phase A and Invert Phase A. Similar depiction could be found for Phase B and Phase C with difference in phase shift. From here, 6 pulse PWMs were created for the switching of the inverter.

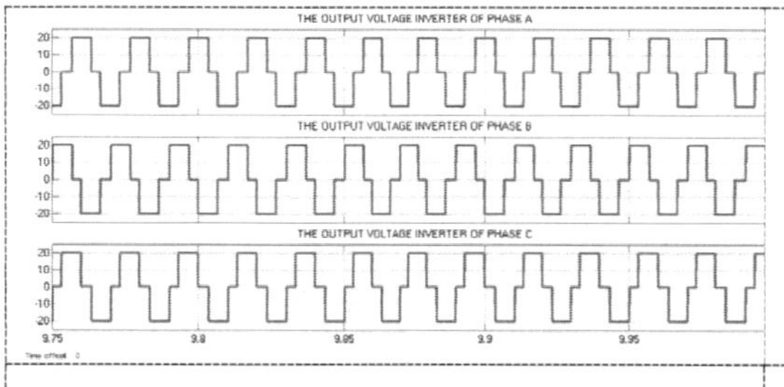

Figure 8(a). The output voltage inverter for three-phases

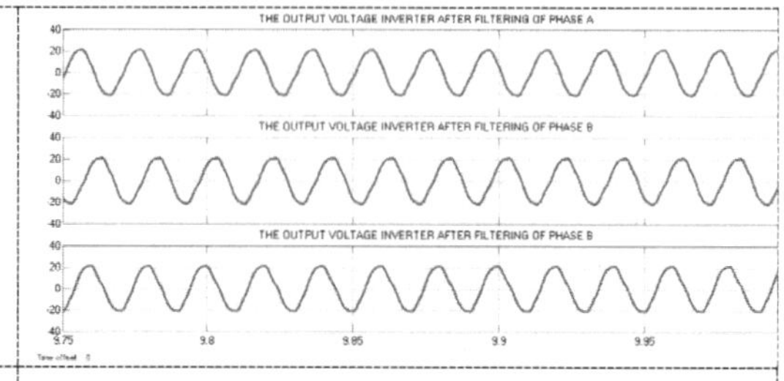

Figure 8(b). The output voltage filter in three-phases (voltage vs time).

The output voltage of the inverter is in nonsinusoidal waveform depicted in figure 8 (a). It is normal and complies with the output voltage of the inverter; the waveform is in square wave. The result was derived from the switching of the MOSFET. Therefore, after placing the filter, the waveform was turned into approximately sinusoidal as shown in figure 8(b).

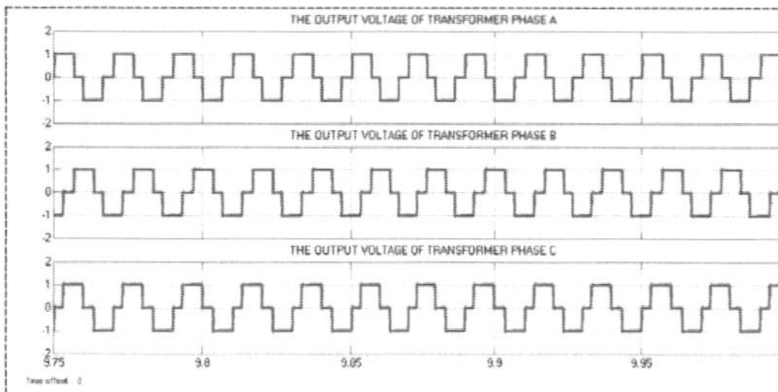

(a)

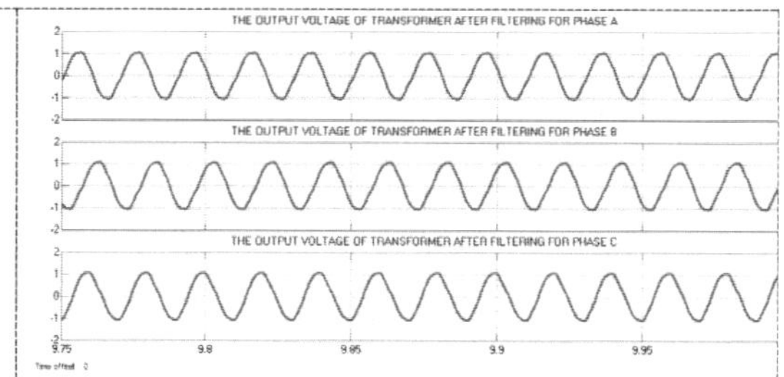

(b)

Figure 9(a) and (b). The output voltage filter of the output transformer (voltage vs time).

The function of the transformer is to scale down the voltage to follow the analog characteristic of Arduino. In the model, the ratio of the transformer is 20:1 hence, 20Vac will be scaled down to 1Vac. The output of the transformer, which is the same pattern as the output of the inverter, is nonsinusoidal as shown in figure 9(a). With the same filter application, figure 9(b) shows the waveform as approximately sinusoidal.

4. Conclusion

This paper introduced hysteresis controller with voltage control technique based on three-phase inverter switching fed to an induction motor. The switching frequency of the inverter is presented using the proposed hysteresis controller in MATLAB Simulink model. The method minimizes the line voltage error of the induction motor preventing the line voltage from being kept low hence will be able to avoid excess current at low speeds. Lastly, implementing the MATLAB Simulink model using a three-phase inverter provides flexibility of control.

5. Future recommendation

Further studies can be carried out by implementing hardware setup to prove the proposed method with real time situation. Arduino can be used to create prototypes, to carry out the duty of controller to control the switching of three phase inverter.

6. Acknowledgement

The author would like to acknowledge Universiti Teknologi MARA (UiTM), Malaysia in providing the facilities and the equipment to conduct this experiment.

References

[1] Farah N, Talib M, Mohd Shah N, Abdullah Q, Ibrahim Z, Lazi J et al 2019 A Novel Self-Tuning Fuzzy Logic Controller Based Induction Motor Drive System: An Experimental Approach. *IEEE Access* **7**:68172-68184.

[2] Zhang J, Li L, Zhang L, Dorrell DG 2017 Hysteresis band current controller based field-oriented control for an induction motor driven by a direct matrix converter. In: *Proc. IECON 2017 - 43rd Annu. Conf. IEEE Ind. Electron. Soc.* **2017** 4633–4638.

[3] Sudheer H, Kodad S F,Sarvesh B 2018 Improvements in direct torque control of induction motor for wide range of speed operation using fuzzy logic *J. Electr. Syst. Inf. Technol.* **5**(3) 813–828.

[4] Ammar A, Bourek A and Benakcha A 2017 Nonlinear SVM-DTC for induction motor drive using input-output feedback linearization and high order sliding mode control *ISA Trans.* **67** 428–442.

[5] Saleh A L, Obaid B A and Obed A A 2018 Motion control of linear induction motor based on optimal recurrent wavelet neural network-PID controller *Int. J. Eng. Technol.* **7**(4) 2028–2034

[6] Braun W D and Perreault D J 2019 A High-Frequency Inverter for Variable-Load Operation *IEEE J. Emerg. Sel. Top. Power Electron.* **7(2)** 706–721.

[7] Vashishtha S and Rekha K R 2018 A survey: Space vector PWM (SVPWM) in 3φ voltage source inverter (VSI) *Int. J. Electr. Comput. Eng.* **8(1)** 11–18.

[8] Kali Y et al 2019 Current control of a six-phase induction machine drive based on discrete-time sliding mode with time delay estimation *Energies* **12(1)** 1–17.

[9] Premkumar K, Thamizhselvan T, Vishnu Priya M, Ron Carter S B and Sivakumar L P 2019 Fuzzy anti-windup pid controlled induction motor *Int. J. Eng. Adv. Technol.* **9(1)** 184–189.

[10] Hamla H, Rahmani L and Belhaouchet N 2019 A modified direct torque control with minimum torque ripple and constant switching frequency for induction motor drives *Int. Trans. Electr. Energy Syst.* **29(12)** 1–20.

[11] Fathi A, Shafiee Q and Bevrani H 2018 Robust frequency control of microgrids using an extended virtual synchronous generator *IEEE Trans. Power Syst.* **33**(6) 6289–6297.

[12] Alla Eddine Toubal Maamar F C, M'hamed H, Taleb R 2018 Analysis and implementation of half bridge series resonant inverter using Arduino *ICCEE'18 Int. Conf. Commun. Electr. Eng. El-Oued, Alger.*

A Study on Different Slot Position for Designing Dual Band Microstrip Antenna for 5G Wireless Communication

N A M Nasir[1], H Ja'afar[1,2*], H Baba[1], N H A Aziz[1]

[1]Antenna Research Center (ARC), Faculty of Electrical Engineering, Universiti Teknologi MARA, Selangor,Malaysia.

[2]Faculty of Electrical Engineering, Universiti Teknologi MARA, Terengganu Branch, 23000 Dungun, Terengganu, Malaysia.

*Corresponding author: hajar_3112@yahoo.com.my

Abstract. In this paper, four antenna designs with different slot positions for dual band frequencies are analysed. A reference antenna operating at 8 GHz is used at a dimension of 30 mm^2 using Rogers RT58880 as the substrate with a dielectric constant of 2.2. Two slots are etched in order to produce a dual band. The antenna resonates at 8 GHz and 15 GHz. Return loss at 8 GHz and 15 GHz are -17.03 dB and -21.37 dB with gain at 7.39 dB at 8 GHz. By using CST Microwave Software, the simulated results are obtained. The slotted antenna produces a dual band antenna that is a suitable candidate for 5G wireless communication tool.

1. Introduction

As 4G has been globally implemented alongside the growth of world population, the need for better quality of life through upgraded and enhanced technology has created a demand that 5G is capable of exceeding. With new emerging ideas of improving people's life, scientists are struggling with the new era challenges that push people to move towards the beyond. With rapid technological progress, the global market has invented a strong desire for businesses to create scientific breakthroughs in nanotechnology that can be utilized in everyday life as well as modern infrastructure usage [1].

Among all antenna types, this research uses micro strip antennas for its advantages and it is widely used for wireless communication because of its uncomplicated design and affordable in terms of the materials used. The basic microstrip patch antenna (MPA) consists of a radiating patch usually made out of copper, a substrate, a ground and its impedance matching are matched at 50 Ω.

Dual band antennas are preferred to wideband antenna for its higher efficiency. Various techniques have been applied to achieve dual band antennas such as coplanar waveguide [2], metamaterials [3] and multi-layered [4] structures but these techniques require a lot of time and is complex to design, hence creating slots on the radiating patch is preferable because its low-complexity in design [5].

In this research, slot positions are studied to determine which position produces suitable results. Double slots are etched in order to produce dual band frequency [6]. Based from R. Parikh findings, arranging the slots into different positions are studied in order to produce multiband frequencies [7]. The antenna is made out of copper as the radiating patch with thickness of 0.035 mm. The area of the antenna is W×L=30 mm^2 by using Rogers RT58880 with a dielectric constant of 2.2 with thickness of 1.575 mm^2. The antenna is designed to work at 8 GHz and 15 GHz suitable for 5G wireless technology.

2. Methodology

The antenna is designed to produce 8 GHz and 15 GHz. The following formulas are used to determine the antenna's dimension.

$$W_p = \frac{v_0}{2f_r}\sqrt{\frac{2}{\varepsilon_r+1}} \tag{1}$$

The length of the patch is calculated

$$L = L_{eff} - 2\Delta L \tag{2}$$

The effective length, L_{eff} calculated by using this formula

$$L_{eff} = \frac{v_0}{2f_r\sqrt{\varepsilon_{reff}}} \tag{3}$$

$$\Delta L = 0.412h\,\frac{(\varepsilon_{reff}+0.3)\left(W/_h+0.264\right)}{(\varepsilon_{reff}-0.258)\left(W/_h+0.8\right)} \tag{4}$$

To calculate the effective dielectric constant ε_{reff}

$$\varepsilon_{reff} = \frac{\varepsilon_r+1}{2} + \frac{\varepsilon_r-1}{2} \tag{5}$$

Calculations are made and is optimised. The results are tabulated in table 1 as a reference.

Table 1. Calculated dimension of the antenna.

Parameter	Value (mm)
Length of patch, *Wp*	15.5
Width of patch, *Lp*	11.6
Length of substrate, *Ws*	30.0
Width of substrate, *Ls*	30.0
Length of feed line, l_f	6.25
Width of feed line, w_f	4.60
Width of half quarter wave port, w_q	0.47
Length of half quarter wave port, l_q	1.59
Length of slot	6.5
Width of slot	0.4

The proposed antenna is designed and simulated using Computer Simulation Technology (CST) Microwave Software. The dimension of the antenna is $\boldsymbol{Wp} \times \boldsymbol{Lp} = \mathbf{30}$ mm^2 by using Rogers RT58880 with a dielectric constant of 2.2 with thickness of 1.575 mm^2. The ground dimension is the same as the substrate of 30 x 30 mm using copper with thickness of 0.035 mm. The width of feedline, $\boldsymbol{w_f}$ and length of feedline, $\boldsymbol{l_f}$ are 4.6 mm x 6.25 mm respectively. 50 Ω matching is achieved by using quarter waveguide feedline with a dimension of $\boldsymbol{w_q}$ and $\boldsymbol{l_q}$ of 0.468 mm x 1.59 mm. Figure 1 show the antenna reference without slot.

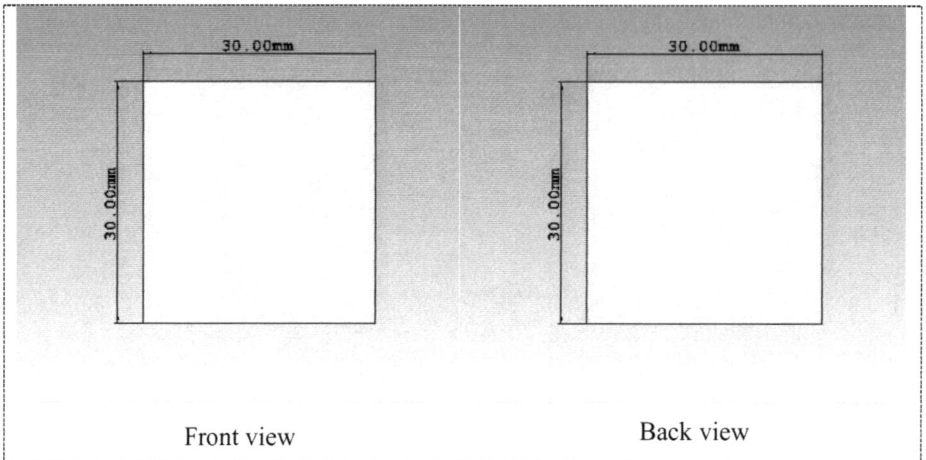

Figure 1. Microstrip antenna patch

3. Results and Discussions

Four antennas are designed and simulated using Computer Simulation Technology (CST) Microwave Software. All four designs, slots are etched on different positions to study the relationship between the slot positions on the radiating patch. The slot dimension is determined by optimization with width and length, 6.5 mm x 0.4 mm. Two slots are etched in order to create dual-band frequency. The designs of the antennas are shown below.

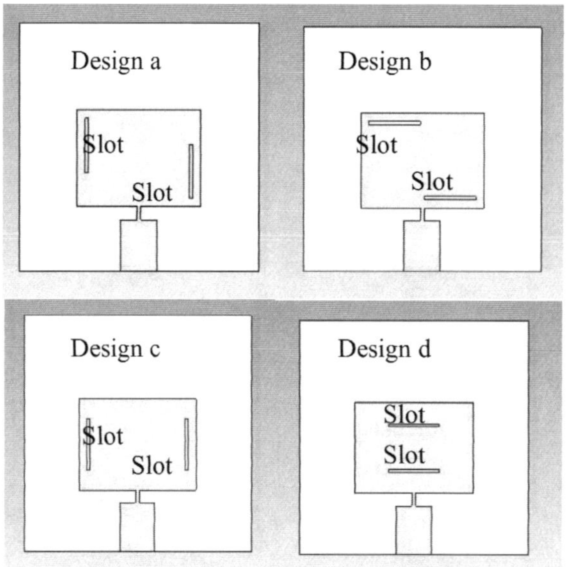

Figure 2. MPA with different slot position

3.1. Antenna Performances

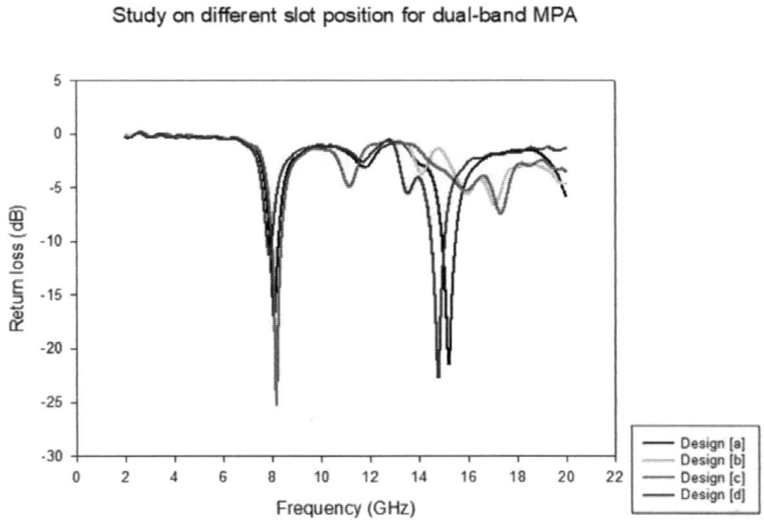

Figure 3. Return loss of MPA with different slot positions

As shown on the graph above, Design [a] and [d] resonates at 8 GHz and 15 GHz with return losses both below -10 dB. For Design [b] and [c], both designs do not produce dual band frequencies, hence they are not preferred unlike Design [a] and [d].

Table 2. Return loss S11 for simulated antenna design

Simulated Antenna	Frequency (GHz)	Return loss (dB)	Bandwidth (MHz)
Design [a]	8	-17.03	366
	15	-21.37	546
Design [b]	8	-24	380
Design [c]	8	-25	394
Design [d]	8	-10.0	192
	15	-11.0	517

3.2. Surface current and gain

Further study on the antenna's surface current and gain to determine which design has better results in terms of slot positions.

IOP Conf. Series: Materials Science and Engineering 1176 (2021) 012014 doi:10.1088/1757-899X/1176/1/012014

Table 3. Surface current density and gain at 8 GHz of simulated results

Simulated antenna	Surface current distribution at 8 GHz	Gain at 8 GHz (dB)
[a]	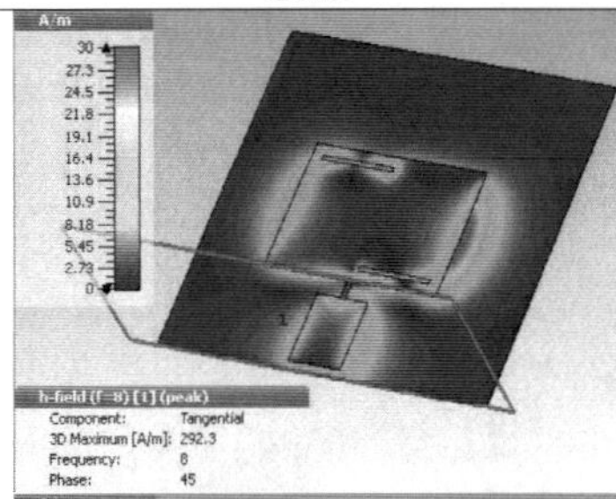	7.39
[b]	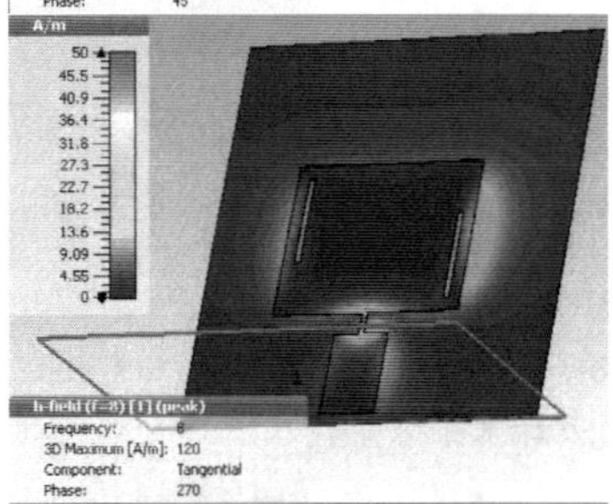	7.58
[c]	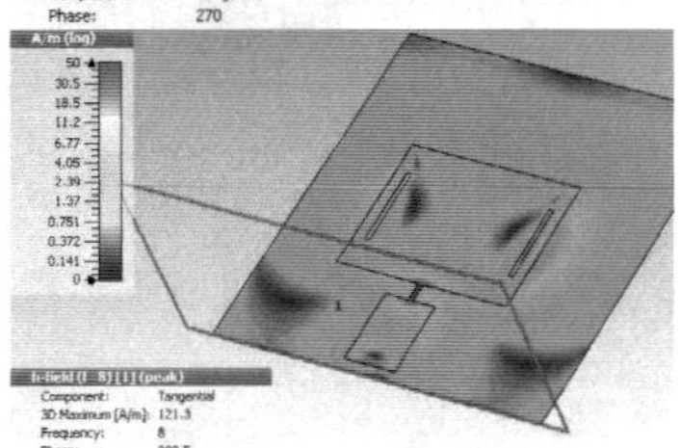	7.46
[d]	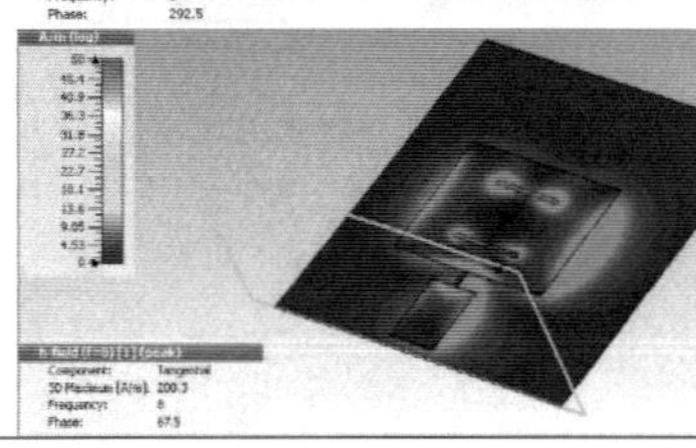	7.57

The results shows that the slot shows that the current is more concentrated on the sides of the antenna patch compared to the other edge of the antenna. With the presence of the slots etched on the radiating patch, it helps to control the current distribution. Design [a] has the lowest gain at 8 GHz compared to order designs. Improving the gain by adding stubs will help to increase the gain.

4. Conclusion

An antenna that operates at 8 GHz and 15 GHz is achieved by etching slots on the radiating patch. Different slot positions give out different results that determines whether it is best suited for better performances. Positioning the slots at the antenna sides will allow current to concentrate at the edge of the antenna for better current distribution. Design a is preferred for its return loss both at 8 GHz and 15 GHz (-17.03, -21.37) dB respectively, as well for its gain at 8 GHz at 7.39 dB. Further investigation will be appreciated in order to improve the antenna performances in terms of return loss and gain. The antenna is suitable for 5G telecommunication applications.

ACKNOWLEDGEMENTS
The authors would like to express their gratitude to University Teknologi MARA for funding this research and to the utmost thanks to the members of Antenna Research Centre (ARC), Faculty of Electrical Engineering, Universiti Teknologi MARA for their shared knowledge and guidance.

References

[1] J. G. Andrews et al. 2020 What Will 5G Be? **32(6)** 1065-1082
[2] T. Hariyadi 2013 A Coplanar Waveguide (CPW) Wideband Octagonal Microstrip Antenna *Int. Conf. of Inf. and Comm. Techn. (Indonesia)* 340-343.
[3] S. Al Nahiyan, A. R. Salehin, G. M. H. Shuvho, S. M. Liaqat, and H. Sagor 2017 Dual Band Operation with Dual Radiation Pattern for Rectangular Microstrip Patch Antenna Loaded with Metamaterial *IEEE Region 10 Humanitarian Technol. Conf. (R10-HTC)* 21-23.
[4] H. Yon, A. H. Awang, M. T. Ali, S. Subahir, and S. N. Kamaruddin 2017 Comparative Analysis for Multilayer Stacked Substrates Microstrip Patch Antenna *IEEE Asia-Pacific Conf. on App. Electromagn. (APACE)* 34-37.
[5] M. Kumar 2018 Introducing Multiband and Wideband Microstrip Patch Antennas Using Fractal Geometries : Development in Last Decade *Wirel. Pers. Commun.* **98(2)** 2079-2105.
[6] D. Lodhi, M. M. Sharma, and R. P. Yadav 2017 Analysis of Dual Band Double Slot Loaded Microstrip Patch Antenna *Inter. Conf. on Comput., Commun and Elect.* 286-288.
[7] Rina Parikh, Bharati Singh 2018 Effects of Slots On Resonant Frequencies of a Microstrip Patch Antenna *Inter. Conf. on Comput. Commun. Contr. and Automat.* (ICCUBEA).

Electronic Voting Machine (EVM) Using Infrared Detection Technique on Arduino Platform

S I Ismail[1*], S A C Kar[1], R Abdullah[1], M A H M Amri[1], M Z Ismail[1] and N N Nik M Zaki[1]

[1]Faculty of Electrical Engineering, Universiti Teknologi MARA Cawangan Terengganu Kampus Dungun, 23000, Dungun, Terengganu, Malaysia

*Corresponding author: syila5416@uitm.edu.my

Abstract. The voting system is a set of rules that determine how people's intentions may be manifest and how decisions can be achieved. Normally, vote papers are placed in the voting box and then will be sorted and calculated manually. But the problem is the use of manpower in calculating the vote manually may lead to error in the counting of votes and it takes a long time to complete the calculation process. For this purpose, an electronic voting machine (EVM) was introduced, where in this work it replace the conventional voting method in which the vote card will be calculated automatically. This paper focusing on implementation of embedded system which will help in the election process, where the Arduino UNO uses as a controller. This system uses a simple counting system based on the detection of black mark on the vote card using infrared (IR) sensor. Then, LCD 16x2 will display the number of votes for the respective candidate and the servo motor will activate to sort the valid and invalid vote card into 2 compartments in the voting box.

1. Introduction

There has been widespread concern about the trust issues on counting the amount of vote card during on the election because it is counted manually. Although the discussion of this problems have taken place in Malaysia, the deployment can go a long way in solving the problems of electoral process while others think only by reducing the level of corruption in the electoral systems can reduce the rigging and other serious problems.

Traditional voting process can be divided into three different phases. The first phase is authentication phase. In this phase, voter authenticates himself or herself by showing his or her voting card, this step is public and verified by the presiding officer. At the end of authentication process, presiding officer give a vote paper to voter to cast his or her vote. Second phase is voting phase. Voters do their candidate selection in a protected booth by writing it with pen on the vote paper. Then the vote paper will be folded and inserted into the voting box. All valid and invalid votes are mixed up in the box. The final phase is vote counting phase. At the end of voting time, the presiding officer collect the voting box containing all vote papers and submit it to the counting center. The voting boxes are opened and votes are counted and the results are then announced.

There have been several studies on methods used to improve elections. There are many ways to track the votes cast by voters. One of it used a set of switches to represent the candidates. The voters select the candidate by pressing the keypad 3x4 provided by the election committee [1]-[3]. Thumb impression for each people has different pattern. A few researches have been done about the fingerprint detection in election process. To maximize the effectiveness and accuracy, the fingerprint is used for voter's identification to avoid fake and repeated voting [4-8].

Although many voting prototypes have been developed, however no prototype has been established for the purpose of separating the invalid votes. This paper presents the development of electronic voting machine which used Arduino UNO as the controller. The IR sensors were used to detect the candidate that has been selected by the voters. The counted votes were displayed by using 16x2 LCD display. This EVM also provide sorting features between valid and invalid votes.

2. Research Method

The development of an Electronic Voting Machine (EVM) consists of two parts which is hardware and software implementation. Figure 1 shows the EVM block diagram for this project. There are 3 infrared (IR) sensors used in vote card's detection. An additional pushbutton is used to send the signal to LCD to display the name of the candidate who received the most votes at the end of the election process, thus becoming the winner in the election. Every time the IR activated, the counter accumulates value for the respective candidate will be increased. The value will display using 16x2 LCD display. Servo motor is used for sorting the vote card into two compartments provided which is valid and invalid vote card storage. All input and output devices controlled by Arduino UNO controller. The 3 IR sensors have been used to represent the three election candidates. Counting the number of votes for each candidate and sorting the votes paper are the two main scopes in designing the project algorithm. The invalid votes stored in box 1, while the valid votes stored in box 2.

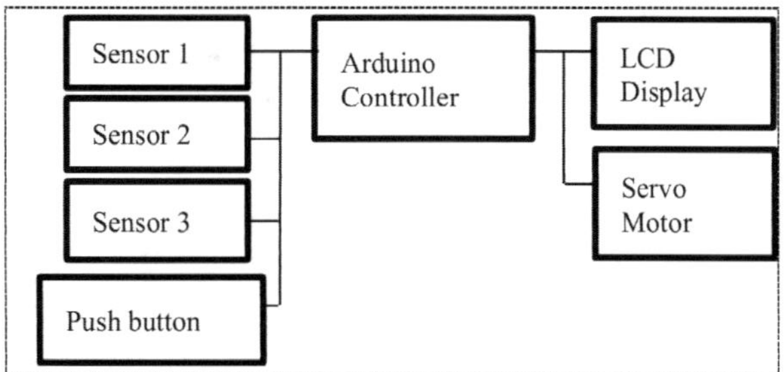

Figure 1. EVM block diagram.

2.1. *Process Flowchart*

Figure 2 and figure 3 shows process flowchart for 2 situations that can occur in the voting process. Condition A activated when more than 1 sensor is active at the same time and Condition B activated when only one sensor is active. If voters make an invalid vote which mean they tick more than 1 candidate, condition A will be triggered. The invalid cards will goes to box 1 when the motor rotates 45 degrees angle in clockwise direction. Meanwhile, for valid vote, condition B is triggered. The counting increases on that particular candidate selected and LCD will display the current votes. The vote card then will be stored into box 2 when the motor triggered and rotates in the other direction with 45° angle. The sorting process is depends on the servo motor rotation. The motor will rotate 45^0 clockwise for invalid vote storage and 45^0 anticlockwise for valid votes storage.

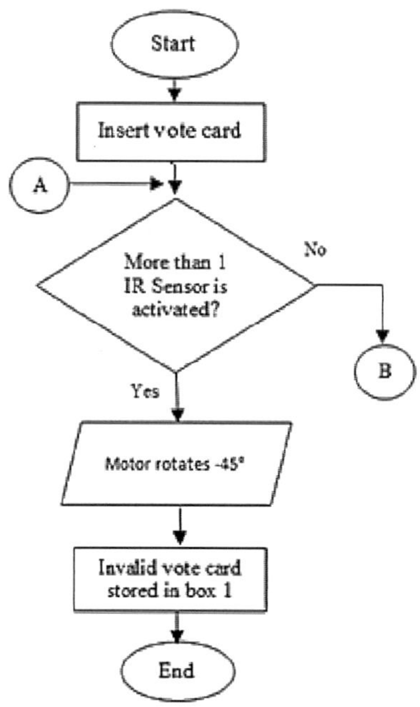

Figure 2. Condition A-Activates more than one IR sensor.

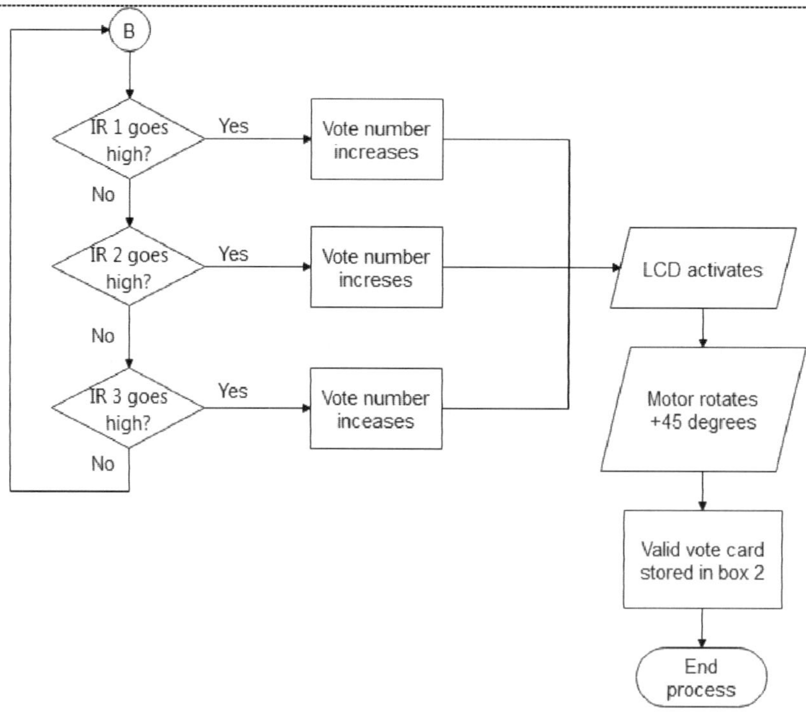

Figure 3. Condition B - Activates only one IR sensor.

IOP Conf. Series: Materials Science and Engineering 1176 (2021) 012015 doi:10.1088/1757-899X/1176/1/012015

2.2. *IR Detection Technique*

The working principle for IR sensors by using a specific light sensor to detect a selected light wavelength in the Infra-Red (IR) spectrum. When an object is close to the sensor, the light from the LED will be reflected back into the light sensor. This type of sensor can then be used to measure how "bright" the object is. Figure 4 shows the white surface will reflects more IR light rather than a darker surface. During the election process, all voters will be provided with a black marker or black sticker to choose their candidates by putting a mark on their selected candidate on the vote card. For this project, the IR will activate when it detect the low intensity which coming from a darker surface, and deactivates at high intensity. The vote card then will be inserted in EVM slot card for detection process. Figure 5 shows the vote card is marked by the voters where candidate A has been selected.

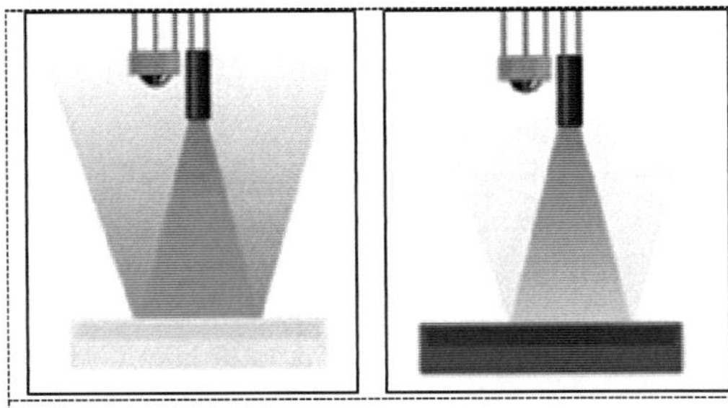

Figure 4. Reflection of IR light on white surface and black surface.

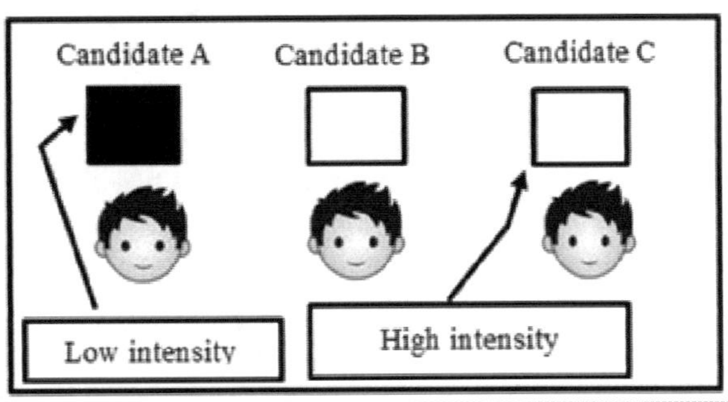

Figure 5. Sample of vote card.

3. Result and discussion

Input circuit consist of three IR sensors. The signal coming from IR sensor will transfer to the Arduino UNO for execution process. The IR sensor which in analog signal will then convert to digital signal by comparator IC LM324 and then connected to the digital pin at Arduino UNO.

3.1. *Simulation*

Figure 6 shows the sensor was activated and increased counter value by 1 for the first candidate. Servo motor will rotate 45° anticlockwise for vote card sorting process. LCD display the count votes for candidate no 1 (ZAM)) and the other two candidate, PISS and NIK shows 0 votes respectively. Figure 7 shows the simulation when it uses two IR circuits to represent the two candidates. When 2 dark surfaces were detected on the vote card, the servo motor rotates 45° angle in clockwise direction. It means that there was invalid vote was detected. The output LCD does not display the invalid vote and only valid votes will be counted and display on the LCD.

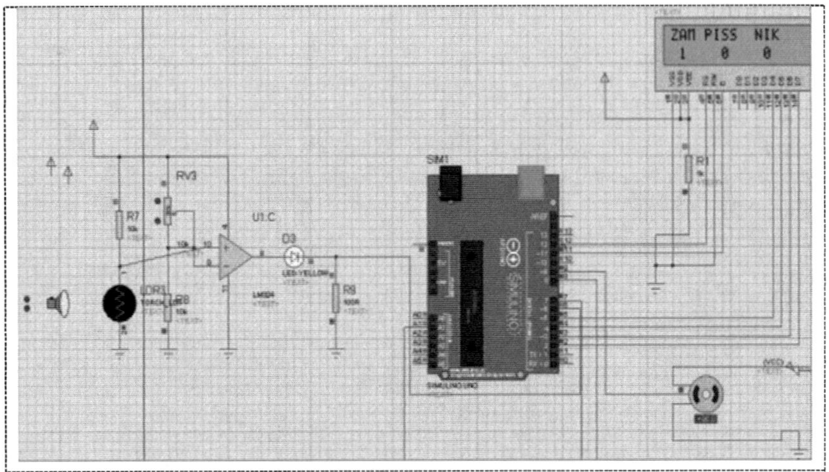

Figure 6. Only one IR sensor was activated (valid vote)

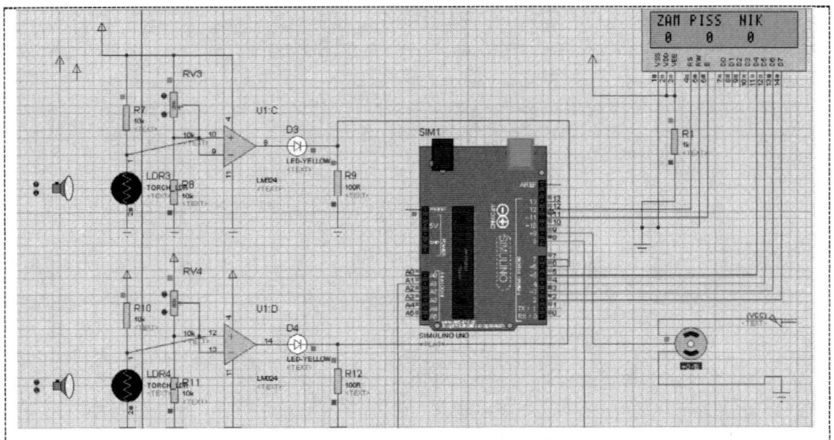

Figure 7. More than 1 IR sensor were activated (invalid vote)

3.2. Prototype Setup

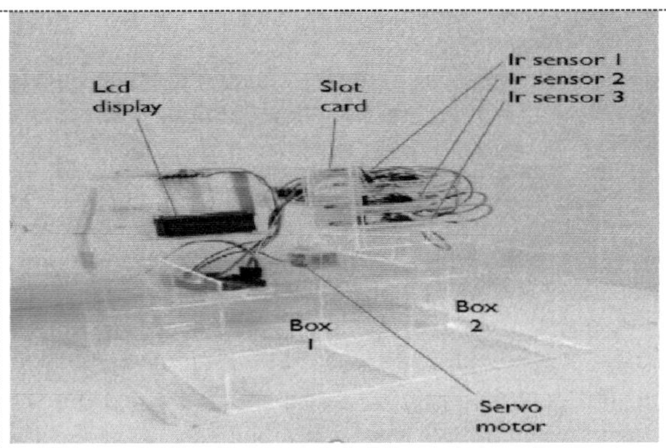

Figure 8. EVM prototype

The prototype as in figure 8 shows two boxes were placed to sort the vote card. Box 1 is for the vote card that has more than one mark (invalid vote). Box 2 is for vote card that have only one mark (valid vote). The result button and LCD display were mounted on the side of the prototype. The servo motor is located precisely under the slot card which acts as the card holder. IR sensors were positioned side by side toward the slot card. Each time a vote card inserted into the slot card, the metal pedal holds the card from falling into the box. After the IR activated, the servo motor rotates accordingly. Then, the card falls into the storage boxes.

4. Conclusion

In conclusion, this developed method can facilitate the election process in which the system can separate valid and invalid votes and also can calculate the number of votes at the same time. This prototype is functioning according to the scope of the design. It is suggested that for future development, the machine can be embedded with an identification process for each voter before the voting process starts. It can track the number of voters who have voted in the respective locations. This developed system would minimize the electoral staff and saves the time.

References

[1] Meher S S 2014 Implementation of Electronic Voting Machine using AT89C51 and Proteus. *Int. J. Adv. Eng. Glob. Technol.* **2** 101-105.

[2] Ghosh M S Development of Microcontroller Based Electronic Voting Machine (EVM). *IOSR J. Electr. Electron. Eng.* (IOSR-JEEE) 9(2) 01-03.

[3] Elkind E, Faliszewski P, Skowron P and Slinko A 2017 Properties of Multiwinner Voting Rules *Social Choice and Welfare* **48(3)** 599-632.

[4] Agrawal S, Majhi P and Yadav V 2014 Fingerprint recognition based electronic voting machine *Nat. Conf. on Syner. Trends in Eng. and Techno.* (STET-2014).

[5] Gelb A and Clark J 2013 Identification for development: the biometrics revolution *Center for Global Development Working Paper* **315.**

[6] Ghosal A, Chakraborty S, Maity S, Paul A and Roy P 2017 Low-cost and secured biometric system-based EVM with instant counting by wireless transmission of data to a central place. *Int. Conf. on Advancement of Comput. Commun. and Electr. Techn. (ACCET 2016)* 139.

[7] Prasad R M, Bojja P and Nakirekanti M 2016 Aadhar Based Electronic Voting Machine Using Arduino. *Int. J. of Comput. Appl.* **145(12)** 39-42.

[8] Jamkar A, Kulkarni O, Salunke A, Pljonkin A 2019 Biometric voting machine based on fingerprint scanner and arduino *2nd Int. Conf. on Intell. Commun. and Comput. Techn. (ICCT)* 322-326.

Water Quality Monitoring System for Fisheries using Internet of Things (IoT)

N Ya'acob[1,2*], N N S N Dzulkefli[1], A L Yusof[1,2], M Kassim[1], N F Naim[1] and S S M Aris[1]

[1]Faculty of Electrical Engineering, Universiti Teknologi MARA, Shah Alam, 40450 Shah Alam, Selangor, Malaysia

[2]Wireless Communication Technology (WiCoT), Faculty of Electrical Engineering, Universiti Teknologi MARA, 40450 Shah Alam, Selangor, Malaysia

*Corresponding author: norsuzila@uitm.edu.my

Abstract. Fish production has significant impacts on the investment and operating costs of the modernization of fish ponds. Water quality monitoring system may help to overcome all the problems by using the implementation of the internet of things (IoT) technology. The project aims in detecting the presence of fishes using an ultrasonic sensor and investigating the parameters of pH and temperature sensors. The system is used pH and temperature sensor as the factors affecting the quality of water and fish growth, the ultrasonic sensor for detecting the presence of fishes, Blynk application through a smartphone for notifications, NodeMCU ESP8266 as its central controller and wireless fidelity (WI-FI) as the communication technology. A sample of three different measurement times in the early morning, midday, and evening based on temperature, pH levels and the fish distance show that this farm has high-quality water. From the measurement at the field test, temperature value increase in the midday, pH level remains constant in the three different times and the fish distance underwater always changes as a fish needs to find oxygen and a portion of food. As a result, a farmer need not hire workers at their site, consequently, drive down operating costs and improve efficiency.

1. Introduction

Fisheries aquaculture remains an important source of food, nutrition, income and livelihoods for hundreds of millions of people around the world. According to The State of World Fisheries and Aquaculture 2018 [1], the world fish supply in 2016 reached 171million tons, of which 88 percent was utilized for direct human consumption, thanks to relatively stable capture fisheries production, reduced wastage and continued aquaculture growth. Fisheries are one of the key Sustainable Development Goal SDG 14 which is to conserve oceans, seas and marine resources while supporting those ecosystem services that are of crucial importance for humans. More sustainable use of resources, changes in production patterns and improved management of human activities will help to reduce negative impacts on the environment and allow the benefit of aquatic ecosystems for present and future generations. However, efforts need to be made to ensure that as the population in the country grows and demand for food and employment likewise grows, the benefits that fishery resources provide are protected through sustainable management and value addition. All fish depends entirely on the water to live, eat, grow and perform other body functions, regardless of the type of water available or the species were chosen. Therefore, it is no surprise that a fish farming establishment's success is largely due to the quality of its water. Temperature and pH are some of the most important parameters in aquaculture [2]. The fisheries management relies totally on water quality monitoring.

Water is a special kind of environmental resource which is a fundamental basis to sustain the ecosystem on earth. Water as an ecosystem not only supports human life and behaviors but also helps preserve the ecological structure, ecological processes and regional environment [3]. Water is not a renewable source, but it needs to be reused. With the rapid development of the economy, more and more serious problems of the environment arise, water pollution is one of these problems. Recently, an illegal tyre recycling factory had allegedly dumped chemical waste in Sungai Kim Kim in Pasir Gudang Johor on the 6th March 2019 due to this effect many lives get into danger [4]. It contaminates the whole water in the lake. So, it affects thousands of fishes. Routinely monitored parameters of water quality are temperature, pH, and turbidity. It is important to collect this water quality measurement because these factors might affect water quality including pollution, wind, weather and rain runoff may affect fish health [5]. Because of continuous monitoring, the parameters help to reduce the problems arrived in the future. In order to improve the lifestyle which makes it easy and effortless, they need to implement this modern technology which is the Internet of Things (IoT) in every field.

The internet of Things (IoT) is a revolutionary new concept that has the potential to turn virtually anything "smart". IoT is a network of physical objects. The goal of the IoT is to enable things to be connected anytime, anyplace, with anything and anyone ideally using any path or network and any services [6]. IoT is much more than machine-to-machine communication, wireless sensor networks, WI-FI, GPS, microcontroller and microprocessor. IoT is not a single technology, but it is a mixture of different hardware and software technology used to store, retrieve, and process data and communications technology which includes electronic systems. It allows controlling or monitoring any objects remotely, from any part of the world through the existing network. However, the fact remains that there are many challenges and issues related to the use of IoT, this brings advantages and disadvantages to the internet of things (IoT).

In this advanced technology, some projects of remote monitoring and sensing system to improve the quality of water and increase fish farm production. One of the projects that use to detect fishes and the presence of an object underwater by using a waterproof ultrasonic sensor JSN-SR04T. This project uses Arduino as the central performance control system by an additional HC-05 Bluetooth Shield that enables Arduino to communicate with Android smartphones [7, 8]. To monitor water characteristics that influence fish production by using remote monitoring via wireless communication technologies like cellular, LoRaWAN, Wi-Fi or Satellite communication. One of the projects is an automated IoT system for automated fish farming where they monitor the water temperature, pH and water level, using Wi-Fi remote connection [9]. The project developed real-time fish pond monitoring and automation where they measure the water, temperature, pH and DO levels integrated with aerating and water supply pumps using Arduino [10]. Another project developed a control and monitoring system where they monitor the pH, Dissolved Oxygen, Temperature and Salinity parameters using PiLeven microcontroller to communicate with sensors and utilize Raspberry Pi microprocessor as a computer interface [11].

There is also a project designed ZigBee-based Wireless Sensor Network system and the Internet of Thing (IoT) [12] - [15]. Sensing data is transmitted through the ZigBee network, stored in a cloud-based database and the LabVIEW is used for the graphical user interface. This system is a low cost, low power consumption, scalable, versatile, distributed, mobile and accurate. Then, another project proposed a smart water quality monitoring system using a GSM network to transfer sensing data to a cloud server by sending SMS messages to the end-user as a notification [16] – [18]. The disadvantage of this proposed GSM system is the process of notification to the end-user after specific time intervals or after a certain parameter value is reached, and not in real-time.

To overcome this in one sense of real-time monitoring, the project in [19] proposed a system with moving cameras, which constantly monitors the fish farming pond based on Deformable Multiple Kernels. Another project is the proposed technology LoRa-based wireless sensor network and a lightweight Message Queuing Telemetry Transport (MQTT) protocol for exchanging messages between small, embedded devices, mobile devices, and sensors [20]. LoRaWAN can be used to cover large areas, and especially satellite communication where can reach areas that are typically not covered by mobile-cellular networks.

The objectives of this project are to monitor and detect the presence of fishes using an ultrasonic sensor and NodeMCU ESP8266 microcontroller. This system is to automate the monitoring and maintenance of a fish farming pond in remote locations by using remote monitoring via wireless communication technologies. This project also highlights investigating the parameters of pH and water temperature value towards water quality systems. This smart system aims to the reduction of the environmental stress that affects the fish population in the fish farming pond. From this project, the user got notification from the user email when the sensor was inserted into the water and at the same time turn on the Wi-Fi connection.

2. Methodology

2.1. Study Area

Freshwater fish pond in Malaysia has the largest total numbers of anglers and it has been in the angling industry for years. Fish pond in Section 24 Shah Alam is one of the best places to fishing. The field test was conducted in fish pond Section 24 Shah Alam, Selangor which is located on the west coast of Peninsular Malaysia with the latitude at 3°2'10.5864"N and longitude at 101°31'17.04"E. The field test was performed for this project in an outdoor fish pond which is a real-world environment to validate the usability of the platform when given commands via a smartphone application.

The fish pond shown in figure 1 has an area of 6070.3 m^2 and a depth of 4.0 m (typical water level). The pond produces Mekong, Amazon Catfish, Tiger Catfish, Red Tails Catfish, Pacu, Toman, Patin Albino, Patin Choaprhaya, Tongsan, Big Head Carp and Rohu. The fish pond also using a pump to continuously give oxygen to the fishes.

Figure 1. Field Test at Fish Pond Section 24 Shah Alam

2.2. Framework

Figure 2 shows the framework of the system. The framework is divided into two major parts includes the software and hardware design. The software has two developments for the program the progress networking which is database and IoT. Database from Blynk application was used for monitoring fisheries and water quality system and get the notification from user email. Arduino IDE was used for IoT coding to interact with Wi-Fi connection and the mobile application. The hardware used is the microcontroller NodeMCU ESP8266, sensors of temperature, pH and ultrasonic as an input, and actuators of LEDs, LCD, and buzzer as an output of the program.

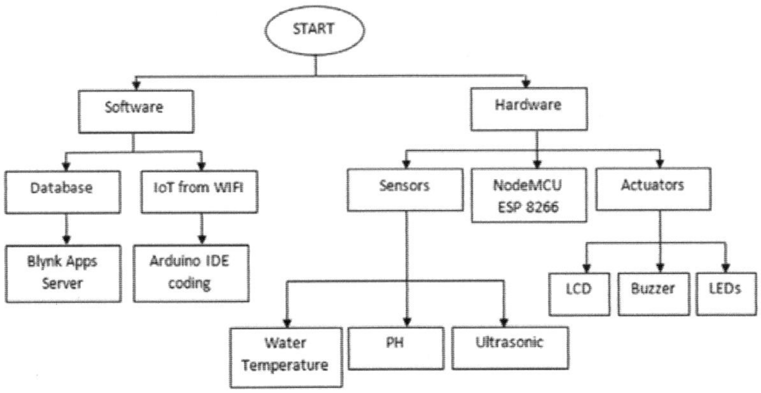

Figure 2. Frameworks of The System

2.3. *Project Overview Concept*

Figure 3 shows the project overview of the system. The sensors will interface with the microcontroller NodeMCU ESP8266 that acts as an open-source IoT platform, it will send the data to the cloud storage which is the Blynk server that has been developed to store the data information. Arduino IDE is a programming language that supports a library for Blynk that can connect any hardware over Wi-Fi. Blynk application is a platform with IOS and Android apps to control NodeMCU ESP8266 through its inbuilt Wi-Fi shield. It can display the digitalized information data from the sensor and convert the data from CSV files to Excel format and store it in a database.

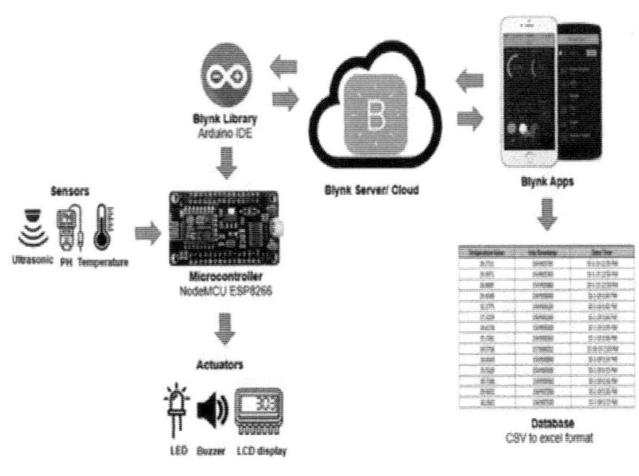

Figure 3. Project Overview of The System

2.4. *Flowchart of Overall System*

Figure 4 shows the flowchart of the system. The flowchart starts by having three sensors (ultrasonic, pH and temperature) detecting the presence of fishes, measuring the hydrogen ion concentration of water and detecting the current temperature of water respectively. The ultrasonic sensor will detect the presence of fishes that move underwater. The range for detecting the fishes is between 0 to 400 cm based on the depth of the fish pond at Section 24, Shah Alam which is 4.0 m. The distance will be display on the LCD. If the distance is out of the range the buzzer will turn on and display "Out of range!" on the LCD. The pH sensor measures the acidity or alkalinity of water. The optimal pH range for fishes is 6.5 to 8.5. The temperature sensor measures the temperature of the water and indicates each stage of temperature level by different colors of LEDs. If the water temperature is below 29 °C, the yellow LED will light up. If the water temperature is between 29 °C to 35 °C, the green LED will light up. This value shows that the normal temperature for fishes to grow. If the water temperature is above 35 °C, the red

LED will light up. Then, all the sensors will be sent and display the data on the Blynk application through a smartphone by using a Wi-Fi connection.

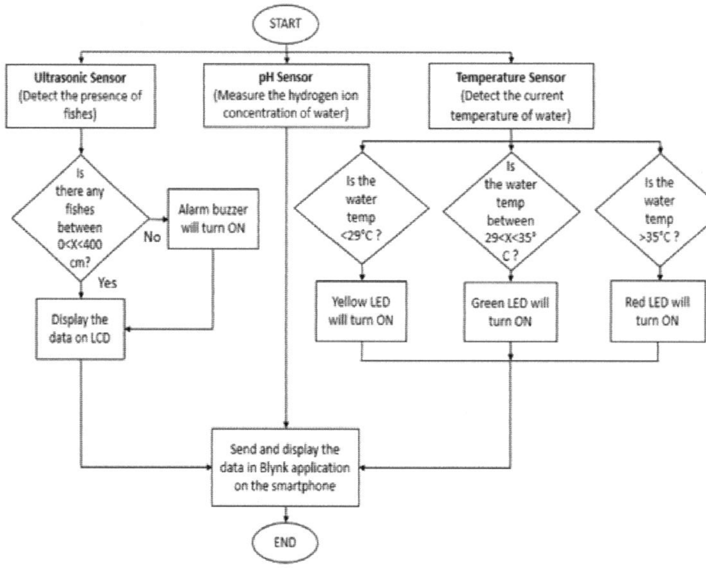

Figure 4. Flowchart of Overall System

2.5. *Circuit Design of the System*

Figure 5 shows the circuit diagram of the system. The components such as the temperature sensor, ultrasonic sensor, pH sensor, buzzer, 3 different light-emitting diodes (LED), a resistor of 100 ohms and the liquid crystal display (LCD) are connected to the different pins on the NodeMCU board.

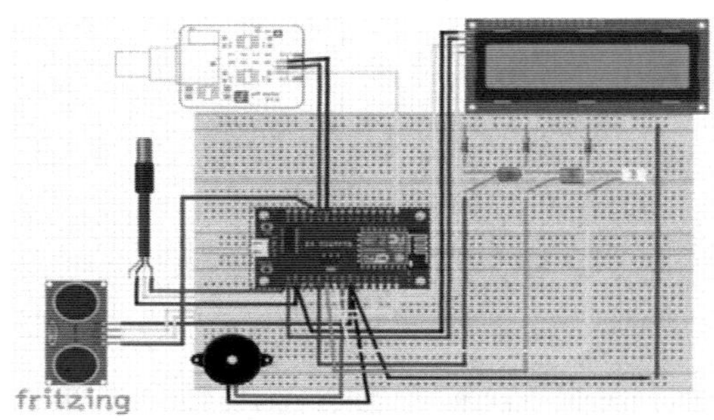

Figure 5. Circuit Design of The System

2.6. *Hardware Development*

2.6.1. NodeMCU ESP8266. Figure 6 shows the NodeMCU which is a development board featuring the popular ESP8266 Wi-Fi chip. As it turns out, it can program the ESP8266 just like any other microcontroller. Its obvious advantage over the Arduino or PIC is that it can readily connect to the Internet via Wi-Fi. The NodeMCU solves this problem by featuring 10 GPIO pins each capable of using PWM, I2C and 1-wire interface [21].

2.6.2. Ultrasonic sensor (JSN SR04T). Figure 7 shows JSN-SR04T ultrasonic sensor distance measurement module that can provide a 20 cm to 600 cm non-contact distance sensing function and ranging accuracy up to 2 mm. This waterproof ultrasonic sensor for locating fishes. The distance range for detecting underwater fish was set from 0 cm to 400 cm based on the depth of the fish pond.

2.6.3. Analog pH sensor (SKU: SEN0161). Figure 8 shows the pH sensor that is to measure water whether is acidic or alkalinity. The pH scale is a logarithmic scale whose range is from 0 – 14 with a neutral point being 7. The value above 7 indicates an alkaline solution and values below 7 would indicate an acidic solution. It operates on a 5V power supply and it is easy to interface with NodeMCU. The optimal pH range for fishes is 6.5 to 8.5. Low and high pH level harms fish especially the young fish in immature stages because they are extremely sensitive to pH levels. The young fish may die when the pH level is below 4 and above 9.

2.6.4. Temperature sensor (DS18B20). Figure 9 shows the temperature sensor, DS18B20. It is a waterproof sensor which detects the temperature of its surrounding. Most of the functions in this development are sensing variables in water, thus the DS18B20 being waterproof helps in detecting the temperature of water without being damaged. It can measure extreme temperature from -55 °C to 125 °C with very good accuracy. The normal range temperature for fishes is 29 °C to 35 °C.

Figure 6. NodeMCU microcontroller

Figure 7. Ultrasonic sensor

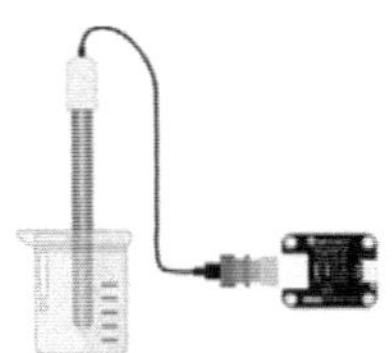

Figure 8. pH sensor

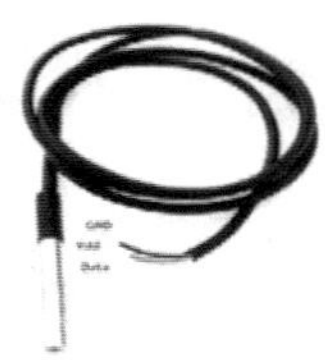

Figure 9. Temperature sensor

2.7. Software Implementation

2.7.1. Arduino IDE. Figure 10 is an Arduino IDE software which is an open-source software used to implement hardware and software. It is used to write and upload programs to Arduino compatible boards, but also, with the help of third-party cores, other vendor development boards. This software allows interference between nodeMCU and the sensors. The language used for this software is C++. The codes were programmed by using Arduino software.

2.7.2. Blynk application. Figure 11 shows the Blynk application. Blynk application is used to connect data in the microcontroller to the smartphone. For this project, the Blynk application connected with nodeMCU to the cloud using a Wi-Fi connection. It displays the data and the graph on the smartphone.

2.7.3. Fritzing. Figure 12 is Fritzing software. Fritzing is used to design the circuit beforehand and is used for simulation before transferring it onto hardware.

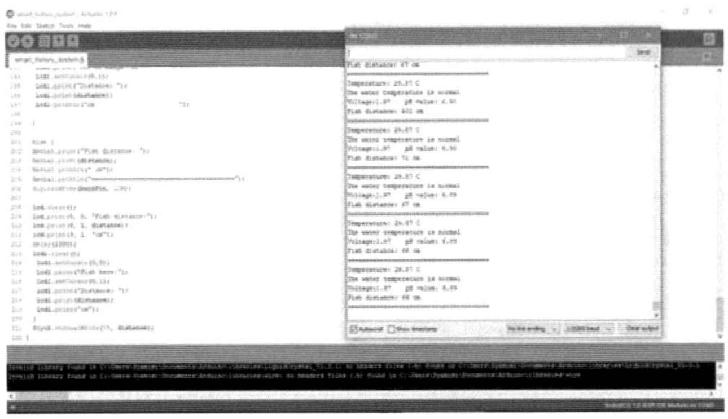

Figure 10. Arduino IDE software

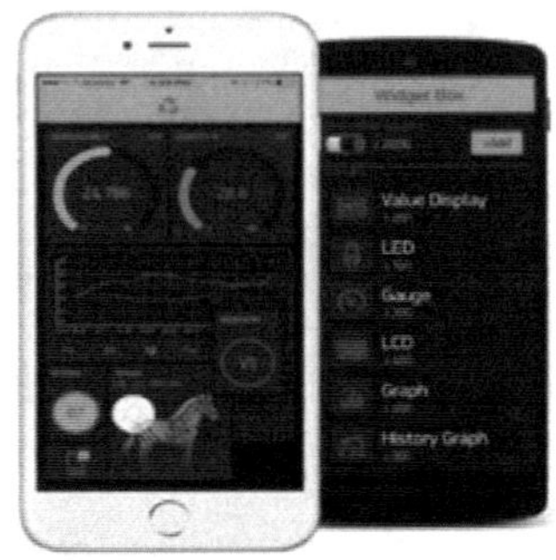

Figure 11. Blynk application

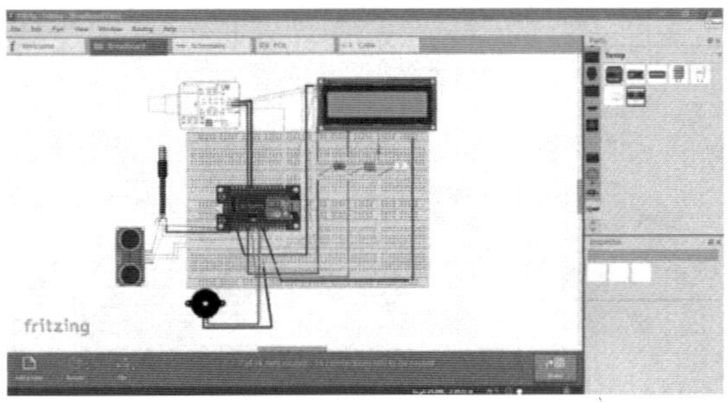

Figure 12. Fritzing software

3. Result and Discussion

The smart fishery and water quality monitoring system using Blynk application to be monitored and linked to one another via Wi-Fi using NodeMCU ESP8266 microcontroller. The measured data were displays through the Blynk application and the result is recorded in the database through a CSV file. The implementation of the project and the prototype are discussed below.

3.1. Hardware Implementation

Figure 13 shows the implementation of the project prototype after testing and troubleshooting. It is equipped with an ultrasonic sensor, temperature sensor and pH sensor as the input. The ultrasonic sensors detect the fishes underwater and at the same time, the temperature and pH sensor measure the water quality that will affect the growth of fishes. The data from monitoring will be sent to user email through a smartphone. The cost is low to build this prototype and easy to bring anywhere.

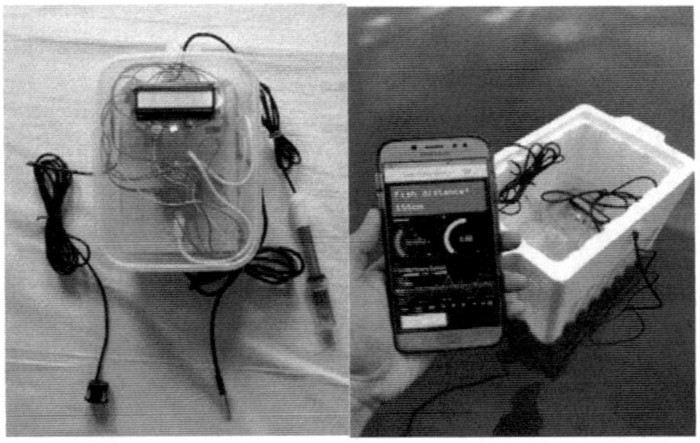

Figure 13. Setting up the hardware and the prototype in the field test at fish pond Section 24, Shah Alam

3.2. Software Implementation

In the software implementation, the data monitoring will be display in two sections which are in the serial monitor on the computer and the Blynk application on the smartphone. Figure 14 shows the data that will display in the serial monitor is the distance of fish, the temperature value and the pH value. If the water temperature is between 29 °C to 35 °C, "The water temperature is normal" will be displayed. If the water temperature is below 29 °C, the serial monitor will notify the user "Warning! The temperature is below 29". If the water temperature is above 35 °C, the serial monitor will notify the user "Warning! The temperature is above 35".

Figure 15 shows the design interfaces for the project using the widgets provided. The widget that is used is LCD that displays the distance of fish, two gauges that display temperature and pH value and a super chart that displays the graph of temperature, pH and ultrasonic. After getting the connection to the cloud using Wi-Fi, the LCD on the Blynk apps will display "Fishery and Water Quality System" in 5 seconds then the display the data monitoring from each sensor. The LCD will display "Out of Range" if the distance is above 400 cm based on the depth of the fish pond.

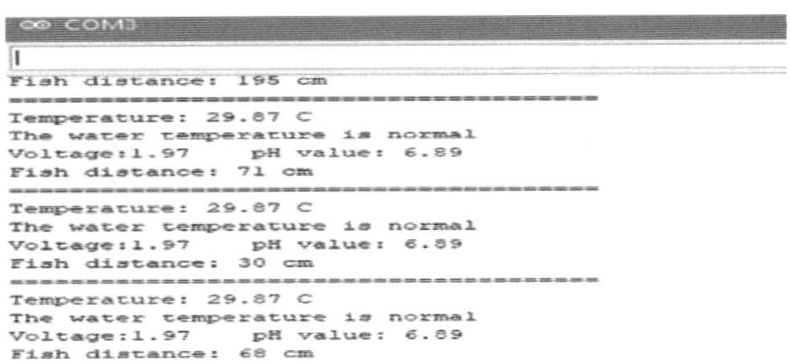

Figure 14. The data monitoring will be display in the serial monitor in the computer

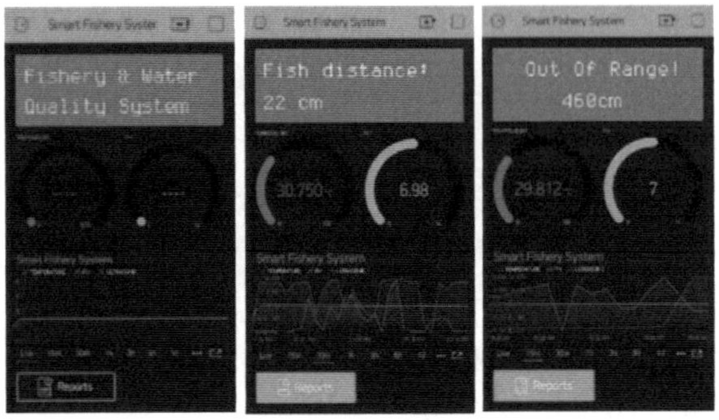

Figure 15. Data monitoring will be display on the smartphone.

3.3. Database

Sample measurements are collected throughout the day at three different times in the morning, noon, and evening for water temperature, pH levels and the fish distance show that this fish pond has high-quality water. From the measurement at the field, temperature value increase in the noon as the sun hits directly to penetrate the water's surface. PH level may rise from a morning value of 6.5 to an afternoon peak of 7 due to photosynthesis which removes carbon dioxide from the water and in the evening, respiration releases carbon dioxide and pH falls continually until daybreak. This condition is normal for the growth and health of fishes. The fish distance underwater always changes as a fish needs to find oxygen and food. As a result, a farmer need not hire a worker at their site, consequently, drive down operating costs and improve efficiency. The database is used to store the data tracked by the history graph. The data which is in a form of a CSV file will be converted into Excel format. All the information is needed to monitor for future improvement of the system.

3.3.1. Early Morning Measurement. Figure 16 shows data and the graph of three sensors in the Blynk application. From this graph, the value of temperature is constant at 29 °C at 10:30 am, the pH value remains constant with 6.9 and the highest distance of fish from prototype implementation and the depth of the fish pond is 370 cm and the lowest distance of fish is 80 cm.

Table 1 shows the list of the data value from the history graph in Blynk application for temperature, pH and fish distance value on 21/11/2019. In the early morning, the measurement value of temperature water is 29°C. The water in the early morning is cold and does not heat up because the sun is low and does not hit the water surface. The pH value remains constant from 6.5 to 6.9. During the day, underwater photosynthesis usually exceeds respiration, so pH rises as carbon dioxide is extracted from the water. The fish become lethargic and tend to be inactive in cold water so that fish swim more in shallow water for meals.

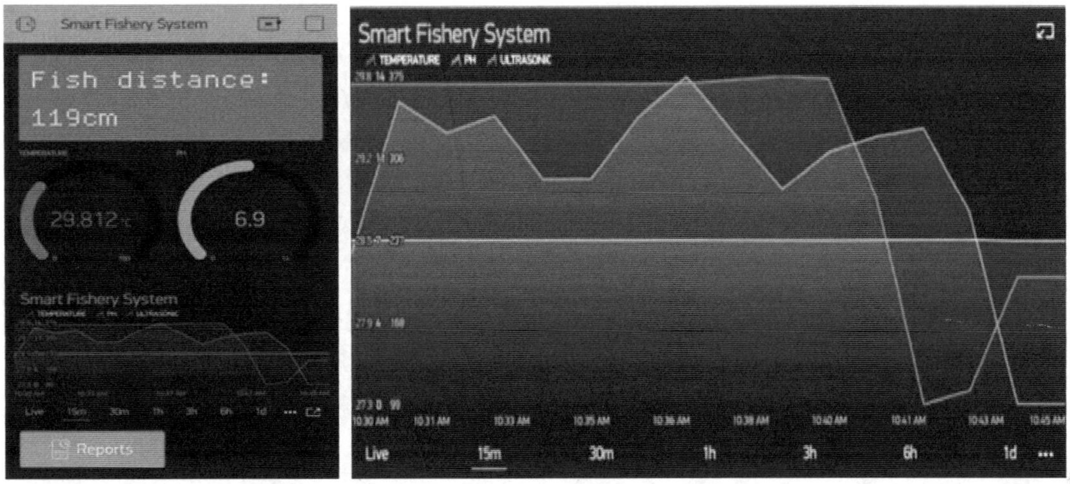

Figure 16. Data displayed and graph of temperature, pH and ultrasonic sensor in Blynk application in early morning

Table 1: The graph in Blynk application for temperature, pH and fish distance value in the early morning on 21/11/2019.

Time	Temp Value (°C)	pH Value	Distance Value (cm)
10:00 AM	29	6.5426	27.0869
10:05 AM	29.1713	6.8037	33.3478
10:10 AM	29.6825	6.8422	29.7826
10:15 AM	29.812	6.9	119.3441
10:20 AM	29.7113	6.8301	234.7826
10:25 AM	29.7284	6.8312	168.2608
10:30 AM	29	6.8677	369.6521
10:35 AM	29.75	6.8392	283.4782
10:40 AM	29.75	6.8337	304.6086
10:45 AM	29.75	6.8332	80.9545

3.3.2. Mid-day measurement. Figure 17 shows the graph of three sensors in the Blynk application. From this graph, the highest value of temperature is 35 °C and the lowest value of temperature is 29 °C the pH value remains constant with 7 and the highest distance of fish from prototype implementation and the depth of the fish pond is 390 cm and the lowest distance of fish is 22 cm.

Table 2 shows the list of the data value from the history graph in Blynk application for temperature, pH and fish distance value on 1/10/2019. During mid-day, the measurement value of temperature water has increased from 29 °C to 35.5 °C. The water at noon is hot than in the early morning because the sun hits directly the surface of the water. When the sun in the peak hour, plants resume photosynthesis and remove carbon dioxide from water, causing pH to rise again. The pH value has increased to 7. The movement of fish increases as the metabolism of fish accelerates, feeding and respiration increases in warm water so that fish swim more on the surface of the water.

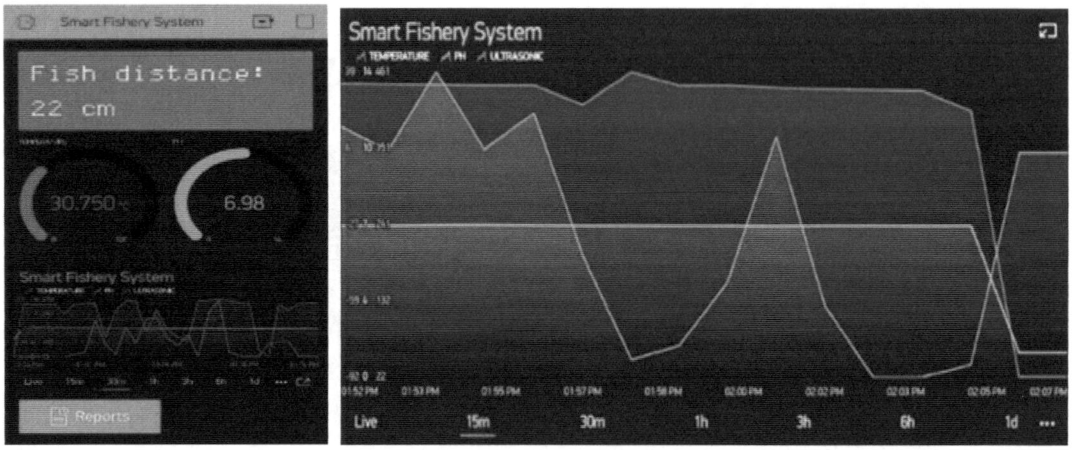

Figure 17. Data displayed and graph of temperature, pH and ultrasonic sensor in Blynk application during mid-day

Table 2: The graph in Blynk application for temperature, pH and fish distance value during mid-day on 1/10/2019.

Time	Temp Value (°C)	pH Value	Distance Value (cm)
12:55 PM	29.2721	6.9717	22.1053
1:00 PM	29.4548	7.1173	34.4444
1:05 PM	34.4176	7.0660	201.2105
1:10 PM	34.5756	6.9725	86.7500
1:15 PM	35.5329	7.0107	47.1579
1:20 PM	29.0031	6.9625	110.1579
1:25 PM	30.6880	7.0235	116.2778
1:30 PM	30.750	6.98	22
1:35 PM	29.4341	6.9568	147.1053
1:40 PM	32.6809	6.9536	45.5714
1:45 PM	34.8882	6.9768	90.2105
1:50 PM	34.2795	6.9340	333.4737
1:55 PM	33.0936	6.9907	390.5263
2:00 PM	33.1560	6.9332	110.3684
2:05 PM	30.1095	6.9470	39.7692

3.3.3. Evening measurement. Figure 18 shows the graph of three sensors in the Blynk application. From this graph, the value of temperature is constant at 30 °C, the pH value remains constant with 6.9 and the highest distance of fish from prototype implementation and the depth of the fish pond is 375 cm and the lowest distance of fish is 126 cm.

Table 3 shows the list of the data value from the history graph in Blynk application for temperature, pH and fish distance value on 20/11/2019. In the evening, the measurement value of temperature water is between 27 °C to 30 °C. The fish pond in the evening during rainy days makes the water became cold. The pH value remains constant from 6.9 to 7. As the sun begins to set in the late evening, photosynthesis decreases and eventually stops, so pH falls throughout the night as respiring organisms add carbon dioxide to the water. The distance of fish always changes as a fish needs to find oxygen and food underwater.

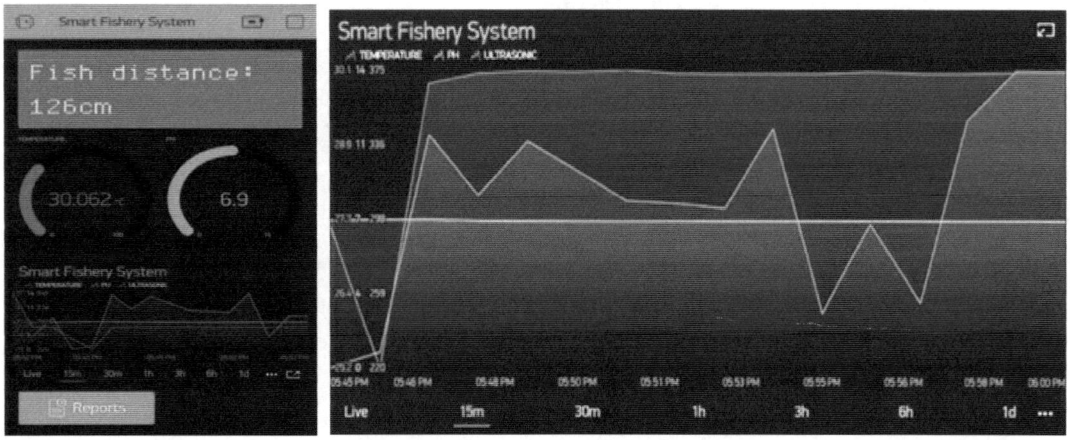

Figure 18. Data displayed and graph of temperature, pH and ultrasonic sensor in Blynk application at noon.

Table 3: The graph in Blynk application for temperature, pH and fish distance value in the evening on 20/11/2019.

Time	Temp Value (°C)	pH Value	Distance Value (cm)
5:25 PM	27.1795	6.9388	268
5:30 PM	30.188	7.0571	290.5909
5:35 PM	30.0626	6.9321	126.3913
5:40 PM	30.1414	6.9161	275.5652
5:45 PM	27.5463	6.9263	259.7391
5:50 PM	30.0849	6.8841	323
5:55 PM	30.062	6.9	248
6:00 PM	30	6.8953	375

4. Conclusion

In this project smart fisheries and water quality monitoring system using IoT in the pond, the objective will focus on the criteria of quality of water and detect the presence of fish was achieved. In this system, it will conclude that earlier prevention of water pollution can help to reduce the amount of fish dead. Monitoring systems using IoT will enable the detection of the quality of water and presence of fish earlier by sending the notification towards the fishermen or community. With technological growth, it will make it easier for fishermen to operate, such as discovering fish as their knowledge of which area that most fish landing based on water conditions. The fishermen can monitor the pond condition using a smartphone and this can help to warn them about pollution of water and at the same can detect the area of fish landing. The water pollution by the human can be prevented earlier thus, reduce the amount of fish dead.

For future works, it can be upgraded from ultrasonic sensors to the sonar with moving cameras based on deformable multiple kernels for more accuracy to detect and track the fishes underwater. Further on, the module for web interface should be developed for easier user interaction. In the future to expand the IoT water monitoring system by adding more parameters for the most secure purpose. The use of the LoRaWAN system as its low-powered sensors and wide coverage area measured in kilometers for future recommendation.

5. Acknowledgment
The authors would like to thank the Faculty of Electrical Engineering, Universiti Teknologi MARA (UiTM) for their valuable support. This research is partly funded by the Malaysian Government through UiTM under 600-IRMI/5/3/BESTARI (122/2018).

References

[1] Marc A B A, Lukey M J and Cerione R A 2018 The State of World Fisheries and Aquaculture, Meeting the sustainable development goals **367.**
[2] Joseph K B 1993 An Introduction to Water Chemistry in Freshwater Aquaculture (NRAC Fact) sheet No.170.
[3] Yang L and Dan S 2011 Assessment on the value of ecosystem service of important protected area for water supply of city *Second International Conference on Mechanic Automation and Control Engineering* 2962–66.
[4] Chee K Y, Shih H T P, Chee S L 2019 Contamination in Pasir Gudang Area, Peninsular Malaysia: What can we learn from Kim Kim River chemical waste contamination *Journal of Humanities and Education Development* **1** 84-7.
[5] Water Quality 2013 Water Quality in Sport fish ponds 165–70.
[6] Keyur K P, Sunil M P, Scholar P, Carloz S 2016 Internet of Things IoT Definition article *Int.J. Elect.Comput. Sci.* **6**(5) 10.
[7] Merna B November 2016 A Waterproof Ultrasonic Sensing System for Locating Fish in Underwater Area **3.**
[8] Bakar S, Ong N, Aziz H, Alcain J, Haimi W and Sauli Z 2017 Underwater detection by using ultrasonic sensor *AIP Conf. Proc.* **1885.**
[9] Kiruthika U, Raja S and Jaichandran R 2017 IoT based automation of fish farming *J. Adv. Res. Dyn. Control. Syst.* **9(1)** 50–7.
[10] Harun Z, Reda E and Hashim H 2018 Real time fish pond monitoring and automation using Arduino *IOP Conf. Ser. Mater. Sci. Eng* **340(1).**
[11] McCann J 2015 Monitoring and control of a marine ecosystem in an aquarium, thesis, (Murdoch University)
[12] Sutar G and Patil P 2013 Wireless Sensor Network System to Monitor the Fish Farm *J. Eng. Res. Appl.* www.ijera.com **3(5)** 194–7.
[13] Encinas Ruiz E, Cortez J and Espinoza A 2017 Design and implementation of a distributed IoT system for the monitoring of water quality in aquaculture *Wirel. Telecommun. Symp.*
[14] Chen H, Sung T and Lin G. Y 2015 Automated Monitoring System for the Fish Farm Aquaculture Environment *Proc. IEEE Int. Conf. Syst. Man, Cybern.* SMC 1161–6.
[15] Parra L, Sendra S, García L and Lloret J 2018 Design and deployment of low-cost sensors for monitoring the water quality and fish behavior in aquaculture tanks during the feeding process, *Sensors* (Switzerland) **18(3).**
[16] Mohammed A and Al-Mejibli I 2018 Smart monitoring and controlling system to enhance fish production with minimum cost *J. Theor. Appl. Inf. Technol.* **96(10)** 2872–84.
[17] Prasad N, Mamun K, Islam R and Haqva H 2015 Smart water quality monitoring system *(2nd Asia-Pacific World Congr. Comput. Sci. Eng. APWC CSE)* no. December 2016.
[18] Murad Z, Harun A, Mohyar N, Sapawi R and Ten S Y 2017 Design of aquaponics water monitoring system using Arduino microcontroller *AIP Conf. Proc* **1885.**
[19] Chuang C M, Hwang N J, Ye H J, Huang C S and Williams K 2017 Underwater Fish Tracking for Moving Cameras Based on Deformable Multiple Kernels *IEEE Trans. Syst. Man, Cybern. Syst.* **47(9)** 2467–77.
[20] Niswar M et al 2019 IoT-based water quality monitoring system for soft-shell crab farming *Proc. - 2018 IEEE Int. Conf. Internet Things Intell. Syst. IOTAIS 2018* 6–9.
[21] Ai-Thinker 2017 ESP-12F WiFi Module WiFi Module 3–11.

Portable Real-Time BMI Nutritional Advice

A D Ab Karim[1], N A Md Lazam[2*], N A Mohd Yahya[3] and S Abdul Rahman[4]

[1,] Student at Computer Engineering University Malaysia of Computer Science & Engineering, Cyberjaya Malaysia

[2,3,4,] Senior Lecturer at Computer Engineering University Malaysia of Computer Science & Engineering, Cyberjaya Malaysia

*Corresponding author: norazlinah@unimy.edu.my

Abstract. Overweight and obesity refer to body weight that is greater than what is considered normal or healthy for a certain height. Your body mass index (BMI) is one way to tell if you are at a healthy weight, overweight, or obese. The BMI is a measure based on your weight in relation to your height. Existing device only measure weight and height require the user to self-calculate their BMI based on the formula related to their weight and height. Furthermore, currently available systems cannot keep track user's previous BMI record and do not provide a guidance or nutritional advice based on user's current BMI. The solution to this is a proposed project which is an IOT based Portable Real-Time BMI nutritional advice that able to calculate user's BMI and keep current and previous user's record. The project aims to help people to monitor their BMI as a way of keeping a healthy weight. The objectives of this project are to develop a portable real-time measuring system to measure height and weight that can auto-calculate BMI and give nutritional advice using web-based application. The sensors used in this system are HX711 Load cell amplifier and Ultrasonic sensor. The system will perform weight and height measurement upon user's command. Data from both sensors will be sent to NodeMCU Wi-Fi module for BMI calculation. Virtual server is used to store the data from the system app. Apart from BMI calculation, this system is able to display user's previous BMI data as a performance tracker in monitoring their BMI. The application is also able to generate a nutritional advice based on user's current BMI.

1. Introduction

Malaysia only views obesity as a risk factor for non-communicable disease (NCDs) rather than a disease even though Malaysia ranked the fattest country in South-East Asia and the sixth in the Pacific Region [1]. World Health Organization (WHO) reported the burden of disease failed to highlight the burden of the disease attributable to overweight and obesity [2,3]. BMI is the convenient rule of thumb used to categorize a person as underweight, normal weight, overweight or obese based on tissue mass and height [4].

The Dietetic and Food Service Department, Ministry of Health Malaysia, is available to provide dietary counselling to patients and provide a meal to all in-patient by outsourcing the services such as manage the dry night rations supply to the staff on duty and monitor meals provided for patients, paramedics, and on-call doctors. Their mission and vision are to promote a healthy lifestyle for Malaysian [5].

Currently, there are many existing systems that can measure weight and height. But a system that can auto calculate user's BMI and store user's record and progress are still missing. Portable real-time

BMI nutritional advice system is designed to help user to monitor their BMI progress. The system will measure user's height and weight and auto-calculate the BMI based on the data receive from the sensors. The system allow user to either save their result or recalculate a new result. The system also will display user's BMI progress for them to easily keep track their performance. The system will provide nutritional advice and healthy tips based on user's current BMI result.

2. Research and Method

2.1. System Overview

Arduino is an open-source physical computing platform which implements the Processing Language on a simple input/output (I/O) board. The Portable Real-Time BMI Nutritional Advice is an Arduino-based system. It is an integration of software and hardware system which connect sensors with Wi-Fi module to a web-based application. This proposed system has two major modules which are hardware connectivity and web application.

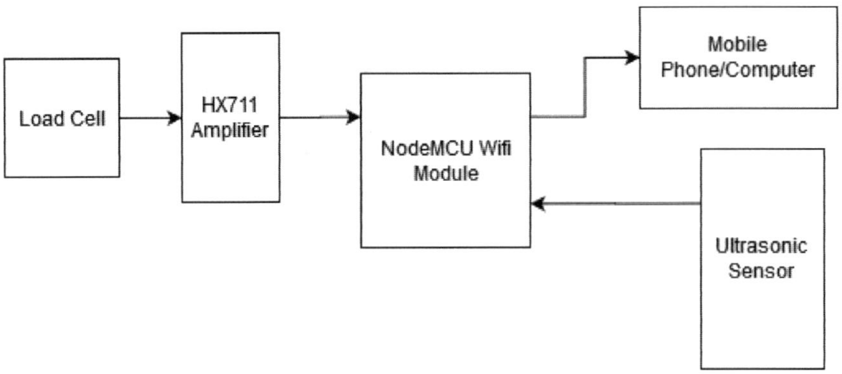

Figure 1. Block diagram of The Portable Real-Time BMI Nutritional Advice System

Figure 1 shows the block diagram of The Portable Real-Time BMI Nutritional Advice System. The hardware connectivity consists of 2 types of sensors that connect to a NodeMCU Wi-Fi module. One of the sensors is a load cell. The load cell is connected to HX711 Amplifier to measure user's weight. Weight measurement using this sensor implement the usage of Wheatstone bridge method [6].

A Wheatstone bridge is a configuration of four balances resistors with a known excitation voltage applied. R1 and R2 compose one voltage divider circuit while R3 and R4 compose the second voltage divider circuit. The output of a Wheatstone bridge is measure at V0. V_{EX} represent the voltage source supply throughout the circuit. The resistor is laid out in a diamond pattern. The circuit has 2 compressions and tangent resistors pair. When load is added onto the circuit, the compression and tangent resistors value will be changed in the opposite direction. The output voltage will be measured between the 2 corners. The output voltage will remain zero when no load is applied in the circuit. With proper calibration, the weight of load can be measured using this circuit configuration. This is illustrated in figure 2.

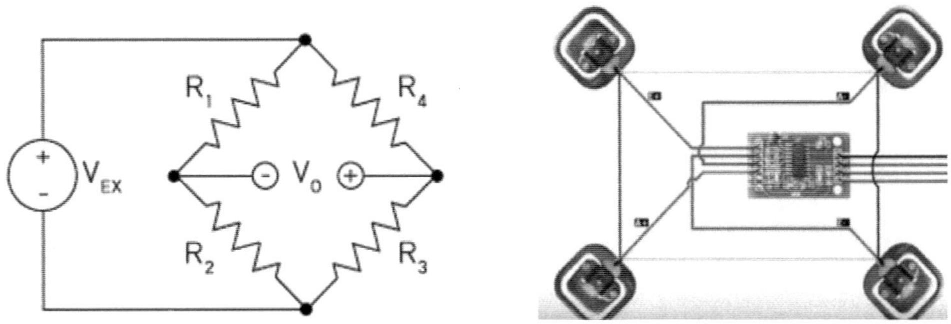

Figure 2. Wheatstone bridge configuration to measure weight

Ultrasonic sensor is used to measure the height of user. The basic principle of ultrasonic distance measurement is based on echo. When sound waves are transmitted the waves are returned to the origin as echo after striking on the obstacle. Time travel of both sounds on outgoing and returning from/to origin after striking on the obstacle are calculated. As the speed of the sound is known, calculation on distance can be done by using the formula in equation 1.

$$Distance\ (cm) = \frac{Time\ x\ 0.0034\ cm/s}{2} \quad (1)$$

By taking multiple readings to different distances, ultrasonic sensor can produce a fairly accurate object location [7,10]. The data receives will be send to the NodeMCU Wi-Fi Module and send to the web application. BMI calculation for this system use a new formula designed by Nick Trefethen [8] as shown in formula (4).

Web application executes three main functions to calculate BMI, to show performance, and to generate nutritional advice. The calculate BMI function is the integrated function connected to the hardware to obtain user's height and weight and auto-calculate user's BMI. The show performance function display user's record in a graphical presentation based on how many measurements they have taken. The last function is generating nutritional advice which offers health tips from authorized dietician based on user's current BMI [9]. Final module diagram is shown in figure 3.

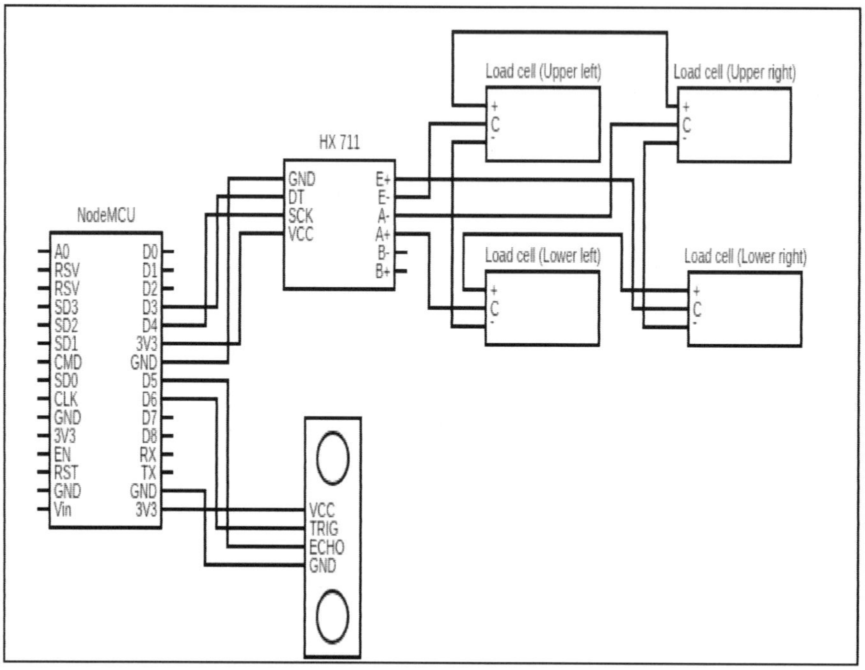

Figure 3. Final module integration for BMI nutritional advice.

The final prototype of this development is shown in figure 4. Four load cells will be embedded inside the scale frame on each edge. A foldable pole will be stored inside the pole case when it is uninstalled. The foldable bracket is installed on the tips of the poles and is positioned at 90° which an ultrasonic sensor is placed to measure user's height.

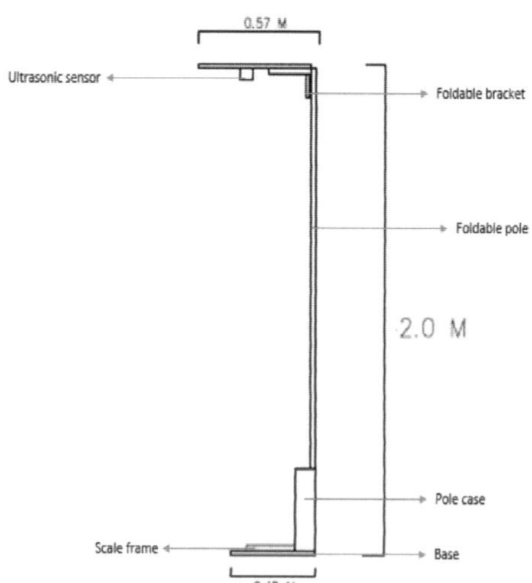

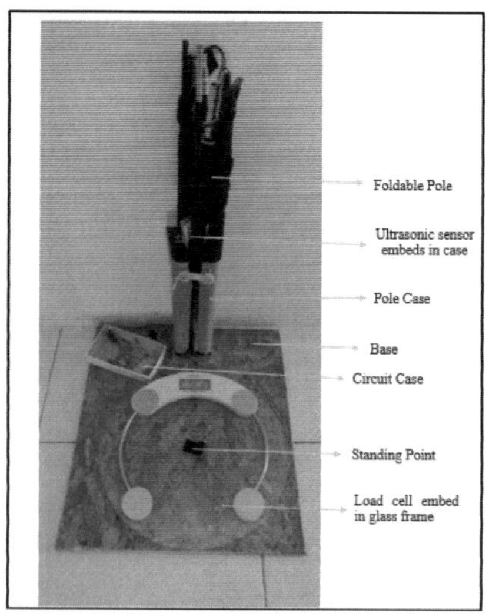

Figure 4. Side and front view of BMI nutritional advice

The system will perform weight and height measurement upon user's command. Then, data from both sensors will be send to NodeMCU Wi-Fi module for BMI calculation. Data received from Ultrasonic sensor is represented as Height using the following formula:

$$Height(m) = 2.15 - \frac{ultrasonic\ sensor\ value}{100} \tag{2}$$

Data received from load cell with HX 711 amplifier is represented as weight using the following formula:

$$Weight\ (kg) = \frac{HX711\ amplifier\ value}{100} \tag{3}$$

Finally, both sensors value recorded resulting in obtaining the BMI value using the following formula.

$$BMI = \frac{1.3\ x\ weight}{height^{2.5}} \tag{4}$$

2.2. System Flow Diagram

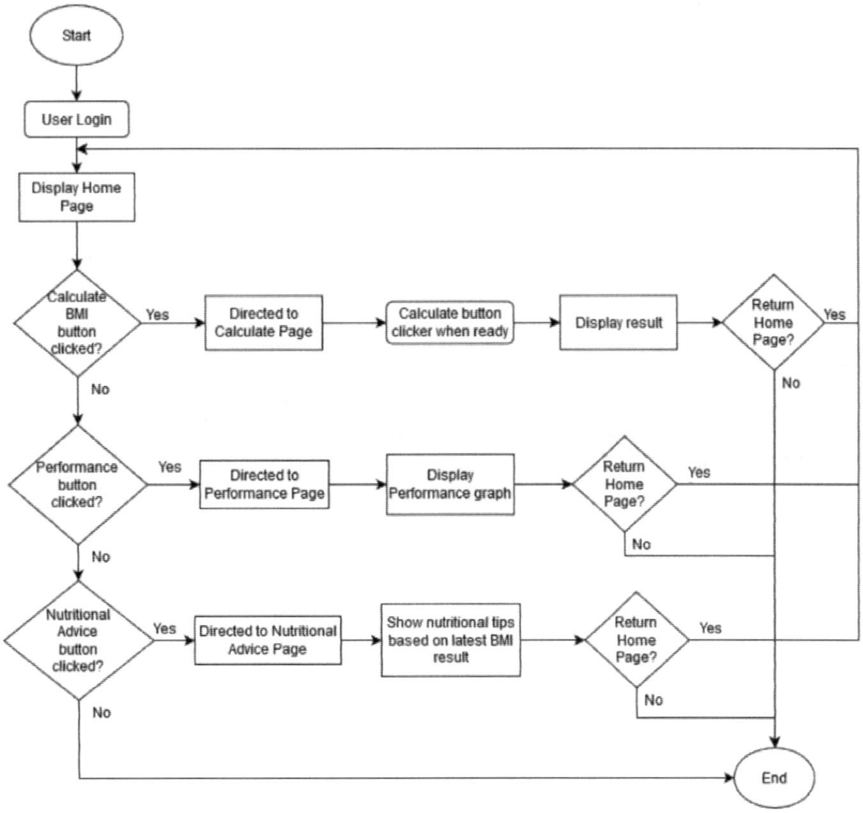

Figure 5. System flowchart

Figure 5 shows the flowchart diagram of the system. Once user browses the system's designated URL, Login page will be displayed. Once user has clicked either one account, they will be directed to the application Home page. On this page, three main buttons will be displayed which is "Calculate BMI", "Performance" and "Nutritional Advice". User can choose which function they wants to perform. For every page that user is directed to, the system will provide a return home page button shown as "Home".

3. Result and Discussion

Figure 6 shows the Homepage of the system. Main Page offers three functions namely Calculate BMI, Performance and Nutritional Advice. The result of user's BMI is shown once the user's height and weight are measured. It summarizes height, weight, BMI and status in one page. The system is able to store the data in MYSQL database if the users wish to save the data. This is shown in figure 7.

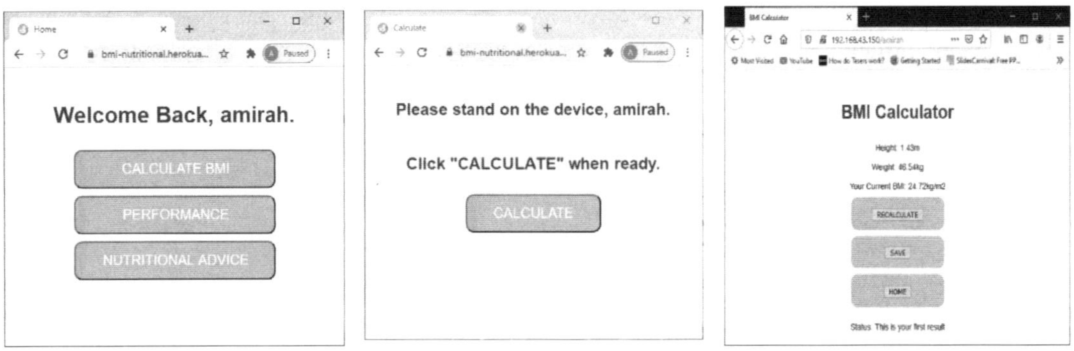

Figure 6. BMI calculation page

	id	user_weight	user_height	user_bmi	created_at
Edit Copy Delete	4	46.54	1.43	24.72	2020-08-05 13:35:31
Edit Copy Delete	3	47.7	1.43	25.36	2020-07-30 06:14:20
Edit Copy Delete	2	49.2	1.43	26.16	2020-07-10 00:00:00
Edit Copy Delete	1	50.2	1.43	27.75	2020-06-30 13:25:50

Figure 7. Database record

User can view their progress, by clicking "Performance" button and it will direct them to a page where the graph is generated based on the user's record. The graph will show user's BMI and date of when the measurement is taken as shown in figure 8.

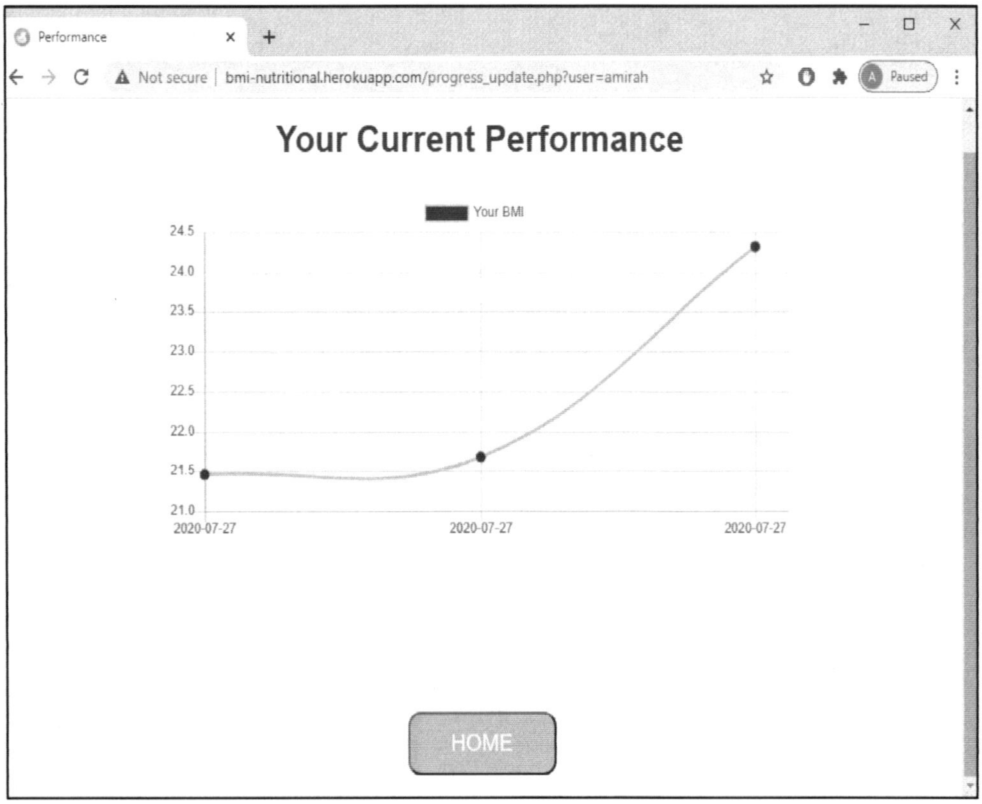

Figure 8. Data represent in graph

A nutritional advice page will be displayed when the user clicks the "Nutritional Advice" button. There are four (4) different health tips based on BMI categorization which are Nutritional Advice for Underweight, Normal, Overweight and Obese. The tips displayed will shows user's current BMI and a list of health tips for user to follow to maintain or improve their healthy weight.

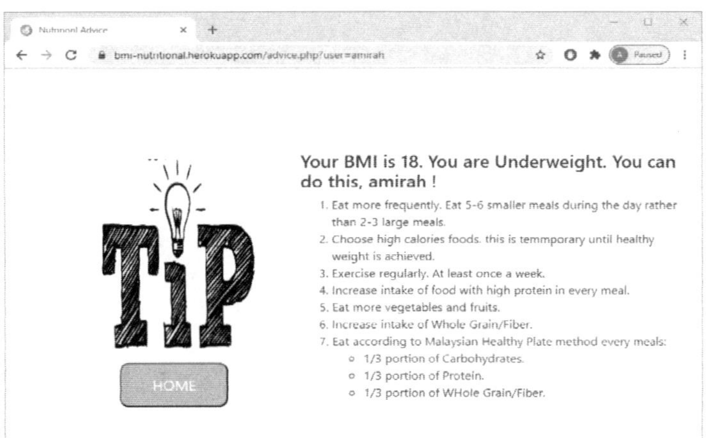

Figure 9. The generated nutritional advice for Underweight

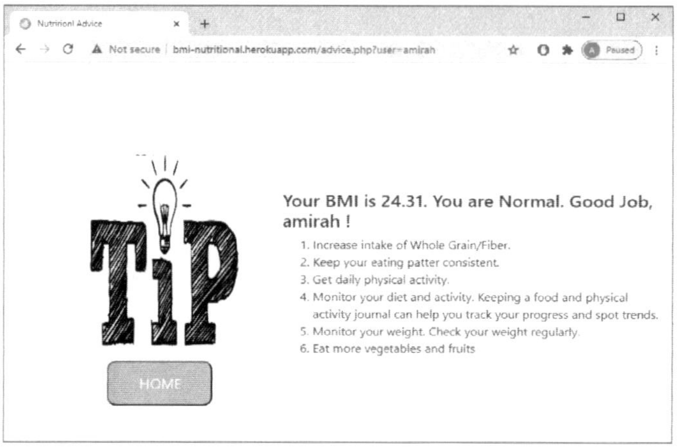

Figure 10. The genrated nutritional advice for Normal Weight

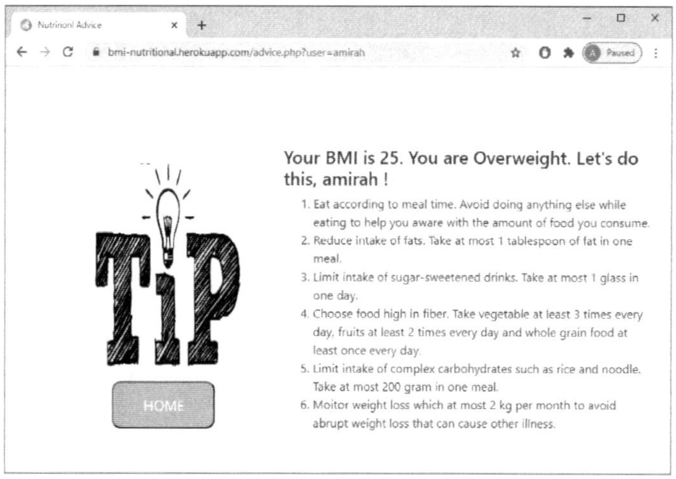

Figure 11. The generated nutritional advice for Overweight

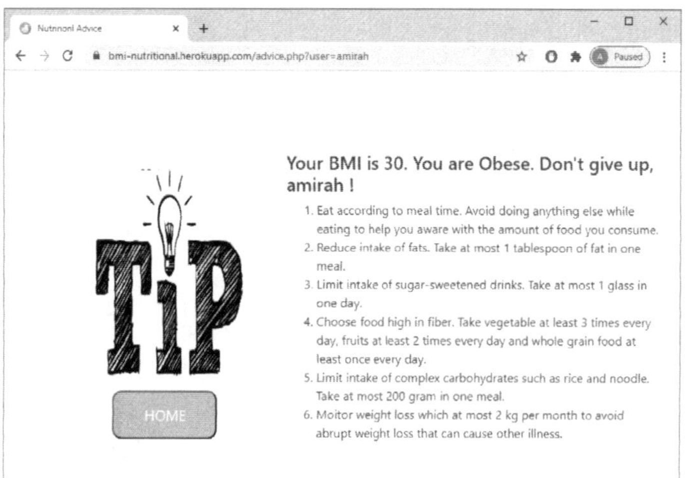

Figure 12. The generated nutrional advice for Obese.

4. Conclusion

The Portable Real-Time BMI Nutritional Advice might be a redundant of the same technologies that have existed. Nevertheless, this system aims to help people to be aware about their BMI (ideal weight) in order to remain healthy. The system is able to integrate two calibrated instrument that measure weight and tall into a comprehensive system that can auto calculate BMI and stores the data into a local database. The data can be retrieved so that users are able to track their progress. The system makes use of Arduino sensor and the add on feature able to display the data in a graphical form. Nevertheless, there are always rooms for improvement. For future work, the system should ideally be equipped with robust operating capabilities such as battery usage instead of power supply and developed in mobile based application for user to easily access. The Arduino has been used widely in IOT application [11-15].

References

[1] Verma R K, Chua G and David S R 2013 Obesity and overweight management in Malaysia and Singapore: *Progress on right track J. Clin. Diagn. Res.***7(12)** 3124-3125.

[2] Mathers C D, Lopez A D, Murray C J L 2006 The Burden of Disease and Mortality by Condition: Data, Methods, and Results for 2001. *Global Burden of Disease and Risk Factors* 45-93.

[3] J-H Park, S-J Yoon, H Lee, et al. 2006 Burden of disease attributable to obesity and overweight *Korea. Int J Obes (Lond)* **330(11)** 1661-1669.

[4] Nuttall, Frank Q 2015 Body Mass Index Obesity, BMI, and Health A Critical Review *Nutrition Today* **50(3)** 117-128.

[5] Hospital Putrajaya Official Website 2019 Dietetic & Food Service. Retrieved 12 December 2019

[6] Müller I, Machado de Brito R, Carlos E P, Brusamarello V J 2010 Load cells in force sensing analysis - Theory and a novel application *IEEE Instrumentation and Measurement Magazine* **13(1)** 15-19.

[7] Mohammad T 2009 Using Ultrasonic and Infrared Sensors for Distance Measurement. *World Academy of Science, Engineering and Technology* **3** 2009-03-26.

[8] Nick, Trefethen 2019 New BMI (New Body Mass Index). *Numerical analysis. Economist*

[9] Sharkawi I, Mohamed Z and Rezai G 2014 Healthy Eating: The Preventive Factors among Malaysians *J. Bus. Econ. Manag.* **2(4)**.

[10] Neeraja Soni C H, Sarita C H, Maheshwari S, Sahu B K, Jain B and Shrivastava G 2017 Distance Measurement using Ultrasonic Sensor and Arduino *Int. J. Eng. Comput. Sci. IJESC* **7(3).**

[11] Hazizan A, Lazam N A M, Hassan N I 2019 Development of Child Safety Car Alert System Using Arduino and GSM Module *IOP Conf. Ser. Mater. Sci. Eng.* **834(1)** 012071.

[12] Badamasi Y A 2014 The working principle of an Arduino *11th Int. Conf. on Elect., Comput and Comput. (ICECCO).*

[13] Bento A C 2018 IoT: NodeMCU 12e X Arduino Uno, Results of an experimental and comparative survey *Int. J. Adv. Res. Comput. Sci. Manag. Stud.* **6(1).**

[14] Louis L 2016 Working Principle of Arduino and Using it as a Tool for Study and Research *Int. J. Control. Autom. (IJCACS)* **1(2).**

[15] Kaswan K S, Singh S P, Sagar S 2020 Role Of Arduino In Real World Applications *Int. J. Sci. Res.* **9(1)** 2277-8616.

IOP Conf. Series: Materials Science and Engineering 1176 (2021) 012018 doi:10.1088/1757-899X/1176/1/012018

High Frequency MOSFET Charging Circuit Using Series Parallel DC-DC Resonant Converter for Wireless Power Application

N A A Nuawi[1], M F M Zin[1]*, E A Kadir[1] and M N Seroji[2]

[1]Faculty of Electrical Engineering, University Teknologi MARA, Terengganu (Dungun Campus, 23000 Dungun, Terengganu, Malaysia)

[2]Faculty of Electrical Engineering, University Teknologi MARA, Shah Alam (40450 Shah Alam, Selangor, Malaysia)

*Corresponding author: mazratul204@uitm.edu.my

Abstract. This paper proposes a new circuit topology of high frequency switching DC-DC Series Parallel Loaded Resonant Converter (SPLRC). The circuit is proposed to be used for a wireless power transmission application that required fast-charging operation. Most of the components used here are low-cost and commercially available in the market. Then, the purpose circuit is transformed into a different equivalent circuit that can efficiently analyze AC and DC modes. For each model, the operation for the converter is first described and explained. A full detailed circuit simulation is then carried out to find out the workability of the purpose converter. From the simulation, optimization is carried out. Here the components value is obtained. MATLAB R2018A is used for simulation purposes rather than being tested again using PSIM software to further prove the validity of the model. All the finding is explained and discussed, here, from the simulation result, it is confirmed that a high-frequency switching circuit is operated as expected.

1. Introduction

Nowadays, high frequencies driver in wireless power transfer is widely used, especially since the advent of amplitude-modulated radio in the early 20th century. Small and medium-sized autonomous electronic devices that involve periodic battery recharging have broad applications in portable communication devices, office equipment, pharmacy, automobile, and so on. Nevertheless, due to the presence of friction contacts, the use of conventional conductive power interfaces affects the efficiency of devices that restrict their use or renders them totally unacceptable, as in implants [1].

Wireless power always about freedom and convenience from power cords; it also influences the way suppliers can manufacture devices. An efficient wireless power transfer method would enable advances in such diverse areas as embedded computing, mobile computing, sensor networks, and micro robotics [2]. To operate wireless power transmission, the signal's frequency to be transmitted and received needs to be high compared to the standard frequency of 50Hz. Most hard switching converters have low-efficiency difficulties, with bulky magnetics resulting in a low frequency of switching [3]. It seems high power losses is depending on the high switching technique [4]. Hard-switching in DC-DC converters is prone to high switching losses as it requires switching on and off with the full load current. Nevertheless, switches only turn on and off in soft-switching resonant DC-DC converters when the voltage across or the current through them is zero, thus reducing or eliminating switching losses [5].

Wireless power transfer has been acknowledged as one of the marginal methods to transfer power quickly and conveniently. Another work has proposed a low-cost project. Previous researchers have presented a variation of technique and approach as a solution to the problem. Mainly in this project, an advanced power converter will be developed in wireless power transfer. The driver consists of a High-frequency DC-AC converter with a design example of a 240 V supply than using an adapter to produce 24 V DC input. Generally, the energy dissipation in the power inverter increases with the operating frequency [6]. After that, a switching frequency of 120k Hz and 600 W power output converter will be used as the second stage of an AC to DC converter. Hence, an advanced topology of the DC to AC converter for the transmitter side and an AC to DC converter topology for the receiver side will be designed.
The research study focused on a high-speed switching circuit using high capacity gate and low voltage value Mosfet, which commercially available in the market. The aim is to develop and design a high-frequency Mosfet driver that can drive Mosfet, mainly within class E configurations. Compared to other switching power amplifiers, Class-E has fewer components, yielding high reliability [7]. Furthermore, to develop the proposed high-speed switching circuit and test for wireless power transfer application.

2. Basic Principal of Resonant Converter

2.1 Series Loaded Resonant Converter (SLRC)

For SLRC, the resonant current rather than the voltage is rectified and smoothed to form DC output in figure 1. It is again assumed that the Mosfet body diodes include the half-bridge antiparallel diodes. Since the rectifier is current-fed, the output filter only consists of a shunt-connected capacitor. The current-fed happens because the filter must have the capacity to absorb abrupt changes in the rectifier output current. It is assumed that there is negligible ripple voltage across the output smoothing capacitor. The sinusoidal rectifier input current defines that the operation of the circuit. The converter also a strong candidate with the capability for current source output behavior with a high impedance output. Nonetheless, SLRCs behave differently with constant current input compared to traditional constant voltage input[5]. Meanwhile, the converter's power throughput is controlled by operating the converter closer to, or further from, the natural resonant frequency, thereby effectively varying the impedance of the resonant tank.

2.2 Parallel Loaded Resonant Converter (PLRC)

Based on figure 2, the PLRC showed that the MOSFET body diodes form the half-bridge antiparallel diodes. Previous studies showed huge potential developments to capable of operating under zero-voltage switching (ZVS) turn-on of switches and zero-current switching (ZCS) turn-off of diodes and works in a narrow switching frequency [8]. The sinusoidal capacitor voltage is rectified by the diode bridge and smoothed by the L-C filter to form the DC output. The rectifier and smoothing circuit operation is identical to a mains-fed equivalent, although the input frequencies are very different. DC load voltage, V_o has a value of $(2 / \pi)$ V_c, where V_c is the sinusoidal V_{in}'s peak.

$$V_O = (\frac{2}{\pi})V_C \tag{1}$$

Rectifier current, Irect is a square wave of $\pm I_o$, which is in phase with V_c, while output current is the DC load current. Therefore, current in the resonant tank is a sum of sinusoidal current I_{xy} and the rectifier's square wave current.

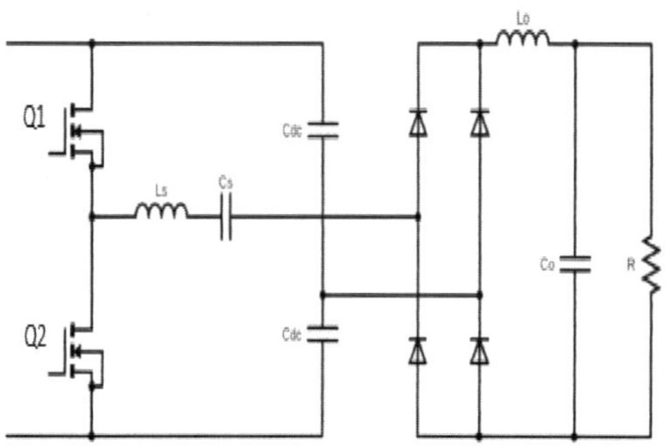

Figure 1. Series Loaded Resonant Converter.

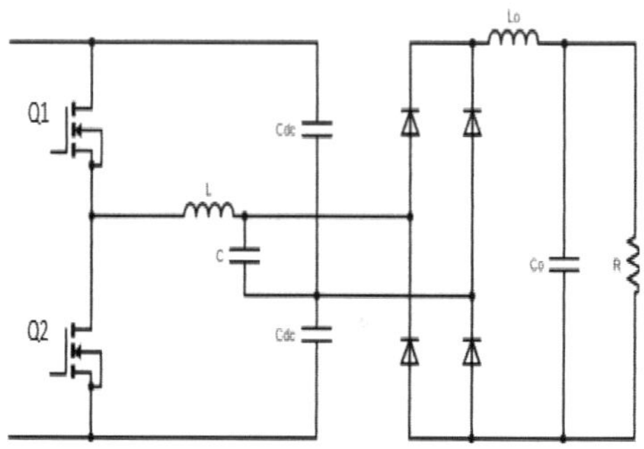

Figure 2. Parallel Loaded Resonant Converter.

2.3 Series Parallel Loaded Resonant Converter (SPLRC)

To overcome the limitations of the basic SLRC and PLRC, the combined series-parallel-loaded converter, as shown in figure 3, is frequently used. The converter is produced by splitting the resonant capacitor into two components C_r, and C_p. The load is connected in parallel with C_f. The principal features of the SPLRC are a voltage-fed rectifier circuit with an L-C filter. The filter capacitor ripple current may be controlled by the filter inductor's appropriate choice, and therefore the circuit is suitable for high output applications. Next, at heavy load, the parallel capacitor Cp is effectively shorted by the rectifier circuit's relatively low input resistance. L-Cr dominates the resonant tank's operation, and the converter takes on the characteristics of a series-loaded converter. In particular, the resonant current I_{xy} falls as the load is reduced, resulting in good part-load efficiency.

Figure 3 shows that the basic series-parallel loaded resonant converter used for the proposed project. Only a fundamental component is considered for the analysis as in figure 5, so equivalent ac resistance, R_{eff} is;

$$R_{eff} = \frac{\pi^2}{8}R \tag{2}$$

Where R is load resistance.

Furthermore, I_{xy} is insensitive to input voltage for a fixed output voltage and current. At light load, Cp is no longer effectively shorted since the rectifier input resistance is increased. The converter, therefore, takes on the characteristics of a parallel-load converter - most importantly, that the output voltage may be regulated at the light and zero loads.

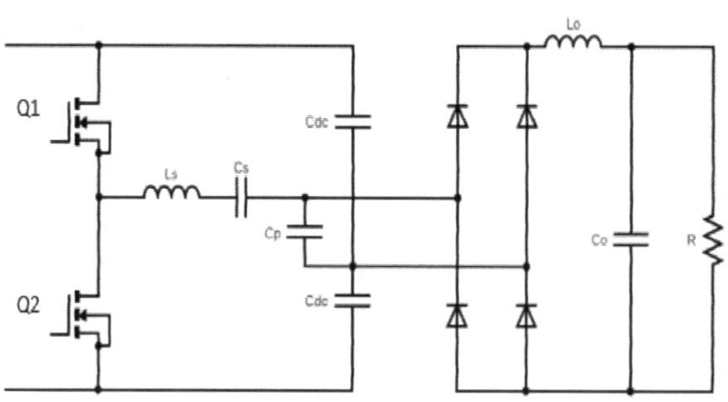

Figure 3. Series Parallel Loaded Resonant Converter

3. Design of Converter

AC analysis or fundamental frequency analysis (FFA) is the most popular, simple, and effective method of analyzing DC-DC resonant converters [4]. In this method, the energy transfer from source to the load is assumed to be contributed only by the fundamental component while harmonics are neglected. Though this method is only approximate and not accurate, it is preferred owing to its simplicity, the lesser computation time for solving, and practical ease of implementation. Hence, researchers continue to use this method for analyzing resonant converters [4], [5]. The assumption made in the converter:

1. All the diodes and switches are ideal, and switches communicate linearly.
2. The passive components behave linearly.
3. All resonant components and filter capacitors assumed lossless.

The converter used by following the specifications bellow:

- $V_{in} = 240$ V
- $V_{dc} = 24$ V
- $f_s = 120$k Hz
- $P_o = 600$ W
- $C_{DC} = 500\mu$ F
- $C_o = 500\mu$ F
- $L_o = 1\mu$ H
- Power Switches = MOSFETs

Figure 4 shows the proposed converter of DC-DC SPLRC for wireless power transmission. The design of a resonant converter often starts by designing the resonant network.

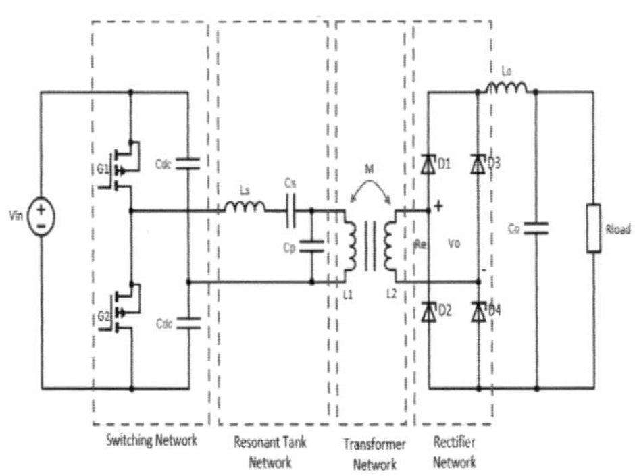

Figure 4. Simulation for proposed converter using SPLRC.

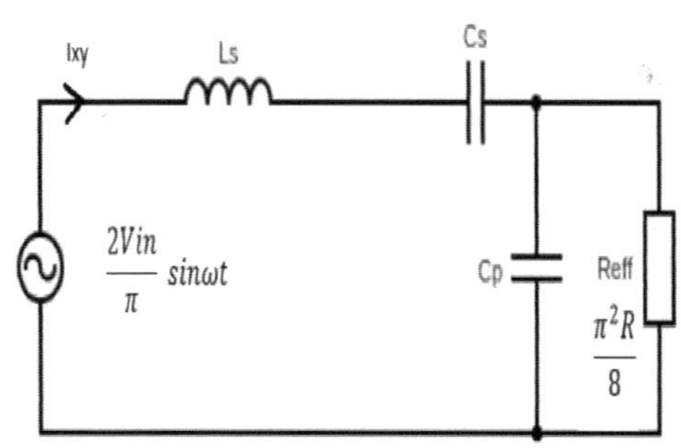

Figure 5. Fundamental frequency equivalent circuit for the half-bridge of SPLRC.

Resonant frequency, ω_o:

$$\omega_o = \frac{1}{\sqrt{L_s C_s}} = \frac{\omega_s}{\omega_n} \tag{3}$$

- L_s = Inductor series of a resonant tank
- C_s = Capacitor series of a resonant tank
- ω_s = Switching frequency

Where normalized frequency, ω_n:

$$\omega_n = \frac{\omega_s}{\omega_o} \tag{4}$$

Quality factor, Q, of the SPLRC:

$$Q = \frac{\omega_o L_s}{R_l} \tag{5}$$

As component stresses are directly a function of Q, so value of Q should be chosen as low as possible.

$$\left|\frac{V_o}{V_{in}}\right| = \frac{0.5}{\left|\frac{\pi^2}{8}\left[1 + \frac{C_p}{C_s}\left[1 - \omega_n^2 + jQ\left[\omega_n - \frac{1}{\omega_n}\right]\right]\right]\right|} \tag{6}$$

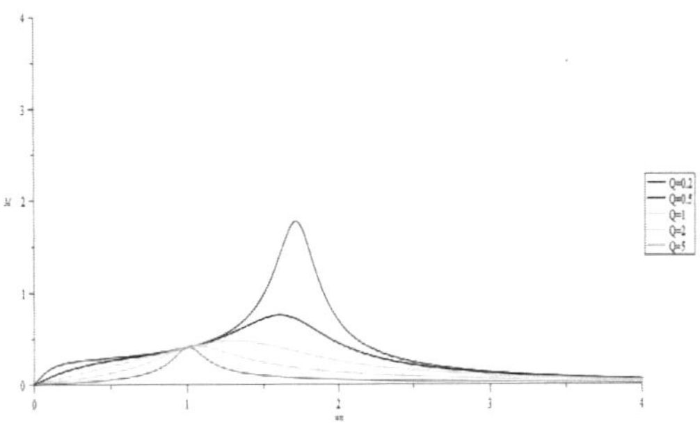

Figure 6. Gain curve for SPLRC for $C_p/C_s = 0.5$.

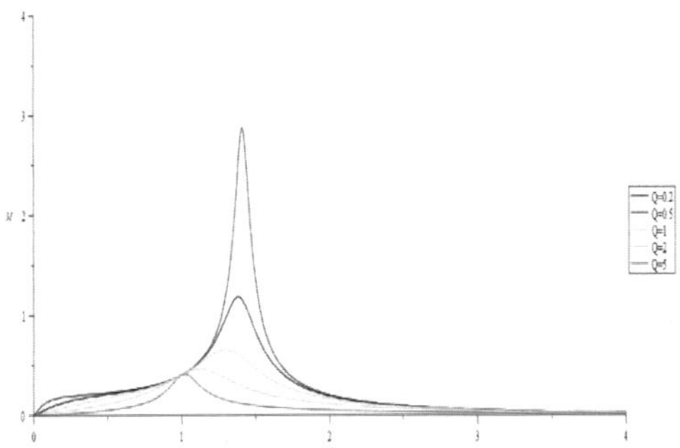

Figure 7. Gain curve for SPLRC for $C_p/C_s = 1$.

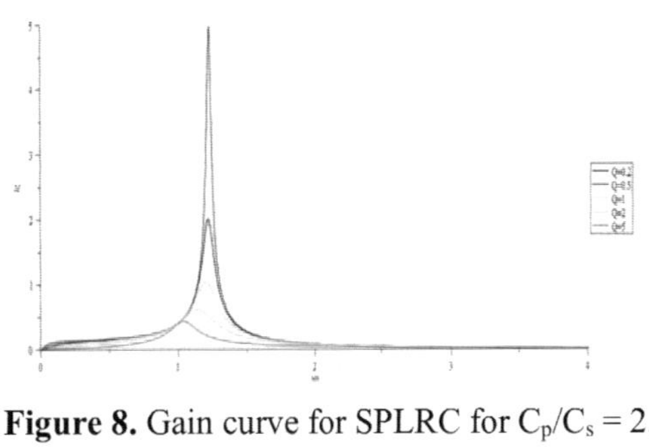

Figure 8. Gain curve for SPLRC for $C_p/C_s = 2$.

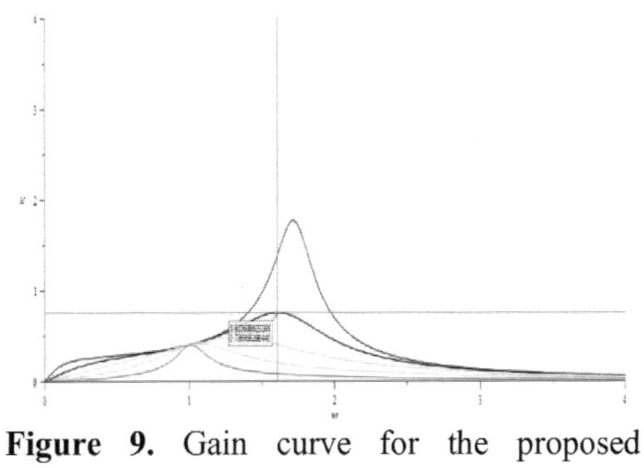

Figure 9. Gain curve for the proposed converter by using $C_p/C_s = 0.5$

The quality factor to design the proposed converter can be analyzed from figure 6 to figure 9, gain versus normalized frequency, for SPLRC when the value of C_p/C_s = 0.2, 1, 2. Thus, the value of normalized frequency, ω_n for the proposed design, can be obtained from the converter's gain value. Since the value of gain for the converter is 0.7, the value for ω_n is 1.6037 for Q = 0.5, as shown in figure 9. The lower value of Q corresponds to light load condition and higher value for full load condition.

4. Result and Discussion

This section represents the simulation of the proposed SPLRC. DC power supply of 240 V was step down by using adapter to 24 V at 50 Hz is used as shown in figure 10 by using PSim. The simulation work on suggested SPLRC is performed using 120k Hz of switching frequency, f_s. The value of Q is 0.5 with ω_n of 1.6037 as stated from previous section. The components value for the design was calculated by using MATLAB R2018A as well as experimental components value are shown in table 1.

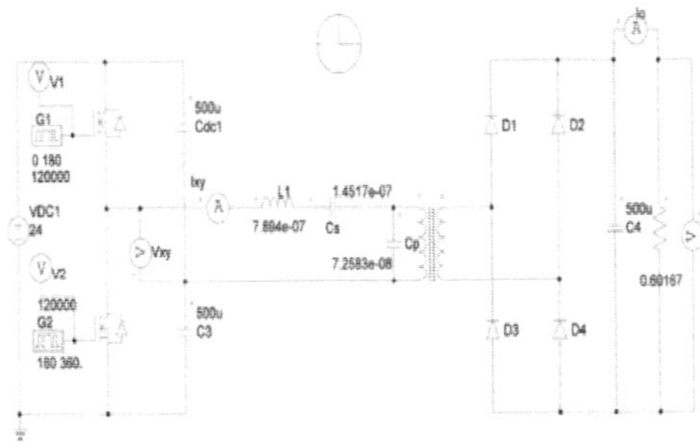

Figure 10. Simulation of SPLRC by using PSIM.

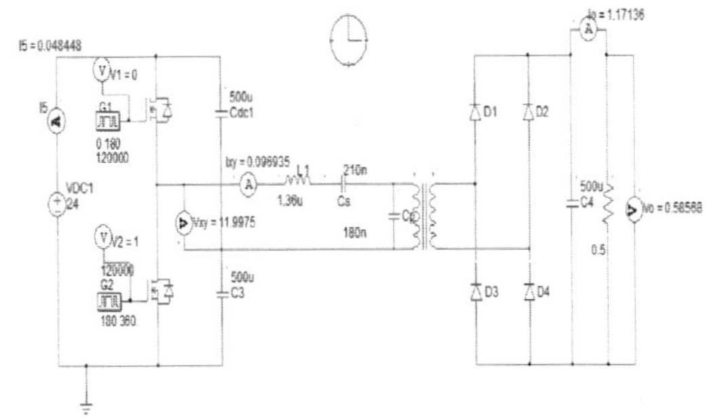

Figure 11. Experimental simulation of SPLRC by using PSIM.

Table 1. Components value of SPLRC from calculation and simulation

	Value (Calculated)	Value (Simulation)
$\mathbf{V_s}$	240 V	240V
$\mathbf{V_{in}}$	24 V	24V
$\mathbf{V_o}$	19 V	19V
$\mathbf{P_o}$	600 W	600W
Q	0.5	0.5
$\mathbf{F_s}$	120000 Hz	120000 Hz
ω_s	753.982 kHz	753.982 kHz
M	0.79167	0.79167
ω_n	1.6037 Hz	1.6037 Hz
R	0.60167 Ω	0.5 Ω
$\mathbf{R_e}$	0.74228 Ω	-
ω_o	470.152 kHz	-
$\mathbf{C_s}$	0.1452 μF	210nF
$\mathbf{C_p}$	0.7258 μF	180nF
$\mathbf{L_s}$	0.7894 μH	1.36μH
$\mathbf{C_o}$, $\mathbf{C_{DC}}$	500 μF	500 μF

A straightforward technique to understand resonant topologies' characteristics is by looking at their gain curve plot, as discussed in section B part III. The curves show that if the resonant converteroperates above the resonant frequency (the right side of the gain curve), the controller won't regulate the output voltage. Once the opposite condition happens, the same is true. The curvesalso imply that Q decreases as the load increases. The gain curve becomes flat, and, hence, major frequency modifications are expected to ensure the desired output voltage. Therefore, open circuit control is unlikely even if there is no resonant peak or selectivity. The high Q value isdesirable to achieve better control in the series resonant converter. The trade-off, however, is that if Q is too high, then the output control mechanism becomes very non-linear. The minorshift in the switching frequency will produce a significant change in the output voltage, and the operatingfrequency also could be shifted quite near to the resonant peak by the circuit. The condition destabilize the power supply in turn.

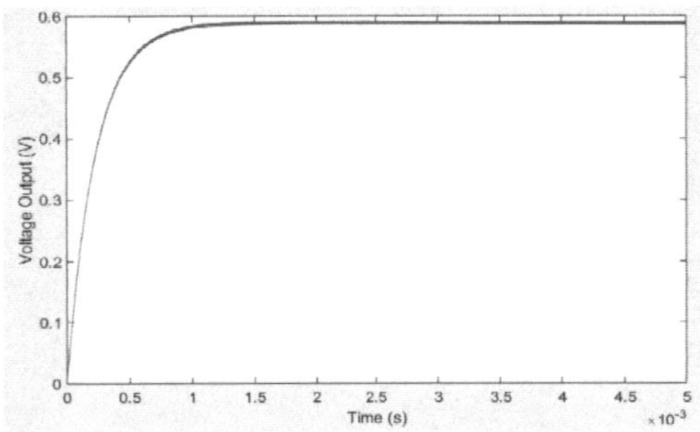

Figure 12. Output Voltage.

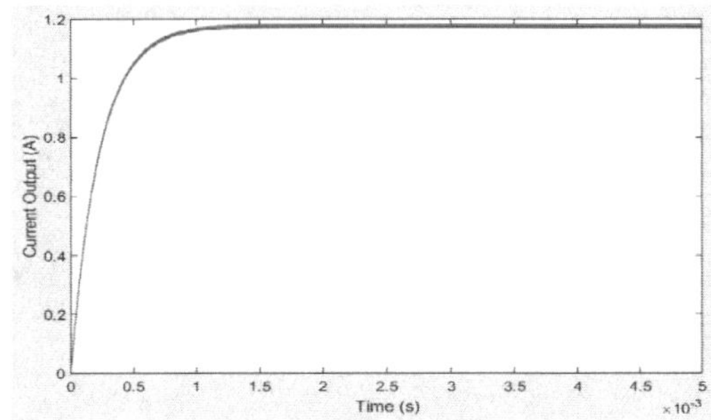

Figure 13. Output Current.

The converter will work with a very low Q to avoid the non-linear gain curve characteristics, so the curve should be flat at virtually any frequency, as shown in figure 12. Even more observation of figure 12 poses the issue of regulating the output voltage; how can you control the output when thevoltage stays the same at either frequency? The proposed design solves this problem. A parallel capacitor is placed after series of inductors and capacitors at a resonant tank. To control the output voltage, the filter capacitor ripples current is governed by the appropriate choice of the filter inductor. Therefore the circuit is suitable for high output applications, as shown in figure 13. It takes on a parallel-load converter's characteristics, most notably that the output voltage may be regulatedat the light and no load. In the case of the two switches or Mosfets operate above the resonant frequency,Fo the resonant tank now acts as inductive impedance resulting in the resonant current, I_{xy} lags the resonant voltage, V_{xy} as illustrated in figure 14.

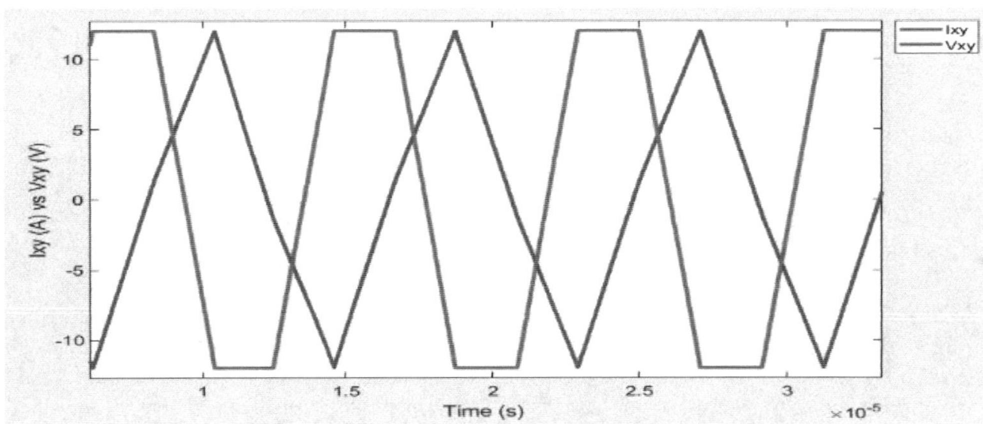

Figure 14. Resonant tank current, I_{xy}.

Due to the resonant current I_{xy}, the antiparallel diode D1 conducts negatively prior to transistor Q1 conduction during turn-on. Since D1 is already on-state, the system Q1 voltage is zero, so the transistor can operate with zero-voltage across it. The mechanism's voltage increases to the off-state level so that the incoming diode is forward-biased until the current can fall. The transistor's turn-off condition is similar to the conventional square wave converter as the loss exists at turn-off in switching devices. The results of simulation of SPLRC using calculation and experimentalshown as table 2 for voltage output, Vo and current output, Io.

Table 2. Results of the simulation using PSim

	Simulation of SPLRC from calculation	Experimental simulation of SPLRC
V_o	0.005	0.58568
I_o	0.008	1.17136

5. Conclusion

A new circuit topology of high frequency switching DC-DC Series Parallel Loaded ResonantConverter (SPLRC), including wireless power transmission application, was proposed in this paper.The converter switches operated in constant voltage with constant high frequency. Wireless power transmission is one of the famous areas of research nowadays. The implementation of an advanced converter for wireless power transfer is significant to society and the economy. Hence it is a concern that the proposed advance high-frequency DC-DC converter or inverter topologyfor the transmitter side could successfully develop to transfer the power with high efficiency to the load. Nevertheless, another concern is to ensure the synthesizedadvanced AC-DC converter or rectifier for the receiver part can execute as a complete wireless power transmission interface that can feed the DC load with improved power conversion efficiency by DC source.

Acknowledgements

Financial support from Ministry of Higher Education & Research Management Institute (RMI), Universiti Teknologi Mara (UiTM) Fundamental Research Grant Scheme (FRGS) No: 600-IRMI/FRGS 5/3 (045/2019), is gratefully acknowledged for implementation of this project.

References

[1] Pavlov G, Vinnichenko I, Pokrovskiy M, and Tarabanov N 2019 Electromagnetic Processes in Serial-To-Parallel Resonant Converter for Contactless Charging of Electric Vehicle Battery *2019 IEEE 39th Int. Conf. Electron. Nanotechnology, ELNANO 2019 - Proc* 668–673.

[2] Stancil D D and Goldstein S C 2009 Magnetic Resonant Coupling As a Potential Means for Wireless Power Transfer to Multiple Small Receivers *IEEE Trans. Power Electron.* **24(7)** 1819–1825.

[3] Patii D, Rathore A K, Srinivasan D, and Panda S K 2014 High-frequency soft-switching LCC resonant current-fed DC/DC converter with high voltage gain for DC microgrid application *IECON Proc. (Industrial Electron. Conf.* 4293–4299.

[4] Ghalib M A and Abdalla Y S 2013 Design and Implementation of a Pure Sine Wave Single Phase Inverter for Photovoltaic Applications *Int. Conf. Informatics, Electron. Vis.* 1–8.

[5] Khalil-Abaker M, Shi J and Kalam A 2016 Design of a 100W bi-directional LCC series-parallel resonant DC-DC converter *2016 Australas. Univ. Power Eng. Conf.* 1–5.

[6] Jiang C, Chau K T, Liu C and Lee C H T 2017 An overview of resonant circuits for wireless power transfer *Energies* **10(7)** 1–20.

[7] Chen W, Chinga R A, Yoshida S, Lin J, Chen C and Lo W 2012 A 25.6 W 13.56 MHz wireless power transfer system with a 94% efficiency GaN Class-E power amplifier *IEEE MTT-S Int. Microw. Symp. Dig.* 25–27.

[8] Kim M, Noh S and Choi S 2015 New parallel loaded resonant converter with wide input and output voltage range *9th Int. Conf. Power Electron. - ECCE Asia "Green World with Power Electron. ICPE 2015-ECCE Asia* 515–520.

IOP Conf. Series: Materials Science and Engineering 1176 (2021) 012019 doi:10.1088/1757-899X/1176/1/012019

High Selectivity Hydrogen Gas Sensor Using Pd/ZnO Tapered Optical Fiber

N A M Yahya[1, 3*], M R Y Hamid[2], B H Ong[2], M A Mahdi[3], M H Yaacob[3]

[1]University Malaysia of Computer Science and Engineering, Cyberjaya 63000, Malaysia

[2]Nanotechnology and Catalysis Research Centre, Institute of Advanced Studies, University of Malaya, Kuala Lumpur 50603, Malaysia

[3]Wireless and Photonics Network Research Centre, Universiti Putra Malaysia, Serdang 43400, Malaysia

*Corresponding author: akmar.yahya@unimy.edu.my

Abstract. Besides so many great potentials of hydrogen (H_2) applications, such as primary reagent used for the industrial chemical processes and a potential clean energy source in future, handling it can cause potential hazardous situations due to its highly flammable and explosive nature. Therefore, hydrogen leakage detection is crucial to control the chemical reaction process running smoothly, thereby ensure industrial manufacturing safety in routine life. For this reason, this paper will be reported on tapered optical fiber coated with Pd/ZnO that is sensitive and highly selective towards H_2 gas. The sensor developed comprises a tapered optical fiber and ZnO nanoparticles incorporated using a chemical bath deposition technique. The synthesized ZnO was inspected by FESEM, RAMAN, TGA and UV-VIS to confirm its material properties. The sensing response towards H_2, CH_4 and NH_3 was conducted to observe its sensitivity and selectivity towards the gases. It was found that the Pd/ZnO tapered optical fiber sensor demonstrated high selectivity and sensitivity on H_2 gas compared to CH_4 and NH_3 with 86.6% increment of response. The repeatability test was also conducted to re-confirm the reliability of the sensor developed. The test measured standard deviation of the number of cycles repeated which yield 0.02902 or 97.1% to produce the same output or responses. These results perhaps are useful, especially in the automotive industry for advanced combustion monitoring and emission control.

1. Introduction

Zinc oxide (ZnO) is an n-type semiconductor with a wide bandgap of the II-VI group. It is an inorganic compound with white color and insoluble in water. For past decades, ZnO has been comprehensively studied for its excellent chemical, electrical, physical and optical properties for various applications. This includes solar cell, gas sensor, photocatalyst, ultra-violet (UV) laser and light-emitting diode (LED) [1]. ZnO is favorable for gas sensor application because it can detect toxic and harmful substances [2] aside from its chemical sensitivity to volatile and other radical gases. Moreover, it is more abundant, low cost and non-toxic [3].

ZnO thin films can be easily synthesized and deposited on different substrates using various technique, for instance, hydrothermal, sputtering, spray pyrolysis, chemical/physical vapor deposition, sol-gel and chemical bath deposition [4]. Many works reported on ZnO nanostructures such as nanobelts, nanowires, nanobridges, nanonails, nanorods, nanoribbons, nanotubes, and whiskers. This nanostructure is providing a superhigh surface-to-volume ratio that contributes to high sensitivity for chemical/gas sensing application [1]. Research on ZnO as a sensing layer for electrical gas sensor application is well established, particularly in detecting toxic gas and inflammable such as methane (CH_4), nitrogen dioxide (NO_2), carbon monoxide (CO) and hydrogen (H_2) [5].

Numerous works have been published and reported regarding ZnO as electrical sensors. ZnO nanostructures were grown by spray pyrolysis method by D.E. Motaung et al. [6] for H_2 detection. The sensor developed is based on resistivity change when exposed to H_2 at room temperature. The sensitivity of the ZnO nanoparticle sensor was more significant compared to columnar structures. Das et al. [7] reported Schottky diode type hydrogen sensor high performance using ZnO nanowires with platinum (Pt) as its electrodes. The sensor developed has shown good sensitivity and fast response of 2500 ppm H_2 at room temperature. The recent development on gold nanoclusters assembled on ZnO nanostructures film for H_2 gas sensor at low operating temperature was reported by Pichitchai Pimpang et al. [8]. The sensing material was prepared by thermal oxidation and deposited on a glass substrate with different temperature oxidation. The sensor has shown better response for Au assembled on ZnO as compared to without Au. The optimum response was obtained at 200 °C for 1% H_2 concentration. It is well known that electrical sensor is vulnerable to electromagnetic interference (EMI) thus affecting the signal responses, and worst of all, possible sparks can trigger an explosion. On the contrary, optical sensors using optical fiber that has unique properties such as lightweight, immune to EMI and able to withstand a high temperature environment can be utilized as a gas sensor [9]. By modifying the cladding part of optical fiber, one can turn into a sensor that is sensitive to the surrounding.

There is lesser number of works published on the optical fiber-based gas sensor using ZnO. Besides, only one work that used cerium (Ce), lithium (Li) and aluminum (Al) doped nanocrystalline ZnO on cladding modified optical fiber for gas ammonia (NH_3), ethanol (C_2H_5OH), and methanol (CH_3OH) (ranging from 0-500 ppm). It was discovered that the spectral intensity was linearly increased with concentration for ammonia but not for methanol and ethanol. The Ce doped ZnO has shown more sensitivity towards these gasses compared to Al and Li doped zinc oxides [8]. Another work on side-polished single-mode optical fiber (SMF) grown with ZnO nanostructures fabricated by A. Og. Dikovska et al. [11] for ammonia sensing. Rana Tabassum and Banshi D. Gupta [12] reported their work on H_2 sensor utilizing palladium (Pd) and ZnO coated on multimode plastic-clad silica optical fiber. The sensor used a surface plasmon resonance approach to detect the shift in resonance wavelength upon H_2 exposure. Yahya et al. has reported H_2 responses for different thickness of ZnO coated onto tapered optical fiber [13]. Pd has been opted as a coating layer due to its high catalytic activities, making it ideal for H_2 sensing and other combustible gases A very thin layer of catalytic metal integrated with metal oxide promotes chemical reactions by reducing the activation energy between the layer and the gas [14-15]. In this work, the selectivity and sensitivity of the developed sensor towards different gases were conducted.

2. Methodology

The optical fiber used is multimode type that is tapered using Vytran GPX-3400 optical glass fiber processor. The tapering process heat the optical fiber and stretch to the preferred dimension profile. For this work, the dimension profile is fixed to waist diameter of 20-um, up/down taper of 2-mm and 10-mm in length.

Chemical bath deposition technique is adopted to deposit ZnO onto the tapered optical fiber. Firstly, the process starts with surface activation on the tapered optical fiber before growing the ZnO nanostructures. The surface activation used potassium permanganate ($KMnO_4$) mixed with droplets of *n*-butanol. Then, the tapered optical fiber is dipped into the deposition solution and later placed in the water bath with a temperature of 85 °C for 10 min. For synthesizing ZnO, a solution was prepared by mixing 1 ml of 1 M zinc sulfate ($ZnSO_4$) as a precursor with 1.5 ml of 4 M ammonium hydroxide (NH_4OH). Subsequently, 2 ml of ethanolamine (C_2H_7NO) as a complexing agent was added into the solution. The mixture later sonicated for 10 min to ensure the solution dispersed homogeneously. Finally, 15.5 ml of DI water was filled in the deposition solution and stirred well. The deposition solution was then placed in the water bath and the activated tapered optical fiber was dipped in the solution for 20 min. Next, the samples were annealed at 200 °C for an hour. The Pd coating was done using DC sputter with 5 nm thickness. The ZnO synthesized was characterized Field Emission Scanning Electron Microscope (FESEM), Raman Scattering, Thermogravimetric Analysis (TGA) and UV-Vis Spectroscopy (UV-Vis).

Figure 1 shows the gas sensing setup comprises of HL-2000 Ocean Optics USA tungsten halogen light source, spectrophotometer to observe the optical absorbance spectrum ranging 200 nm to 1100 nm, sealed gas chamber equipped with heater, gasses cylinder (H_2, CH_4, NH_3 and pure synthetic air) and Alborg controller. The sensor sample is placed in the gas chamber as to conduct the testing.

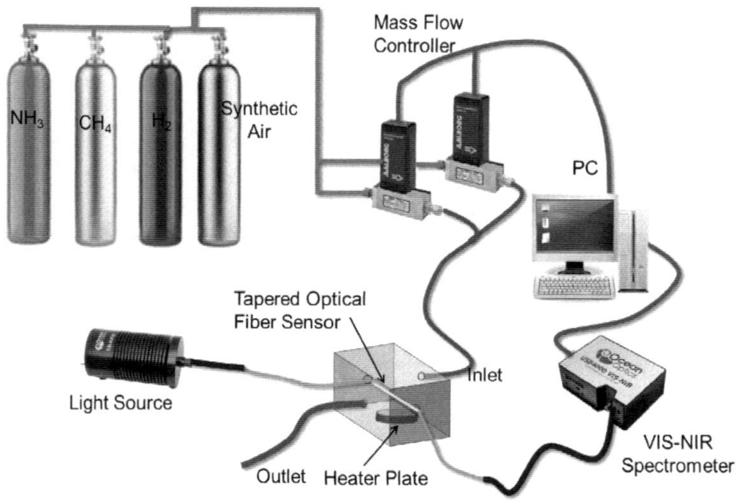

Figure 1. Gas sensing experimental setup.

3. Results and Discussion

The surface morphology of the ZnO coated on tapered optical fiber was examined using FESEM as illustrated in figure 2. Image (a) and (b) show the optical fiber before and after ZnO coating. It can be seen that ZnO coating is well coated around the optical fiber. The morphology of the synthesized ZnO observed to be nanoflowers structure (c) ranging from 60 nm to 150 nm in size (d). The nanostructure size is really important as it will provide a high surface area to volume ratio which yields a large number of surface atoms. This will lead to insufficiency of surface atomic coordination and high surface energy. Thus, the surface becomes highly active. This will also promote further adsorption of gas molecules when sensing.

Figure 2. FESEM image of (a) bare optical fiber, (b) optical fiber coated with ZnO and (c) nanoflowers structure of ZnO (d) particle size distribution

Figure 3 shows Raman spectra of synthesized ZnO sample, excited with a 532 nm laser. A few peaks were identified from the spectrum at 323 cm^{-1}, 438 cm^{-1}, 570 cm^{-1} and 1099 cm^{-1}. A_1 and E_1 modes are polar and split into transverse optical (A_1TO and E_1TO) and longitudinal optical (A_1LO and E_1LO) components. In contrast, E_2 mode comprises two modes of low (heavy Zn sub-lattice vibration) and high frequency (oxygen atoms vibration) phonons. The peak at 323 cm^{-1} and 570 cm^{-1} correspond to acoustic overtone and E_1 low mode vibration. The peak at 438 cm^{-1} corresponds to E_2 high mode dominates in the non-resonant Raman scattering spectra indicating good crystal quality. It is also found that peak 1099 cm^{-1} is due to the acoustic combination of A_1 and E_2. The peaks stated above are significant, suggesting that the ZnO has good crystallinity and high purity. Based on [16], good crystallinity will provide more surface area, hence, strong gas molecules-sensing layer interaction.

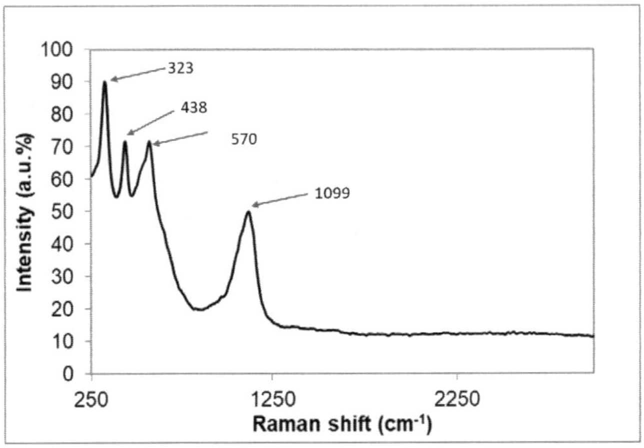

Figure 3. Raman spectra of synthesized ZnO, excited with a 532 nm laser.

TGA measurement was performed in a dynamic nitrogen atmosphere with sample sizes of 21 mg and a heating rate of 10 °C/min. The TG plot in figure 4 displays a weight loss of 0.27% (0.059 mg) between 30 °C and 200 °C, due to the evaporation of the water absorbed on the material's surface. The loss of weight continues about 1.89% between 200 °C to 530 °C, indicating the loss of intercalated water and hydroxyl ions physically adsorbed in the material. The temperature tested was 30 °C to 600 °C in which the total weight loss of the sample was only 0.477 mg from the initial value of 21 mg. The loss is insignificant, showing the thermal stability and high purity of synthesized ZnO nanostructures [17].

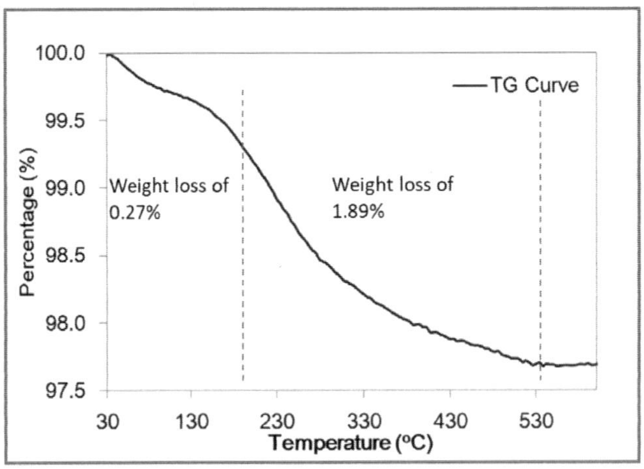

Figure 4. TG curves of the synthesized ZnO

The absorbance spectrum obtained from the UV-Vis measurement is used to plot $(\alpha h\nu)^2$ versus $h\nu$ so the optical band gap can be calculated. The optical band gap (E_g) value can be measured by taking the tangent to the curve cross the $h\nu$ axis, as demonstrated in figure 5. From the graph, the estimated optical band gap for the synthesized ZnO is 3.1 eV. It was reported in [18] that narrowing the optical band gap will provide a high concentration of oxygen deficiency. This deficiency served as the active sites for adsorbed oxygen to interact with the targeted gas molecules. Hence, the sensitivity of the sensing material can be increased.

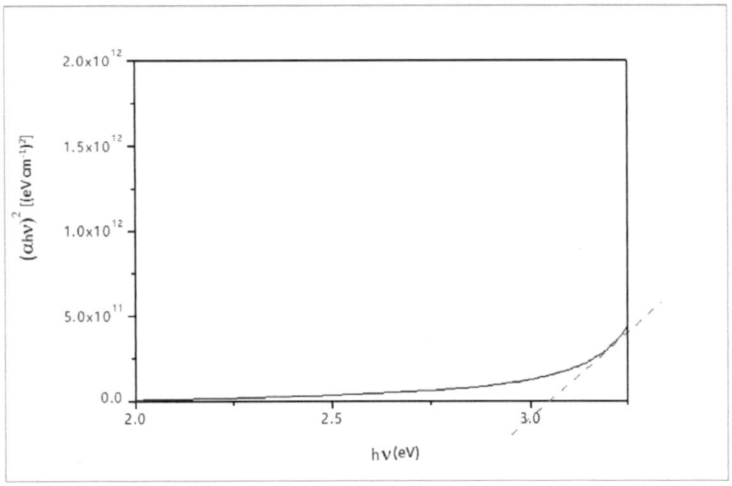

Figure 5. Estimated optical band gap of ZnO nanoflowers (3.1 eV).

The sensing responses are monitored through optical absorbance that exhibits when the developed sensor was exposed to 1% of H_2, CH_4 and NH_3 gasses. The cumulative change in the absorbance can be integrated over a specific wavelength range, defining the dynamic responses. Figure 6 shows for both ZnO and Pd/ZnO sensors on the cumulative absorbance response against operating temperature when exposed to 1% of H_2. The ZnO sensor only exhibits a low absorbance response at 200 °C, and the absorbance change decreases as temperature increases. On the other hand, the Pd catalyst coated on top of the ZnO layer has provided some gas sensing enhancement. Pd/ZnO tapered optical fiber sensor exhibits absorbance change even at low temperature. It was found that the optimum operating temperature that displays the maximum absorbance response is at 180 °C. Not only the absorbance change has increased, but also the optimum operating temperature also has decreased to the lower point with the effect of Pd as a catalyst in H_2 sensing.

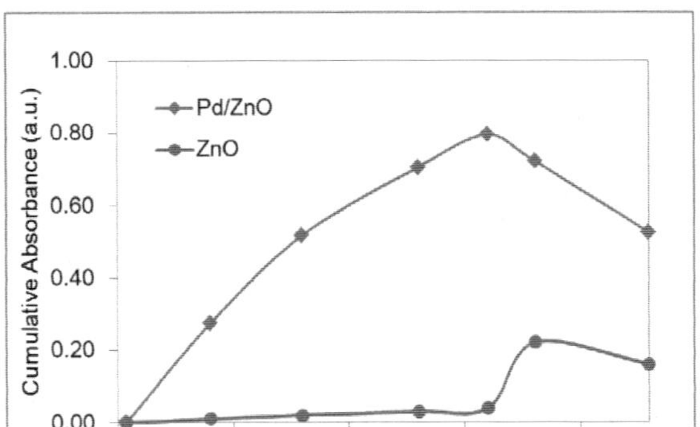

Figure 6. Change of cumulative absorbance versus operating temperature for ZnO and Pd/ZnO coated on tapered optical fiber when exposed to H_2 of 1.0% in air.

From figure 7, it was observed that there is no distinguish absorbance increase when the sensor is exposed to CH_4 and NH_3. However, a distinct absorbance increase can be seen from the initial condition of synthetic air to 1% H_2 exposure at 700 to 1000 nm range of wavelength. This absorbance change is due to H_2 adsorption onto the ZnO sensing layer in which the Pd thin layer dissociates H_2 into H^+ and reacts with chemisorbed oxygen (O^- and O^{2-}) hence producing H_2O molecules [19]. The generated electrons reduced the Zn^{2+} in the center of ZnO crystal lattice to Zn^{+}. Therefore, any changes in sensing layer properties will also change the absorption of light, thus changing absorbance magnitude.

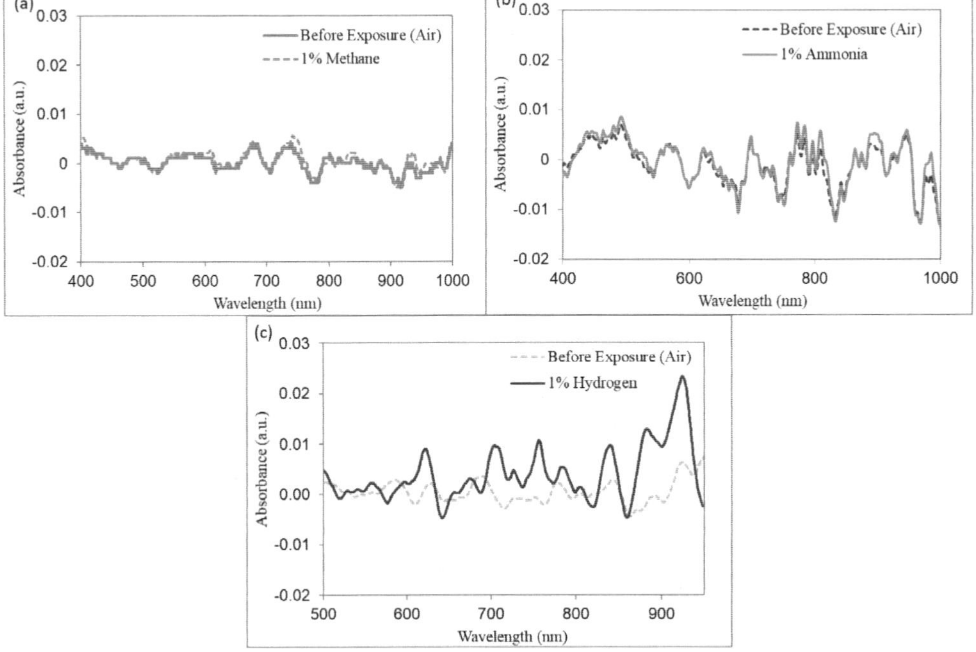

Figure 7. Absorbance response versus wavelength of Pd/ZnO coated tapered optical fiber sensor when exposed to 1% concentration of (a) CH_4, (b) NH_3 and (c) H_2 at 180 °C operating temperature.

The Pd/ZnO sensors were tested towards different gases to examine their selectivity performance. The gasses selected were ammonia (NH_3) and methane (CH_4). The measurement was performed by exposing the sensor to synthetic air and selected gas alternately at 180 °C operating temperature. Figure 8 shows the dynamic responses of cumulative absorbance increase when these gasses exposed to the sensor. Note that an increase of 2.435 upon exposure of 1% H_2, 0.237 and 0.133 for CH_4 and NH_3, respectively. It was observed that the absorbance increase for H_2 is 86.8% that is distinguishable compared to CH_4 and NH_3, which are 8.45% and 4.74%, respectively. The sensor developed has shown high selectivity towards H_2 gas and less sensitive to NH_3 and CH_4. According to [20], CH_4 gas is a stable gas that requires very high energy to dissociate H from C. High operating temperature is needed to enhance the sensitivity towards this gas. The sensor is less sensitive towards NH_3, probably because of Pd coating since the Pd is more suitable in dissociating the H_2 gas. This confirms that the produced sensors portray a good selectivity for H_2 compared to the cases of NH_3 and CH_4.

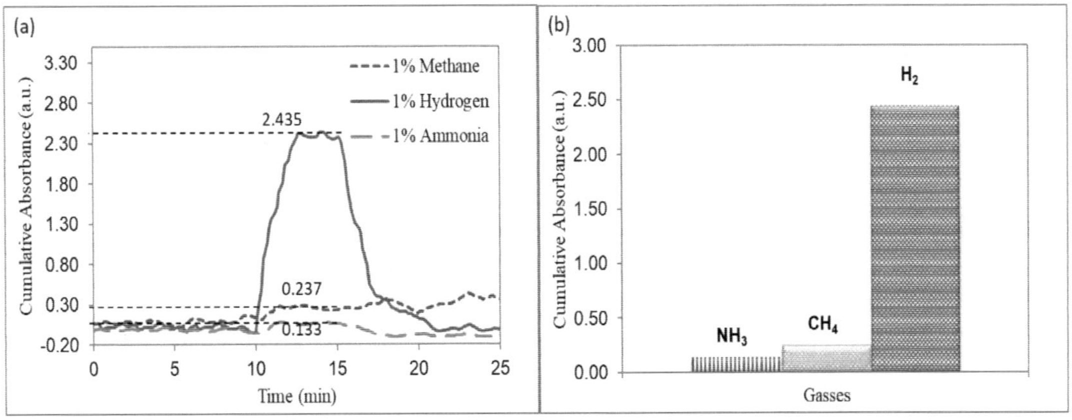

Figure 8. (a) Dynamic response for Pd/ZnO coated tapered optical fiber sensor when exposed to 1% concentration of H_2, NH_3 and CH_4 at 180 °C and (b) comparison bar graph for selectivity test.

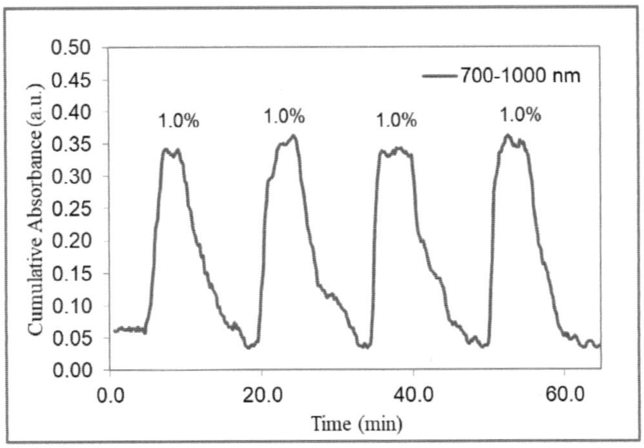

Figure 9. Four (4) cycles repeatability test for 1% of H_2

Figure 9 displays the repeatability of the Pd/ZnO tapered optical fiber sensor when exposed to four (4) cycles of 1% concentration of H_2 gas. From the graph, it is observed that the sensor able to exhibit the same output response for every cycle tested. The repeatability is measured by the standard deviation of the number of cycles repeated, which yield 0.02902 or 97.1% to produce the same output or responses.

4. Conclusion

The developed Pd/ZnO tapered optical fiber sensor has shown high selectivity towards H_2 gas (86.8%) compared to CH_4 (8.4%) and NH_3 (4.7%), respectively. On top of that, the developed sensor showcases good repeatability to produce the same responses with a standard deviation of 0.02902 or 97.1%. The synthesized ZnO exhibits nanoflowers structures with good crystallinity and purity based on material characterization performed. Hence, the increase of the film surface area provided strong gas molecules-sensing layer interaction. The narrowing optical band gap contributes to a high concentration of oxygen deficiency. This deficiency served as the active sites for adsorbed oxygen to interact with the targeted gas molecules. As a result, the sensitivity of the sensing material can be increased too. The future work can be improved further by incorporating other nanomaterials onto a gas sensor platform in this work. The selectivity towards other gases can be achieved by merely integrating a chemo resistive type of semiconducting metal oxide sensors to the developed sensor.

References

[1] Gasparotto G, Mazon T, Gasparotto G, Zaghete M A, Perazolli L A and Varela J A 2014 *Adv. Mater. Res.* **975** p 189–193.

[2] Feng W, Yang X, He Z, and Liu M 2021 *J. Phys. D: Appl. Phys.* **54** p 135105-135109.

[3] Renganathan B and Ganesan A R 2014 *Opt. Fiber Technol.* **20(1)** p 48–52.

[4] Hendi A A and Alorainy R H 2014 *Superlattices Microstruct.* **66** p 23–32.

[5] Roy S, Saha H and Sarkar C K 2010 *High Sensitivity Methane Sensor by Chemically* **3** p 605–620.

[6] Motaung D E, Mhlongo G H, Kortidis I, Nkosi S S, Malgas G F, Mwakikunga B W, Ray S S and Kiriakidis G 2013 *Appl. Surf. Sci.* **279** p 142–149.

[7] Das S N, Kar J P, Cho J H, Lee T Il, Moon K J and Myoung J M, 2010 *J. Phys. Chem. C* **114(3)** p 1689–1693.

[8] Pimpang P, Zoolfakar A S, Rani R A, Kadir R A, Wongratanaphisan D, Gardchareon A, Kalantar-zadeh K and Choopun S 2017 *Ceram. Int.* **43** p S511–S515.

[9] Chen K, Yuan D, Zhao Y 2021 *Optics and Laser Technology* **137** p 106808-106828.

[10] Wang C, Yin L, Zhang L, Xiang D and Gao R 2010 *Sensors (Basel)* **10(3)** 2088–106.

[11] Dikovska A O, Atanasova G B, Nedyalkov N N, Stefanov P K, Atanasov P A, Karakoleva E I and Andreev A T 2010 *Sens. Act. B Chem.* **146(1)** p 331–336.

[12] Tabassum R and Gupta B D 2015 *Appl. Opt.* **54** 5 p 1032.

[13] Yahya N A M, Hamid M R Y, Ong B H, Rahman N A, Mahdi M A and Yaacob M H 2020 *IEEE Sens. J.* **20(6)** 2982-2990.

[14] Del Orbe D V, Yang H, Cho I, Park J, Choi J, Han S W, Park I 2021 *Sensors and Actuators: B-Chemical* **329** p 129129-129137.

[15] Lin Y-M, Gao J-J, Chen K-P, Huang C-H, and Huang C-S 2021 *IEEE Sensors Journal* **21(3)** 2798-2804.

[16] Jiménez I, Arbiol J, Dezanneau G, Cornet A and Morante J R 2003 *Sensors Actuators, B Chem.* **93(1–3)** p 475–485.

[17] Mutasim I K, Maha M A, Aisha M A and Joselito P L 2014 *Arab. J. of Chem.* **7(6)** p 1178-1184.

[18] Tan C H, Tan S T, Lee H B, Yap C C and Yahaya M 2016 *AIP Conf. Proc.* **1784.**

[19] Banerjee N, Roy S, Sarkar C K and Member S 2013 *IEEE Int. Conf. on Nanotech. (Beijing)* p 682–685.

[20] Basu S and Basu P K 2009 *J. Sensors* **2009.**

IOP Conf. Series: Materials Science and Engineering 1176 (2021) 012020 doi:10.1088/1757-899X/1176/1/012020

Students' Performance Prediction using Artificial Neural Network

N Ahmad[1*], N Hassan[1], H Jaafar[1] and N I M Enzai[1]

[1]Faculty of Electrical Engineering, University Teknologi MARA Cawangan Terengganu
Kampus Dungun, 23000, Dungun, Terengganu, Malaysia

*Corresponding author: norha291@uitm.edu.my

Abstract. This paper proposed a method for predicting diploma students' performance of Faculty of Electrical Engineering (EE), Universiti Teknologi MARA (UiTM) Terengganu. Data of 59 first semester students from Electrical Engineering (EE) were obtained to predict students' academic performance. A predictive model based on a machine learning technique was employed. The predictive model utilizes Artificial Neural Network (ANN) technique that was developed to predict the actual performance of first semester students based on Sijil Pelajaran Malaysia (SPM) results, first semester results, and the interest level of the participants towards EE course. The findings have shown that the developed model could accurately predict the actual result of the first semester students successfully with minimal errors.

1. Introduction

Educational quality is essential for a country's development. With the use of systems for admission, academic information, learning management system and e-learning, the data amount in education domain has been increasing rapidly. The data collected from students are usually used only for decision making. However, most of the data remain unused due to its complexity and large volume of data sets. Therefore, there is a need for analysis of educational data, especially to predict the students' performance.

It is necessary to build a system that can continuously keep track of the students' academic performance and accurately predict their future performance; such as when are they likely to graduate and what are their estimated final grade point averages (GPAs).

In this paper, we propose a method for predicting diploma students' performance of Faculty of Electrical Engineering, Universiti Teknologi MARA (UiTM) Terengganu. The scope of this paper is the prediction of students who could achieve Dean's List (GPA of 3.5 and above) but the general framework can be used for other performance prediction tasks.

The rest of this paper is organized as follows. Section 2 discusses the related work. Section 3 details the methodology. Section 4 describes results and discussion and section 5 concludes the paper.

2. Literature Review

There are various factors and parameters which could affect the students' academic performance. The factors could be internal or could be caused by surrounding or external.

Social media such as Facebook has been investigated by [1] in term of its impact towards academic performance. As social media has become inevitable, its usage can't be avoided even though there's concern that it could cause distraction in learning process. A systematic review of Facebook usage impact was conducted using Mixed Method Appraisal Tool. The findings indicate both positive and negative effects of Facebook usage on the academic performance of university students. However, the work outlined mechanisms to better utilize Facebook for improving students' performance academically.

Factors of resilience and engagement of students in relation to their academic performance are studied in [2]. 748 full-time first-year students of the Administration and Business Management degree at two public Business faculties in Northern Spain were involved in the study. The study aims to figure out whether the dimensions of resilience and engagement play important part in ensuring their ability to survive and continue their study until the end. The findings indeed verified that two dimensions of resilience and engagement predict the students' academic performance. Interestingly, gender also affects the difference in importance of these variables.

Based on the hypothesis that Emotional Intelligence (EI) could affect academic performance, [3] conducted a meta-analysis by searching nine databases for related researches on the topic of interest: ProQuest Dissertations, ERIC, Google Scholar, ISI Web of Science, Medline, and Theses, PsycINFO, ScienceDirect, PubMed, and Scopus. Students with higher EI are expected to be more able to control the negative emotions, manage the social surroundings, and more competent in humanities subjects. The Inclusion and Exclusion Criteria for the meta-analyses are: (a) written in English; (b) published in either a test manual or journal article; (c) direct appraisal of academic performance indices or self-reported. It is found that EI has moderate impact towards producing better academic performance, in particular for skills-based EI tasks.

The knowledge, attitudes and behavioural skills (KAB) model is the basis of [4] in predicting the student academic performance. To enhance the modelling accuracy, variable selection and dimensionality reduction techniques are proposed. A real data set of ten factors that are related to both attitude and skill-related behaviours of 320 first-year students was utilized to conduct evaluation. The prediction accuracy of student achievement was improved by employing the proposed methodologies.

Physical activity, aerobic fitness, and motor skills are believed to have some effect towards executive functions and academic achievement. Cross-sectional data from 697 children from 57 schools in Norway was analysed using a linear mixed model. Even though the impact of physical activity on academic achievement was found to be weak, the results still indicate positive relationship [5].

Motivational beliefs and learning strategy factors in enhancing student learning outcomes are studied by [6]. This work also adopts Motivated Strategies for Learning Questionnaire (MSLQ) as their methodology like [7]. SPSS 17.0 was utilized to perform analyses of data obtained from 323 participants that comprised of Liberian junior and senior high school student. The results indicate that the students' motivational beliefs and learning strategies give moderate impact on their learning outcomes. Rather, the improvement of education quality is more required by the students.

To study the effect of family background factors combined with students' previous academic performance on their tertiary first year performance, machine learning algorithms: WEKA and multilayer perceptron have been employed by [8]. This work constructs a model of students' academic performance to act as building block for other students which consequently forecasts the overall performance. The research sample involved students' enrolment forms from Babcock University, Nigeria. Among the extracted information are age, gender parent occupations and education, parents' marital status, religion, and previous educational performances of related subjects. Family background factors as well as culture and region were found to play significant role in the first-year students' performance.

In the era of Internet and mobility, impact of mobile phone usage has been the research focus of [9]. The research motivation stems from the idea that mobile phone usage during class could distract the students even though mobile phone has its benefits in making the teaching and learning process more interesting. Structured questionnaires and additional interviews were conducted among 45 students of Federal Polytechnic, Ilaro to obtain their personal perspectives. Based on the responses and data analysis, mobile phones give significant negative impact as the students' attention deteriorated which leads to poor academic performance. However, if the mobile phone usage is managed properly, the academic performance of students can actually be improved.

The effect of students' exposure to green space and vegetation was also explored by [10]. The relation of tree cover, tree diversity, and tree species composition with students' academic performance was examined in 387 schools across the Toronto District School Board (TDSB). School-level geospatial and tree inventory data and statistical analysis were employed as research techniques. Schools which showed the highest level of external challenges experienced the most profound effect of tree cover and species composition on student performance [10].

With the objective of assisting in improving students' performance based on predicted grades, machine learning techniques were utilized by [11] to predict students' grades in 24 different courses of different semesters. 225 Electrical Engineering undergraduate students were involved in the study. Collaborative Filtering (CF), Matrix Factorization (MF), and Restricted Boltzmann Machines (RBM) techniques were used for systematic analysis and discussed for students' grade prediction. Out of the three techniques, RBM showed the best results in students' performance prediction.

Another promising technique, Artificial Neural Network (ANN) is utilized by [12-14] for students' performance prediction. ANN emulates the biological processing ability by determining the upcoming events based on past data [15].[12] forecasted the undergraduate students' CGPA of a Chinese university based on their entrance examination results and socio-economic background using ANN. Among the findings are: female students scored better and rural or urban areas did not affect the scores. Meanwhile, [13] took into account factors such as high school score, scores of science and technology related subjects, CGPA obtained during first year, and gender of students in the Faculty of Engineering and Information Technology in Al-Azhar University of Gaza to feed the ANN model. The performance of more than 80% students was able to be predicted correctly. Blended learning course is selected by [14] for ANN implementation. The ANN model considers four learning activities, namely email interaction, wiki content collaboration, files view measurement and online quizzes. Correct classification rate (CCR) of 98.3% was obtained for students' performance prediction.

Based on the reviewed literature, there are many works that have been done with regards to prediction of student academic performance. This indicates that the need for such works are really crucial in ensuring that the students will be able to perform from the very beginning of their study. ANN technique in particular has garnered wide attention due to its success in prediction, classification, and forecasting in various areas including education, which is the focus of this paper.

3. **Methodology**

To predict students' performance, a predictive model based on a machine learning technique was employed. The predictive model utilizes Artificial Neural Network (ANN) technique that was developed to predict the actual performance of first semester students. The details of the prediction model will be discussed in the next section.

3.1 Data collection and Neural Network Model Development

The flow of the overall proposed system could be seen in figure 1. The work comprised data collection, data pre-processing, network/model development, and performance validation. Data of 59 first semester students from Electrical Engineering (EE) were obtained for this study. A survey that consists of Sijil Pelajaran Malaysia (SPM) results, first semester results, and the interest level of the participants towards EE course were collected. These collected parameters will be set as input and target for ANN topology to develop the prediction model shown in table 1. The input variable factors were transformed into a format that is suitable for neural network analysis. The domain of the input variables used in this study is shown in table 1. The reasons for choosing those parameters as the input data/independent parameters for modeling are because all the chosen subjects are mandatory to be eligible for enrollment in EE course, and students' interest could be considered as a potential indicator to acquire a good result.

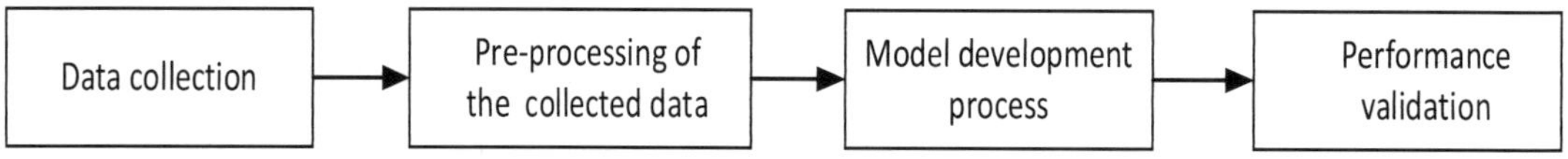

Figure 1. Block diagram of the students' performance prediction

Table 1. Input and Output variable

Input variable	Domain	Output variable
Additional Mathematics Mathematics Physics	A+ = 6, A= 5, A-= 4, B+ = 3, B = 2, C = 1	First-semester CGPA
Interest in Electrical Engineering courses	Really interested = 5, Interested = 4, Neutral = 3, Less interested = 2, not interested = 1	

For efficient training process, it is recommended to pre-process the samples so that the input and target dataset will fall in the same range of values and become reliable. A reliable dataset is important because it can affect the accuracy and performance of the developed model. Furthermore, a dataset could also affect the learning process of the model during training. To get a clean dataset, two types of data pre-processing techniques were utilized in this study, namely data selection and data inspection. For the model development process, ANN was adopted to develop a model that can predict the potential of a student to acquire Dean List status for actual result in the first semester. The structure of the model is depicted in figure 2. ANN is a universal function approximator that generally consists of an input layer, several hidden layers, and an output layer. The networks are very sensitive to the training hyperparameters such as number of the hidden layer, number of hidden neuron in one hidden layer, and activation function. Proper hyper-parameters must be chosen to produce a high-quality network. ANN with one hidden layer trained with backpropagation technique is developed in Matlab. The initial layer where the input comes into ANN is called the input layer. It is immediately followed by one hidden layer while the last layer where the output is produced is called the output layer. To develop the prediction model, the collected data will be divided into two parts. The first part is the training and learning process which is to create a nominal model with regards to the loaded data. The second part is the efficiency evaluation of the developed model and the data for this process which is called unlearning data or unseen data.

In the training phase, data was divided into three groups: 70% of the data were used to train and construct the network, 15% to validate the network that is developed by the training data set, and the remaining 15% will be utilized to reconfirm the prediction ability of the developed network. Several values of the hidden neuron are varied by using systematic trial and error method in training state to develop a reliable network. To achieve an optimal and desirable set of weights to pre-defined network architecture, learning and training processes are needed. These processes are based on an optimization approach to adjust the network weights and bias concerning the set of input-output. In this study, the network is trained by Lavernberg-Marquardt (LM) Back Propagation algorithm, since it offers a more efficient training scheme compared to gradient descent [16]. This training method falls under the variation of Newton's method which updates weights and biases value based on LM optimization. The optimal hidden neurons value for this work is 20.

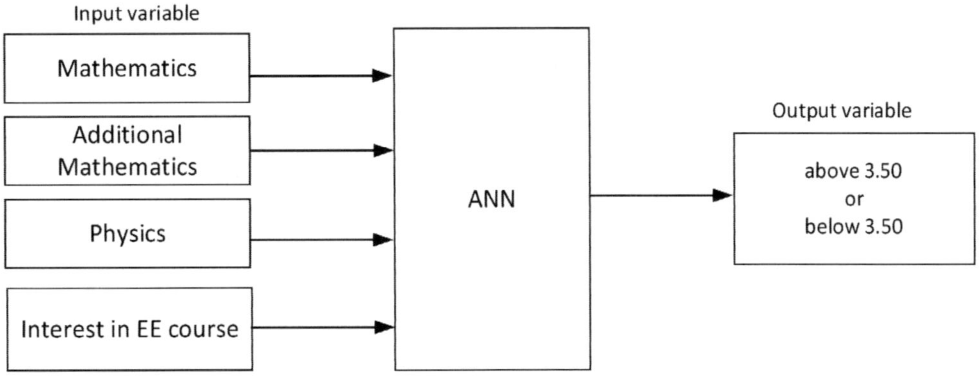

Figure 2. Neural Network Structure for students' performance prediction

4. Results and Discussion

Four components must be considered to evaluate the performance of a model in the training state or during the model development process: confusion matrix training, Receiver Operating Characteristic (ROC), cross entropy performance, and error histogram plot.

A confusion matrix is known as a contingency table or an error matrix. It is a specific table layout that allows visualization of the performance of an algorithm as supervised learning. For confusion matrix mapping shown in figure 3, the network outputs could be said as an accurate model since it shows the higher numbers of correct responses in the green regions. The number of incorrect responses is lower as shown in the red regions. The lower right blue region indicates the overall accuracy of the developed model which is 93.2%.

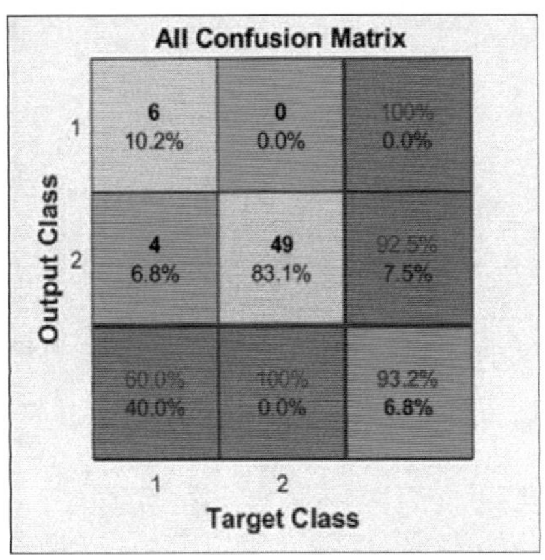

Figure 3. Confusion matrix

Receiver Operating Characteristic (ROC) shown in figure 4 is a plot of the true positive rate (sensitivity) versus the false-positive rates (1- specificity) as a threshold is varied. The plot shows that the model almost had a perfect test for the training phase since the value is approaching 100% sensitivity and 100% specificity. Thus, it is coherent with the performance mapping in the confusion matrix.

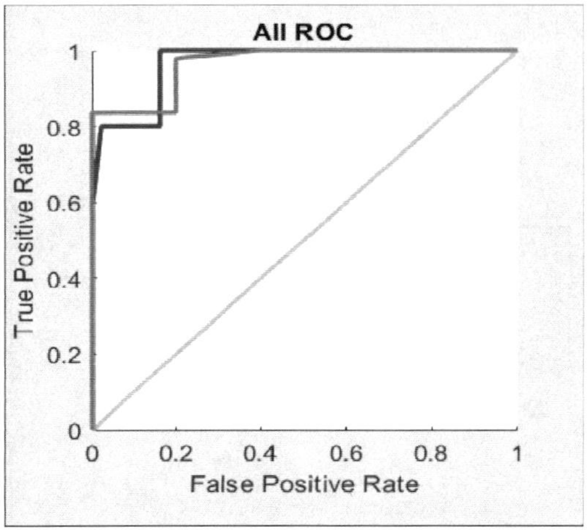

Figure 4. Receiver Operating Characteristic (R0C)

Figure 5 and figure 6 show the cross-entropy result and error histogram respectively. Cross-entropy is a loss function to optimize the classification model and the best cross-entropy performance for this model is 0.10664. The lower the cross-entropy value, the better the resulting model. In the error histogram, the significant error mapping range is between – 0.5752 and 0.5752, and most of the errors are tabulated at the center of the plot which is near zero value. Based on the above-mentioned model performance evaluator, it indicates that most of the training data are successfully discriminated during the training process.

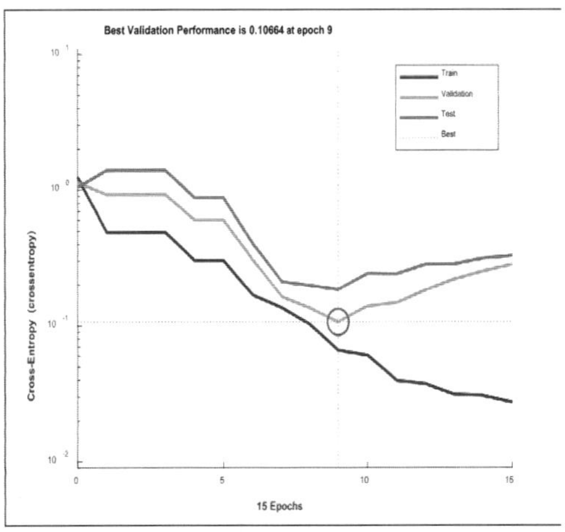

Figure 5. Cross entropy network performance

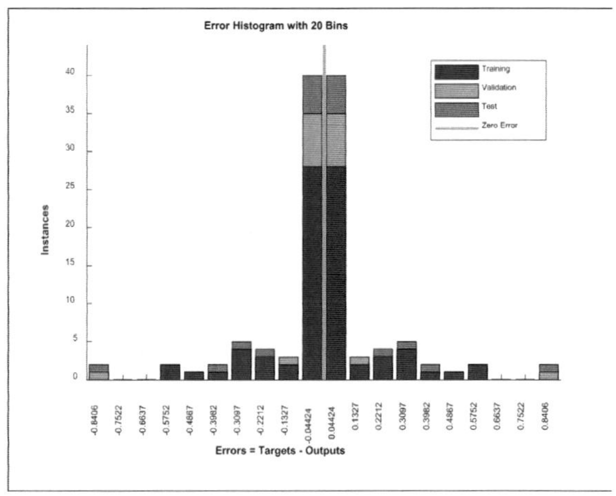

Figure 6. Error histogram

To make sure the developed model is robust and reliable; an additional step or process is conducted. During this process, the network is introduced to the data that is never encountered during the training process. The comparison of the real output of unlearned (data that has never been seen by the network during the training process) data set and predicted output from the developed model is shown in table 2. The errors between the real value and predicted value of five students are calculated to see the network prediction ability when dealing with the unseen data set. All errors are shown in table 2 which depicts low values. This means that the target output is discriminated very well.

Table 2. The validation process for five students

Student	Input [add math, math, Physic, interest]	Real output	Prediction output	error
Student 1	5,5,5,5	1	0.9320	0.068
Student 2	3,3,3,5	1	0.6069	0.393
Student 3	3,2,2,4	0	0.0066	0.0066
Student 4	2,2,2,2	0	0.0066	0.0066
Student 5	2,5,3,4	1	0.9071	0.0929

5. Conclusion

This research aims to predict the students' performance using a supervised learning technique which is ANN. Based on the students' SPM results and interest in EE course, the findings have shown that the developed model could accurately predict the actual result of the first semester students successfully with minimal errors. This framework is also expected to be adopted as a tool to make an early prediction of Dean List candidates from the first semester students. Although the developed model shows reasonably positive results, further studies and investigation are necessary to understand the correlation between cause and effect of every single parameter involved in this study. For future work, input variables like family factors and financial issue of students can be added. This research could contribute to education quality management through systematic students' performance monitoring.

References

[1] Nadeem S, Dhool M A, Butt R M, Rasheed A and Bashir A 2020 A Systematic Review on the Impact of Facebook Usage on Academic Performance *Rev. Manag. Sci.* **2(1)** 17-39.

[2] Ayala J C and Manzano G 2018 Academic performance of first-year university students: The influence of resilience and engagement *High. Educ. Res. Dev.* **37(7)** 1321-1335.

[3] MacCann C, Jiang Y, Brown L E, Double K S, Bucich M and Minbashian A 2020 Emotional intelligence predicts academic performance: A meta-analysis *Psychol. Bull.* **146(2)** 50.

[4] Al-Sheeb B A, Hamouda A and Abdella G M 2019 Modeling of student academic achievement in engineering education using cognitive and non-cognitive factors *J. Appl. Res. High. Educ.* **11(2)** 178-198.

[5] Aadland K N, Moe V F, Aadland E, Anderssen S A, Resaland G K and Ommundsen Y 2017 Relationships between physical activity, sedentary time, aerobic fitness, motor skills and executive function and academic performance in children *Ment. Health Phys. Act.* **12** 10-18.

[6] Gbollie C and Keamu H P 2017 Student academic performance: The role of motivation, strategies, and perceived factors hindering Liberian junior and senior high school students learning *Educ. Res. Int.* **2017** 1-11.

[7] Pardo A, Han F and Ellis R A 2016 Combining university student self-regulated learning indicators and engagement with online learning events to predict academic performance *IEEE Trans. Learn. Technol.* **10(1)** 82-92.

[8] Goga M, Kuyoro S and Goga N 2015 A recommender for improving the student academic performance *Procedia Soc. Behav. Sci.* **180** 1481-1488.

[9] Soyemi J, Oloruntoba S and Okafor B 2015 Analysis of mobile phone impact on student academic performance in tertiary institution *Int. J. Emerging Technol. Adv. Eng.* **5(1)** 361-365.

[10] Sivarajah S, Smith S M and Thomas S C 2018 Tree cover and species composition effects on academic performance of primary school student *PLoS One* **13(2).**

[11] Iqbal Z, Qadir J, Mian A N and Kamiran F 2017 Machine learning based student grade prediction: A case study *arXiv preprint arXiv* ***1708****.08744*

[12] Lau E, Sun L and Yang Q 2019 Modelling, prediction and classification of student academic performance using artificial neural networks *SN Appl.Scien.* **1(9)** 1-10.

[13] Abu-Naser S S, Zaqout I S, Abu Ghosh M, Atallah R R and Alajrami E 2015 Predicting student performance using artificial neural network: In the faculty of engineering and information technology *Int. J. Hybrid Inf. Technol.* 221-228.

[14] Zacharis N Z 2016 Predicting student academic performance in blended learning using Artificial Neural Networks *Int. J. Artif. Intell. Appl.* **7(5)**17-29.

[15] Arsad P M and Buniyamin N 2013 A neural network students' performance prediction model (NNSPPM) *IEEE Int. Conf. on Smart Instrum. Measure. and Appli. (ICSIMA)* 1-5.

[16] Munakata T 2008 *Neural Networks: Fundamentals and the Backpropagation Model:* Springer p 7-36.

IOP Conf. Series: Materials Science and Engineering 1176 (2021) 012021 doi:10.1088/1757-899X/1176/1/012021

The development of IoT Smart House Automation System controlled wirelessly via OpenHab central server

F N M Redzwan[1*], A H F Suhaimi[1] and A K Husaini[1]

[1]Faculty of Electrical Engineering, University Teknologi MARA Cawangan Terengganu, Kampus Dungun, 23000, Dungun, Terengganu, Malaysia

*Corresponding author: fatimahredzwan@gmail.com

Abstract. Smart home technology, also often referred to as home automation, provides homeowners security, comfort, convenience, and energy efficiency by allowing them to control smart devices, often by a smart home app on their smartphone or another networked device. A part of the internet of things (IoT), smart home systems and devices often operate together, sharing consumer usage data among themselves and automating actions based on the homeowners' preferences. This smart home automation system will be a server-based system using the OpenHab central server that can connect and control the house appliances by using an application that can be installed on the smartphone. By using this smart house system, the user can control their home appliances from distance. It can reduce the cost and provide better time management of the household.

1. Introduction

Nowadays, technology becomes a vital part of our daily life. It is included in most life aspects. One of the emerging technologies is smart home technology. This technology will influence the structure of the houses in the coming ten years. Smart homes promote comfort, luxury, entertainment, security, and world peace. Altering an existing home to accommodate changing needs can cost up to three times more than including the same features in the initial design. Current existing smart homes need special kinds of appliances to deal with. These appliances should be equipped with a network adapter since they should connect to a wired network. Smart homes also should contain a pre-installed wired infrastructure to provide a means of communication between the appliances. Smart home technology originally depends on the wired network infrastructure but after the huge improvements in that technology, it also supports wireless communication over multiple frequencies to control the system [1-3].

Several researchers studied and developed the home automation system using any technologies. The researchers in [4] reported that the home automation system that has been built by them is controlled via based on the Android app communicating with the micro-web server providing more than the switching functionalities. The Arduino Ethernet is used to eliminate the use of a personal computer (PC) keeping the cost of the overall system to a minimum while voice activation is incorporated for switching functionalities.

The Arduino Ethernet is used to eliminate the use of a personal computer (PC) keeping the cost of the overall system to a minimum while voice activation is incorporated for switching functionalities [5]. The studies in [6-8] have presented Bluetooth based home automation systems using Android Smartphones without Internet controllability. The devices are physically connected to a Bluetooth sub-controller which is then accessed and controlled by the Smartphone using built-in Bluetooth connectivity. However, due to the limited range of operation (maximum up to 100 m), the system is unable to cope with mobility and can only be controlled within the vicinity. The paper in [9] mentioned that the smart home automation system can also be controlled using Android Application on the smartphone via Wi-Fi connectivity.

This project proposed a smart home automation system that can control household devices wirelessly through the smartphone via the OpenHab server. This Openhab server helps to give the command wirelessly to the home appliances. This smart home system is user friendly to people even the kids starting at the age of six years and above can also know how to use it correctly.

The smart home automation system is mainly focused on maintaining the security of the house when the household is away from home, for example going for a long holiday or go back to hometown for a long term. This system will keep the security of the house from thieves and the users can monitor their home security by observing it through their smartphone.

Other than that, this system also can help to control the home appliance whether to turning it off condition or it on condition. This can be controlled also by using a smartphone via IoT technology compared to the conventional way where the users need to press the switch button.

2. The methodology of IoT Smart Home Automation System

2.1. Overall System Flow Chart

Figure 1 shows the overall flowchart of the smart home automation and security system controlled via OpenHab server directly to the user's smartphone and any communication devices. According to this flowchart, the main microcontroller of the system is Raspberry Pi and connected to the Node MCU as the second main component. The function of Node MCU is to act as the open-source firmware and development kit that helps to develop the IoT product. It includes firmware that runs on the ESP8266 Wi-Fi and to the OpenHab server. The command from the users will goes to the server and receiving at the Node MCU and then, it is will wirelessly control the home appliance such as fan, lamp, and other home appliances.

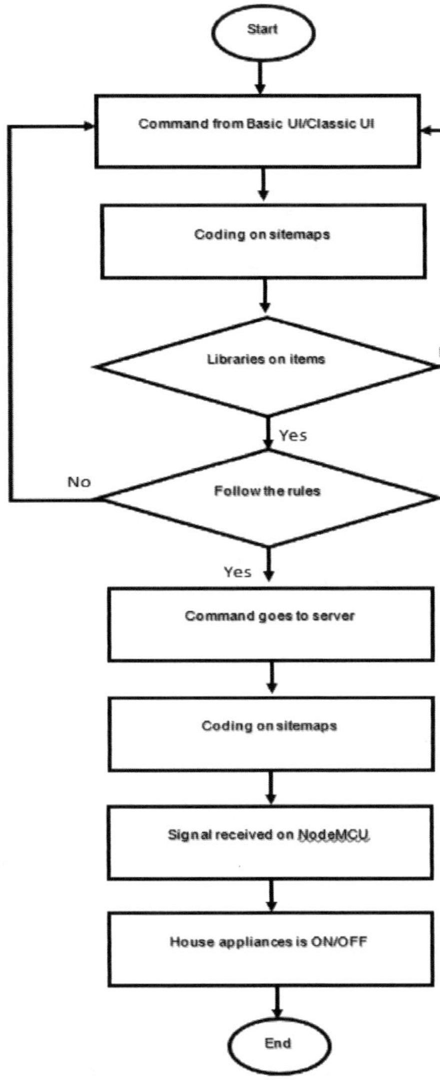

Figure 1. Flow chart of the overall process of the smart home automation system

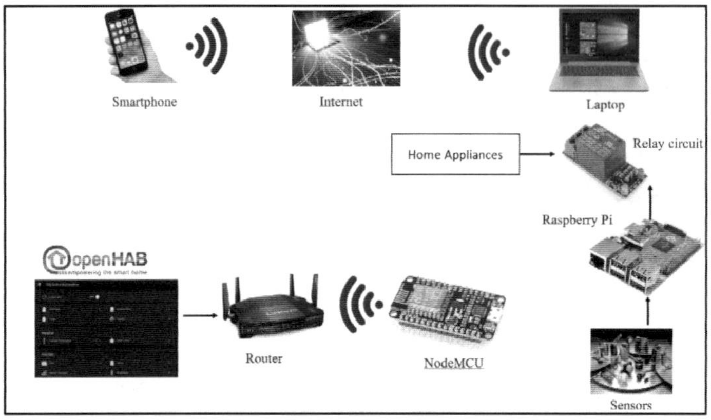

Figure 2. System architecture

From the system architecture shown in figure 2, the system consists of two main parts: the wireless control system and the home automation network system. The smartphone and laptop or computer act as the wireless controllers, and it is connected wirelessly to the network directly or via the Internet in enabling the remote control of home appliances. The Raspberry Pi used in this project and act as the microcontroller and connect to the OpenHab server. This OpenHab acts as the central server for the system. The function of the modem router can log the IP address of the different devices connected to it and besides, it can be operated as the central connection point for all the devices and appliances in the house. The NodeMCU wifi module is used to communicate the signal wirelessly from the router to the OpenHab central server.

The user can access this central server with an *email address* and a *password* registered to it. The connection can be made via two methods either from the web interface or any smartphone applications such as Android or IoS. The users can make the command to the server after gaining access. The receptiveness of the designed architecture was necessary because, in smart homes, appliances must respond to either switching commands or voice commands.

2.2. Block Diagram of the IoT Home System

Figure 3 below represents the block diagram of this IoT home system. Based on the figure, there are three main parts in this block diagram where the left-hand side is the input process, the center part is the main microcontroller (Raspberry Pi) that act as the brain of the system and give the command to the devices. And the right-hand side is the outputs obtained in this system to control the home appliances for security and electrical energy reduction purposes.

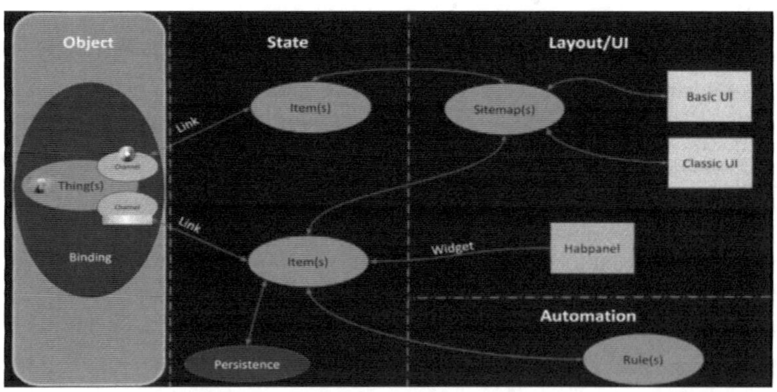

Figure 3. The block diagram of the smart home automation systems.

2.3. Schematic Diagram

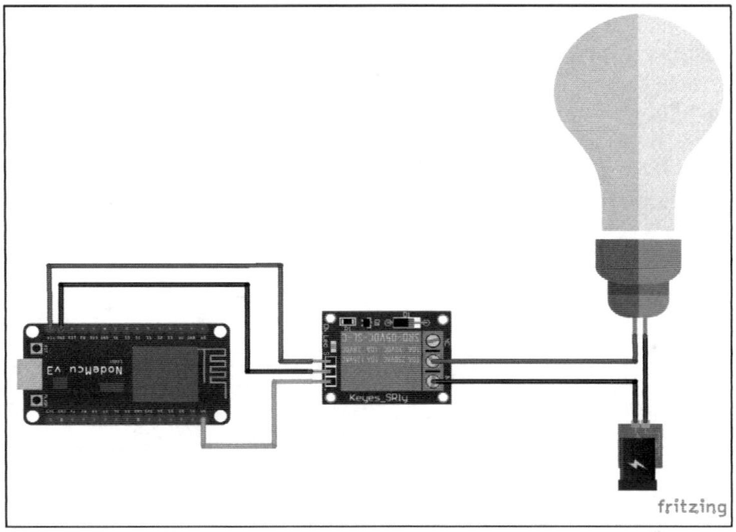

Figure 4. The schematic diagram simulated on fritzing circuit simulation software

Figure 4 represents the sample of the schematic diagram and one of the outputs that can be controlled wirelessly using this smart home system. In this schematic diagram, the main "brain" of the component is the Raspberry Pi and connected to the NodeMCU. As cannot be seen in this figure is the server that connects the signal wirelessly and is called as OpenHabian or OpenHab.

2.4. User Interface of the System

The user interface of the home automation system developed using the OpenHab server is presented in figure 5. The OpenHab is chosen compared to others server like SmartThings, Blynks Application and ThingsSpeak are due to the user-friendliness of its interface; it offers the user the capability to modify the interface based on their preference [10], low-security issues, more compatible with any devices; it is open-source and can be integrated with smart devices and NodeMCU based sensors, and reasonable cost.

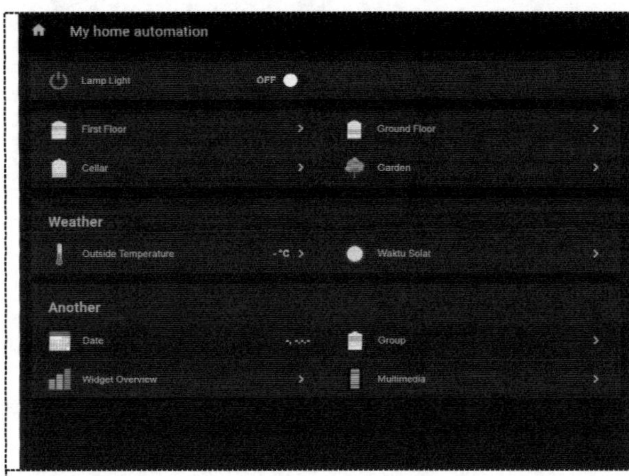

Figure 5. The user interface of the OpenHab server.

2.5. Circuit Testing and Troubleshooting

This kind of project will reduce the cost of the household due to that they can monitor their home appliances away from their home. It also can help the household to decrease their electric bill accordingly because they can switch off their fan and lamp wirelessly through the mobile devices. Figure 6 shows the schematic diagram of the IoT Smart Home System. This circuit has been designed using the Proteus simulation software tools. Then, this schematic diagram has been converted to a Printed Circuit Board (PCB) layout to proceed with the fabrication process as shown in figure 7.

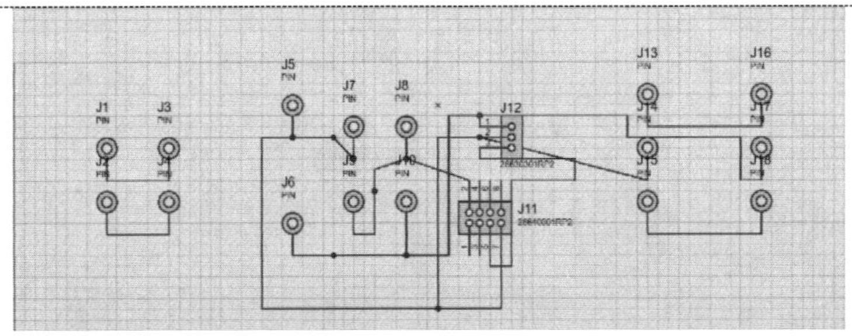

Figure 6. Schematic diagram of IoT Smart Home System designed with Proteus simulation software.

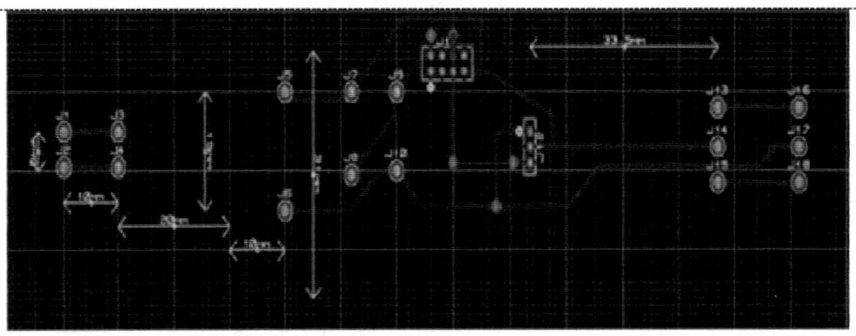

Figure 7. Printed Circuit Board Layout on Proteus ARES

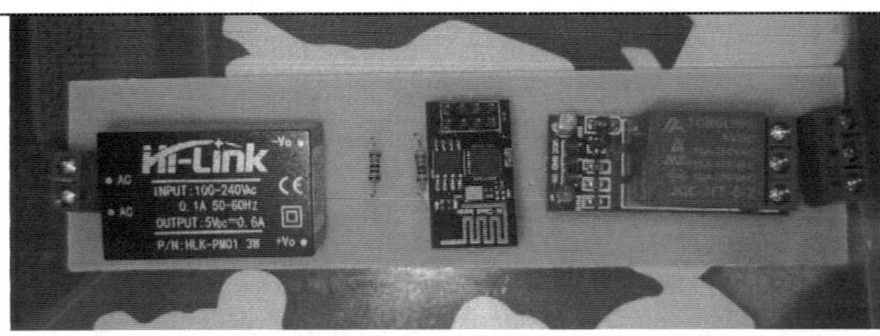

Figure 8. The circuit was fabricated on the Printed Circuit Board (PCB) through the etching process.

Figure 8 shows the final fabricated circuit printed on the PCB board through the etching process. As can be seen, all the components have been arranged according to the PCB layout. The middle component is NodeMCU that will connect the signal from Raspberry Pi to the OpenHab server wirelessly through the same internet connection.

The Raspberry Pi is used in this project to host a lightweight distribution with OpenHab. It is installed with the Mosquitto MQTT broker and it can be communicated with the network by connecting either via Ethernet cable or wirelessly connected to the router.

3. Results and Discussion

3.1. Software Simulation Test

The simulation for this system is constructed and simulated using Proteus Professional 8 software. To simulate the circuit, first, the library of the software must be added with the Arduino Library. After that, the simulation can be done by picking all the components and the script for the circuit was loaded to the Arduino board in the Proteus.

3.2. Hardware Implementation Test

In this project, the real standard hardware of the AC fluorescent lamp (220 V to 240 V) is used to verify the results. The condition of this lamp can be controlled wirelessly through the OpenHab server either in the ON state or OFF state. The OpenHab server has been installed on the laptop and also the smartphone. The list of outputs obtained according to the OpenHab server is demonstrated in figure 9 to 12. As can be seen in figure 9, the command on the OpenHab server is set to OFF conditions in controlling the switching of the AC fluorescent lamp and the value of MQTT is set to '1'. Then, the result of the lamp is shown in figure 10, where the lamp is turned OFF following the command encoded from the OpenHab server.

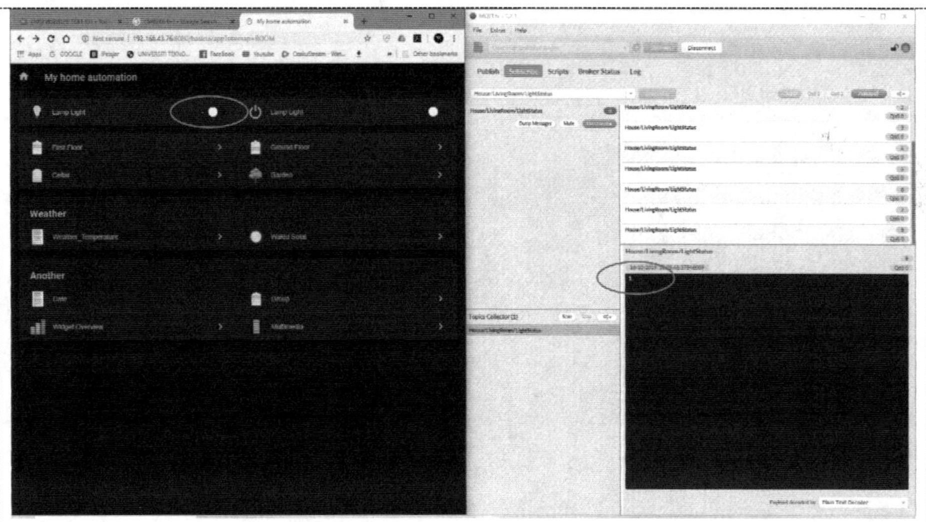

Figure 9. The command on the OpenHab server in controlling the switching of the home appliances (OFF conditions).

IOP Conf. Series: Materials Science and Engineering 1176 (2021) 012021 doi:10.1088/1757-899X/1176/1/012021

Figure 10. The AC fluorescent lamp is turn OFF according to the command set in the OpenHab.

Figure 11 shows the command on the OpenHab server in controlling the AC fluorescent lamp is in ON conditions and figure 12 shows the lamp is turned ON according to the command set in the OpebHab server that has been connected wirelessly through the same network via NodeMCU. It can be seen that, the value of MQTT is set to '0' value to indicate the ON state condition for the command in the OpenHab central server.

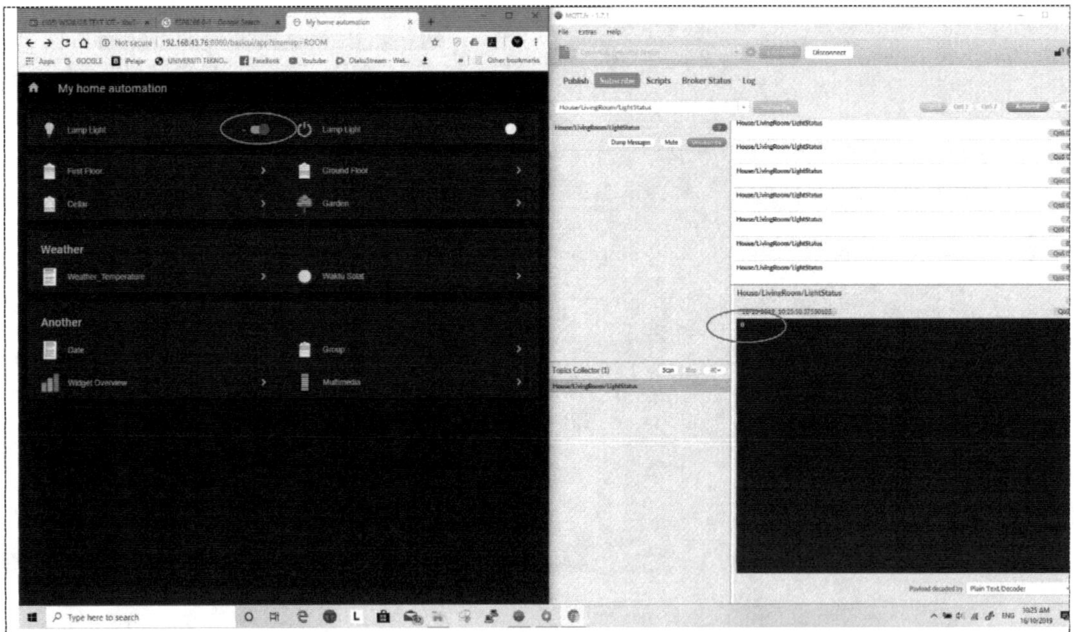

Figure 11. The command on the OpenHab server in controlling the switching of the home appliances (ON conditions)

IOP Conf. Series: Materials Science and Engineering 1176 (2021) 012021 doi:10.1088/1757-899X/1176/1/012021

Figure 12. The AC fluorescent lamp is turned ON according to the command set in the OpenHab.

3.3. Notification results on the smartphone

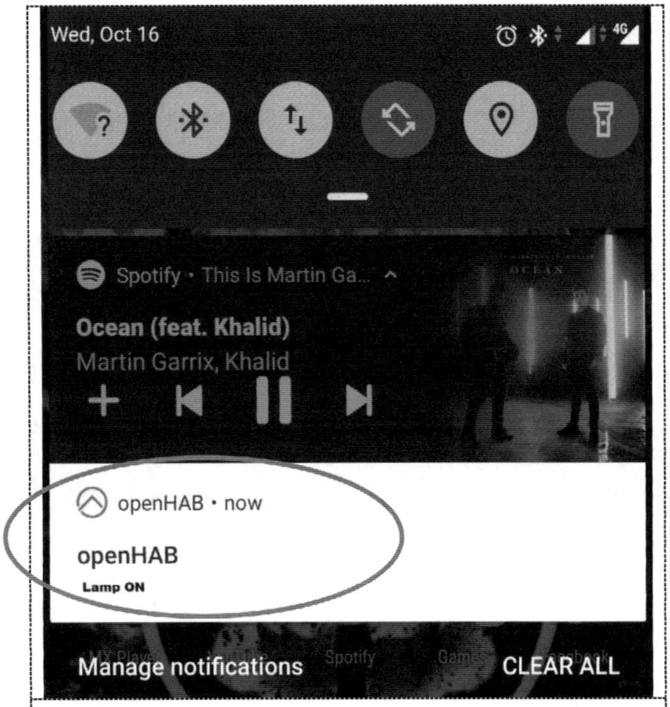

Figure 13. Alert notifications are received on the smartphone via the OpenHab Apps.

Figure 13 shows the alert notification received on the user smartphone via the OpenHab application that have been installed in the Android Operating System smartphone. The message and command from the OpenHab central server will be sent directly and wirelessly using the OpenHab application. The message then will be appeared on the user's smartphone and stated the condition of the lamp either in ON state or OFF state.

4. Conclusion

In conclusion, this project will make some changes to the daily life and perspective to the people how important the maintain the security and controlling the condition of the home appliances. This system could provide homeowners security, comfort, convenience, and energy efficiency. These home appliances can be controlled control via the smart devices such as using smart home applications on their smartphone or other networked devices. Besides, this smart house will be a server-based system that connects and control the house appliances by using OpenHab application that can be installed in the user's smartphone. In the nutshell, this project can help users to control their home appliances far from distance and thus reduce the electricity and the electrical energy consumptions.

Acknowledgments

The authors would like to show their appreciation to the Faculty of Electrical Engineering, Universiti Teknologi MARA Cawangan Terengganu for providing the facilities and equipment to carry out this project.

References

[1] UKEssays 2018 *The Smart House Project Information Technology Essay.*
[2] Labs A C 2018 *IoT Based Street Light Monitoring System.*
[3] Giaimo C, Door Sensor Secrets: What They Are, How They Work & 6 Unexpected Uses 24 July 2013. [Online]. Available: https://simplisafe.com/blog/door-sensor-secrets
[4] Shiu K 2014 Ubiquitous Smart Home System Using Android Application *Int. J. of Comp. Net. & Comm.* **6** 33-34.
[5] Abu M and Kamarulzaman W 2018 Development of home automation systems by using android apps. *ARPN J. Eng. Appl. Sci.* **13(8)** 2809-2814.
[6] Hisham A B and Ishak M 2014 Bluetooth-Based Home Automation System Using an Android Phone *J. Teknol.* **70** 57-61.
[7] Oo Z, Theint W and Maung T 2019 Bluetooth based home automation system using android and Arduino *Proc. Conf. on Comput. Appl. and Res.* 1-3.
[8] Kulkarni M 2019 Home Automation using Bluetooth Control Technology *Int. J. Res. Appl. Sci. Eng. Technol.* **7** 2136-2140.
[9] Aravindhan R, Ramanathan M, SanjaiKumar D and Kishore R *2017 Int. J. of Eng. and Tech.* **4** 2542-2545.
[10] Ramljak M 2017 Security Analysis of Open Home Automation Bus System *40th Int. Conv. Inf. and Commun. Technol. Electron. and* Microelectron. *(MIPRO)* 1242-1250.

Design and Fabrication of Kayak Carrier

H Ghazali[1*], N H Mas'aud[1] and N A Rahman[1]

[1]Faculty of Mechanical Engineering, University Teknologi MARA (UiTM), 23200 Bukit Besi, Terengganu, Malaysia

*Corresponding author: hanipah@uitm.edu.my

Abstract. Kayak has become the most popular outdoor activities among the community of UiTM Cawangan Terengganu, Kampus Bukit Besi. It is due to the location of this campus is situated near Tasik Puteri. During kayak activities, the users need to bring the kayak from campus to the lake. The distance from the storage area to the lake is about 800 m. A minimum of two persons is required to carry the kayak due to its size and weight. Where the average recreational kayak weighs 16kg for a single kayak and 27kg for a tandem kayak. Therefore, this study is focused on designing the kayak carrier to overcome these problems. There are several stages involved in producing the kayak carrier, which are concept generation and evaluation, CAD design, fabrication, and testing. In a conclusion, by using this kayak carrier, only one person needed to carry the kayak from the storage area to the lake, and the traveling time also can be reduced as the person need to cycle instead of walking.

1. Introduction

A kayak is a vehicle used for transporting across the water. It is distinguished from canoeing by the sitting position of the paddlers and the number of blades on the paddle. Canoe-like boat in which the paddler sits facing forward, legs in front, using a double-bladed paddle to pull front-to-back on one side and then the other in rotation. Most kayaks have closed decks, although sit-on-top and inflatable kayaks are growing in popularity as well [1]. According to Hudson et al. [2], a kayak is economical transportation and has a small structure in size therefore it can be transported effortlessly using a car or manageable trailer from one location to another.

Kayaking is one of the most popular outdoor activities at the University of Technology MARA (UiTM), Terengganu Branch, Bukit Besi Campus due to its location next to Puteri Lake, Bukit Besi, Terengganu. Figure 1 shows the kayaking activity at Puteri Lake, Bukit Besi, Terengganu. It is customary for students to shuttle the kayak from the storage located on campus to Puteri Lake that takes 800 meters of travel. The kayak that available on the UiTM campus weights almost 21kg and usually needs two persons to carry it on their shoulders. From the observation, after done with the kayaking activity some difficulties occur to transport back the kayak to campus manually as there is a lack of energy left.

Figure 1: Kayaking activity at Puteri Lake, Bukit Besi, Terengganu.

Industriales et al. [3] have stated that the European Conference of Ministers of Transport (ECMT) defined a trailer as "a vehicle without own propulsion to transport goods, intended to couple with an automobile, except semi-trailers", where the dissimilarity between trailer and semi-trailer is the existence of the front axle. There are few studies on a trailer that has been done by previous researchers. Naukowe et al. [4][5] studied the strength analysis of the frame trailer used to carry personal watercraft such as boat and water scooters. The method used to analyze the strength analysis was the numerical mathematical method known as the finite element method (FEM). Ladra et al. [6] used the same method of FEM to study the strength analysis on the prototype of a multi-tasking trailer. The analysis was developed using the SolidWorks program and performed in the Simulation Module. The development of this prototype focus on specific design criteria for the trailer system functional in many tasks such as space, multifunctional folding frame, and backlight of the sliding back panel. A similar research project of trailer development has been studied by Agrait et al. [7]. The project is subjected to security and durability where it minimizes the disposal of the dysfunctional trailers that lead to contamination due to material selection.

In recent work, focus on Computer-Aided Design (CAD) developed by SolidWorks as a tool to visualize the outcomes of the intended product before continuing with the product fabrication. However, to complete the design projects well, the generation process and concept selection are required as it will affect the cost, quality, and performance of the final product. For the concept generation process, the Morphological Method is used. This method applies the following process in generating feasible concepts, which; breaking down a problem into sub-functions, generating an option solution for each sub-function, and selecting and combining the suitable option solution into an alternative concept solution. This method is advantageous in terms of generating novel concepts and able to represent a wide range of concepts which allows the unexpected matching of components to be considered [8].

Concept selection is the method used to determine the most suitable concept that follows all the specifications given in the decision-making criteria before proceed to the design stage process that allows a reduction to cost and time implication [9][10]. According to Augustine et al. [11], the right selection of design concepts will eliminate the cost of redesign the final product and escalate the productivity in the project development process. This researcher has listed the possibility of multi-criteria decisions to aid concept selection namely as Quality Function Deployment (QFD), Analytic Hierarchy Process (AHP), Pugh Matrix, multi-criteria optimization, and fuzzy logic methods. In this paper, Pugh Matrix or also known as Pugh Chart is selected as a systematic approach in selected the best design from the three proposed design concepts.

Many studies on Pugh Matrix have been done previously by other researchers. Thakker et al. [12] study the Pugh method in the analysis concept of designing an impulse turbine using a 3DCAD environment. It was found out that this method successfully delivered the evaluation of alternate design by compared the impulse turbine rotor concepts in the product development process. Seperamaniam et al. [13] also used this method in developing the conceptual design of a hydrostatic bearing pad in a heavy-duty automotive application and effectively obtain the optimum design. Another study was done by Silaen et al. [14] as they used the Pugh concept to select the best design of lecture chairs at Sebelas Maret University and turn out the results show good time management throughout the assembly process.

This paper highlights the design and fabricate process of the kayak trailer which is also known as a kayak carrier using the Morphological Method and Pugh Chart as for the concept generation and selection method as for the decision-making criteria activity before proceeding with the fabrication process.

2. Research Method

The process of fabrication of kayak carrier is shown in figure 2. It started with concept generation and evaluation, CAD design, fabrication of the product, and testing.

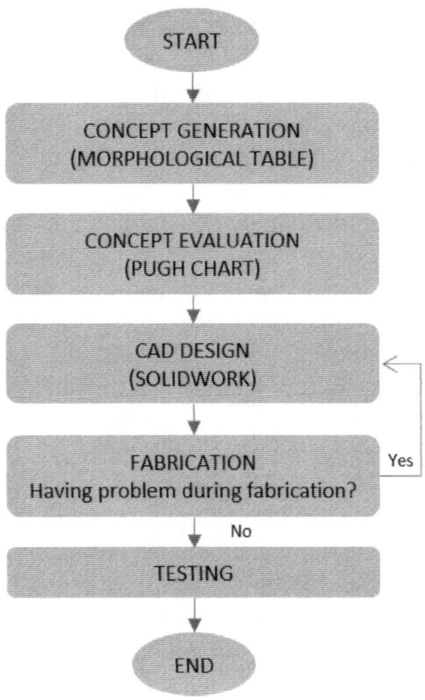

Figure 2: Flowchart of kayak carrier fabrication

First, the process started with the concept generation of the product by using the morphological table. It is based on function analysis. This table will list down all the sketch ideas for each function. According to the function of each part of the product, it is then selected to come out into several design concepts using a different option. There are at least, three combination concepts that will be produced during this stage.

Then, the next step is concept evaluation. The design concept produced will be then evaluated to select the best design. Several factors need to be considered in selecting the design concept. Since there is limited raw material provided by the organization, it is required to consider the availability of each part. It might be available from hardware or need to be fabricated based on the dimension. Other than that, in terms of ease of installation, durability, aesthetics value, safety, etc. Pugh Chart is used in concept evaluation. All criteria are having a weightage to get the score for each design concept. Design concepts with the highest score will be selected.

The selected concept is then being created in a 3D design by using a Computer-Aided Design (CAD) software, Solidworks 2013 x64 Edition. This software can produce a detail drawing of each part as well as the assembly drawing of a product with a specific dimension. With that, the blueprint of the product will be then produced and endorsed by the supervisor before entering the workshop, to proceed with the fabrication process.

The fabrication process includes cutting the raw materials, machining process, joining process either permanently or temporarily joined, and finishing process which is required for the aesthetical value and safety purpose. Therefore, by referring to the blueprint, the fabrication process will be managed properly as the detail dimension already stated, and the preparation of material will be easier, and the number of waste material also can be reduced. Lastly, the prototype product will undergo the product testing process to make sure that the product is safe for the users.

3. Results and discussion

3.1. Concept Generation

Product concept generation is a process that starts with a list of parameters based on customer's needs and specifications. Based on the requirements, concept generation helps to pinpoint a variety of possible solutions and ideas that answers those needs. A morphological table is a table based on the function analysis. On the left side of the table, the sub-functions are listed, while on the right side, different types which can be used to perform the functions listed are drawn. It is a visual aid used to come up with different ideas. The morphological table is used in concept generation and Table 1 shows the morphological table of concept generation for kayak carrier. Based on the table, there are six sub-functions listed in our product design, kayak carrier. The six sub-functions are considered such as kayak stands, rear tire, front tire, chassis, chassis connector, and bicycle connector. There is three options solution for each sub-function. For example, three shapes of kayak stands are considered in our design which are rectangle bar, semi-circle bar, and V-bar.

Three different design concepts have been designed using different option solutions provided in Table 1, to produce the three best designs. We used a bicycle to pull the kayak carrier as a solution to avoid the users carrying the kayak on their shoulders. Figure 3a shows the design concept 1. The mechanism for this design concept is the kayak is placed on the V-bar stand. This design does not require a belt to grip the kayak because the shape of the V-bar is similar to the shape on the bottom surface of the kayak. Figure 3b shows the design concept 2. The mechanism for this design concept is the kayak is placed on the rectangle stands, which are located at the rear and front of the chassis. Then the kayak is held by the belt to avoid sliding or moving. Figure 3c shows the design concept 3. The mechanism for this design concept is the kayak is placed at the rear semi-circle stands only. Then the kayak is held by the belt to avoid sliding or moving.

Figure 3a: Design concept 1.

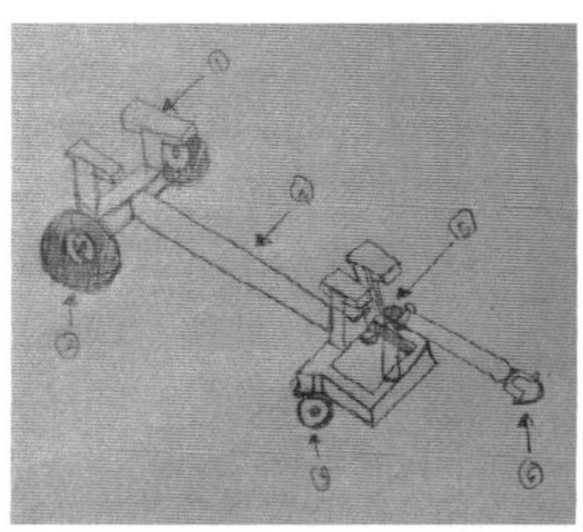

Figure 3b: Design concept 2.

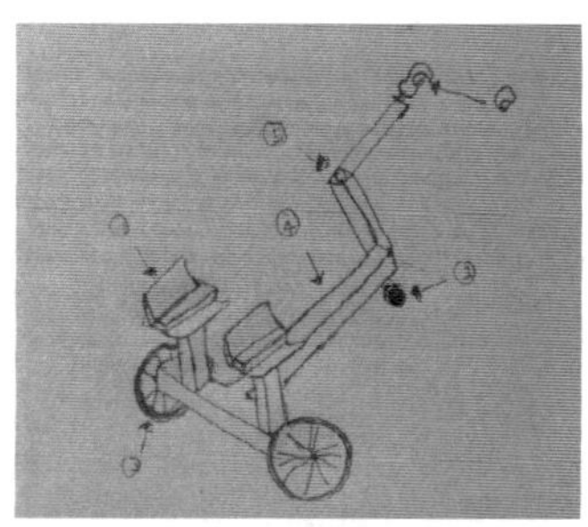

Figure 3c: Design concept 3.

After the three design concepts were designed, the reason for choosing the material or shape was determined and listed in the table. Table 2 shows the list of parts for design concept 1, Table 3 shows the list of parts for design concept 2, and Table 4 shows the list of parts for design concept 3. For example, in Table 2, we choose the V-bar shape for kayak stands because the design can hold the kayak from sliding or moving, the go-kart tire is chosen for the rear tire because the design gives more stability for a variant type of road surface, the double trolley tire is chosen for front tire because the shape can support rear tire to make sure the carrier is stable, and so on.

Table 1: Morphological table of concept generation for kayak carrier

Sub-function	**Option solution** 1	2	3
Kayak Stands	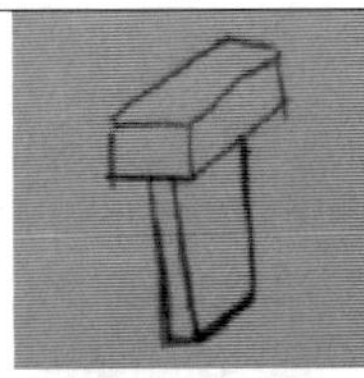 Rectangle bar	 Semi-circle	 V-bar
Rear Tire	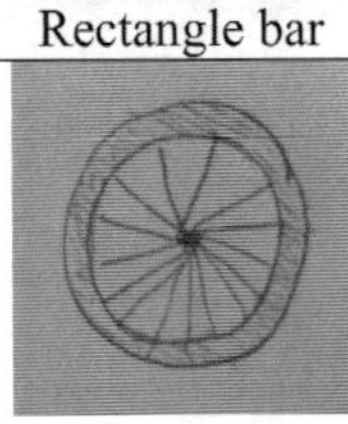 Bicycle tire	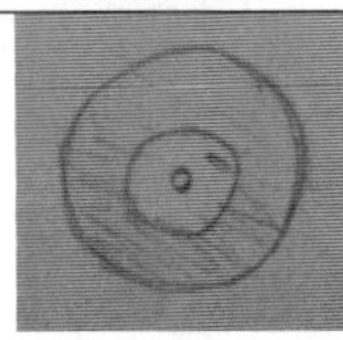 Wheelbarrow tire	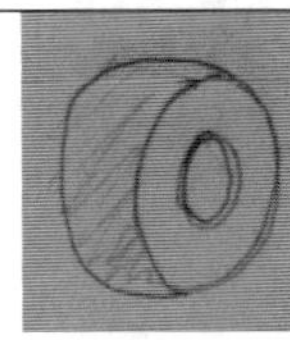 Go-kart tire
Front Tire	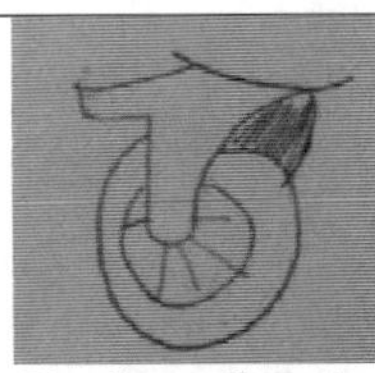 Single trolley tire	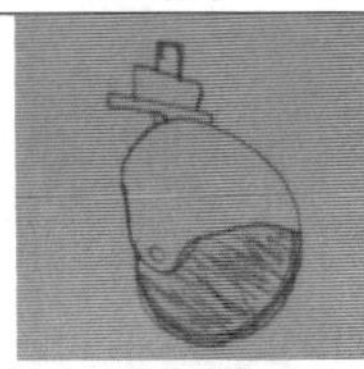 Round trolley tire	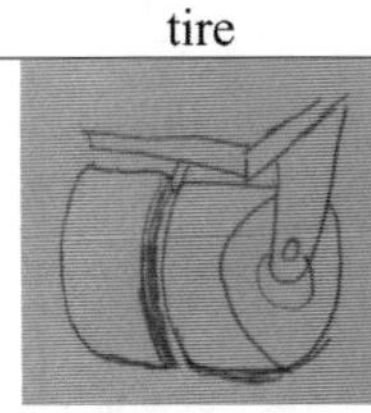 Double trolley tire
Chassis	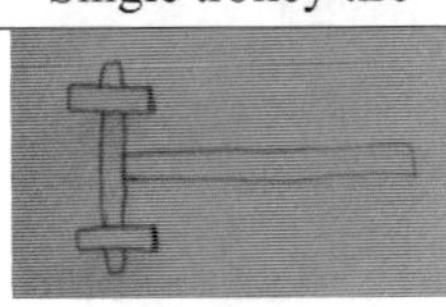 T-shape	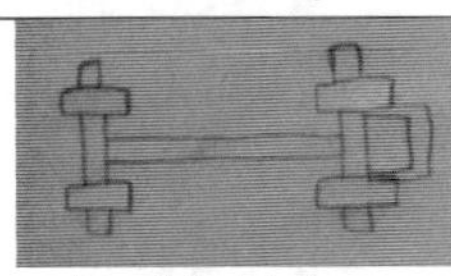 Double T-shape	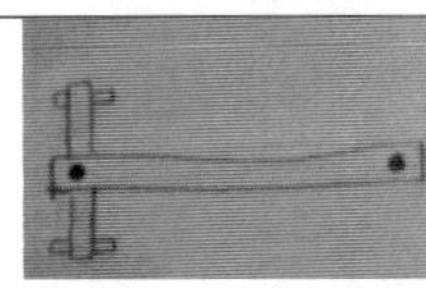 Single straight shape
Chassis Connecter	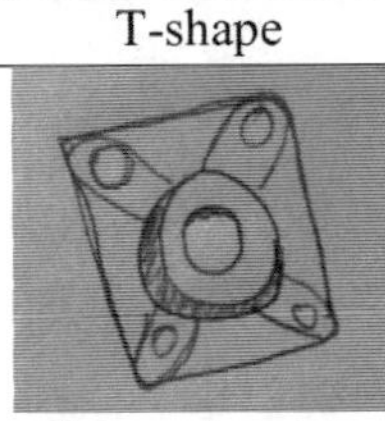 Square bearing	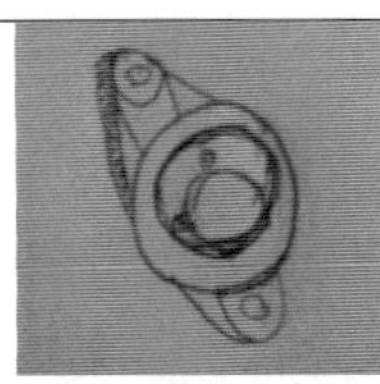 Round bearing	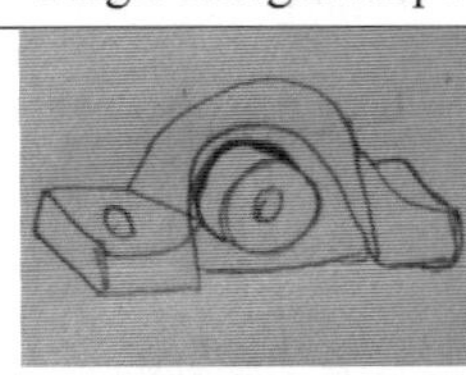 Rectangle bearing
Bicycle Connecter	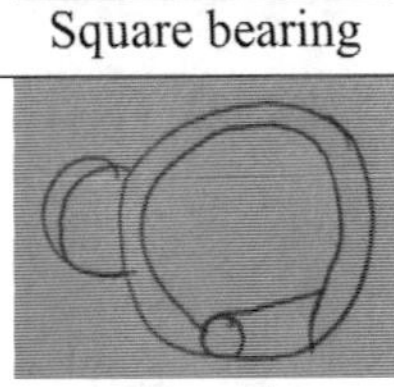 Round	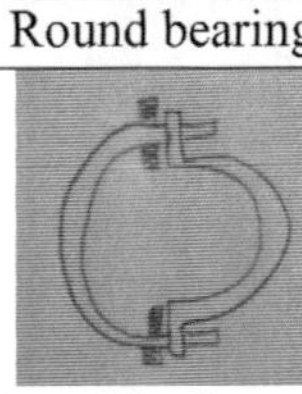 Double C-type	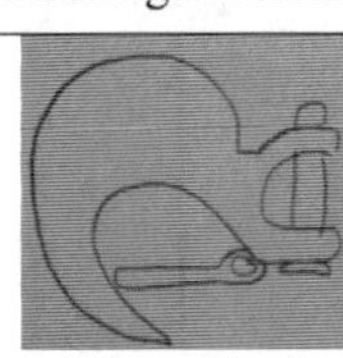 Moon Hook

Table 2: List of part for Design Concept 1

No	Sub-function	Option function	Reason
1	Kayak stands	V-bar	Can hold the kayak from sliding or moving.
2	Rear tire	Go-kart tire	Give more stability for a variant type of road surface.
3	Front tire	Double trolley tire	Support rear tire to make sure the carrier is stable.
4	Chassis	Single straight shape	The design is simple and reduces the load that kayakers need to carry.
5	Chassis connecter	Round bearing	Durable.
6	Bicycle connecter	Round type	Easy to make a turn.

Table 3: List of part for Design Concept 2

No	Sub-function	Option function	Reason
1	Kayak stands	Rectangle bar	Can carry many types of kayak.
2	Rear tire	Wheelbarrow tire	Easy to maintain.
3	Front tire	Single trolley tire	Quick response when changed direction.
4	Chassis	Double T-shape	Better hold of the kayak at its place.
5	Chassis connecter	Rectangle bearing	Can take out if bearing broken and easy to install.
6	Bicycle connecter	Double C-type	A variety of types of a bicycle can use.

Table 4: List of part for Design Concept 3

No	Sub-function	Option function	Reason
1	Kayak stands	Semi-circle	Can hold the kayak from sliding or moving.
2	Rear tire	Bicycle tire	Lighter compared to other designs.
3	Front tire	Round trolley tire	Wireless tire so no issue of puncture.
4	Chassis	T-shape	This chassis design for a kayak that has a length of fewer than 2.2 meters.
5	Chassis connecter	Square bearing	Able to carry more weight compared to other designs.
6	Bicycle connecter	Moon hook	The carrier can pair with many types of vehicles such as a motorcycle.

3.2. Concept Evaluation

A Pugh chart is a tool used for evaluating multiple options against each other and to consider the pros and cons of each design concept relative to design criteria. The Pugh Chart is used to compare the design concepts 1, 2, and 3. Table 5 shows the Pugh Chart of kayak carrier. The design criteria are listed in the left column. Eight criteria are considered, which are durable, portable, affordable, aesthetics, easy to use, weight, safety, and adjustable. Each criterion is given weightage according to how important it is. We use a scale of 1 to 3 where the larger scale indicates that the criteria are more important. Design 1 is chosen as a datum and set zero (0) scale. Then, each design is compared to the datum and evaluated using a scale [-, 0, +]. The total points for each design concept are calculated by multiplying the +'s and

- 's by their weights. Subsequently, the net score is calculated by subtracting the + points to – points. The design with the highest number is the best.

Based on the table 5, the best design concept selected is Design 2 with a total 4 net score. The most important criteria that have been observed before selecting the best design is how the design help kayakers carry the kayak without using lots of energy. Design 2 has fulfilled these most important criteria like easy to use, adjustable, and portable.

Table 5: Pugh chart of kayak carrier

Description		**Design Concept**		
				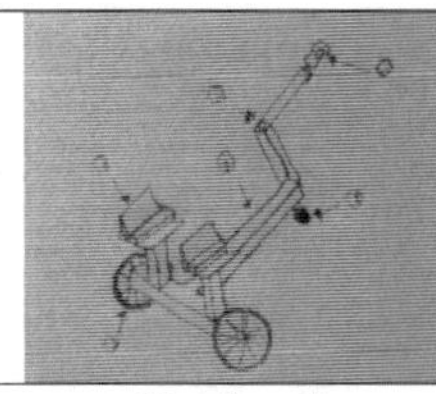
Criteria	**Weight**	**Design 1 (Datum)**	**Design 2**	**Design 3**
Durable	2	0	0	-
Portable	2	0	+	+
Affordable	2	0	-	+
Aesthetics	1	0	-	+
Easy to Use	3	0	+	-
Weight	2	0	0	+
Safety	3	0	0	-
Adjustable	2	0	+	0
	+	0	7	7
	0	17	7	2
	-	0	3	8
	Net Score	0	4	-1

Scale Indicator:
[+] = greater than datum
[0] = as per datum
[-] = lesser than datum

3.3. *Detail Design*

Solidworks 2013 x64 Edition software is used to draw the kayak carrier. Figure 4 shows the assembly drawing of the kayak carrier. The overall dimension of this kayak carrier is 2300 x 600 x 250 mm (LxWxH).

IOP Conf. Series: Materials Science and Engineering 1176 (2021) 012022 doi:10.1088/1757-899X/1176/1/012022

Figure 4: An assembly drawing of kayak carrier

3.4. Fabrication

The process of fabrication is included prepare material and components, measure raw material, cut the raw material, drilling process, welding process, and assembly of the material until the kayak carrier is completed. Figure 5 shows the fabrication process to produce the kayak carrier.

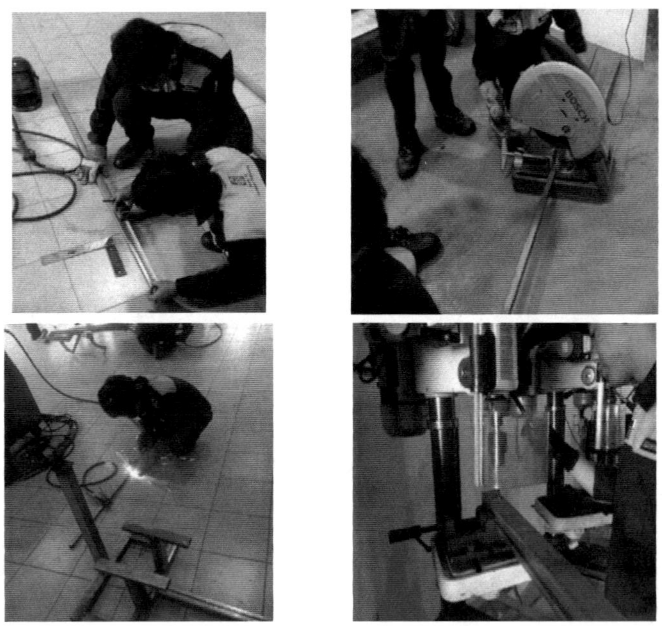

Figure 5: The fabrication process to produce the kayak carrier

3.5. Testing

After finished the fabrication and assembly process, the kayak carrier is tested to ensure a kayak can fit onto the kayak carrier. Figure 6 shows the picture of a kayak carrier that is attached to a bicycle and carries a kayak. From the test run, several issues were found, and the team has discussed and proposed solutions. Table 6 shows the three main issues of the kayak carrier according to its priority and the proposed solution to overcome.

Figure 6: The kayak carrier

Table 6: Three main issues of the kayak carrier according to its priority and the proposed solution to overcome.

No	Main Issues	A proposed solution to overcome
1	The kayak was off from its place when the bicycle was pedaled especially when cycling on a bumpy road.	To design proper kayak holder according to kayak shape e.g. protruded bar.
2	The Double-C type connector was unable to hold the kayak carrier when the bicycle braking causing the carrier to hit the rear bicycle's tire.	To use a better and reliable connector.
3	A biker will have trouble and feel heavier on pedaling when making turning at a corner.	To use a bigger tire at the front of the kayak carrier.

4. Conclusion

The design of the kayak carrier was completed as per the planned design. Fabrication processes were carried out accordingly. Then, the product has undergone testing to ensure all the safety aspects and the objectives of this project were achieved. Based on that, it is concluded with this kayak carrier, only one person needed to carry the kayak instead of two persons. For the next project, it is proposed to design a kayak carrier which able to carry two kayaks at one time.

Acknowledgments

Special thanks to UiTM Terengganu Branch, Bukit Besi Campus for the facilities and equipment provided to complete this project.

References

[1] Berger K 2009 Different Strokes for Canoeing Folks, Backpacking and Hiking *Scounting Mag.* May - June Issue.
[2] Hudson S 2007 Kayaking *Water-Based Tourism, Sport, Leisure, and Recreation Experiences* **1989** p 171–186.
[3] Industriales E D E I and Rubio A 2018 Grado en Ingeniería Mecánica Design and development of a kayak trailer.
[4] Naukowe Z, Śląskiej P and Transport S 2017 Series Transport DOES ELECTRIC CAR PRODUCE EMISSIONS *Scientific Journal of Silesian University of Technology* **94**.
[5] Silvia M-B 2017 Strength analysis of the frame of a trailer *Sci. J. Silesian Univ. Technol. Ser. Transp.* **96** pp 105–113.
[6] Ladra P and Posiadała B 2019 Modeling and Strength analysis of the prototype of the multi-tasking car trailer *Lect. Notes Mech. Eng.* **01014** pp 419–426.

[7] Design M C, David A and Advisor F 2007 *Final Project : Boat ' s Trailer Main Axis* Academic UPRM.

[8] Börekçi N A G Z 2018 Design Divergence Using the Morphological Chart *Des. Technol. Educ.*, **23** p 62–87.

[9] Honkala S, Hämäläinen M and Salonen M 2007 Comparison of Four Existing Concept *Int. Conf. Eng. Des.* p 1–11.

[10] Okudan G E and Shirwaiker R A 2006 A multi-stage problem formulation for concept selection for improved product design *Portl. Int. Conf. Manag. Eng. Technol* **6** pp 2528–2538.

[11] Augustine M, Yadav O P, Jain R and Rathore A P S 2010 Concept convergence process: A framework for improving product concepts *Comput. Ind. Eng.* **59(3)** p 367–377.

[12] Thakker A, Jarvis J, Buggy M, and Sahed A 2009 3DCAD conceptual design of the next-generation impulse turbine using the Pugh decision-matrix *Mater. Des.* **30(7)** p 2676–2684.

[13] Seperamaniam T, Jalil N A A and Zulkefli Z A 2017 Hydrostatic Bearing Design Selection for Automotive Application Using Pugh Controlled Convergence Method *Procedia Eng.* **170** p 422–429.

[14] Silaen D T H, Herdiman L and Susmartini S 2020 Design of assembly tools at woodworking and upholstery workstations with PUGH's Method to improve productivity *J. Phys. Conf. Ser.*, **1450**(1).

Palm Oil Fuel Ash and Mussel Shell Powder as Supplementary Cementitious Materials in Non-Load Concrete Brick

A Z M Zahid[1*], M M Zin[1], N Z Saifullizam[1], N Khalid[1], S F Kamaruddin[1] and W S W Chik[1]

Faculty of Civil Engineering, University Teknologi MARA (UiTM), 81750 Pasir Gudang, Johor, Malaysia

*Corresponding author: azyan8410@uitm.edu.my

Abstract. The pozzolanic properties of the industrial by-product palm oil fuel ash (POFA) has given this material a unique characteristic that can be used as a cement substitute. Meanwhile, the abundance of waste seashells which contain an enormous amount of calcium oxide (CaO) available around the world has led this material to be another potential waste material as substitutes for conventional materials in concrete. The percentage amount of POFA used as a replacement was constant at 20% from the amount of cement. On the other hand, mussel shell powder (MSP) was replaced by 2%, 4%, 6% and 8% with the decrement of 2% of OPC used for each of the design. The physical and chemical properties of the materials and the mechanical properties of concrete brick were identified for 7, 14 dan 28 days. The inclusion of POFA and MSP in the concrete brick mix in this study reduced the compressive strength of the concrete in the early days and enhanced at the late stages. Apart from that, the compressive strength of the concrete brick mixtures decreased along with the increasing percentage of MSP as the increment content of CaO from MSP reduced the hydration process. The optimum percentage of MSP replacement in this study is 4% with 20% replacement of POFA and 76% OPC. The application of these materials as supplementary cementitious materials in the construction industry, particularly in producing non-load concrete brick will simultaneously encourage reducing potential pollutions and promote sustainability.

1. Introduction

The cement industry is currently facing several challenges including the increasing demands of cement production and the necessity of diminishing energy consumption during the production of Portland cement clinker. However, the biggest challenge faced in this industry is the environmental concerns regarding greenhouse gas emissions. The cement industry is responsible for 5% to 8% of CO_2 emission in the world [1-3] and approximately 2% of the world energy consumption is attributed to the process of cement production [3,4]. These challenges have driven the cement industry to continuously expand and improve its cement formulations and production methods and meet the sustainable development goal. This includes globally reduce the consumption of non-renewable material and maintaining natural resources concepts of recycling and sustainability.

The pozzolanic properties of by-product palm oil fuel ash (POFA) has given this material a unique characteristic that can be used as a cement substitute in the mortar and concrete mixes [5]. As one of the largest producers of palm oil with 28% of world palm oil production and 33% of world exports [6], Malaysia has continuously increased its production, lead to the excessive amount of POFA produced and hence, create a large environmental-load [7]. In order to minimize the environmental load caused by this by-product, numerous investigations have been carried out to study the potential of POFA as fertilizer for agricultural purposes [8]. However, the insufficient nutrients required for fertilizers causes POFA to be mostly dumped in an open field close to palm oil mills, creates a lot of environmental pollutions and attracts criticism due to its carcinogenic and bioaccumulative effects [9-11]. To reduce the potential pollutions and health hazards caused by this agricultural by-product, as well as to reduce the production cost of concrete, POFA often being utilized as one of the supplementary cementitious materials in the concrete mix. The high silica oxide content in POFA which met the pozzolanic properties criteria has increased the potential of this agricultural waste to be utilized as cement replacement or as filler to produce a durable and strong concrete [12,13].

Meanwhile, the abundance of waste seashells available around the world has led this material to be another potential waste material as replacements for the existing conventional materials in concrete. As mentioned by [14,15], the largest producer of shellfish in the world, China, has disposed of about 10 million tonnes of waste seashells including oyster, clam, scallop, and mussel shells in landfills every year. However, only a small fraction has been reused for other purposes such as fertilizers and handicrafts due to the limitation on the amount that can be used, the problem of soil solidification and economic problems [15,16]. The remaining large quantity of waste seashells generated hence has created problems with illegal dumping into public waters and reclaimed land [15]. The waste seashells can cause foul odours or the microbial decomposition of salts into gases, such as hydrogen sulfide (H_2S), ammonia (NH_3) and amines if left untreated for over a long period [15,17]. According to [3], the uses of waste seashell in the construction sector vary from the replacement of natural aggregates in producing concretes and mortars to adding filler in the manufacturing of Portland cement. Many previous studies have shown that the chemical compositions of seashells include more than 90% calcium carbonate ($CaCO_3$) [16-21]. This composition is similar to the content of $CaCO_3$ in the limestone dust used in Portland cement production [22]. Hence, seashell can be considered as a prospective substitute for conventional limestone in the cement industry.

However, there is inadequate literature available to conclude the effect of using POFA and waste seashell as cement replacements on its workability. Therefore, in this study, the effect of POFA and a small fraction of mussel shell powder (MSP) on concrete properties were explored under several proportions and curing time in developing concrete to be applied for non-structural building element, particularly as a non-loading brick.

2. Research Method

2.1. Materials

With a specific gravity of 2.983, Ordinary Portland Cement (OPC), a type of cement that has been produced worldwide was used in this study. The sand with a specific gravity of 2.7 also was used as fine aggregate with particle size ranged 0.3 and 4.75 mm. To prepare the concrete brick mix, tap water was used and taken from the concrete laboratory. The cementitious materials as substitution used in this study were POFA and mussel shells.

2.1.1. POFA. POFA is a solid waste product obtained from the burning palm oil husk or fibre in a form of ash. The combustion usually takes place in a palm oil mill boiler. With a planted area of 332.34 hectares, Keck Seng Oil Palm Estate, a plantation under Keck Seng (M) Berhad company located in Masai, Johor has been selected as the source of by-product POFA in this study. It is then was air-dried for 24 hours to reduce the moisture. Then, POFA was oven-dried at a temperature of 105 – 110°C for another 24 hours. Initially, POFA obtained from the mill was coarser than cement size, hence it was sieved at the size of 90 μm for the purpose of supplementary cementitious material.

2.1.2. Mussel Shells. Perna viridis (family: Mytilidae) is known as the Asian green mussel, Philippine green mussel or green-lipped mussel in certain parts of the world [23]. According to [24], Johor has been the largest producer of *P. viridis* mussels in Malaysia and majorly harvested at the adjacent to the Pasir Gudang Seaport and industrial zones, such as Kampung Pasir Puteh [23,25]. Due to the abundance of *P. viridis* mussels in this area, Perkampungan Orang Asli Kampung Pasir Puteh has been selected as the sampling collection site in this study. *P. viridis* mussel is a wedge-shaped shell that can be found in saltwater and freshwater locale. This mussel is either wild-caught or farmed to meet the demand of the market that requires its flesh as a food source. However, mussel shells do not have any direct application and unrecyclable, hence often ended up in landfill or dumped in the sea. The collected mussel wastes were washed and cleaned by scrubbing off any debris such as barnacles, sand and mud spot that attached to the shells. Fully cleaned shells were then dried in the oven at 105 – 110°C for 2 hours. The oven-dried shells were crushed and finely ground before being sieved at the size of 90 µm, equivalent to cement particle size. Finally, the mussel shell powder (MSP) underwent a calcination process at a temperature of 800°C for 2 hours [26] (figure 1).

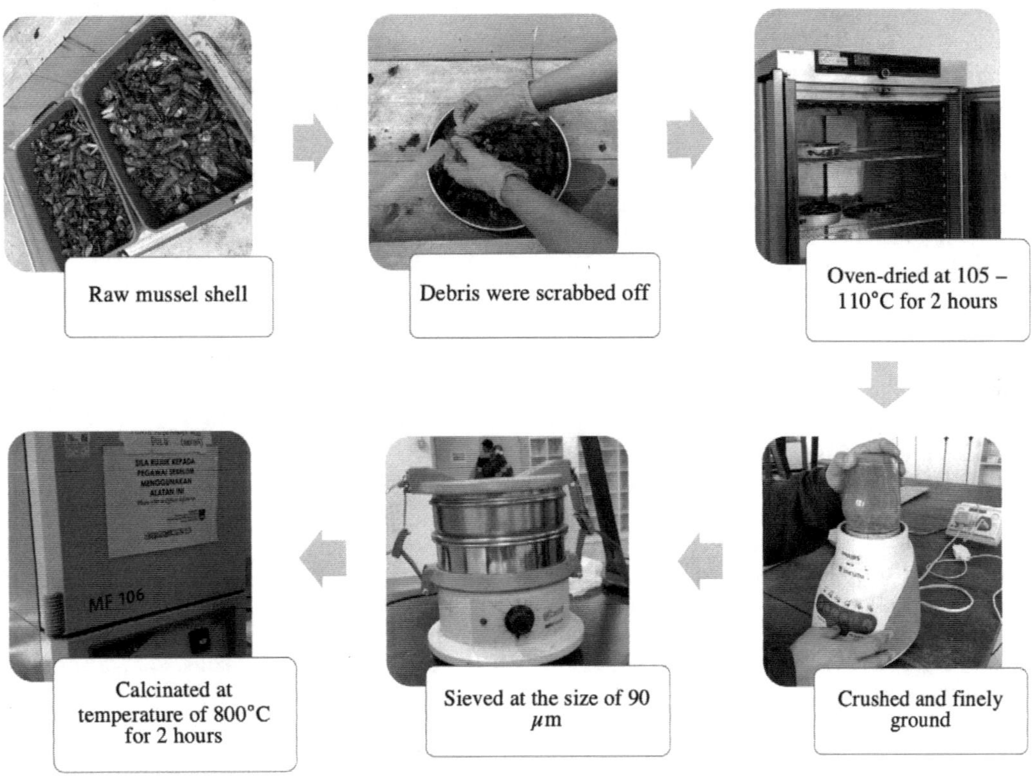

Figure 1: Preparation of MSP

2.2. *Mix Design*

The brick moulds were designed based on the standard brick size of 215 mm x 102.5 mm x 65 mm (length x depth x height) [27]. The total volume of each brick is 0.00158 m^3 including 10% of wastage consideration. The design of the non-load bearing brick was based on JKR standard specification (JKR 20800-0183-14) with a ratio of 1: 6 (cement to sand) [27]. The mix proportion was prepared based on the proportion of cement replacement with POFA and MSP (Table 1). A water/binder ratio of 0.5 was used in all mix designs. Table 1 shows the percentage and weight of materials used as partial OPC replacement. The percentage amount of POFA used to replace cement was set constant at 20% from the amount of cement, while the amount of MSP increased by 2% with the decrease of 2% of OPC used for each design. As mentioned by [28], the amount of POFA may be used up to 20% without any considerable reduction in the workability. To assess the viability of the result, control specimens with 100% OPC also had been prepared. All prepared samples were tested for compressive strength on day-7, day-14, and day-28 of room temperature.

Table 1: Quantity (kg/m^3) and percentage (%) of material proportions used in brick samples

Sample	Cement/material proportion						Sand
	OPC		POFA		MSP		
	%	kg/m^3	%	kg/m^3	%	kg/m^3	kg/m^3
M1	100	426.68	0	0	0	0	2317
M2	78	332.81	20	68.75	2	6.08	2317
M3	76	324.28	20	68.75	4	12.16	2317
M4	74	315.75	20	68.75	6	18.25	2317
M5	72	307.21	20	68.75	8	24.33	2317
M6	70	298.68	20	68.75	10	30.41	2317

2.3. Characterization of Physical Properties and Chemical Compositions

2.3.1. Particle Density. Particle density is a density of solid particles, expressed in units of grams per cubic centimetre (g/cm^3). The particle density of OPC, POFA and MSP were determined according to BS EN ISO 17892-12:2018 [29]. The determination of density was done by a small pycnometer method whereas the material was prepared for three sets to get the average value of density. All data were tabulated and calculated based on BS EN ISO 17892-12:2018 [29].

2.3.2. Fineness. The fineness of POFA is important as it can be used as a supplement or replaces the cement. This method is required to check and control any presence of coarse POFA particles that may affect the mixing process. The amount of SiO_2, the important chemical composition in concrete, increases with the increase in fineness of POFA [28]. During the test, the fineness of POFA was measured by sieving it on a standard sieve size of 90 μm, and the proportion of POFA of which the grain sizes are larger than the specified sieve size was then determined.

The fineness of MSP is also important as it affects hydration rate and sequentially, the strength of the concrete. A satisfactory value of fineness indicates whether green mussel shells can be adopted as a partial replacement for cement. This method is required to determine any presence of coarse green mussel shells particles that may affect the mixing process and indirectly controls the quality of the production process. The fineness of green mussel shells was measured and determined by the same procedures as POFA. According to BS EN 196-6:2018, the value of fineness is equivalent or should not exceed 10% [30].

2.3.3. Chemical Composition. X-Ray Fluorescence (XRF) provides an insight into chemical properties of POFA and MSP by estimating the proportion of each element accurately, whereby the elemental ratios acquired is converted into weight fractions of each oxide present in these two materials. The chemical composition determined from the procedures designates the suitability and stability of POFA and MSP, whether both can act as well as cement.

2.4. Mechanical Properties of Concrete Containing POFA and MSP

In this study, the samples of 215 mm x 102.5 mm x 65 mm in sizes were tested for compressive strength test, ultrasonic pulse velocity (UPV) test and rebound hammer test. All samples including the control mix were tested on day-7, day-14 and day-28 after curing.

2.4.1. Compressive Strength Test. Compressive strength is considered as the utmost necessary engineering property among other mechanical properties to determine the capacity of the concrete structure to resist failure under compression.

2.4.2. UPV Test. UPV test is conducted to determine the quality of the concrete by analyzing its uniformity, cracks, voids, defects, and cavities [31,32] and is assessed by determining the velocity of an ultrasonic pulse penetrating through the inner concrete structure. A low UPV value implies the presence of voids inside the concrete structure, which influence the compressive strength and quality of concrete as cited by [33].

2.4.3. Rebound Hammer Test. Rebound hammer is another non-destructive measurement that considered a reliable method in identifying deficiencies in concrete. It mainly provides an in-situ rapid indication of elastic properties and strength, mainly surface hardness and penetration resistance.

3. Results and discussion

3.1. Physical Properties and Chemical Composition

The physical properties and chemical composition of OPC, and the substitute materials, POFA and MSP are presented in table 2 and table 3 respectively.

3.1.1. Particle Density. The physical properties of POFA are strongly influenced by the burning environment, particularly the burning temperature [34]. The colour of POFA collected was dark grey to black due to unburned carbon content left at relatively low burning temperature. Based on the results obtained, the specific gravity of OPC and POFA used in the study was found to be 2.983 and 2.403 respectively as tabulated in table 2. Thus, it can be seen that the specific gravity of POFA was 19.4% lower and lighter than OPC. The grinding process has increased the specific gravity of POFA because a decrease in porosity would result in a reduction in particle size.

On the other hand, mussel shells are almost always in the form of a wedge or symmetrical shell. The shells are usually much longer and wider as shown in figure 1. As presented in table 3, the colour of MSP turned grey after the calcination process. The result obtained shows that the specific gravity of MSP used in the study is 2.126, 28.7% lower and lighter than that of OPC. This convinces that the MSP is comparable to the specific gravity of OPC. They can thus be combined to produce another material such as blended cement.

3.1.2. Fineness. Fineness is an essential asset for the hydration and pozzolanic reaction for rapid development of the brick strength. Due to this, the materials used must adhere to cement fineness in order to produce uniform hydration distribution and strength growth [35]. Table 2 depicts the fineness of OPC, POFA and MSP used in this study. On this matter, the OPC used in this research has a small grain size fineness of approximately 99.72% with respect to the 90 μm sieve. This result was supported by [36] which reported that POFA was much finer than cement as a measure in a specific surface area. Meanwhile, the particle size of MSP was marginally larger than cement at 99.50% referring to the percentage passing on the 90 μm sieve.

Table 2: Physical properties of OPC, POFA and MSP

	Physical properties			
Material	Density (kg/m^3)	Specific gravity	Fineness (%)	Colour
OPC	2983	2.983	0.28	Grey
POFA	2403	2.403	0.08	Dark grey/black
MSP	2126	2.126	0.5	Grey

3.1.3. Chemical properties. X-Ray Fluorescence (XRF) test provides insight into chemical compositions determined from the procedures designates the suitability and stability of POFA and MSP, whether both can act as well as OPC. The chemical analysis of OPC, POFA and MSP used in this study are shown in Table 4. The chemical analysis revealed that POFA consisted of SiO_2, Al_2O_3, Fe_2O_3 less than 70% and high in calcium (>5% calcium oxide, CaO) which may be classified as Class C pozzolan as indicated in ASTM C618 [37]. Based on table 4, all these ashes have an almost similar chemical composition and complimenting one another. The distinction in the chemical composition of POFA was usually due to the burning situation and the source of materials. This outcome was comparable to [38] who found that a pozzolanic material had little or no cementing properties. However, when it is at a fine size particle, with the presence of moisture, it can react with calcium hydroxide at ordinary temperatures to provide the cementing properties. Thus, the brick produced from the partial replacement of cement with POFA has reaction by silicate, SiO_2, from POFA and slaked lime, $Ca(OH)_2$, from OPC to form calcium silicate hydrate (C-S-H) which is responsible for the compressive strength.

The temperature of the waste green mussel shell preparation may decide its crystallinity level, amorphosity and compound nature [39,40]. As revealed in table 4, the results of the XRF test indicate MSP mainly consists of CaO with several minor percentages of other chemical compositions. The descriptions coincided with the findings of the [16-21]. Through the calcination process, $CaCO_3$ can be converted into CaO. The percentage of CaO depends on the method of cleaning and calcination temperature that could increase both pH and the nature of compound especially calcite or CaO peaks [39,41]. Besides, the presence of silica in the mussel shell is within the optimum acceptable condition or recommended range by ASTM C618 [37], and not exceeding 3.0%. This indicates their propensity to be pozzolanic and could also be a potential source for replacement of conventional quarried limestone that can found in the OPC. These findings also align with the findings reported by other researchers [3,20,29] as well.

Table 3: Chemical compositions of OPC, POFA and MSP

	Chemical composition								
Material	SiO_2	Al_2O_3	Fe_2O_3	CaO	MgO	TiO_2	K_2O	SO_3	P_2O_5
OPC	20.690	3.253	4.077	66.562	0.999	0.307	0.787	3.195	0.155
POFA	64.542	10.343	4.956	20.066	8.001	0.345	31.161	19.729	5.192
MSP	1.077	-	0.348	97.742	-	0.028	0.059	-	0.290

3.2. Mechanical Properties of Concrete Brick

3.2.1. Compressive Strength. The compressive strength of concrete brick samples for each mix at curing ages of 7, 14, and 28 days is shown in figure 2. The compressive strengths of day-7 for M2, M3, M4, M5 and M6 mixtures were recorded lower than the control mix M1 by approximately 29.2%, 27.1%, 60%, 64.3% and 50%, respectively. Day-14 shown increment for control mix M1, M2, M3, M4, M5 and M6 mixtures with 4.9, 3.5, 3.9, 2.4, 2.0 and 2.7 MPa. The compressive strength of the concrete bricks containing POFA and MSP continued to increase for M2, M3, M4, M5 and M6 to 3.6, 5.5, 3.3, 3.2 and 3.1 MPa on the 28th day, as presented in figure 2. Based on the findings, all concrete brick mixes shown improvement on day-28 compared to the condition on day-7 and day-14. The M3 mixture exceptionally showed a high compressive strength on day-28, gives 1.85% higher than the control mix M1 and has reached minimum permissible average compressive strength, 5.2 MPa, based on the JKR standard specification for brick [27]. This shows that the optimum percentage of MSP replacement in this study is 4% with 20% replacement of POFA and 76% OPC. This finding is parallel to that obtained by [42] that used ground cockle seashell as a partial cement replacement. The cement replacement level of MSP must not exceed 15% as suggested by [43].

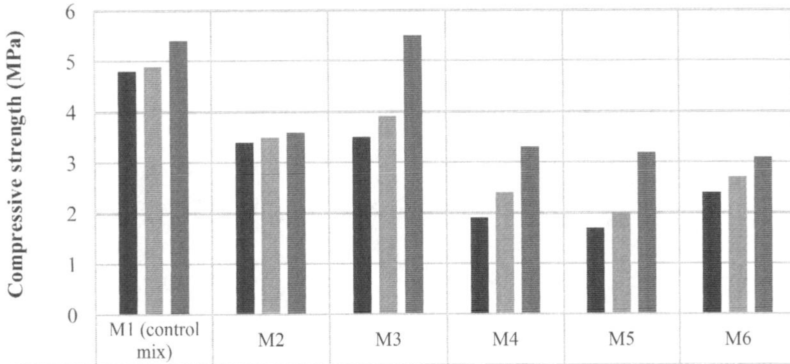

	M1 (control mix)	M2	M3	M4	M5	M6
■ Day-7	4.8	3.4	3.5	1.9	1.7	2.4
■ Day-14	4.9	3.5	3.9	2.4	2.0	2.7
■ Day-28	5.4	3.6	5.5	3.3	3.2	3.1

Figure 2: Compressive strength test on concrete brick samples containing POFA and MSP

The inclusion of POFA and MSP in the concrete brick mix reduced the compressive strength of the concrete at an early age and improved at the late age, and parallel with the findings recorded by [33] in examining nano-POFA and eggshell powder in concrete. According to [28], the ability of finer POFA particles to fill the void in the concrete structure might have increased the compressive strength at an early age. Meanwhile, at the later age, the presence of SiO_2 in the POFA reacted with the CaO contained in the MSP, formed further calcium silicate hydrate (C-S-H) and advanced the interfacial bonding in the concrete and hence, increased the strength [28]. However, the compressive strengths of the concrete brick mixtures in this study were constantly declined as the percentage of MSP increased by 2% each. The increment content of CaO from MSP reduced the hydration process and hence decreased the strength of concrete [15,44]. This is agreed by [43] who studied cockle shell ash replacement for cement and filler in concrete. Similar findings were reported by [45] concerning the compressive strength obtained in the concrete mixtures when the percentage of waste materials replacement increased. The strength also decreased as a result of the reduced OPC content when the MSP was used as a cement replacement [15,45].

The main component of MSP is $CaCO_3$, which has a similar composition as limestone, can be categorized as inert material in concrete. At a high temperature, the $CaCO_3$ can be converted into CaO that is useful in enhancing the reactivity in the concrete with the presence of pozzolans [15], such as POFA in this study. Despite limited studies have been done on the combination of MSP along with pozzolanic material by the previous researchers, [46] found that the compressive strength of the concrete was increased when fine oyster shell powder was used as addition up to 20% in cement-fly ash brick mixtures. The presence of oyster shell powder in the cement-fly ash brick mixtures aided the pozzolanic reaction with flu ash as the calcium hydroxide concentration increased [15,46].

3.2.2. Ultrasonic Pulse Velocity. The day-7 UPV values recorded between 4528 and 4871 m/s range, while the UPV values for day-28 for all mixes were recorded between 5340 and 8590 m/s, which indicate good quality concrete. The wave velocity obtained for each mixture increased with age. From figure 3, the maximum UPV recorded was 8590 m/s for M3 at an age of 28 days. This high wave velocity obtained indicates a dense internal structure of the concrete brick. The replacement level of 20% of POFA and 4% of MSP might give a high pozzolanic reaction of fine POFA and MSP to improve the interfacial C-S-H bonds and fill the internal pores. Hence, increase the quality and strength of the concrete. Similar findings were reported by [47] using a reference sample of waste oyster shells as controlled low-strength materials. Based on figure 3, the overall trends associated with the wave velocity from the UPV test and compressive strength obtained in figure 2 was parallel in this study.

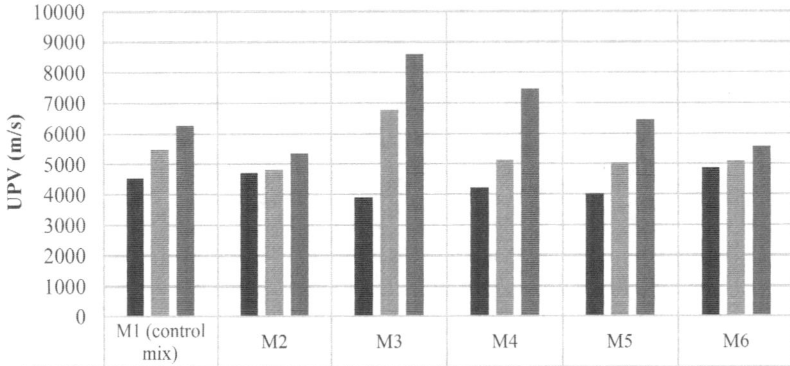

	M1 (control mix)	M2	M3	M4	M5	M6
■ Day-7	4528	4706	3907	4211	4023	4871
■ Day-14	5491	4831	6778	5155	5047	5119
■ Day-28	6263	5340	8590	7457	6438	5577

Figure 3: UPV test on concrete brick samples containing POFA and MSP

3.2.3. Rebound Hammer. As illustrated in figure 4, the early age of curing, RI value ranged between 27.8 and 41.3. The trends show a slight increment towards the late age in each of the mixtures, in line with the increase of compressive strength and UPV values as shown in previous figure 2. Although the RI values have not coincided with the overall trends of compressive strength and UPV values in all samples, however, the device is not expected to give an accurate measure of concrete compressive strength [48].

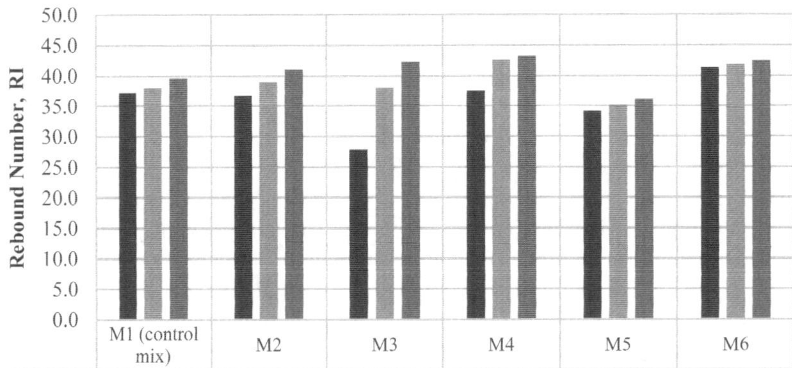

	M1 (control mix)	M2	M3	M4	M5	M6
■ Day-7	37.2	36.7	27.8	37.6	34.1	41.3
■ Day-14	38.0	38.9	38.0	42.6	35.2	41.7
■ Day-28	39.6	41.1	42.2	43.3	36.1	42.4

Figure 4: Rebound hammer test on concrete brick samples containing POFA and MSP

3.3. Microstructural Properties of Concrete

3.3.1. Scanning Electron Microscopy (SEM) of POFA and MSP. Figure 5 shows the SEM images of POFA particles. The particles were irregular in shape and having porous surfaces texture. In addition, there was no agglomeration of POFA particles after heat treatment. The descriptions were in agreement with the results of [7] and [49]. The morphology of green mussel shell calcined at 800°C was analysed by SEM as shown in figure 6. The MSP revealed a typical layer. With the calcination temperature of

IOP Conf. Series: Materials Science and Engineering 1176 (2021) 012023 doi:10.1088/1757-899X/1176/1/012023

800 °C, natural shell microstructures were significantly showed layered to porous [50,51]. The calcined waste shells were irregular in shape, and some were attached as aggregates [50].

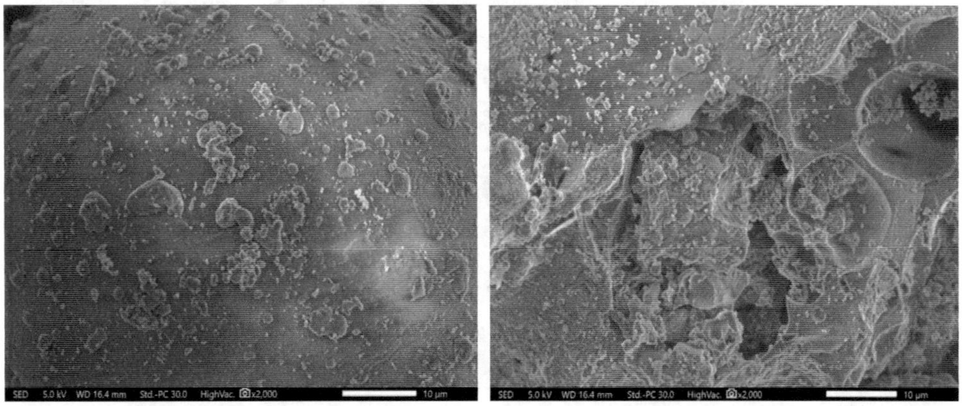

Figure 5: SEM images of POFA (2000x magnification)

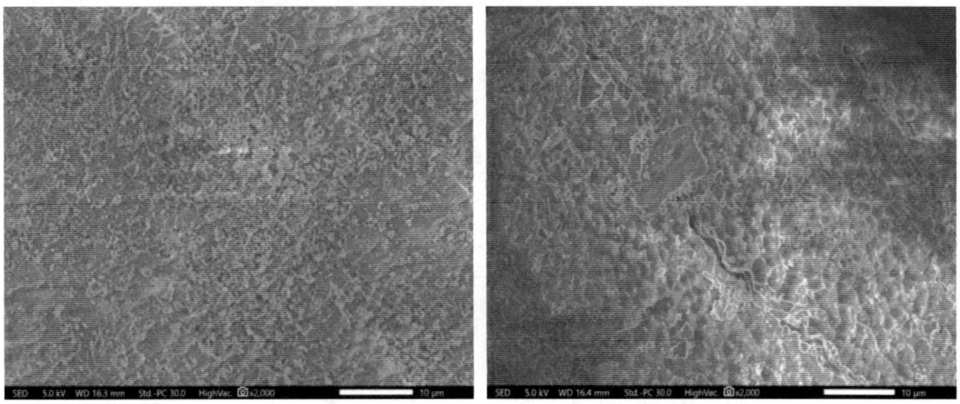

Figure 6: SEM images of MSP (2000x magnification)

3.3.2. SEM of Brick. The SEM images of brick M3 (figure 7) were taken considering its highest strength based on the compression strength test result has been conducted and recorded in figure 2. The SEM shows a flaky yet compact surface of the brick sample with minimal voids. The materials were well combined and bonded together after 28 days and the mixture reached a high compressive strength for a non-load bearing brick.

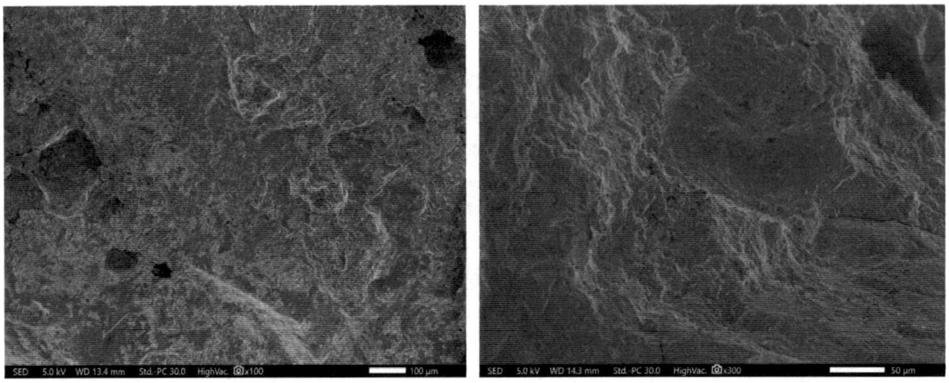

Figure 7: SEM images of brick M3 (100x magnification, left and 300x magnification, right)

4. Conclusion

Based on the assessed physical and chemical properties of OPC, POFA and MSP and mechanical properties of non-load bearing brick containing these materials as cement replacements, and the following conclusions were obtained.

- The specific gravity of POFA and MSP was 19.4% and 28.7% lower and lighter than cement. This convinces that these materials are comparable to the specific gravity of OPC and can be mixed to produce another substance such as blended cement.
- In comparison to OPC, the particle size of POFA was marginally finer than cement at 99.92% corresponding to the percentage passing on the 90 μm sieve. The main component of POFA is SiO_2.
- The result of the XRF test indicated that MSP mainly consists of CaO with minor percentages of other chemical components.
- The ability of finer POFA particles to fill the void increased the compressive strength in the concrete brick mix at an early age. Meanwhile, the formation of additional C-S-H bonds from the presence of SiO_2 in the POFA that reacts with the CaO contained in the MSP at the later stages enhanced the interfacial bonding in the concrete and hence, increased the strength.
- Except for M3, the compressive strength of the concrete brick mixtures decreased along with the increasing percentage of MSP. The increment content of CaO from MSP reduced the hydration process, thus decreased the strength of concrete.
- The UPV values obtained from the UPV test increased with age. The high pozzolanic reaction of fine POFA and MSP to improve the interfacial C-S-H bonds and fill the internal pores. Hence, increased the quality and strength of the concrete.
- The rebound hammer test showed a slight increment towards the late age of all mixtures.
- The optimum percentage of MSP replacement in this study is 4% with 20% replacement of POFA and 76% OPC.

As mentioned by many previous researchers, introducing POFA in the mix designs had extensive outcomes on concrete properties, exceptionally at a late age. Meanwhile, due to its filler effect, the MSP could add a positive influence but reduced the mechanical properties of the structure. A suitable portion of MSP, however, may improve the interfacial C-S-H bonds and increase the strength of the concrete. Based on the findings, it is certainly interesting to determine the effect of concrete brick on the insulation properties, especially on thermal and sound, for further research in the future. With the high production of POFA and abundance disposition of mussel shells are forecasted to increase in the future, the combine application of these two as supplementary cementitious materials in the construction industry, particularly in producing non-load concrete brick will reduce the environmental glitches and promote sustainability.

Acknowledgements
Utmost gratitude to the Faculty of Civil Engineering, UiTM Johor Branch, Pasir Gudang Campus for the support towards completing this study. Special thanks to Keck Seng (M) Berhad, Masai and the residents of Kampung Pasir Puteh, Pasir Gudang for the cooperation and the material provided, as well as the involving parties for assistance, providing insight and expertise with us.

References
[1] Celik K, Jackson M D, Mancio M, Meral C, Emwas A H, Mehta PK and Monteiro P J M 2014 High-volume natural volcanic pozzolan and limestone powder as partial replacements for portland cement in self-compacting and sustainable concrete *Cem. Concr. Compos.* **45** 136–147.
[2] Ulusu H, Aruntas H Y and Gencel O 2016 Investigation on characteristics of blended cements containing pumice *Constr. Build. Mater.* **111** 11–19.
[3] Soltanzadeh F, Emam-Jomeh M, Edalat-Behbahani A and Soltan-Zadeh Z 2018 Development and characterization of blended cements containing seashell powder *Constr. Build. Mater.* **161** 292–304.
[4] Ghiasvand E, Ramezanianpour A A and Ramezanianpour A M 2014 Effect of grinding method and particle size distribution on the properties of Portland-pozzolan cement *Constr. Build. Mater.* **53** 547–554.
[5] Awal A S M A and Nguong S K 2010 A Short-Term Investigation On High Volume Palm Oil Fuel Ash (POFA) Concrete *35th Conf Our World Concr Struct (Singapore)* 1–9.
[6] MPOC 2021 Malaysia Palm Oil Industry Available at: *http://mpoc.org.my/malaysian-palm-oil-industry/* Accessed on 15 Feb 2021.
[7] Abdullah K, Hussin M W, Zakaria F, Muhamad R and Hamid Z A 2006 POFA: A Potential Partial Cement Replacement Material in Aerated Concrete *6th Asia Pacific Struct. Eng. Constr. (APSEC 2006) (Kuala Lumpur, Malaysia)* 132–140.
[8] Yin C Y, Kadir S A S A, Lim Y P, Syed-Ariffin S N and Zamzuri Z 2008 An investigation into physicochemical characteristics of ash produced from combustion of oil palm biomass waste in a boiler *Fuel Process Technol.* **89** 693–696.
[9] Sumadi S R and Hussin M W 1995 Palm oil fuel ash (POFA) as a future partial cement replacement material in housing construction *J. Ferrocem.* **25** 25–34.
[10] Foo K Y and Hameed B H 2009 Value-added utilization of oil palm ash: A superior recycling of the industrial agricultural waste *J. Hazard. Mater.* **172** 523–531.
[11] Subramaniam V, Ma A N, Choo Y M and Sulaiman NMN 2008 Environmental performance of the milling process of Malaysian palm oil using the life cycle assessment approach *Am. J. Env. Sci.* **4** 310–315.
[12] Munir A, Abdullah, Huzaim, Sofyan, Irfandi and Safwan 2015 Utilization of palm oil fuel ash (POFA) in producing lightweight foamed concrete for non-structural building material *Procedia Eng.* **125** 739–746.
[13] Awal A S M A and Hussin M W 1997 The effectiveness of palm oil fuel ash in preventing expansion due to alkali-silica reaction *Cem. Concr. Compos.* **19** 367–372.
[14] Yao Z, Xia M, Li H, Chen T, Ye Y and Zheng H 2014 Bivalve shell: Not an abundant useless waste but a functional and versatile biomaterial *Crit. Rev. Env. Sci. Tec.* **44** 2502–2530.
[15] Mo K H, Alengaram U J, Jumaat M Z, Lee S C, Goh W I and Yuen C W 2018 Recycling of seashell waste in concrete: A review *Constr. Build. Mater.* **162** 751–764.
[16] Yang E I, Yi S T and Leem Y M 2005 Effect of oyster shell substituted for fine aggregate on concrete characteristics: Part I. Fundamental properties *Cem. Concr. Res.* **35** 2175–2182.
[17] Yoon G L, Kim B T, Kim B O and Han S H 2003 Chemical-mechanical characteristics of crushed oyster-shell *Waste Manag.* **23** 825–834.
[18] Mo K H, Alengaram U J, Jumaat M Z, Yap S P and Lee S C 2016 Green concrete partially comprised of farming waste residues: A review *J. Clean. Prod.* **117** 122–138.

[19] Safi B, Saidi M, Daoui A, Bellal A, Mechekak A and Toumi K 2017 The use of seashells as a fine aggregate (by sand substitution) in self-compacting mortar (SCM) *Constr. Build. Mater.* **78** 430–438.
[20] Martínez-García C, González-Fonteboa B, Martínez-Abella F and Carro- López D 2017 Performance of mussel shell as aggregate in plain concrete *Constr. Build. Mater.* **139** 570–583.
[21] Cuadrado-Rica H, Sebaibi N, Boutouil M and Boudart B 2016 Properties of ordinary concretes incorporating crushed queen scallop shells *Mater. Struct. Constr.* **49** 1805–1816.
[22] Shad M W, Gregory C, Frank X W, Damian S and Thomas J K 2012 Effects of lead on Na+, K+-ATPase and hemolymph ion concentrations in the freshwater mussel Elliptio complanata. *Environ. Toxicol.* **27** 268–276.
[23] Mahat N A, Muktar N K, Ismail R, Abdul Razak F I, Abdul Wahab R and Abdul Keyon A S 2018 Toxic metals in Perna viridis mussel and surface seawater in Pasir Gudang coastal area, Malaysia, and its health implications *Environ. Sci. Pollut. Res.* **25** 30224–30235.
[24] DOF Malaysia 2013 *Annual Fisheries Statistics* (Putrajaya: Department of Fisheries)
[25] Yap C K, Peng S H T and Leow C S 2019 Contamination in Pasir Gudang Area, Peninsular Malaysia: What can we learn from Kim Kim River chemical waste contamination? *J. Humanit. Educ. Dev.* **1** 84–87.
[26] Nordin N, Hamzah Z, Hashim O, Kasim F H and Abdullah R 2015 Effect of temperature in calcination process of seashells *Malaysian J. Anal. Sci.* **9** 65–70.
[27] JKR Malaysia 2014 Standard Specifications for Building Works 2014 (JKR 20800-0183-14)
[28] Thomas B S, Kumar S and Arel H S 2017 Sustainable concrete containing palm oil fuel ash as a supplementary cementitious material – A review *Renew. Sustain. Energy Rev.* **80** 550–561.
[29] BS EN ISO 17892-12 2018 *Geotechnical investigation and testing. Laboratory testing of soil. Determination of liquid and plastic limits* (London: British Standards Institution)
[30] BS EN196-6 2018 *Methods of testing cement. Determination of fineness* (London: British Standards Institution)
[31] Komloš K, Popovics S, Nürnbergerová T, Babál B and Popovics J S 1996 Ultrasonic pulse velocity test of concrete properties as specified in various standards *Cem. Concr. Compos.* **18** 357–364.
[32] IAEA 2002 *Guidebook on non-destructive testing of concrete structures* (Vienna: International Atomic Energy Agency).
[33] Hamada H, Tayeh B, Yahaya F, Muthusamy K and Al-Attar A 2020 Effects of nano-palm oil fuel ash and nano-eggshell powder on concrete *Constr. Build. Mater.* **261** 119790.
[34] Safiuddin M, Salam M A and Jumaat M Z 2011 Utilization of palm oil fuel ash in concrete: A review *J. Civ. Eng. Manag.* **17** 234–247.
[35] Binag N D 2016 Powdered shell wastes as partial substitute for masonry cement mortar in binder, tiles and bricks production *Int. J. Eng. Res. Technol.* **5** 70–77.
[36] Hussin M W, Muthusamy K, Arifin M A M, Ariffin N F and Lim N H A S 2016 P*ofa Concrete Theory and Application 1st Edition* (Pahang: Penerbit Universiti Malaysia Pahang).
[37] ASTM C618 2019 *Standard Specification for Coal Fly Ash and Raw or Calcined Natural Pozzolan for Use in Concrete* (Pennsylvania: ASTM International)
[38] Sooraj V M 2013 Effect of Palm Oil Fuel Ash (POFA) on Strength Properties of Concrete. I*nt. J. Sci. Res. Publ.* **3** 1–7.
[39] Tayeh B A, Hasaniyah M W, Zeyad A M and Yusuf M O 2019 Properties of concrete containing recycled seashells as cement partial replacement: A review *J. Clean. Prod.* **237** 117723.
[40] Ez-Zaki H, Diouri A, Kamali-Bernard S and Sassi O 2016 Composite cement mortars based on marine sediments and oyster shell powder. *Mater. Constr.* **66.**
[41] Wan Mohammad W A S, Othman N H, Wan Ibrahim M H, Rahim M A, Shahidan S and Rahman R A 2017 A review on seashells ash as partial cement replacement. *IOP Conf. Ser.: Mater. Sci. Eng.* **271** 012059.

[42] Olivia M, Mifshella A A and Darmayanti L 2015 Mechanical properties of seashell concrete *Procedia Eng.* **125** 760–764.
[43] Othman N H, Bakar B H A, Don M M and Johari M A M 2013 Cockle shell ash replacement for cement and filler in concrete *Malaysian J. Civ. Eng.* **25.**
[44] Lertwattanaruk P, Makul N, Siripattarapravat C 2012 Utilization of ground waste seashells in cement mortars for masonry and plastering. *J. Environ. Manage.* **111** 133–141.
[45] Khankhaje E, Rafieizonooz M, Salim M R, Mirza J, Salmiati and Hussin M W 2017 Comparing the effects of oil palm kernel shell and cockle shell on properties of pervious concrete pavement *Int. J. Pavement Res. Technol.* **10** 383–392.
[46] Li G, Xu X, Chen E, Fan J and Xiong G 2015 Properties of cement-based bricks with oyster-shells ash *J. Clean. Prod.* **91** 279–287.
[47] Kuo W T, Wang H Y, Shu C Y and Su D S 2013 Engineering properties of controlled low-strength materials containing waste oyster shells *Constr. Build. Mater.* **46** 128–133.
[48] Sanchez K and Tarranza N 2014 Reliability of Rebound Hammer Test in Concrete Compressive Strength Estimation *Int. J. Adv. Agric. Environ. Eng.* **1.**
[49] Brown O R, Yusof M B B M, Salim M R Bin and Ahmed K 2011 Physico-chemical properties of palm oil fuel ash as composite sorbent in kaolin clay landfill liner system *2011 IEEE 1st Conf. on Clean Energy and Technology (Kuala Lumpur, Malaysia)* 269–274.
[50] Buasri A, Chaiyut N, Loryuenyong V, Worawanitchaphong P and Trongyong S 2013 Calcium oxide derived from waste shells of mussel, cockle, and scallop as the heterogeneous catalyst for biodiesel production. *Sci. World J.* **2013** 1–7.
[51] Geist J, Auerswald K and Boom A 2005 Stable carbon isotopes in freshwater mussel shells: Environmental record or marker for metabolic activity? *Geochim Cosmochim Acta.* **69** 3545–3554.

Effect of Petroleum-Based Oil Towards Physicochemical Properties of Natural Rubber-Based Magnetorheological Elastomers

N M Zaki[1*], N A Yunus[1], M S Yusoff[1], F Ahmad[1] and N Wahid[2]

[1]Department of Mechanical Engineering, Universiti Teknologi PETRONAS, 32610, Seri Iskandar, Perak, Malaysia.

[2]Faculty of Electrical Engineering, University Teknologi MARA (UiTM), 23000 Dungun, Terengganu.

*Corresponding author: zatie_nori@yahoo.com

Abstract. This study presents an investigation of the effect of magnetorheological elastomers (MREs) using petroleum based-oil (PBO) as dispersing aids. The inhomogeneous dispersion and mobility of magnetic particles within the elastomer matrix result in poor MREs performance. Isotropic natural rubber (NR) based MREs were fabricated by homogeneously mixed with the addition of different ratios of naphthenic oil (NO) to aromatic oil (AO), whereby NO 1(100:0), NO 2 (70:30), NO 3 (50:50), and NO 4 (30:70), correspondingly. The cure characteristics, morphological, magnetic properties and thermal analyses were conducted. Based on the results, NO 1 leads the best results for microstructure observation and magnetic properties of the MREs. Meanwhile, NO 2 mixed with AO showed an improvement in curing characteristics and thermal analyses compared to other samples. In summary, the ratio of NO 1 and NO 2 is considered the preferable dispersing aids for MREs compared to NO 3 and NO 4 due to the structural compound in the oil that enhanced the characteristics of the NR-based MREs.

1. Introduction

A magnetorheological elastomer (MRE) belongs to smart material groups, which their rheological properties can be regulated continuously, reversely and rapidly controlled by an external magnetic field within an elastomer matrix [1]. These benefits provide a large variety of engineering fields in several applications including prosthetic devices, spring elements, automotive bushing, medical devices, and vibration absorbers [2-4]. MREs can be fabricated in two different states either isotropic and anisotropic which are known as with or without magnetic field is applied during the curing process, respectively. MRE is a composite material that consists of the main components such as matrix, magnetic particles and additives where all have shown great influences on the properties of MREs. The researchers found out that the selection of matrix for MREs is an essential part. Most MREs research groups used different types of rubber materials such as natural rubber (NR), silicone rubber (SiR) and polyurethane rubber (PU) as matrix [4]. In this study, NR was chosen as an alternative matrix. NR is a non-magnetic viscoelastic material and is well-known to exhibit an ideal matrix for excellent MREs performances, such as good dynamic performance, high elasticity and abundant availability [6].

Conventionally, dispersing aid such as silicone oil is used in the rubber processing for SiR to help the dispersion of filler and additive during that process. In this work, petroleum based-oil (PBO) was

IOP Conf. Series: Materials Science and Engineering 1176 (2021) 012024 doi:10.1088/1757-899X/1176/1/012024

chosen as dispersing aid for unsaturated rubber matrix such as NR to enhance the rubber's flexibility, reduce the viscosity of the matrix and improve processability [9]. The most commonly used PBO as dispersing aids is aromatic oil (AO) which has good compatibility with the rubber. The main characteristics of AO as dispersing aid such as good compatibility with the rubber and acts as a softening agent to improve the workability of rubber composition [7]. Based on these properties, AO may have the potential to improve the MR performance of MREs when mixed with other PBO [6]. Aziz et al. [5] have discovered other several oils which include epoxidized palm oil (EPO), petroleum based-oils (PBO) involving naphthenic oil (NO) and light mineral oil (LMO) as additives for NR-based MREs. These oils showed a potential towards improving the dispersibility of magnetic particles within MREs performance. Based on their findings, the agglomeration of carbonyl iron particles (CIPs) still occurs in the MREs sample, where it causes a decrease of crosslink density and leads to poor MR performance. The mixture of NO:AO may reduce cohesive forces between polymer chains due to a high level of unsaturated rings and high reactivity with the rubber [6]. The characteristics of PBO have shown a significant effect on the process of rubber compounding and mechanical properties of the vulcanizate [8]. The addition of the proposed additives is expected to enhance the dispersion ability of the magnetic particles and consequently improves the MRE performance [1,4,8]. Therefore, in this work, the evaluation of different ratios of NO:AO towards the performance of MRE are experimentally investigated and evaluated. NR-based MRE results in terms of cure characteristics, micrograph, thermal analysis, and magnetic properties were compared and discussed.

2. Research Method

2.1. Materials Preparation and Fabrication of MREs samples

The main components of MREs are elastomer matrix, magnetic particles and additives. Generally, the fabrication process of isotropic MREs consists of two stages: mixing and curing. Natural rubber (NR) was used as a matrix and prepared by the Malaysian Rubber Board. The CIPs were purchased from Sigma Aldrich, Germany and used as a magnetic particle in this study. The petroleum based-oil (PBO), which includes naphthenic oil (NO) and aromatic oil (AO) were purchased by the Malaysian Rubber Board. Zinc oxide (ZnO) and stearic acid were used as an activator. Meanwhile, carbon black, sulphur and cyclohexyl benzothiazole-2-sulphenamide (CBS) were added as the function as reinforcing agents, crosslinking agent and accelerator, respectively. The schematic diagram of the fabrication process is shown in figure 1. The first stages of the mixing process were carried out using the conventional double-roll mill. The addition of ZnO, stearic acid, carbon black, NO and AO in the rubber. Then, sulphur and CBS were added and finally, the addition of CIPs was fixed at 60 wt%. Subsequently, the measurement of the curing time for isotropic MREs was determined using the Monsato Moving Die Rheometer (MDR 2000) at 150°C without the presence of a magnetic field. The final MREs compound was fabricated into 1.0 mm thickness. The ratios of NO to AO are shown in table 1 and the MREs samples were named as sample NO 1, NO 2, NO 3 and NO 4.

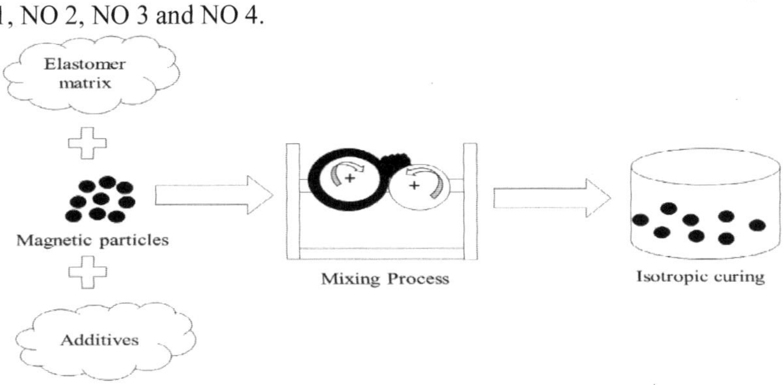

Figure 1: A schematic diagram of the fabrication process of MREs

Table 1: Compound ingredients for different ratios of NO to AO

Sample	Ratio of PBO	Amount (phr)	
		Naphthenic oil	Aromatic oil
NO 1	100:0	10	0
NO 2	70:30	7	3
NO 3	50:50	5	5
NO 4	30:70	3	7

*Phr = parts per hundreds of rubbers

2.2. Characterizations of MRE samples

The cure characteristics were examined using a moving die rheometer (MDR) at a constant temperature of 150°C for 30 minutes. The parameters of cure characteristics were obtained such as cure time (t_{90}) and scorch time (t_{S1}). The vulcanization time of the MRE sample corresponds to the optimum cure time (t_{90}) was calculated according to Equation (1).

$$t_{90} = M_L + \frac{90}{100}(M_H - M_L) \quad (1)$$

M_L : Minimum torque value
M_H : Maximum torque value

The morphology of the MRE samples was carried out through a Field Emission Scanning Electron Microscope (FESEM). The cross-section of samples was determined with a magnification of 500x at an accelerating voltage of 5kV during the experiment process. To prevent charging during the observation process, all samples were coated with gold. The magnetization curves of MR materials were measured at room temperature using a vibrating sample magnetometer (VSM). All the samples have been varied and were conducted under the magnetic field of -8000 to 8000 Oe. Thermal characteristics of the NR-based MREs were performed via thermogravimetric analysis (TGA) in a nitrogen atmosphere. The mass of each MREs sample was measured in the range of 8-10 mg based on standard sample preparation. The samples were heated at a temperature of approximately 25 to 800°C under a heating rate of 10°C/min.

3. Results and Discussion

3.1. Cure Characteristics

Figure 2 shows the cure curves of NR-based MREs with different ratios of NO:AO. The torque values increase with the increase of NO contents in NR-based MREs. It can be seen that reversion curing curves were acquired for the sample NO 1, NO 2, NO 3 and NO 4. Reversion curves occur at the final stage when some compounds are overheating during the test that corresponds to the break down of rubber networks. As compared to NO 1, NO 2 and NO 3, NO 4 shows the most reversion where torque decreases with time as shown in figure 2. This is probably due to the breakdown of the ring structure in oil between NR molecular chains.

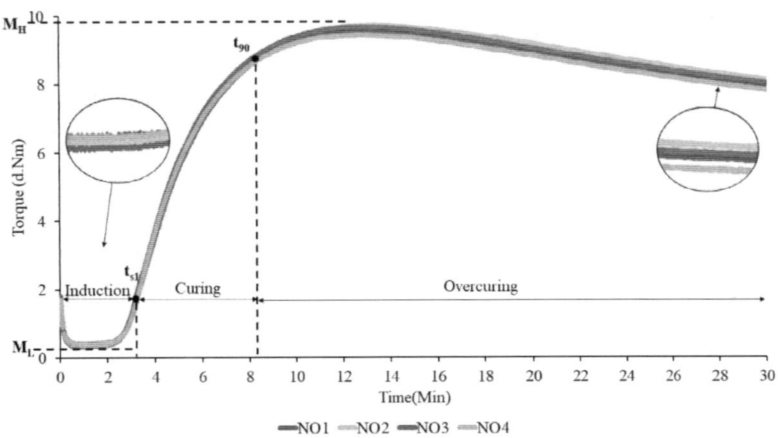

Figure 2: The curves torque versus time with different ratios of NO:AO

The different ratios of NO:AO affects the cure characteristics of the NR-based MREs including the value of scorch time (t_{S1}) and the optimum cure time (t_{90}). According to table 2, NO 2 demonstrated the highest cure time in which the value is 3.79 minutes. This could be due to the addition of AO within the samples and consequently delay the curing process. Chandrasekara et al. [10], reported that NR compound containing AO has the highest cure time during the vulcanizing process due to the energy consumption as well as showed good processing safety.

Table 2: Cure characteristics of the NR-based MREs

Sample	t_{S1} (min)	t_{90} (min)
NO 1	3.42	8.64
NO 2	3.41	8.79
NO 3	3.43	8.75
NO 4	3.49	8.54

3.2. Microstructure Observation

Figure 3 (a)-(d) shows the microstructure of isotropic MREs samples with different ratios of NO:AO. All MREs samples exhibit an isotropic condition and show a homogenous distribution of CIPs within the MREs. There were significant differences in the microstructure of MREs samples when AO was added with the NO. According to figure 3 (a), a small agglomeration is observed in MREs sample containing NO without mixing with AO and the void still occurs in the sample. It can be seen that CIPs can blend well with the matrix and dispersing aids. This is due to the high content of NO that is compatibility with NR and more stable as compared to AO. Meanwhile, in Figure 3 (b), (c) and (d), the MREs samples formed large aggregations and void in the presence of more AO. The unsaturated rings in the AO could be the reason for the large agglomerations that occurred.

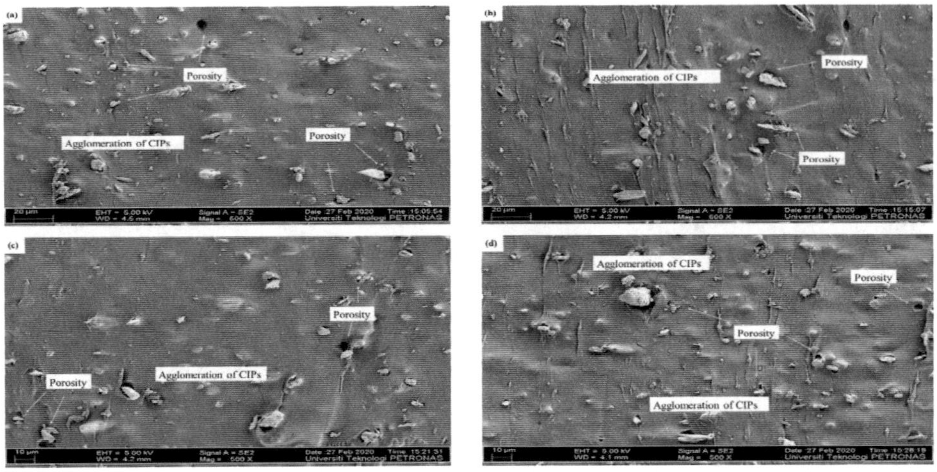

Figure 3: The microstructure of MREs at different ratio of NO:AO. (a) NO 1, (b) NO 2, (c) NO 3 and (d) NO 4

3.3. Magnetic Properties

Figure 4 illustrates the magnetization curves of NR-based MREs with different ratio of NO:AO, while table 3 presents the magnetic parameters that had been obtained from figure 4. Based on the magnetization curves shown in figure 4, the saturation magnetization of NO 1 has the highest magnetization curves. It could be due to the high content of NO which the particles homogeneously dispersed in the samples. Hence, the NO 1 has high compatibility with the rubber as compared to the NO 2, NO 3 and NO 4. The magnetization curves of NO 3 showed the lowest in figure 4. This is referred to the movement of the particles, where the ratio of NO:AO is 50:50 had reached saturation magnetization easily. Besides, the saturation magnetization value for NO 3 was insignificant with the NO 4.

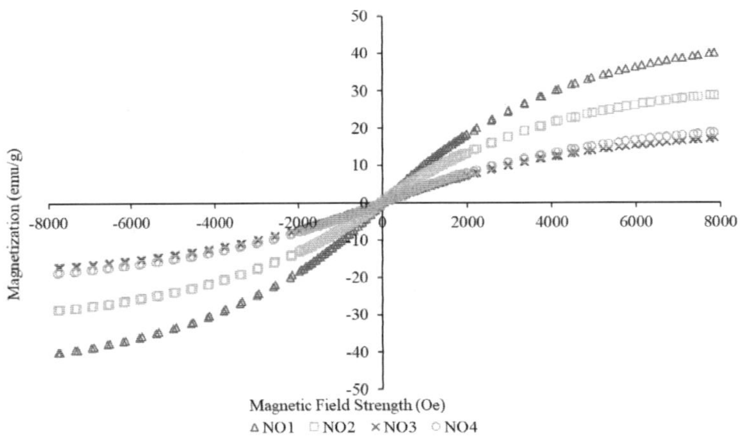

Figure 4: Magnetization curves of NR-based MREs with different ratio of NO:AO

Table 3 represents the saturation magnetization (M_s), retentivity magnetization (M_R) and coercivity (H_c) at different ratio of PBO in which the values are obtained from figure 4. It can be seen that the rise of NO content contributes to the enhancement of magnetic properties of the NR-based MREs. This is proven by the increasing of M_s with NO content as referred in table 3. The NO 1 showed the highest magnetic saturation and followed by NO 2, NO 4 and NO 3 with 40.06, 28.73, 18.74 and 17.23 emu/g, respectively. The compatibility and interaction between matrix and filler are the predominant factors in improving the performance of MREs. By using the different ratio of NO:AO, the magnetic properties had significantly changed. It can be seen the MREs with NO 1 has the highest magnetic saturation as stated in table 3.

Table 3: Magnetic properties of MREs with different ratio of NO:AO

Samples	M_s (emu/g)	M_R (emu/g)	H_c (Oe)
NO 1	40.06	0.18	16.38
NO 2	28.73	0.13	15.57
NO 3	17.23	0.075	16.77
NO 4	18.74	0.078	16.36

3.4. Thermal Characteristics

The thermal stability of a polymer can be evaluated by using TGA. The effect of NR-based MREs with different ratio of NO:AO was measured using thermogravimetric analysis in a nitrogen atmosphere are demonstrated in figure 5. The thermal decomposition of MREs samples consists of two types which are known as the first stage and second stage shown in figure 5. The first stage of thermal decomposition begins at a temperature between 260 to 343°C while the second stage occurred at 315 to 465°C. The initial decomposition of all the samples occurred during the first stage. Afterwards, TG patterns were dramatic decrease during the second stage. According to figure 5, NO 1 and NO 4 showed higher weight loss as compared to other samples. This is due to the presence of different ring structure in oil. The NO consists of a saturated ring structure while AO has an unsaturated ring structure. These rings were reacted with CIPs via a chemical reaction and consequently cause the corrosion occurred and affect the CIPs in the MREs samples at the high temperature. In addition, the presence in the matrix of different structural NO and AO rings could occupy the space between the rubber chains. However, due to the labile structures in the rubber chains, the material easily decomposed at low temperatures decreases so that it breaks easily.

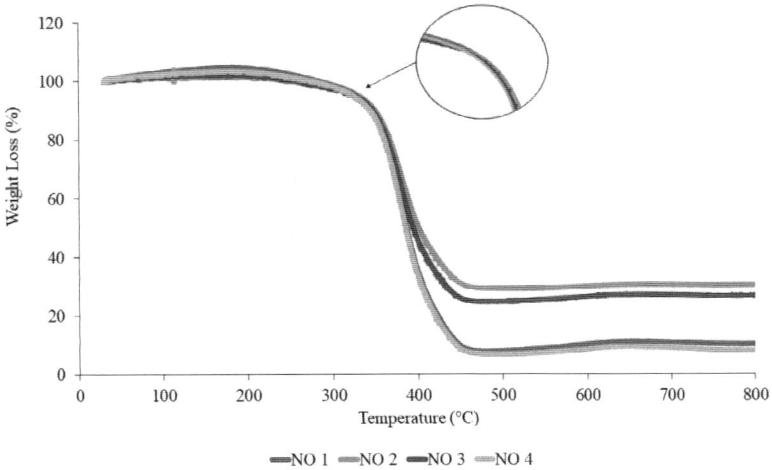

Figure 5: TGA curves of NR-based MREs with different ratio of NO:AO

The degradation temperature of MREs with different ratios of NO:AO obtained from thermograms are summarized in table 4. T_{onset} represents the temperature at which the NR matrix begins to degrade while the T_{end} corresponds to the end of degradation. It can be seen that the T_{onset} for NO 1 is lower than the other samples as shown in table 4. T_{onset} from NO 1 to NO 2 increased from 260 to 300°C and then decreased from 290 to 280°C with an increase in AO content. High NO content has delayed degradation.

The effect of NO on the thermal decomposition of MREs could be due to the presence of a saturated ring which delays the degradation of the MREs. The saturated ring structure in the oil may take longer to break down the rubber chains. The addition of dispersing aids could occupy the space between the rubber chains. Other than that, it assumes that the different ratios of NO to AO, have had a significant impact on the characteristic NR temperature and weight loss.

Table 4: Thermal degradation temperature of MREs with different ratios of NO:AO

Samples	T_{onset}(°C)	T_{end}(°C)	Ash at 800°C (%)
NO 1	260	450	10
NO 2	300	465	30
NO 3	290	457	26
NO 4	280	454	8

4. Conclusion

In this work, the influences of different ratios of NO to AO as dispersing aids with 60 wt% CIPs on the performance of isotropic NR-based MREs was investigated. The results revealed that scorch time decrease and optimum time increased for the NO 2. The results of morphology revealed the formation of CIPs in an isotropic state. The addition of PBO as dispersing aids allowed the molecular chains of rubber to glide easily and consequently reduce the gap and agglomeration of magnetic particles. The NO showed an improvement in morphology observation and magnetic properties due to the stronger bonding of NR and CIPs which favorable for MREs applications since MREs are exposed to various magnetic field strength. Sample NO 1 has indicated the best ratio for morphology observation and magnetic properties. In a nutshell, NO has a great influence on MRE properties of NR-based MREs due to the good compatibility between rubber and oil and could be a promising material for certain MRE applications. Studied mechanical properties such as tensile properties and damping performance could be considered for future work particularly for implementation of MRE in automotive and civil industries.

Acknowledgements

The authors gratefully acknowledge support from the Yayasan Universiti Teknologi PETRONAS (YUTP-FRG-015LC0-135) and the Ministry of Education Malaysia under FRGS Grant (015MA0-086)

References

[1] Kwon S H, Lee J H, and Choi H J 2018 *Materials (Basel).* **11** 6.
[2] Jung H S, Kwon S H, Choi H J, Jung J H, and Kim Y G 2016 *Compos. Struct.* **136** 106 112.
[3] Yunus N , Mazlan S A, Ubaidillah, Aziz S A A Shilan S T and Abdul Wahab N A 2019 *Int. J. Mol. Sci.* **20** 3 1–19.
[4] Ahmad Khairi M H, Mazlan SA, Choi S, Aziz S A A, Mohamad N, Hapipi N M and Nordin N 2019 *Adv. Eng. Mater.* **21** 1800696.
[5] Aziz S A A, Mazlan S A, Ismail N I N, Ubaidillah, Choi S B, Nordin N A and Mohamad N 2018 *Smart Mater. Struct.* **27** 11 117002.
[6] Kang S S, Choi K, Nam J and Choi H J, 2020 *Materials (Basel).* 1**3** 20 4597.
[7] Li X, Tan S, Liu G, Hoch M and Zhao S, 2016 *J. Macromol. Sci. Part B Phys.* **55** 5 494–502.
[8] Yunus N A, Mazlan S A, Ubaidillah, Choi S B, Imaduddin F, Abdul Aziz S A and Ahmad Khairi M H 2016 *Smart Materials and Structures* **25** 10 1–11.
[9] Haslina N, Abdul N, Ismail H, Abdullah M K and Shuib R K 2020 *J. Appl. Polym. Sci.* **137** 13 48492.
[10] Chandrasekara G, Mahanama M, Edirisinghe D and Karunanayake L 2011 *J. Natl. Sci. Found. Sri Lanka* **39** 3 243–250.

Fire-Resistant Properties of Modified Epoxy Based Intumescent Fire-Retardant Coating: Hybridizing of Oyster Shell Powder with Glass Frit

Y X Lee[1*], F Ahmad[1], S Kakooei[1], Y Azmi1, and S Kabir[1]

Mechanical Engineering Department, Universiti Teknologi PETRONAS Bandar Seri Iskandar, 32610, Perak, Malaysia.

*Corresponding author: yuan_20001965@utp.edu.my

Abstract. Fire retardant coatings play vital role in safety of construction nowadays, to delay the fire propagation and provide more time for evacuation. The current research work aimed to study the use of modified siloxane epoxy binder to develop intumescent coating. Polyamide amine was used as a hardener. Three different sets of formulations were developed by using three different fillers, namely the oyster shell powder (OSP), glass frit (GF) and combination of both oyster shell powder and glass frit (OG). All three set of formulation also includes the basic ingredients of intumescent coating. These samples were used to study the synergistic effects of fillers on thermal performance of the coating as well tested for char expansion, heat shielding, char morphology and composition. Fire resistant testing was conducted by using a Bunsen Burner and GF samples had found to have minimum substrate temperature of 175.3°C. The char morphology was analysed via Scanning Electron Microscope (SEM) which confirmed the adhesion between the matrix with fillers. Furnace test was carried out to investigate the char expansion of OSP/GF/OG with the modified siloxane epoxy composition and the maximum expansion was OSP 1% with intumescent factor of 7.78. XRD was conducted to evaluate the residual compounds in the char and FTIR was utilized to analyse the functional group where O-H, C-N, O=C=O and C=C bonds were found in the degradation compounds in the char residual of the coatings.

1. Introduction

The inclination usage of durable and environmentally friendly materials with wide range of application in global industry boosts the development of composites [1]. Recent development in polymer composites targeted biowaste as reinforcement to be added into polymer to improve the properties of polymer. Researchers believe that some of the biowaste containing high amount of minerals that can be useful as a reinforcement. Utilization of biowaste reduces the pollution caused by waste dumping and it also initiates a new way to enhance the current technology of waste management [2].

Intumescent fire-retardant coating is often being stressed out in the construction industry [3]. Mechanism of fire retardant can be divided into three categories: condensed phase mechanism, gas phase mechanism and combination of both mechanisms. The condensed phase mechanism involves the decreasing rate of degradation of coating material and modification of the mechanism of degradation to produce less flammable volatile products. Meanwhile, the gas phase mechanism involves the releasing of non-flammable compounds evolved by additive on heating which dilutes the flammable volatile product feeding the flame [1].

Polymer such as polypropylene, polyurethane and epoxy often become polymer intumescent coating. Polydimethlysiloxane (PDMS) offers heat stability, cold flexibility and chemical resistance which are vital for various applications, especially for cable applications [4]. Pyrolysis of PDMS generates amorphous silica residual, forming a protection barrier which provides a shielding effect from the heat by means of delaying or preventing the heat transfer to the polymer [5]. Besides, researchers believe epoxy are the most versatile thermosetting polymers which offering appropriate properties in terms of chemical, corrosion, adhesion as well mechanical [6]. Modified epoxy resins by mixing PDMS with epoxy believed to offer great dimensional stability, mechanical properties and chemical properties, which are an excellent material for intumescent coating.

Oyster shell powder and glass frit had been targeted to become a filler to be added into epoxy resin. Oyster shell is identified as waste product from mariculture [7]. It has become a major problem in coastal areas, especially in south east Korea. According to Yoon et al. [8], oyster shell comprises up to 95% of calcium carbonate and some other minerals and substances. Calcium carbonate will release a great amount of carbon dioxide upon decomposition. Researches, which compared calcium carbonate extracted from oyster shell powder with commercial calcium carbonate, have shown both material exhibits have similar properties [9]. Rehman Sheh et al. [10] stated that oyster shell powder used for reinforcement as a filler in polypropylene had successfully increased fire retardant of polypropylene. Carbon dioxide is considered a non-flammable gas, which will dilute the temperature of the flame during combustion. Glass frit containing a high amount of silica will form char during combustion. The formation of char will act as a barrier to prevent heat from reaching the coated material [11].

Most of the studies focus on the combination of thermosetting and thermoplastic polymer with other biowaste such as rice husk ash. No study has been reported on combination of epoxy resin with oyster shell powder and glass frit. This project aims to produce an intumescent fire-retarding coating material to synthesize siloxane-based epoxy resin for development of fire-retardant intumescent coating using oyster shell powder and glass frit. The performance of coating for building elements is characterized to impart flame resistance performance.

2. Research Methodology

The materials used in this study were PDMS and epoxy to serve as matrix. Phosphoric acid was used as the catalyst while polyamine act as the hardener. The binding agent used in this study was Bisphenol A. The filler, namely GF obtain from market and processed by utilize facility in university whereby the OSP were obtained from Xi'an Biotechnology. The basic ingredient incorporated in the intumescent coating were melamine, ammonium polyphosphate, boric acid and expandable graphite.

2.1. Glass Frit Powder Preparation

The glass bottles were break into small pieces by using hammer. The smaller pieces of the waste glass were then undergoing grinding process using grinding machine. The process took 1 minutes to acquire powder. The waste glass powder was then sieved using sieving machine and retained at 212 μm sieve.

2.2. Coating Fabrication

Epoxy weighted according to table 1 is heated at 60°C using a hotplate and stirred with 40 rpm for 30 minutes by using a stirrer. The heating and stirring process is continued for 10 minutes after PDMS is added into the epoxy. The bisphenol A is then added into the epoxy resin and stirred for 10 minutes. 6 to 7 drops of phosphoric acid are added in the epoxy resin. The rotational speed of stirrer is increased to 70 rpm for 30 minutes.

All coating formulation is mixed and grind for 1 minute using mixer grinder except for expandable graphite (EG). The EG is then mixed manually with the grinded mixture. The mixture is mixed with epoxy resin by using stirrer for 15 minutes. The hardener is added into the mixture. The rotational speed of stirrer is then reduced to 40 rpm for 5 minutes. The mixture is then applied to steel plates. It was then left for 2 days for drying.

Table 1: Control weightage of intumescent coating

Formulation no.	OSP (wt.%)	GF (wt.%)	APP (wt.%)	MEL (wt.%)	Boric Acid (wt.%)	EG (wt.%)	Epoxy (wt.%)	PDMS (wt.%)	Harderner (wt.%)
OSP 1%	1	-	11.36	5.5	11	5.5	30.63	13.13	21.88
OSP 2%	2	-	11.36	5.5	11	5.5	30.17	12.93	21.55
GF 1%	-	1	11.36	5.5	11	5.5	30.63	13.13	21.88
GF 2%	-	2	11.36	5.5	11	5.5	30.17	12.93	21.55
OG 1%	0.5	0.5	11.36	5.5	11	5.5	30.63	13.13	21.88
OG 2%	1	1	11.36	5.5	11	5.5	30.17	12.93	21.55

3. Characterization

3.1. Fire Test
The fire test will examine the ease of ignition and ability of fire endurance of material under certain temperature according to standard ASTM E119. Flame spread of material also can be examined to investigate burning rate of material across material surface. A portable Bunsen burner was used to burn the coating applied on the steel plate. The distance between the Bunsen burner and the coated steel plate is 7cm. Three thermocouples are attached to the back of steel plate. The thermocouple is then connected to Anarittsu Data logger, Input Channel 6 Model AM 8000K with Anarittsu software. The temperature of the steel plate was recorded for every 30 seconds at an interval of 1 min.

3.2. Furnace Test
Furnace test is also a type of fire resistance test, which is designed to simulate particular fire threats on structures. Furnace test in this project was used box furnace (Carbolite Gero RWF 1100) to examine the char expansion rate, which is a mechanism for increasing the performance of char to serve as a barrier. The heating rate of the furnace was set at 20°C/min for 2 hours. The char was measured by using digital vernier caliper and average values were recorded. The thickness of samples was measured and recorded for before and after. Intumescent factor (IF) was calculated from the thickness obtain.

$$\text{Intumescent Factor (IF)}=\frac{(d_2-d_0)}{(d_1-d_0)} \quad (1)$$

where d_2 = expanded char thickness; d_1 = coating thickness; and d_0 = steel plate thickness.

3.3. Scanning Electron Microscope
Scanning electron microscope (SEM) generates a variety of signals on a material's surface in a sample by using a focused beam of high energy electrons. The signals will reveal information about the sample including external morphology, chemical composition, structure and orientation of sample. SEM is used to evaluate char morphology of the sample after fire test.

3.4. X-Ray Diffraction (XRD)
X-Ray Diffraction is a frequent used non-destructive test to analyze the crystalline materials. In this project the scan range of XRD is set from 10° to 90° with step size 0.01. The results plotted in graphs is interpret using software HighPro Plus.

3.5. Fourier-Transform Infrared Spectroscopy (FTIR)

FTIR is an analytical technique used to identify organic, polymeric and inorganic materials. Infrared is used in FTIR to scan test samples to observe the chemical properties of the samples in the range of 4000cm^{-1} to 400cm^{-1}.

4. Results and Discussion

4.1. Heating Shielding

The substrate temperature for each formulation was plotted after the fire test results were obtained as shown in figure 1. From figure 1, the rapid increase of temperature in first 10 minutes of the fire test due to formation of char was yet to take place. The graphs reach their peak and temperature of substrate gradually decrease due to the formation of char initiated and release of non-flammable gas such as carbon dioxide to dilute the temperature of the flame.

Comparison had been made between 3 fillers; glass frit is found to have the best performance in retarding the heat energy transmission. Sample GF 1% with final temperature 175.3°C has the best performance among other fillers and followed by sample OG 1% and GF 2% with approximately 177°C. This result indicates glass frit performs well in fire retarding properties. We can also observe that oyster shell powder has the least performance among all three fillers. This result may be due to the calcium carbonate which only contribute carbon dioxide where glass frit form silica when decomposed. Silica exists to prevent heat from reaching substrate whereby carbon dioxide dilutes the temperature of the flame while release to surrounding. Other than that, from the plotted graph, we can obtain that combination of oyster shell powder with glass frit able to reduce the peak of temperature due to quick formation of char and release of non-flammable gas.

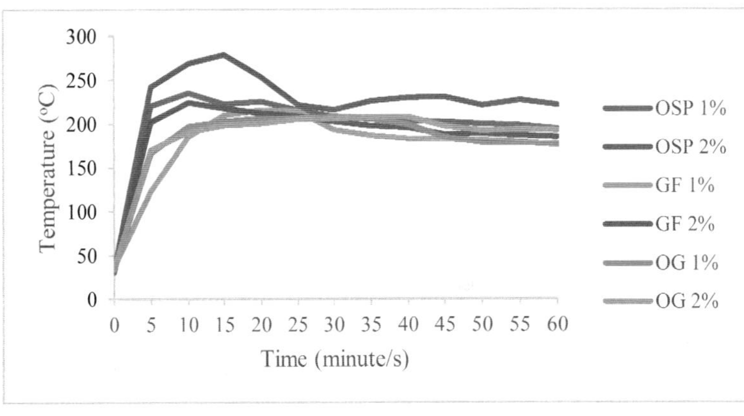

Figure 1: Fire test result

4.2. Char Expansion

All the samples were sent for furnace test. Table 2 shows the intumescent factor obtained for studied formulations. OG samples showed the least intumescent factor while oyster shell powder shows the best intumescent factor. Among all the sample, modified epoxy resin intumescent coating with 1% of OSP filler showed the highest intumescent factor of 7.78 followed by OSP 2% with 7.06 and GF 2% with 6.84. This result may due to the decomposition of calcium carbonate in OSP samples which contribute carbon dioxide which serve as blowing agent to expand char effectively whereas glass frit only form silica when decompose to serve as physical barrier.

Table 2: Thickness and intumescent coating factor of samples

Formulation no.	Steel Plate Thickness (mm)	Coating Thickness (mm)	Expanded Char Thickness (mm)	Intumescent Factor
OSP 1%	1.5	4.52	25.01	7.78
OSP 2%	1.5	3.35	14.56	7.06
GF 1%	1.5	5.00	25.41	6.83
GF 2%	1.5	4.98	25.32	6.84
OG 1%	1.5	5.32	25.65	6.32
OG 2%	1.5	5.36	24.89	6.06

4.3. Char Morphology
Multiple structures consisting of micro- and nano-scale particles were observed in the low-magnified SEM image. Figure 2 showed the char of the samples which the expandable graphite expands after combustion. From figure 2(a) and figure 2(b), there were void gaps between the structure which interfere the fire-retardant properties of the samples [13]. This result also proven the factor that temperature of OSP is higher than other 2 fillers during fire tests. Both GF 1% and GF 2% showed better adhesion compare to OSP as less voids were found [14]. Bubbles were found in figure 2(c) and figure 2(d) due to the emission of N_2, ammonia and CO_2 gases. The bubbles formed improved fire-retardant properties as it increases the difficulty of the heat from reaching the substrate. From figure 2(e) and figure 2(f) we found OG samples showed less adhesion compare to GF but better adhesion than OSP where few voids gaps can be observed on the microstructure of OG.

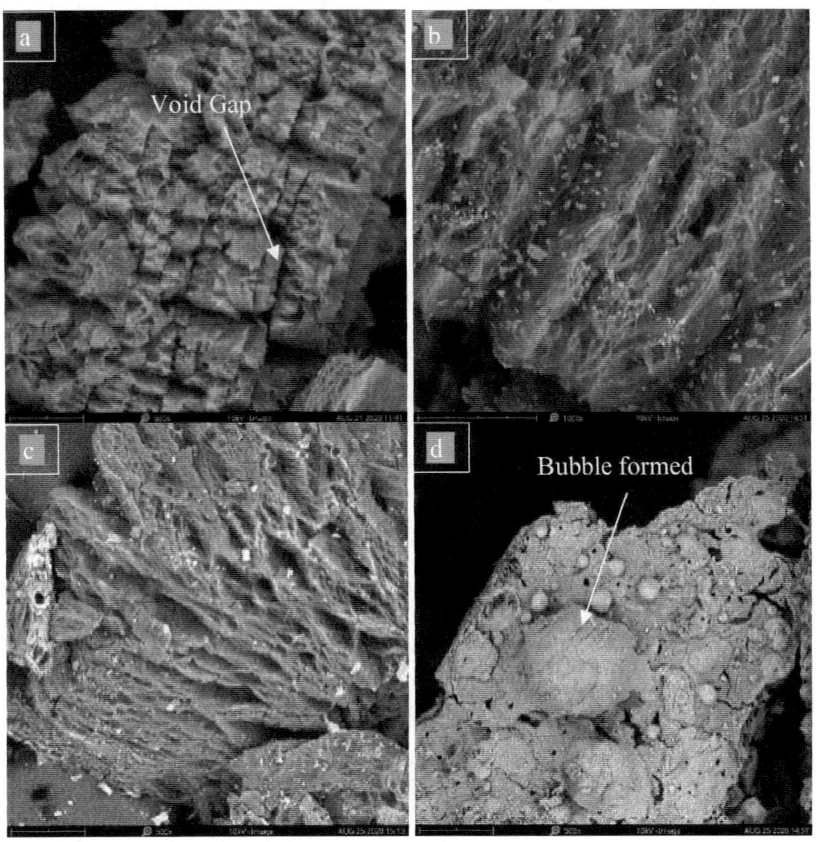

Figure 2: Microstructure of char (100μm) (a) OSP 1%, (b) OSP 2%, (c) GF 1%, (d) GF 2%, (e) OG 1%, (f) OG 2%

4.4. Analysis of Char Composition

The XRD results were plotted in graphs and interpret using software HighPro Plus. For the formulations for all samples, the major peaks around $2\theta = 24^{o}$, 40°, 48.6^{o} represent the boron phosphate. The boron phosphate formed due to the reaction between boron acid with phosphoric acid [15]. Based on the Figure 3(a), the highest peak of $2\theta = 27^{o}$ and other major peaks at $2\theta = 44.6^{o}$, 54.8^{o} represent the carbon element. This result may due to the decomposition of calcium carbonate contain in OSP. Through data interpretation using software, we obtain that for OSP 1% contain 56% of carbon and 44% of boron phosphate whereas OSP 2% contain 61% of carbon and 39% of boron phosphate. From figure 3(b), major peaks at $2\theta = 14.9^{o}$, 26.5^{o}, 27.8^{o} represent the silicon oxide. This result indicates the presents of silica in glass frit. The data shows the 59% of boron phosphate and 41% of silicon oxide in GF 1% whereas GF 2% contain 33% of boron phosphate and 67% of silicon oxide. Based on the graph 3(c), the major peaks at $2\theta = 14.7^{o}$, 26.5^{o}, 27.9^{o} shows the presents of silicon oxide. The result of OG is similar to GF as OG is the combination of OSP and GF.

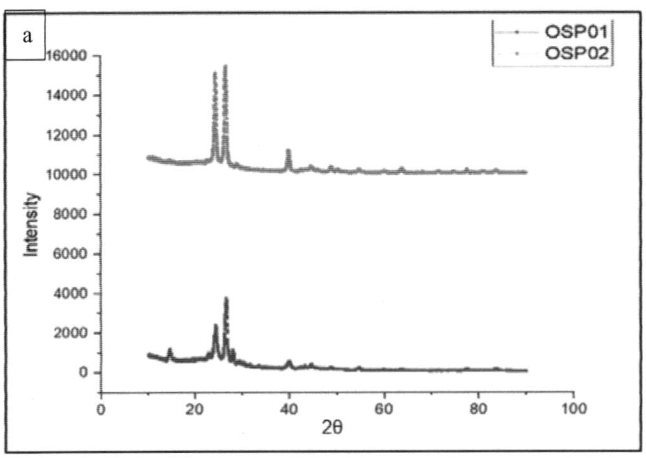

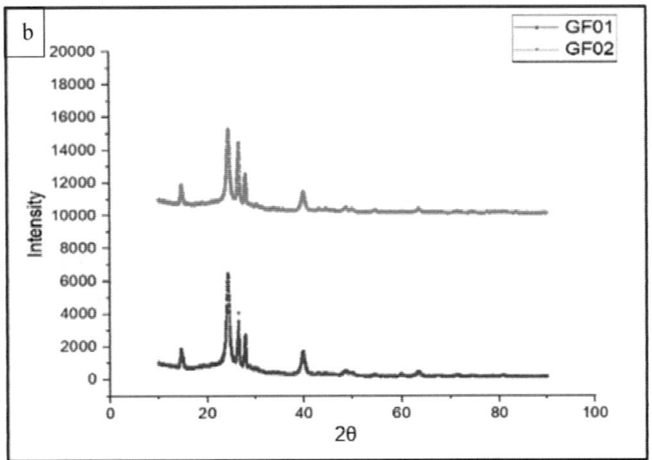

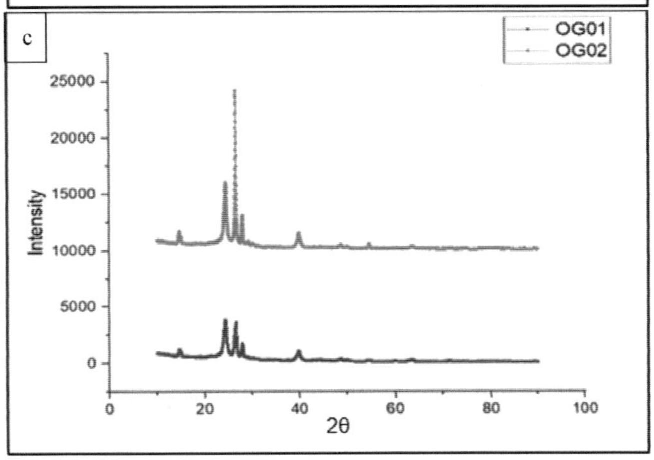

Figure 3: XRD spectra of (a) OSP samples, (b) GF samples, (c) OG samples

4.5. Analysis of Functional Group of Char

The char of samples is characterized by FTIR. These samples showed similar properties as the same formulation of the intumescent coating was used. Based on the Figure 4 and with the aid of FTIR table, we can analyze that at the wavelength range of 3040cm^{-1} to 3300cm^{-1}, we obtained wide peak absorption of alcoholic groups which the O-H bond stretching [16]. At the range of 2300cm^{-1} to 2380cm^{-1} wavelength, we obtained the carbon dioxide bond stretching in a small wide peak. We obtained that O-H bond bending which might represent the alcohol at the range of 1320cm^{-1} to 1480cm^{-1}. Whereas at the range of 1000cm^{-1} to 1200cm^{-1}, we confirmed the C-N bond stretching, which indicates the presents of amine. Lastly, at the range of 800cm^{-1} to 960cm^{-1}, we obtained the bending of C=C bond.

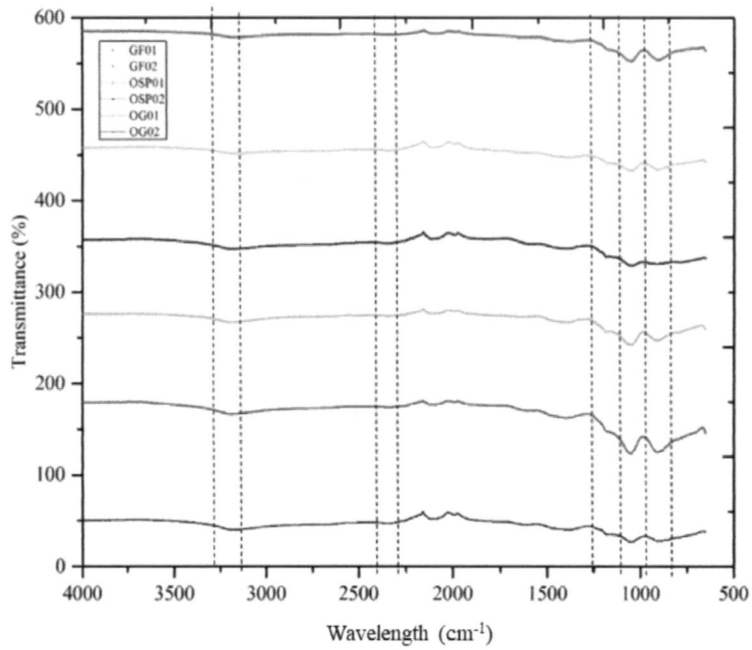

Figure 4. FTIR spectra of samples

5. Conclusion

The incorporation of oyster shell powder and glass frit incorporate with modified siloxane-based epoxy, either alone or together, improves fire retardant properties of the modified siloxane-based epoxy. The highest intumescent factor obtained for OSP, GF and OG samples were 7.78, 6.84 and 6.32 respectively, indicating the OSP samples were excellent in expanding the char. Minimum backside temperature of the substrate obtained was 175.3°C for GF samples, 194.6°C for OSP samples and 177.5°C for GF samples from the lab scale jet fire test. Among all three types of samples, GF samples showed the best performance in enhancing fire retardant properties. GF samples were found to decrease the temperature of substrate gradually after the char formation took placed. XRD confirmed the presence of boron phosphate in all samples which the main source of char formation. Carbon elements were found in coating with filler of oyster shell powder due to decomposition of calcium carbonate. XRD also confirmed the presence of silicon oxide in the coating with glass frit, which are found to contribute effectively to fire retardant. FTIR data shows the bonds O-H, C-N, O=C=O and C=C bonds presence in the samples indicates the presence of amorphous carbon which comes from amine and graphite functional group, the existence of water and contribution of calcium carbonate and silica towards char formation. The SEM confirmed the good adhesion of the filler with modified siloxane epoxy with GF composition which had least void gaps. OG samples were successful in reducing the peak of the temperature but weakens the adhesion of the coating which confirmed through analysis of char morphology.

ACKNOWLEDGEMENTS

The first and second authors acknowledges the financial support from the Ministry of Higher Education Malaysia through Fundamental Research Grant Scheme (FRGS) (CS: 015MAO-081).

References

[1] Bar M, Alagirusamy R and Das A 2015 Flame Retardant Polymer Composites *Fibers and Polymers* **16** 705.

[2] Safiuddin M, Zamin M J, Salam M, Islam M and Hashim R, Utilization of solid wastes in construction *International Journal of the Physical Sciences* **5** 1952-1963.

[3] Sami U, Faiz A, Azmi S, Bustam M, Girma G and Qandeel G 2017 Effects of ammonium polyphosphate and boric acid on the thermal degradation of an intumescent fire retardant coating *Prog. Org. Coat.* **109** 70-82.

[4] Adly N Y, Hassani H, Tran A, Balski M, Yakushenko A, Offenhausser A, Mayer D and Wolfrum B 2017 Observation of chemically protected polydimethylsiloxane: towards crack-free PDMS *Soft. Matter* **13** 6297.

[5] Genovese A and Shanks R A 2008 Fire performance of poly(dimethl siloxane) composites evaluated by cone calorimetry *Compos. Part A* **39** 398-405.

[6] Henri V, Reza M S, Krzystof F and Jose-Marie C L 2017 Flame retardant epoxy/halloysite nanotubes nanocomposite coatings:Exploring low-concentration threshold forflammability compared toexpandable graphite as superiorfire retardant *Prog. Org. Coat.* **119** 8-14.

[7] Tsou C H, Wu C S, Hung W S, Guzman M R D, Gao C, Wang R Y, Chen J, Wan N, Peng J Y and Suen M C 2019 Rendering polypropylene biocomposites antibacterial through modification with oyster shell powder *Polymer* **160** 265-271.

[8] Yoon G L, Kim B T, Kim B O and Han S H 2008 Chemical-mechanical characteristics of crushed oyster-shell *Waste Management* **23** 825-834.

[9] Hamester M R R, Balzer P S and Becker D 2012 Characterization of Calcium Carbonate Obtained from Oyster and Mussel Shells and Incorporation in Polypropylene *Materials Research* **15** 204-208.

[10] Rehman A U S, Prabhakar M, Lee D W, Kim B S and Song J I 2014 Development and Characterization of Oyster Shell Powder Filled Polypropylene Composite *Composites Research* **27** 201-206.

[11] Soheir T Y Flame Retardants: Additives in Plastic Technology *Polymers and Polymeric Composites: A Reference Series* (Egypt, Springer-Verlag GmbH Germany) pp 2-27.

[12] Long M Y, Peng S, Deng W S, Miao X R, Wen N, Zhou Q N, Yang X J and Deng W L 2017 A robust superhydrophobic PDMS@$ZnSn(OH)_6$ coating with under-oil self-cleaning and flame retardancy *J. Mater. Chem. A Mater.* **5** 22761-22771.

[13] Manilal M, Sourav S K, Naresh M C and Banerjee P 2019 Amine cured double Schiff base epoxy as efficient anticorrosive coating materials for protection of mild steel in 3.5% NaCl medium *J. Mol. Liq.* **278** 521-535.

[14] Faiz A, Eza Z S, Ullah S and Norlaili A 2018 Effects of nano-sized boron nitride on thermal decomposition and water resistance behaviour of epoxy-based intumescent coating *J. Anal. Appl. Pyrolysis* **132** 171-183.

[15] Hatem A, Rao T N, Song J I, Yarasani P, Faheem A,Botsa P, Asma A A, AL-Anazy M M and Ifseisi A A 2020 Influence of Magnesium Aluminate Nanoparticles on Epoxy-Based Intumescent Flame Retardation Coating System *Coatings* **10** 968.

Study on the Comfort of 3 Local Hijab Sport Brand on Muslim Woman during Physical Activities

A Baharudin[1,2*], A H A Hafidz[1], N M Roslan[1], I N Zahir[1], N A Norjeferi[1], A Musa[1], M R Ahmad[2]

[1]Universiti Teknologi MARA (UiTM), 72000 Kuala Pilah, Negeri Sembilan, Malaysia.

[2]Faculty of Applied Sciences, University Teknologi MARA (UiTM), 40450 Shah Alam, Selangor, Malaysia.

*Corresponding Author: aznînbahar@yahoo.com

Abstract. This study goals are to identify the most comfortable sport hijab from 3 local brands. The selection of the sport hijab from the brand are identify through a preliminary survey distributed to 90 students from Faculty of Sport Science UiTM Negeri Sembilan. From the survey, types of sport hijab and the length of sport hijab are identified. The purchased sport hijab is then gone through the fabric identification and testing included weighing, burning test and water impact penetration test. The fabric for the sport hijab labelled as Sample A is Nylon, Sample B as Polyester and Sample C as Acrylic. Among this, it is found that Sample B gives the best comfort according to all 3 selected respondent who wear the sport hijab in the physical test activities. The physical test activities are done by the 3 participants running on a treadmill for 20 minutes at 7.5km/h with warming up and warming down at 2.5km/h for 2 minutes. The results obtained from the comfort evaluation were then analyzed using Analysis of Variance (ANOVA).

1. Introduction

Sport is one of the areas where wearing a hijab can pose a range of practical challenges. These challenges can be linked to the potential physical limitations of modest clothing that might restrict movement or be uncomfortable to wear during strenuous physical exercise. Hijab in sport have taken a lot of toll, athlete dropping from competitions like Noor Abukaram, 16, have been banned from participating a running event in Ohio for district level in 2019 [1]. Kalsoom Abdullah, a weightlifter for American Open tournament, 2011[2], just because they are wearing hijab and refuses to take it off to participate the game. This situation might link to the understanding of Islam and there has been no test done on hijab and confirmed that it is safe to be worn during the game [3]. In recent years, the clothing industry has increasingly responded to Muslim women's desire to find practical solutions that address the difficulties associated with wearing a hijab during physical exercise. This has, for example, led to the development of the sports hijab [4]. When working on the design of hijab sportswear, producers took on board the concerns competing athletes face when wearing a hijab, examining the ways these affected their performance. For instance, Nike produce The Pro Hijab that stretches to adapt to the wearer's head, and its cut is optimised for sports [5].

Every woman who has worked out in a hijab has gone through the unpleasant experience of the material sticking to their necks, the endless flapping around and the ugly sweat stains. Our everyday chiffon, cotton or viscose hijabs make us look and feel amazing on other occasions. But when it comes to breaking out a sweat, these mentioned materials are not suitable. Thus, a proper fabric selection for sport hijab is needed. In a fabric the structure and its thickness can influence the heat and moisture transfer and hence thermal comfort [6]. This is because fabric structure and thickness affect the air and moisture air permeability, which play a significant role in heat and moisture transfer [7] which every active sports female need to wear hijab that has a wicking property to ensure that they will not be fainted when doing activities due to ineffective of heat transfer.

In ensuring the comfort of the hijab during a sport activity, a simple test can be done to imitate the actual high performance sport activities like running, cycling, sports games and marathon, by increasing a body core temperature by $1.5^0C - 2.0^0C$ where a thermophysical comfort can be measure as the high metabolic system in our body able to generate heat of approximately 800 – 1300W [8]. In this study, a preliminary survey is given out to 90 Muslim women who actively do sport to identify the best style of sport hijab worn by the athletes. Then a test is done with 3 local sport hijab brands to identify the style and the material used suitable for sport hijab in Malaysia.

2. Methodology

2.1 Methods Flowchart

Figure 1 shows the summary of the process flowchart done in the study.

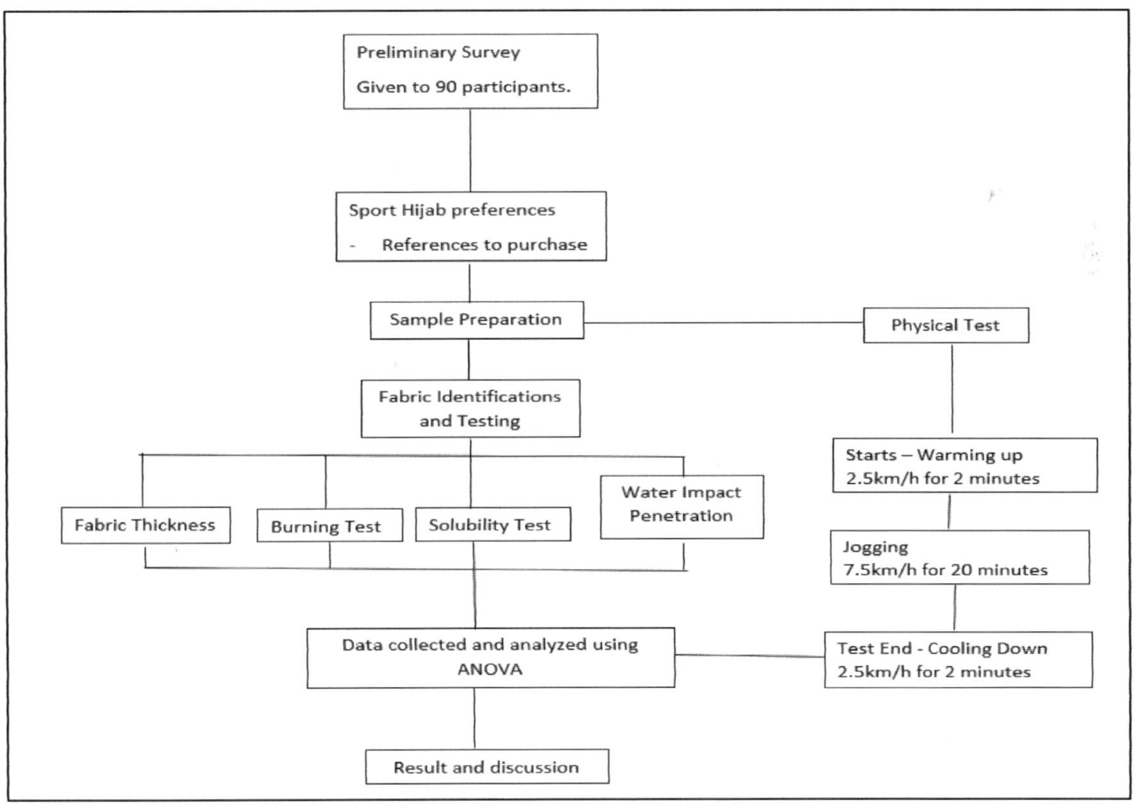

Figure 1: Process flowchart

2.2 Survey
Sports Science program students from Universiti Teknologi MARA, Negeri Sembilan are the focal point in this study because the study needs to have an avid sports person or a person who do sports on daily basis. Aside from that the second most important criteria in respondent is that they are all Muslim. Hence, this study is done by distributing a preliminary survey to 90 students from sport science background. The questions are about general knowledge, perception and the properties of the sports hijab that are usually worn to do sports activities. Textile terms are explained to the participants to give conscious on the questions given. the objective of the preliminary survey is to identify the preferences of the athlete or a person who regularly do sport and wear sport hijab, their styles of wearing the hijab and material choose for the sport hijab. This in the end will help the selection of buying readymade sport hijab from the market.

2.3 Fabric Identification and Testing
In this section the Samples of sport hijab fabric need to be identified. There are three ways of identifying the materials according to the American Society for Testing and Materials (ASTM) which are Burning Test, Solubility Test (ASTM D276-12) and Water Impact Penetration Test (AATCC 42-2007). In burning test, a strip of fabric from all of the sample will be burns and the behaviour of the burning are monitored. How the fabric ignitions, the residues either it leaves a grey ash (natural fibres) or a black beaded (Man-Made Fibres), afterglow and the smells of the burning samples. As for the solubility test, a chart from ASTM is follow. This chart will characterize the chemical that able to soluble the fabric material.

The water impact penetration is conducted by mounting tautly the sample, face-up on an angled surface. A pre-weighed sheet of blotter paper is mounted behind the specimen. A volume of water is sprayed against the face of the specimen then the blotter paper is then re-weighed to determine water penetration. The increase in mass of the blotter paper indicates the mass of water that passed through the test specimen. Lower test results indicate less penetration and better water repellency or resistance of the fabric.

Hence by conducting the fabric identification and test, types of fabric used in the sport hijab either it is man-made fibres, natural fibres or blended of both will influence the comfort of the sport hijab. This include the wicking properties, thermoregulatory, sweat management and tactile of the fabric surface [9]. Measuring fabric thickness of the sports hijab samples are also done in the fabric testing.

2.4 Physical Test wearing Samples A, B, and C
A comfort wear tests was conducted to evaluate the comfort based on the sensory of the human body. Referring to figure 2 and figure 3 different sports hijab samples are bought from 3 different brands which are labelled as Sample A, Sample B and Sample C. The samples are given to 3 participants and they are required to sweat under a controlled environment and activity. The activity is done on a treadmill in a gym with the set surrounding temperature of 28°C. Participants are required to warm up by walking at the speed of 2.5 km/h for 2 minutes, then they are required to jog in 20 minutes at increased speed of 7.5 km/h. Lastly, cooling off are done by walking at the speed of 2.5 km/h for 2 minutes. The evaluations are done in a form of Likert-survey where the participants rate the scale of satisfaction from 1-10 based on the properties that are focused such as wicking, perspiration, breathability, and the hand of the sports hijab sample. The result from the survey is collected and analysed by using Analysis of Variance (ANOVA) under SPSS software.

IOP Conf. Series: Materials Science and Engineering 1176 (2021) 012026 doi:10.1088/1757-899X/1176/1/012026

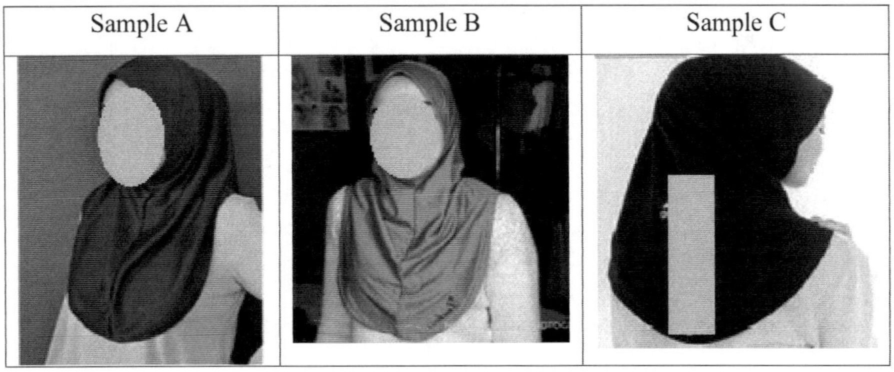

Figure 2: Sport Hijab from 3 different local brand being label as Sample A, Sample B and Sample C

3. Result and Discussion.

3.1. Evaluating General Information

Based on the survey answered by 90 respondents, all the female respondents do sports in daily bases. Figure 3 shows that majority of the respondent (93.3%) wears hijab during sports activities.

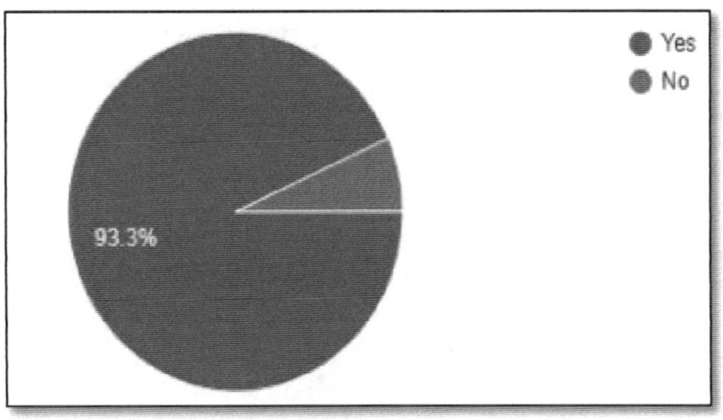

Figure 3: Percentage of respondents who wears hijab during sport activity

Since the respondents are from Science Sports program the respondents are active in sports, time of playing sport in hours per day also influenced by the respondents which different courses have different time playing sport in hours. As refer to a graph of figure 4, most of the respondents spent time on performing sports activities for 2 hours per day. As the sport hijab worn should always maintained it comfort, not sticking onto the neck, good thermal transfer and have a proper wicking property, 2 hours of sport activities will give the best impression of comfort while wearing hijab during sports.

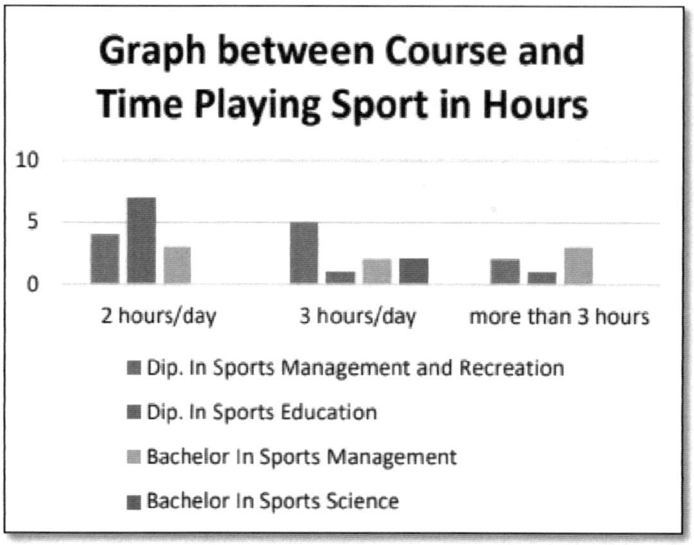

Figure 4: Sports durations according to courses in Sports Science Programs.

Through the preliminary survey, three findings are emphasized. Based on the survey distributed, it is found that design influences the selection of sports hijab. As shows in figure 5, as much as 70% from the total respondents prefer the hijab with awning. This is due to women personality that wanted to look nice in most of the time. So, the hijab with awning is beautifully shaped when it is worn and maintained its shape thorough out the activities.

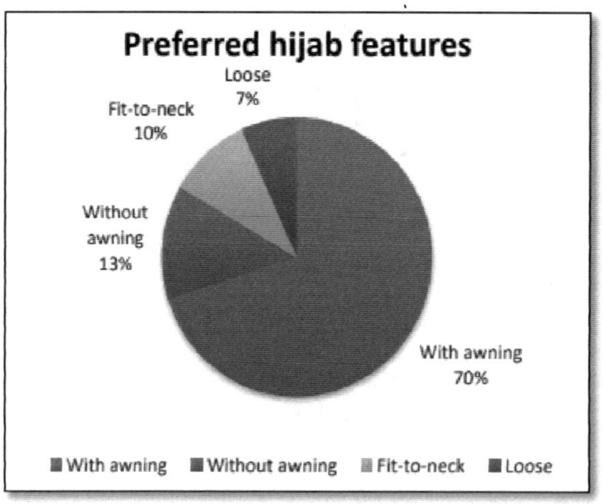

Figure 5: Percentages of preferred hijab features for sport hijab.

In addition, based on the figure 6 it shows that 66.7% of the participants preferred to tied-at-the-back when wearing "bawal" and "shawl" as for their sport hijab style specifically while performing sport activities. Meanwhile, instant hijab with 66.7% preferred to let-off their sport hijab. This is because it will not hinder the movement of the respondents when performing sport activities compared to the other styling in hijab. Nevertheless, different people have different preferences of comfort on choosing styling of hijab. Figure A, B and C shows the preferred styling based on the survey. The length of the hijab also influences the movement of the participants when doing sport activities. As refer to figure 7, shows that majority of the participants with 73% preferred medium length of hijab. The long hijab with 40cm length covering the chest area and shoulder area. This makes it difficult for the respondent who wears the hijab during sports to raise their hand in an impulse for example playing badminton or receiving a ball in a netball game. As for the short hijab which is 15cm length does not properly covers the respondent chest. In a study, a physical comfort also plays an important role to influence a person to feels comfortable [10]. Thus, the most suitable length of hijab when doing sports is medium length which is at 30cm.

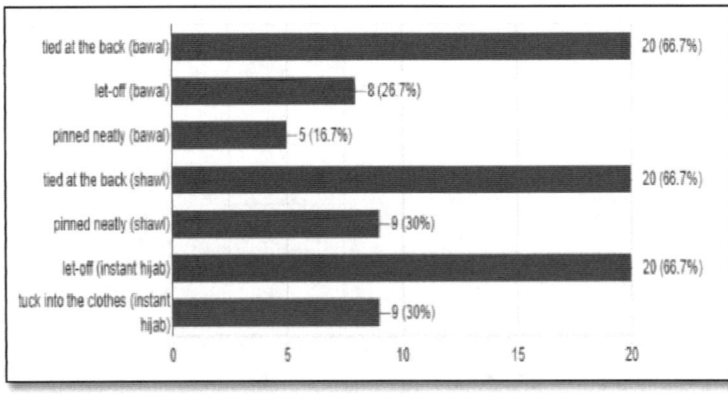

Figure 6: Percentage of preferred hijab styling for sport hijab.

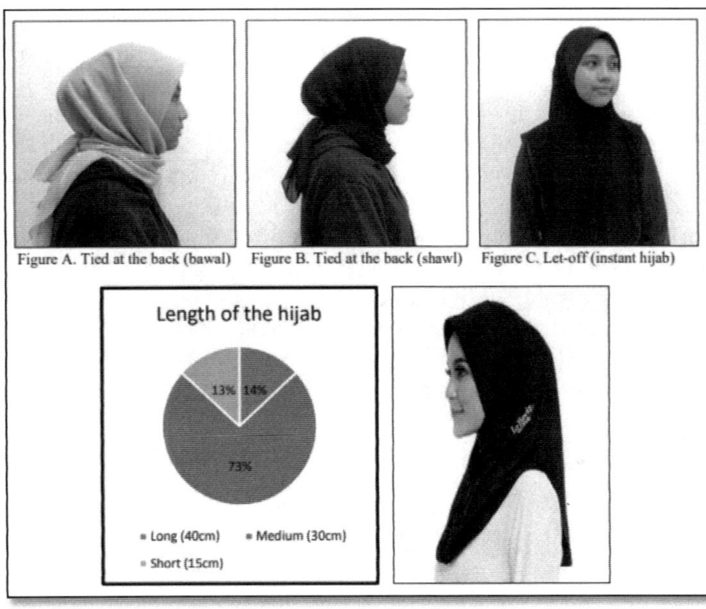

Figure 7: Example of variety of hijab length and the pie chart of respondent selection on the hijab length for sport hijab.

3.2. Fabric Identification and Testing.

3.2.1. Fibre Identification. Table 1 shows the thickness of the fabric samples from the local brand sport hijabs. From the table, Sample B has the thickest value with the thickness of 0.44 in millimeters (mm) compared with the other two samples. The standard that was used to measure thickness was ISO 534. In the recent studies discuss the higher is the thickness, increases the resistance of heat transfer and moisture due to presence of a large amount of excess air in between the fabric structure compared to the fibre content **[11]**. Hence it can be said that Sample B with the highest thickness, means that the sport hijab able to retain heat and most suitable to be use in the set temperature of 28^0C. Nevertheless, the comfort does not rely solely on the thickness, as it also depends on the material used and the design of the sport hijab.

Table 1: Samples thickness in millimetres.

Sample	Thickness (mm)
A	0.36
B	0.44
C	0.39

On the other hand, burning test results can be seen in table 2. It shows the types of material for each sample consists of nylon, polyester and acrylic. Based on the theoretical of the burning test, nylon fabric supposedly has smells like plastic when burnt and produce a celery-like smell. Meanwhile, the polyester smells slightly sweet and a combination of smell with a chemical odour. The acrylic burns with a strong flame and smells like acid and have smell of a chemical odour. Hence, the burning test results of the three samples is accurately have similarities with the theoretical from the aspect of its material characteristics. Fibre identification was checked according to the standard method of AATCC Test Method 20A.

From the burning test result, it is confirmed that all the samples are man-made fibres. Man-made fibres (MMF) today are the best selection for sport attire especially for sport hijab. Among the material used are from polyester and nylon. Though acrylic is also MMF, but the development of acrylic is to imitate wools **[12]**. Acrylics have an exceptionally good drapability, but some of the material if it is made into a garment, it will show or stick to the body silhouette. Where else, polyester and nylon fabric have a similar property as light in weight, versatile when wet and dry, have good stretchiness and moisture wicking properties. Therefore, it can be forecasted based on the material finding, Sample A and Sample B might have a better comfort compared to Sample C.

Table 2: Burning Test results corresponding to the Sample A, B and C fabric.

Sample	Types of Material	Explanations.
A	Nylon	- Burns slowly with melting drip in flame. - Smells like celery. - The residue was hard, tough, grey, and tan bead.
B	Polyester	- Burns slowly, continues to melt and drip. - It has a chemical odour. - The residue is hard, tough, tan bead.
C	Acrylic	- Burns rapidly with hot flame, sputtering, drips, melts. - Smells like acrid.

Table 3 shows the result of water impact penetration test. This can be seen that Sample C has the heavier weight of water contained in blotting paper which is 2.349 g. As in this test, from the table 3, it shows that the highest percentage of water that penetrate the fabric sample to the blotting paper is weighed with 59.11% for the Sample C. The results show that both Sample A and B have a water repellence behaviour.

Table 3: Water Impact Penetration test result.

Sample	Weight of blotting paper before testing (g)	Weight of blotting paper after testing (g)	Weight water (g)	Percentage of water (%)
A	2.364	2.562	0.198	8.37
B	2.022	2.420	0.398	19.68
C	2.282	3.631	2.349	59.11

3.3. Evaluating Comfort from the Outdoor test.

3.3.1. Evaluation on the Comfort. After outdoor test is done, the results from the survey are collected and analysed using Analysis of Variance (ANOVA). The results are showed in table 4. Through the ranking rated by the participants on sample A, B and C, it shows that the 3 different sports hijab samples have the similarity in the properties focused.

Table 4: ANOVA Results showing on the significance of properties for 3 sport hijab brands

Properties	P-Value	Conclusion
Wicking	0.487	Not Significant
Perspiration	0.380	Not Significant
Comfortness	0.520	Not Significant
Breathability	0.864	Not Significant
Hand	0.587	Not Significant

Figure 8 shows the ranking given by participants on wicking, perspiration, breathability, comfort, and hand properties for Sample A. As can be seen in Figure 8, it shows an inconsistency rating among 3 participants. However, from the graph we could see that perspiration and comfort properties have a consistency in value as the different in rating are only 1.

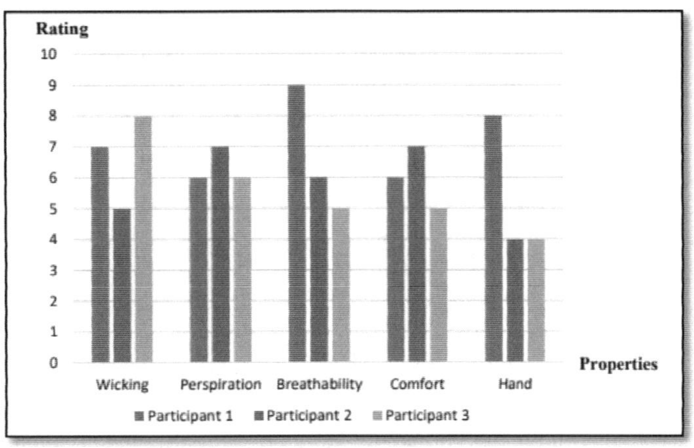

Figure 8: Rating from participant 1, 2, and 3 for Sample A

In figure 9, a consistent graph based on the wicking, perspiration, breathability, comfort, and hand properties are obtained for Sample B. In this graph, the participants prone to have similar feelings on Sample B where the participants rated the properties that exist in Sample B are alike. Thus, Sample B shows a better result in comfort compared to Sample A.

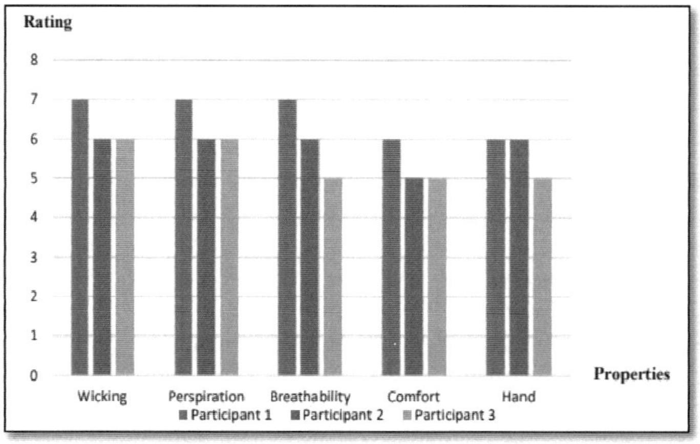

Figure 9: Rating from participant 1, 2, and 3 for Sample B

Figure 10 shows an unstable graph on wicking, perspiration, breathability, comfort, and hand properties for Sample C. It shows more inconsistent graph which is seen to be having a larger different compared to Figure 8. However, the properties are presence in Sample C but it is rated poorly by the participant 3 yet rated well by the participant 2. Hence, the participants may not have similar agreement on the properties based on the participants' feelings.

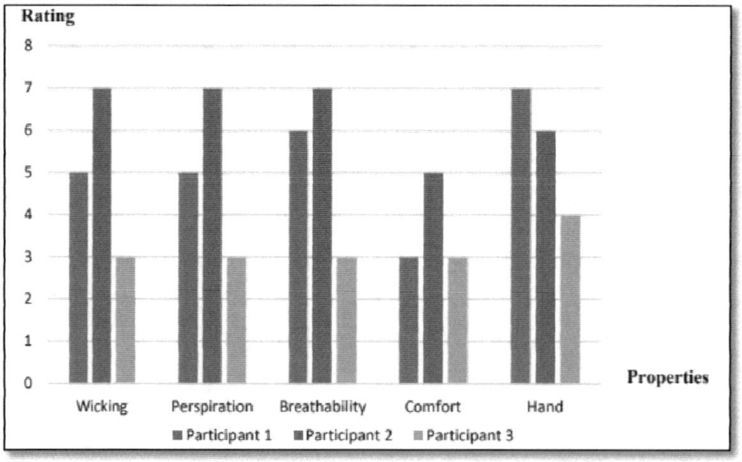

Figure 10: Rating from participant 1, 2, and 3 for Sample C

4. Conclusion.

In conclusion, the first part of the study shows that majority of the respondent preferred to wear an awning hijab and medium length (30cm) from the chin and have an awning feature when doing sport. As for the 3 different brands, Sample B is found to be the most preferable sports hijab based on the rating that is done. This is because, Sample B promotes a medium length of hijab which is 30 cm corresponding result with the survey that have been done with 30 survey respondents in the first part of the study. On the other hand, material for Sample B is polyester where polyester fabric is high moisture wicking ability and breathable. Hence, Sample B is selected among the participants for the best hijab sportswear that can be worn to perform sports activities.

Acknowledgement

The authors would like to thank the Ministry of Higher Education for the funding under the Fundamental Research Grant Scheme (FRGS/1/2018/STG07/UITM/02/1). The authors would also like to acknowledge the assistance given by the Research Management Centre (RMC), Universiti Teknologi MARA.

References

[1] BBC News. *Muslim teen athlete disqualified in Ohio race over hijab* (25 Oct 2019) Retrieved on 5th January 2021. https://www.bbc.com/news/world-us-canada-50186728.

[2] Cristina N *2011 Muslim women Athletes Banned from Tournaments get creative with Hijabs* ABC News Retrieved on 5th January 2021. https://abcnews.go.com/US/athletic-hijabs-female-muslim-athletes/story?id=13829421

[3] Manal H 2015 Jordanian national football Muslimat players: Interrupting Islamophobia in FIFA's hijab ban *Phys. Educ. Sport Pedagogy* **20(5)** 517-531.

[4] Cook K 2018 Uncovering the Evolution of Hijabs in Women's Sport *The Graduate Review* **3** 62-67.

[5] Scarano G Nike Debut Pro Hijab for Muslim Athletes *Sourcing Journal* Retrieved on 5th January 2021 https://sourcingjournal.com/topics/fashion-trends/nike-debuts-pro-hijab-muslim-athletes-63055/

[6] Damayanti and Rahayu N I 2017 Effect of Muslim Women's Sportwear (Jilbab) to Dehydration Level and Thermoregulation After Exercise *IOP Conf. Series: Material Science and Engineering* **180** 012204.

[7] Ho C P, Fan J, Newton E and Au R 2011 Improving Comfort in Clothing *Woodhead Publishing Series*, p 161-181.

[8] Awais M, Sybille K, Wolfling B and Classen E 2020 Thermal Simulation of Close-Fitting Sportwear *Energies* **13** 2419.

[9] Oktay P 2008 Clothing Comfort Properties in Textile Industry *E-Journal of New World Science Academy* **3**.

[10] Yazid M G and Sugiura K 2016 Designing a modest sportwear to provide comfort in daily sport activities for Muslim women wearing hijab *Keio University Graduate School of Media Design. (Master Thesis)*.

[11] Marolleau A, Salaun F, Dupont D, Gidik H and Ducept S 2017 Influence of Textile Physical Properties and Thermo-Hydric Behaviour on Comfort Journal of Ergonomics **7(6).**

[12] Bajaj P and Surya K 2007 Modification of Acrylic Fibres: An Overview *Journal of Macromolecular Science, Part C* **27(2)**.

IOP Conf. Series: Materials Science and Engineering 1176 (2021) 012027 doi:10.1088/1757-899X/1176/1/012027

Piper Nigrum (Green Corrosion Inhibitor) as a Modified Quenchant in Heat Treatment of Ductile Iron

M F Idham[1,2*], S H Nasip[3], N Hazirah[3], B Abdullah[4] and S K Alias[5]

[1]Faculty of Mechanical Engineering, University Teknologi MARA (UiTM), 23200 Bukit Besi, Terengganu, Malaysia

[2]Department of Earth System Science and Technology, Interdisciplinary Graduate School of Engineering Science, Kyushu University, Fukuoka 816-8580, Japan

[3]Faculty of Chemical Engineering, University Teknologi MARA (UiTM), 23200 Bukit Besi, Terengganu, Malaysia

[4]Faculty of Mechanical Engineering, University Teknologi MARA (UiTM), 40450 Shah Alam, Selangor, Malaysia

[5]Faculty of Mechanical Engineering, University Teknologi MARA (UiTM), 81750 Pasir Gudang, Johor, Malaysia

*Corresponding author: faizulidham@uitm.edu.my

Abstract. Corrosion of ductile iron leads to many consequences in all its application aspects, like a structural failure due to the porosity formation, which is consequenced by corrosive chemical exposure. Hence, this project was carried out to modify the water quenchant using *Piper Nigrum* (black pepper) essence, as a green corrosion inhibitor, in promoting corrosion resistance towards austenitized ductile iron in hydrochloric acid (HCl). Furthermore, its effects on the hardness and microstructure of austenitized ductile iron were also investigated. The modified quenchant employed in the austenitizing process was modified by diluting *Piper Nigrum* extract to several concentrations. The corrosion resistance performance of ductile iron quenched by this modified quenchant was tested using the gravimetric method. The samples were exposed to 1 molar (M) of HCl for 2 hours in the method. This study's novelty has proved that the *Piper Nigrum* extract was successfully used as an alternative quenchant to reduce ductile iron's corrosion rate in HCl and its corrosion resistance performance was equipotential to other synthetic inhibitors practiced conventionally. This approach could replace the use of synthetic inhibitor, which is harmful and costly. The gravimetric test showed that the corrosion rate of austenitized ductile iron decreased to 51.2685 mm/year and inhibition efficiency increased to 72.17% as the extract's concentration of quenchant increased to 15%. This extract's presence as a quenchant in the heat treatment process neither improved nor exacerbated the samples' mechanical properties, and it has not changed the behavior of microstructure transformation.

1. Introduction

In industry nowadays, Ductile Iron (DI) is preferable material and extensively utilized as the pipeline material for the water and chemical transportation system, sewer lines, and automotive components due to its versatile properties. Even though the material has excellent mechanical properties, it is not exempted withstanding the corrosion problem, especially this material widely used in transporting and dealing with corrosive chemicals [1]. The acidic and basic mediums lead to a different mechanism of corrosion [2][3][4][5]. Other than that, HCl is most commonly used in industry as an acid cleaner and pickling acid [6][7]. Thus, corrosion has weakened the DI by porosity formation on the material's surface, which resulted from corrosive chemical exposure stemming from this highly corrosive media.

Corrosion is a crucial worldwide problem commonly encountered by many industries and strongly affects the industrial environment. Corrosion is the corroding of material through reduction, oxidation, or other chemical reactions occurred on a material's surface. Expensive preservation and upgrades, plant shutdowns, productivity loss, and vital energy waste are among corrosion consequences [8]. It commonly occurs when reduction and oxidation have taken place on the surface of the material. DI also showed the same problem in its application [9]. Therefore, surface treatment is an alternative to enhance the material's surface, inhibit corrosion, and improve material performance [9][10] [11].

A heat treatment approach is a versatile tool in improving the mechanical and surface properties of a metal. Heat treatment parameters like operating temperature and period, cooling rate and the type of quenchant, and the atmospheric environment influence the microstructure nucleation and mechanical properties of DI [12][13]. Thus, this approach can be implied to treat DI's surface, then inhibit oxidation and reduction that caused corrosion.

There have been numerous studies conducted and reported to diminish metal corrosion attacks. Corrosion rate reduction can be made in many practical ways. The corrosion inhibitor application has driven detailed attention and is among the most economical, popular, and acceptable practices against corrosion [8]. However, most corrosion inhibitors are industrial compounds that are costly and environmentally harmful [14][15]. This phenom happens because the organic inhibitor contains heteroatoms (S, N, O) and aromatic rings in their structures, allowing them to be adsorbed on the metal surface [16]. Consequently, natural sources as green corrosion inhibitors for metal corrosion have nowadays taken part in recent research because it is free from toxic contaminants and cost-effective and environmentally safe [8][14][17][18][19]. It is a synergistic action in a plant extract of individual elements [20]. Besides specific corrosion processes, natural products derived from plant sources and certain non-toxic organic compounds containing polar functions containing nitrogen, oxygen and sulfur atoms in their molecules have been used successfully as inhibitors [21][22]. Since the advantageous outcome given by plant source in inhibiting the corrosion, many researchers extracted the essence from the plant like *Musa Sapientum* (Banana) [8][21][23], watermelon [24][25][26], *Piper Nigrum* (Black Pepper) [1][27][28] and *Cinnamomum Zeylanicum* (Ceylon cinnamon) [29], to be applied as green corrosion inhibitor for metal.

Due to the growing need to invent the new low environmental impact practice incorporated with non-toxic and effective inhibitors in protecting metal from corrosion, this present study was thus conducted to explore the potential of *Piper Nigrum* extract (green corrosion inhibitor) as a quenchant in heat treatment (Austenitizing) process for DI in promoting corrosion resistance towards HCl. Furthermore, its effect on the hardness and microstructure of DI also were evaluated. The *Piper Nigrum* was chosen in this research because this naturally occurring plant product is easy to get, less toxic, eco-friendly, and biodegradable, promising the tremendous effect in inhibiting corrosion [22][30]. A hybrid process approach was invented to achieve the most desirable result where the *Piper Nigrum* extract was applied as modified quenchant in the DI's heat-treatment process. The advantages of mechanical and inhibition properties drawn by the heat treatment process and green inhibitor respectively were envisaged could be combined, thus giving benefits to the material.

2. Research Method

2.1. DI Sample and Specimen Preparations

The DI sample was cast using the standard CO2 sand casting method, and Ferrosilicon Magnesium was added to the molten metal as a graphite nucleating agent. The produced samples were in the form of a double cylindrical shape with 300 mm long and 25±2 mm diameter. The chemical composition of the material, as shown in table 1, was obtained through the spectrometer test.

The specimens dimension of the gravimetric test were prepared in 30 mm diameter and 1 mm thickness. These specimens were ground using metallography silicon carbide sandpaper with 180, 240, 320, 600, and 800 grades subsequently before performing the test.

Table 1. The chemical compositions (wt.%) of DI

Element	C	Si	Mn	P	S	FE
wt. %	3.49	2.62	0.55	0.069	0.0074	Balance

2.2. Piper Nigrum Extraction

The raw *Piper Nigrum* was washed and rinsed with distilled water and dried for 24 hours at room temperature. Then they were ground into a fine powder form before extracted with 2-propanol by using the refluxing method for 2 hours at 85 °C. The resulted solution was filtered to separate the residual. The evaporation process was employed to remove the alcohol in the solution producing the final *Piper Nigrum* extract. The extract was stored in the chiller at 3°C to avoid contamination.

2.3. Quenchant Preparation & Heat Treatment Process

The *Piper Nigrum* extract was diluted by distilled water to produce green corrosion inhibitor modified quenchant with the initial variation concentration of 5, 10, and 15%. These quenchants were employed in a heat treatment process of DI.

In the heat treatment process, the DI specimen were austenitized at 850 °C for an hour in a Carbolite furnace. The DI specimens were then rapidly cooled by quenching in conventional water quenchants, and modified quenchants composed different initial concentrations of *Piper Nigrum* extract to room temperature. The schematic diagram of the heat treatment process is shown in figure 1.

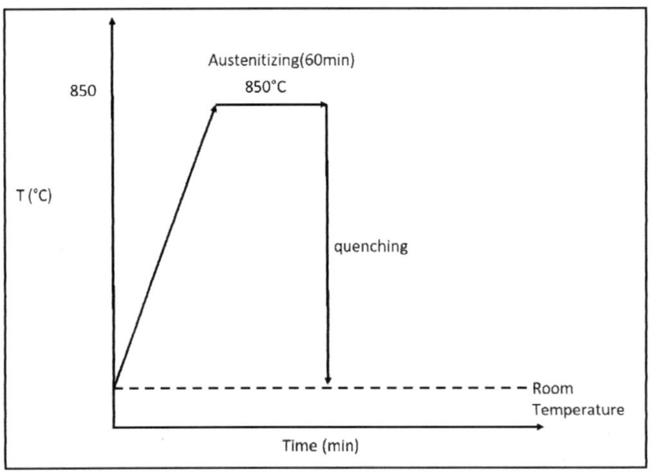

Figure 1: Schematic representation of heat treatment process

2.4. Fourier Transform Infrared Spectroscopy (FTIR) Analysis

Fourier Transform Infrared Spectroscopy (FTIR) analysis was used to characterize the chemical bond and functional group in the *Piper Nigrum* extract. The equipment used for FTIR was VERTEX Series FTIR Spectrometer. For the measurement, Infrared (IR) absorbance spectra were taken between 500 and 4000 cm^{-1} at a resolution of 4 cm^{-1}.

2.5. Gravimetric Test

The Gravimetric test was performed to analyze the corrosion behavior of the sample in acidic media. 1M HCl solution as corrosive media was prepared in the volumetric flask. The as-austenitized samples were wholly immersed in 100 ml of corrosive media and gently shaken in an incubator for 2 hours at a temperature of 30°C. The samples' mass was then measured before and after immersing in the corrosive media using an electronic scale. The density of this material was measured, and it was 7.0887 g/cm³. The corrosion rate (CR) and inhibition efficiency (IE) were calculated by using formulae (1) and (2), respectively, according to the standard of the American Society for Testing Material G1-03.

$$\text{Corrosion Rate (mm/yr)}, CR = \frac{87600W}{PAt} \tag{1}$$

where W is the weight loss (g), ρ is the density (g/cm3), A is the exposed area of the specimen (cm²), and t is the immersion time (h)

$$\text{Inhibition Efficiency (\%)} = IE = \frac{(W_0 - W_i)}{W_0} \times 100\% \tag{2}$$

where W_o is the weight loss (g) in blank etchant W_i is the weight loss (g) in etchant composed of *Piper Nigrum* extract.

2.6. Hardness test
Vickers hardness test was performed in this study to evaluate the effect of quenchant composed *Piper Nigrum* extract on the surface properties of the sample. The test was performed using a *Vickers* hardness tester, where the penetration load was 1000g.

2.7. Microstructure observation
The microstructures of the samples were observed using a series of standard microstructure observation methods. Huvitz HRM-300 microscope and *CapturePro* software were employed to characterize the samples' surface morphology after quenching with conventional and modified quenchants. Specimens were etched by 2% Nital etchant reagent before observing under the microscope.

3. Results and Discussion

3.1. Characteristic of Piper Nigrum Extract
The FTIR spectra of *Piper Nigrum* extract are presented in figure 2. It clearly shows the alcohol group's broad characteristic band at 3369.41 cm^{-1} is suggested to O-H stretching or N-H stretching. The same absorption peaks were also reported by Qiang et al.[19]. The vital characteristic bands found at 1274.88 cm^{-1}, 1193.82 cm^{-1}, and 1151.43 cm^{-1} are associated with Benzoates, Ethanoates, and Propanoates groups, respectively, which is due to C-O stretching vibration. Moreover, the characteristic bands located at 1017.55 cm^{-1} correspond to the C-F halogen group's framework vibration. This analysis proved alcohols, alkaloids, and free compounds (N, O, and F) in *Piper Nigrum* extract. These compounds attribute an inhibitory effect towards corrosion [18]. Furthermore, the carbon-oxygen double bond and carbon-fluoride single bond functional groups attribute the corrosion inhibition efficiency of *Piper Nigrum* extract. All of the functional groups are regarded as excellent cores in corrosion inhibitors.

The presence of heteroatoms, primarily oxygen and nitrogen atoms in these functional groups, gave more evidence on the inhibitory effect of *Piper Nigrum*. Rahuma et al. [28] supported this result, which they used Piper Longum (long pepper) fruit as a green inhibitor. The presence of those atoms met the general consideration of usual corrosion inhibitors [19].

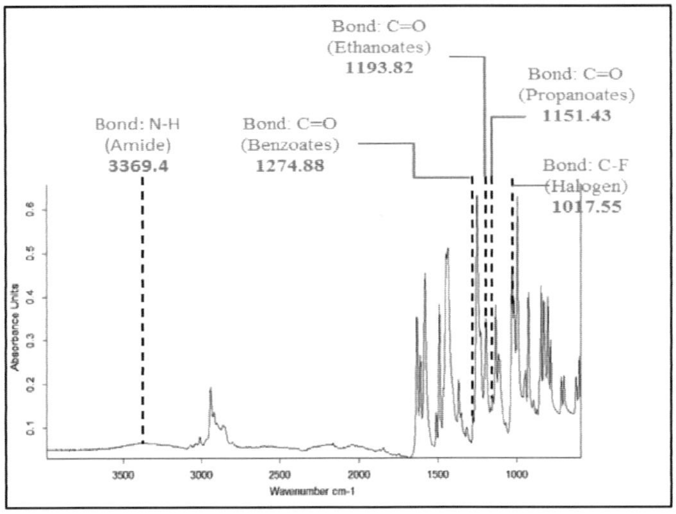

Figure 2. The FTIR spectra of Piper Nigrum extract.

3.2. Effect of Piper Nigrum Concentration on Corrosion Behavior

Table 2 displays the variation in the inhibition efficiency and corrosion rate of the as-austenitized DI sample quenched in quenchant at different *Piper Nigrum* extract concentrations. As-austenitized DI quenched by conventional water quenchant showed a corrosion rate around 189.6962 mm/year. The corrosion rate decreased to 122.0789 mm/year, which showed a 36% reduction when the sample was quenched with a quenchant composed of 5% concentrated *Piper Nigrum* extract. This phenom indicated that the *Piper Nigrum* extract composed in the quenchant effectively reduced the corrosion rate of as-austenitized ductile iron. The corrosion rate decrement increased around 65%, as the concentration of *Piper Nigrum* in quenchant was increased to 10%, and the recorded corrosion rate was 66.6087 mm/year. The inhibitor efficiency increased to 72.17%, and the corrosion rate of the sample had decreased to 51.2685 mm/year as the *Piper Nigrum* extract concentration increased to 15%.

Table 2. The inhibition efficiency and corrosion rate of the as-austenitized DI sample

Sample	Concentration of extraction (%)	Inhibition Efficiency (%)	Corrosion Rate (mm/year)
1	0	0	189.6962
2	5	33.04	122.0789
3	10	65.22	66.6087
4	15	72.17	51.2685

Figure 3 depicts the impact of *Piper Nigrum* extract concentration in quenchant on the as-austenitized DI's corrosion rate. The corrosion process occurred rapidly in the conventional quenchant. It is clearly shown that the corrosion rate steadily decreased with increasing the concentration of *Piper Nigrum* extract to 5% and significantly decreased as the concentration increased to 10%. This concentration can be considered the equilibrium concentration of the modified quenchant since there was no significant corrosion rate decrease as the concentration increased to 15%. The active compounds in *Piper Nigrum* extract reacted with DI's surface during quenching to form a layer that avoided corrosion. The compounds contribute to the phenomenon, including piperine, piperidine, piperittine, and thiamine.

These compounds were adsorbed on the sample surface by displacing water or solvent molecules and form a compact barrier film and hence combating metal dissolution to a considerable extent.

Figure 4 depicts the variation of inhibition efficiency with the concentrations of *Piper Nigrum* extract in quenchant. Contrary to the corrosion rate, increasing the concentration of *Piper Nigrum* extract in quenchant resulted in increasing inhibition efficiency. Both trends suggested that the high concentration of *Piper Nigrum* extract was superior to supplement water quenchant. This pattern was most definitely due to the number of potential reaction sites on the sample surface, limited and protected by complex chemical composition extracts. A similar trend was also reported by Arthur and Abechi [14], where the inhibition efficiency of mild steel in HCl increased as the concentration of plant extract increased.

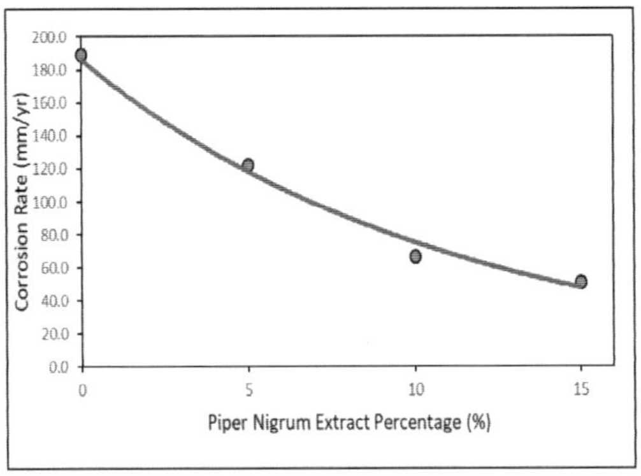

Figure 3. Effect of the *Piper Nigrum* extract concentration in quenchant to the as-austenitized DI's corrosion rate.

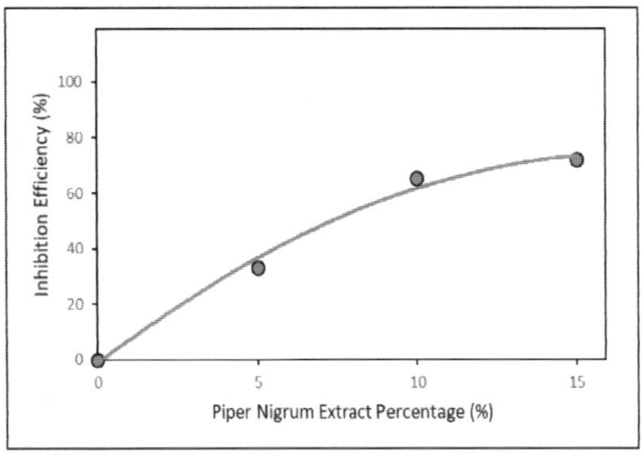

Figure 4. Effect of the *Piper Nigrum* extract concentration in quenchant to the inhibition efficiency

3.3. Ductile Iron Morphology and Characteristics

Microstructure observation was done purposely to evaluate the effect of quenchant composed *Piper Nigrum* extract on the sample's surface. Besides the type of heat treatment process, different quenchant physically and chemically also yields different effects on microstructure and hardenability of a material. The microstructure morphology and transformation of material are essential because they have a close correlation with the material's mechanical properties[11][13].

Figure 5a exhibits the microstructure of as-cast DI. It displays DI's common constituents, comprised of graphite nodules, the sea of ferrite (white region), and the island of pearlite (dark region). On the other hand, figure 5b and figure 5c depict the microstructure of as-austenitized DI samples quenched in conventional quenchant and modified quenchant with *Piper Nigrum*. It is seen that both microstructures consist of martensite and nodular graphite structures. Martensite structure transformed when the austenite structure existed at a temperature higher than eutectoid temperature (Critical), was rapidly cooled (by quenching) to room temperature. The morphology of martensite closely relates to the type of quenchant and cooling method of a heat-treatment process. It is noticed that the distribution of martensite and nodular graphite morphology did not change even though the sample has been quenched in modified quenchant. This result has proven that green inhibitor (*Piper Nigrum*) in quenchant gave a similar effect on the microstructure transformation, as shown in conventional quenchant. The findings likewise suggested that the presence of *Piper Nigrum* extract in water quenchant did not negatively affect the sample morphology behavior and present the almost identical microstructure shown by the sample heat-treated by the conventional method. Since the microstructure is closely related to mechanical properties, this result showed that the mechanical properties of DI unchanged even the quenchant used consisting of *Piper Nigrum* extract. Significant microstructure alteration is sometimes not required to maintain the quality of mechanical properties.

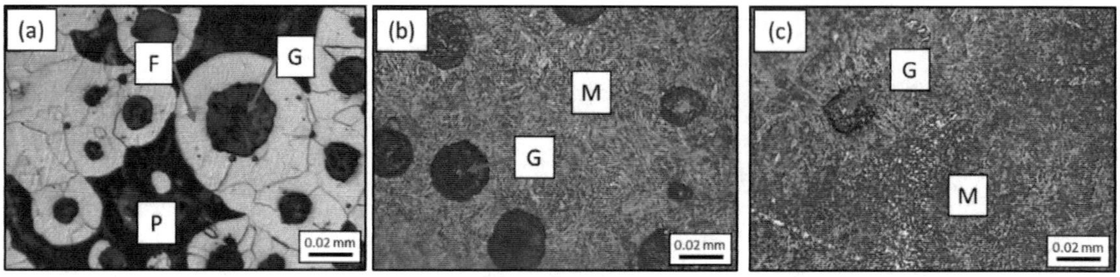

Figure 5. The microstructure of DI: (a) As-cast DI; (b) As-austenitized DI quenched with blank water quenchant; (c) As-austenitized DI quenched with modified quenchant

3.4. Hardness Test

The quenching process and quenchant solution of a heat treatment contribute to the stimulating effect on the material's hardness. Hence, considering the type of quenchant and its element are essential in the hardening process. Vickers hardness test was performed in this study to evaluate the effect of *Piper Nigrum* extract on the surface properties of as-austenitized DI samples. In this analysis, the hardness was measured at several points on the surface of the as-austenitized samples. The average of each sample's readings was then calculated to represent the hardness value of the sample. As shown in figure 6, the graph depicts that the inhibitor's presence did not give a significant consequence on the surface properties of the sample since there is neither significant increment nor decrement is shown. Even though the sample's hardness seemed to decrease as the initial concentration of *Piper Nigrum* extract increases at 10%, the decrease could be considerably low and negligible. The average hardness values of each sample are in the range of 150HV - 167HV. This trend aligns with the as-austenitized samples' morphology behavior, where no significant changes were shown on the microstructure.

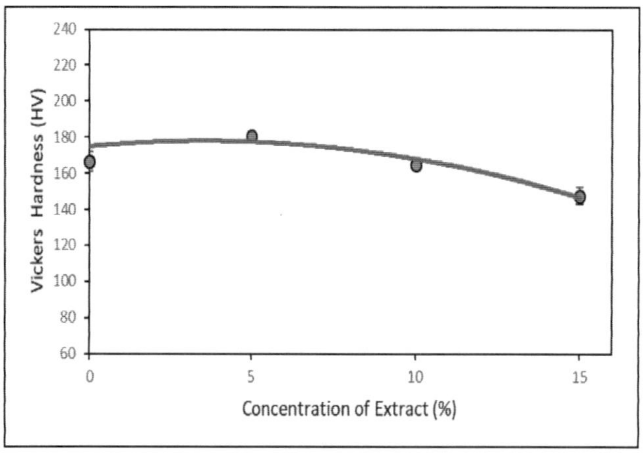

Figure 6. Effect of the *Piper Nigrum* extract concentration to hardness of sample.

4. Conclusion

In a nutshell, this study has proven that the *Piper Nigrum* extract (green corrosion inhibitor) can be used in quenchant as a new alternative to inhibit corrosion of DI in the acidic condition during the heat-treatment process. The increase of the extract concentration in water quenchant produced increased inhibition efficiency and decreased corrosion rate of DI in the acidic condition. From the economic perspective, a 10% concentration of *Piper Nigrum* quenchant is adequate to give excellent inhibition performance of DI in HCl and offer optimum corrosion rate. This extract's presence as a green corrosion inhibitor during the heat treatment process neither improved nor exacerbated the mechanical properties and sustains microstructure transformation behavior.

Acknowledgements
The authors would like to acknowledge Faculty of Mechanical Engineering and Faculty of Chemical Engineering at Universiti Teknologi MARA Bukit Besi Campus for providing technical supports.

References

[1] Idham M F, Masor M F, Mohd N, Abdullah B and Alias S K 2018 Performance of Piper Nigrum (Black Pepper) as Corrosion Inhibitor of Ductile Iron *International Journal of Engineering and Technology* **7** p 115–118.
[2] Yaro A S, Khadom A A and Wael R K 2013 Apricot juice as green corrosion inhibitor of mild steel in phosphoric acid *Alexandria Eng. J.* **52(1)** p 129–135.
[3] Asipita S A, Ismail M, Majid M Z A, Majid Z A, Abdullah C and Mirza J 2014 Green Bambusa Arundinacea leaves extract as a sustainable corrosion inhibitor in steel reinforced concrete *J. Clean. Prod.* **67** p 139–146.
[4] Mourya P, Banerjee S and Singh M M 2014 Corrosion inhibition of mild steel in acidic solution by Tagetes erecta (Marigold flower) extract as a green inhibitor *Corros. Sci.* **85** p 352–363.
[5] B. Tan *et al.* 2021 Papaya leaves extract as a novel eco-friendly corrosion inhibitor for Cu in H2SO4 medium *J. Colloid Interface Sci.* **582** p 918–931.
[6] Boumhara K, Tabyaoui M, Jama C and Bentiss F 2015 Artemisia Mesatlantica essential oil as green inhibitor for carbon steel corrosion in 1M HCl solution: Electrochemical and XPS investigations *J. Ind. Eng. Chem.* **29** p 146–155.
[7] Fouda A E A S, Etaiw S E D H, El-Bendary M M and Maher M M 2016 Metal-organic frameworks based on silver (I) and nitrogen donors as new corrosion inhibitors for copper in HCl solution *J. Mol. Liq.* **213(3)** p 228–234.
[8] Rosli N R, Yusuf S M, Sauki A and Razali W M R W 2019 Musa sapientum (Banana) peels as green corrosion inhibitor for mild steel *Key. Eng. Mater.* **797** p 230–239.
[9] Yang P, Fu H, Absi R, Bennacer R, Lin J and Guo X 2020 Improved corrosive wear resistance of carbidic austempered ductile iron by addition of Cu *Mater. Charact.* **168(100)** p 110577.
[10] Dhilip G and Arul S 2020 Investigation of mechanical properties and wear rate on cryogenically treated austempered ductile iron *Mater. Today Proc.*
[11] Wang B, Qiu F, Barber G C, Pan Y, Cui W and Wang R 2020 Microstructure, wear behavior and surface hardening of austempered ductile iron *J. Mater. Res. Technol.* **9(5)** p 9838–9855.
[12] Idham M F, Abdullah B, Bahari A R, Marwan S H and Yusof K M 2016 Effect of double-quenching heat treatment process on microconstituent and XRD of Ductile iron *ARPN J. Eng. Appl. Sci.* **11(16)** p 9596–9601.
[13] Idham M F, Abdullah B and Yusof K M 2017 Effects of two cycle heat treatment on the microstructure and hardness of ductile iron *Pertanika J. Sci. Technol.* **25(S3)** p 99–106.
[14] Arthur D E and Abechi S E 2019 Corrosion inhibition studies of mild steel using Acalypha chamaedrifolia leaves extract in hydrochloric acid medium *SN Appl. Sci.* **1(9)**.
[15] Sarkar T K, Saraswat V, Mitra R K, Obot I B and Yadav M 2020 Mitigation of corrosion in petroleum oil well/tubing steel using pyrimidines as efficient corrosion inhibitor: Experimental and theoretical investigation *Mater. Today Commun.* **26** 101862.
[16] Tasić Z Z, Petrović Mihajlović M B, Radovanović M B, Simonović A T and Antonijević M M 2018 Cephradine as corrosion inhibitor for copper in 0.9% NaCl solution *J. Mol. Struct.* **1159** p 46–54.
[17] Hoai N T *et al.* 2019 An improved corrosion resistance of steel in hydrochloric acid solution using Hibiscus sabdariffa leaf extract *Chem. Pap.* **73(4)** p 909–925.
[18] Zhang X *et al.* 2020 Evaluation of Idesia polycarpa Maxim fruits extract as a natural green corrosion inhibitor for copper in 0.5 M sulfuric acid solution *J. Mol. Liq.* **318**.

[19] Qiang Y, Zhang S, Tan B and Chen S 2018 Evaluation of Ginkgo leaf extract as an eco-friendly corrosion inhibitor of X70 steel in HCl solution *Corros. Sci.* **133** p 6–16.
[20] Anupama K K, Ramya K, Shainy K M and Joseph A 2015 Adsorption and electrochemical studies of Pimenta dioica leaf extracts as corrosion inhibitor for mild steel in hydrochloric acid *Mater. Chem. Phys.* **167** p 28–41.
[21] Kumar H and Yadav V 2020 Musa acuminata (Green corrosion inhibitor) as anti-pit and anti-cracking agent for mild steel in 5M hydrochloric acid solution *Chem. Data Collect.* **29**.
[22] Prabakaran M, Kim S H, Sasireka A, Kalaiselvi K and Chung I M 2018 Polygonatum odaratum extract as an eco-friendly inhibitor for aluminum corrosion in acidic medium *J. Adhes. Sci. Technol.* **32(18)** p 2054–2069.
[23] Ji G, Anjum S, Sundaram S and Prakash R 2015 Musa paradisica peel extract as green corrosion inhibitor for mild steel in HCl solution *Corros. Sci.* **90** p 107–117.
[24] Odewunmi N A, Umoren S A and Gasem Z M 2014 Watermelon waste products as green corrosion inhibitors for mild steel in HCl solution *J. Environ. Chem. Eng.* **3(1)** p 286–296.
[25] Odewunmi N A, Umoren S A and Gasem Z M 2015 Utilization of watermelon rind extract as a green corrosion inhibitor for mild steel in acidic media *J. Ind. Eng. Chem.* **21** p 239–247.
[26] Odewunmi N A, Umoren S A, Gasem Z M, Ganiyu S A and Muhammad Q 2015 L-Citrulline: An active corrosion inhibitor component of watermelon rind extract for mild steel in HCl medium *J. Taiwan Inst. Chem. Eng.* **51** p 177–185.
[27] Tantawy A H, Soliman K A and Abd El-Lateef H M 2020 Novel synthesized cationic surfactants based on natural *Piper Nigrum* as sustainable-green inhibitors for steel pipeline corrosion in CO2-3.5%NaCl: DFT, Monte Carlo simulations and experimental approaches *J. Clean. Prod.* **250.**
[28] Rahuma M N, Amer H and Alfergani M 2014 Corrosion Inhibition of Mild Steel in 11% Hydrochloric Acid Solutions by Using Black Pepper (*Piper Nigrum*) *Chem. Sci.Trans.* **3(2)** 2.
[29] Chemistry A 2019 Cinnamomum zeylanicum Extract as Green Corrosion Inhibitor for Carbon Steel in Hydrochloric Acid Solutions *Prog. Chem. Biochem. Res.* **2(3)** pp 120–133.
[30] Meena O and Chaturvedi A 2020 Corrosion Inhibition of Aluminium by Extract of Aerial Parts of Phyllanthus niruri in Hydrochloric acid and Study of Extract as Antibacterial **13(7)** p 22–32.

IOP Conf. Series: Materials Science and Engineering 1176 (2021) 012028 doi:10.1088/1757-899X/1176/1/012028

Constructing a Blueprint of a Kolek

N A A Rahman[1], M A Johan[1*], N M K N Ismail[1], and S Ahmad[2]

[1]Faculty of Mechanical Engineering, University Teknologi MARA (UiTM), 23200 Bukit Besi, Terengganu, Malaysia
[2]Faculty of Mechanical Engineering, Universiti Teknologi Mara (UiTM), 40450 Shah Alam, Selangor, Malaysia.

*Corresponding author: azahari287@uitm.edu.my

Abstract. Kolek, a Malay Traditional Boat, is slowly forgotten by the younger generation due to several reasons. In this paper, an in-depth study was conducted to construct a Kolek blueprint for preservation using the Reverse Engineering method on the existing artifact in Malay Traditional Boat Yard located at Pulau Duyung. The study was conducted through several phases namely boat measurement, digitization, and analyses. As a result, a Kolek Lines Plan was produced using Rhinoceros 5 for reference by future generations. Then analyses were carried out using PolyCAD version 10.4 to obtain hydrostatic properties of the Kolek for its stability, and resistance to determine the power requirements needed for the Kolek. The results show the blueprint of the Kolek that can be develop as well as mechanical properties of the Kolek can be obtained using modern naval architecture techniques. This blueprint can be used to build new boats by the next generation using the hybrid method which is a combination of modern and traditional methods.

1. Introduction

The traditional Malay boat is one of the heritages slowly forgotten by Malaysians, especially the younger generation. Malay Traditional boats had existed since ancient times. There are types of boats depending on their functions and unique design. Kolek is one of the traditional Malay boats with unique art of heritage used as a fishing boat. The usage is not only for water transportation but also as symbolic of Malay thought and civilization [1]. The invention of technology and techniques in boat construction reflects Malay society intellectual thought in producing things. However, the technological progress and development of machinery propelled transport system led to the declining usage of a Kolek. [2]

Kolek, the Malay traditional boat, is made of wood and using the traditional method without any design blueprint during the construction stage and alterations made based on the rule of thumbs and trial and error [3]. It is supported by Indonesian researchers [4] that the construction of fishing boats does not use design drawings. The boat draft or design is only imagined in the boat builder's minds and verbally communicated to the boat builder assistants in the form of instructions in building fishing boats. These situations lead to difficulty for the reconstruction and identification of the traditional boat by the future generation, leading it to diminish in the future.

Moreover, the younger generation's lack of knowledge on traditional boats' construction as there is no design blueprint documentation contributed to the diminishing. Several studies on the Malay traditional boat preservation was done these few years by the researchers to resolve this situation. Most of the research focused on the art and decorations of Malay and western traditional boats. There is no in-depth study highlighted on the blueprint of the Malay traditional boat's naval architecture and engineering design aspects. This study proposes to conduct an in-depth study of the mentioned aspect by applying the Reverse Engineering methodology from the existing artifacts. The main objective of this study is to develop a complete design blueprint on Kolek for preservation purposes.

It starts with reproducing the actual shape of the Kolek by measuring the actual boat. This blueprint is expected to support the Malaysia National Culture Policy. One of its strategies is to restore, preserve, and develop culture towards strengthening national culture through joint research, development, education, and cultural expansion and connections. The cultural heritage can enhance national identity and promote the tourism industry as a source of national income. Although few studies on the preservation of traditional boat, the research focused on the art and decoration [5] and traditional western boat, none of them highlighted traditional Malay boats from the engineering perspective.

The traditional shipbuilding in this archipelago area has already existed since the migration of Protomalay [6]. The technique is passed down from generation to generation of shipwright master. In the early days, the wooden ship was widely used to travel by the Malay communities. This seaworthy craftsmanship travelled up to Madagascar island in the west and the Polynesian islands in the east [7]. There are many types of Malay ships. Among them was a big ship or Perahu Besar that can be further categorized, such as Pinis and Bedar of Terengganu type. As Perahu Besar role conquered by the modern ship, most traditional shipyard nowadays tends to focus more on producing the smaller perahu and fishing vessels. [7]

The in-depth study of Malay traditional boats is essential to ensure the preservation of Malay maritime heritage and educate the young generation nowadays, as reported by the researcher. The study focused on Malay traditional boats' art and decoration by using the existing artifacts from the numerous museums in Malaysia [8]. The collection of these data was analyzed based on the forms, motives, and relevance inherent in the traditional Malay boat decoration components. In Malaysia, the fishing boats are the same as traditional Malay boats built without any blueprint design on the construction stage, guidance, and approval from naval architecture [9]. Any modification is only based on the trial and error method to countermeasure any technical issue. [3]

Reverse Engineering, also known as back engineering, is acquiring knowledge of design information by examining a product. For example, dismantling an automobile to study the various components and reproduce them based on the original model. Reverse Engineering is used as a tool that can preserve data by scanning the originals. Reverse Engineering applies to many areas like Architecture, works of art, and old artifacts [10]. In our engineering Industrial commercialized society, Reverse Engineering shows much promise for the application like mass production, and so forth.

Reverse engineering is not only applied to the engineering design aspect; it also further utillizes on the old artifacts, for example, monuments and various musical instruments from the past. The digitalization of cultural heritage with R.E (reverse engineering) methodology has piqued the interest of many researchers from a different field, especially the archaeologist. Digitization helps preserve and can also make digital copies and store the digital data at various locations at a minimal cost, which is not possible with the original.

With the advancements of computer technologies, their applications are used to virtualize the cultural heritage to make digital copies of the models showing its shapes, texture, and type of material used and the reverse engineering methodology to reconstruct the objects lost. The process allows for the reconstruction and preservation of objects that are lost due to unexpected events. [11] As reported by the researchers [12], there are several different methods to record individual ship timbers and ship timber assemblies applied by marine archaeologists. The current ship timber recording methods generally fall under one of four categories, which are 2D scaled drawings, 2D tracing, 3D contact digitizing, and 3D scanning.

2. Research Method

The study was conducted through several phases namely boat measurement, digitization, and analyses. In analyses phase, hydrostatic, stability and resistance analysis were analyzed.

In the boat measurement phase, the activity was conducted manually at a known boat maker yard owned by Haji Abdullah Muda, in Pulau Duyung, Terengganu for the existing Kolek. The Kolek needed to be measured in several parts: stem profile, stern profile, and cross-sections at 5 location as shown in figure 1.

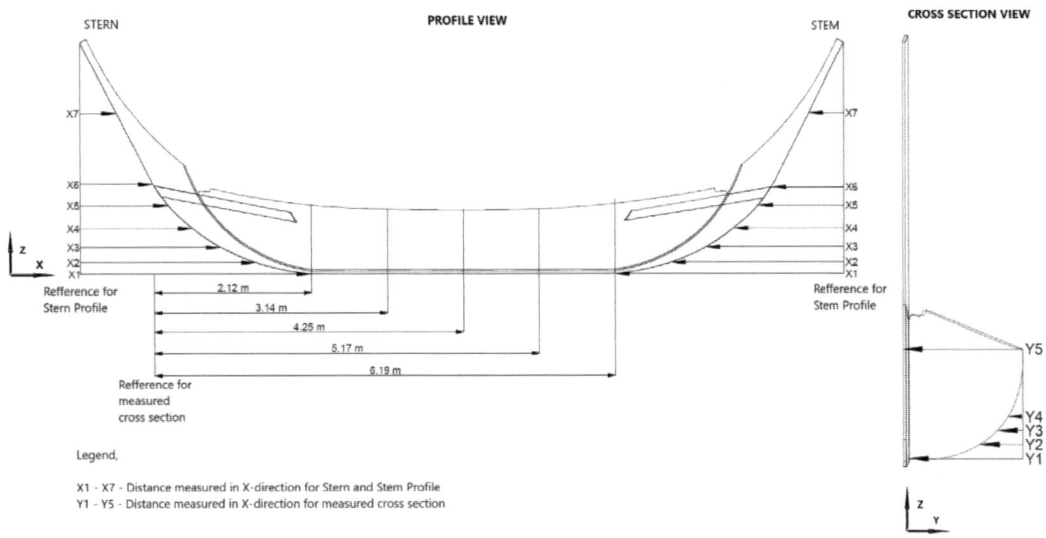

Figure 1: Measurement plan of Kolek

In the digitization phase, the shape of the Kolek hull was drawn using 3-D modeling software called Rhinoceros 5. The Kolek was modelled in the 3-D. The re-drawing process began by placing the stem and stern profiles in the profile view to get the overall length of the Kolek. Next, the 5 measured cross-sections were then positioned along the Kolek length as shown in figure 1. After that, the deck line can be produced by connecting the edges of each cross-section and ending at the stem and stern as shown in figure 2. Then a surface was lofted along those measured cross-sections as shown in figure 3. From this surface, the actual 10 stations were generated using an automatic function in the software.

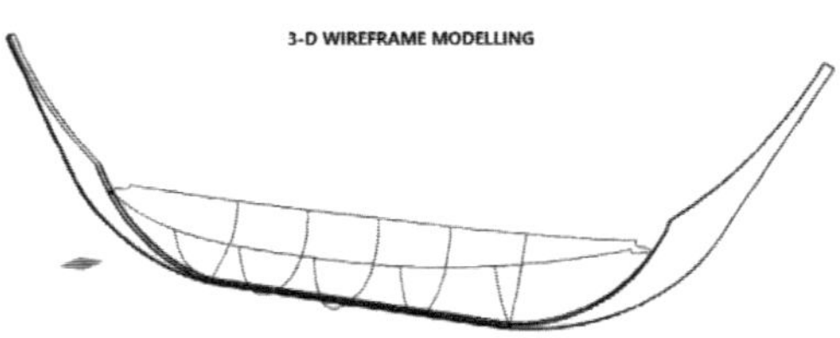

Figure 2: 3-D wireframe modelling of Kolek

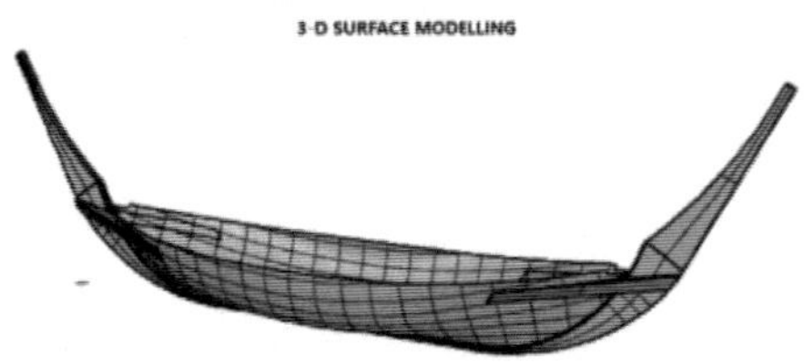

Figure 3: 3-D surface modelling of Kolek

3. Results and Discussion

From the 3-D modelling in digitization phase, a 2-D drawing known as Lines Plan has been produced. This Lines Plan modelling consists of 3 projections: Body Plan, Profile, and Half Breath Plan, to visualize a ship's shape. Figure 4 shows the replica of the Kolek Lines Plan. This Lines Plan helps the younger generation who have carpentry skills to visualize the shape of a Kolek. Today, the skills to visualize the shape of the Kolek are found in the minds of highly skilled boat makers where they are becoming less and less.

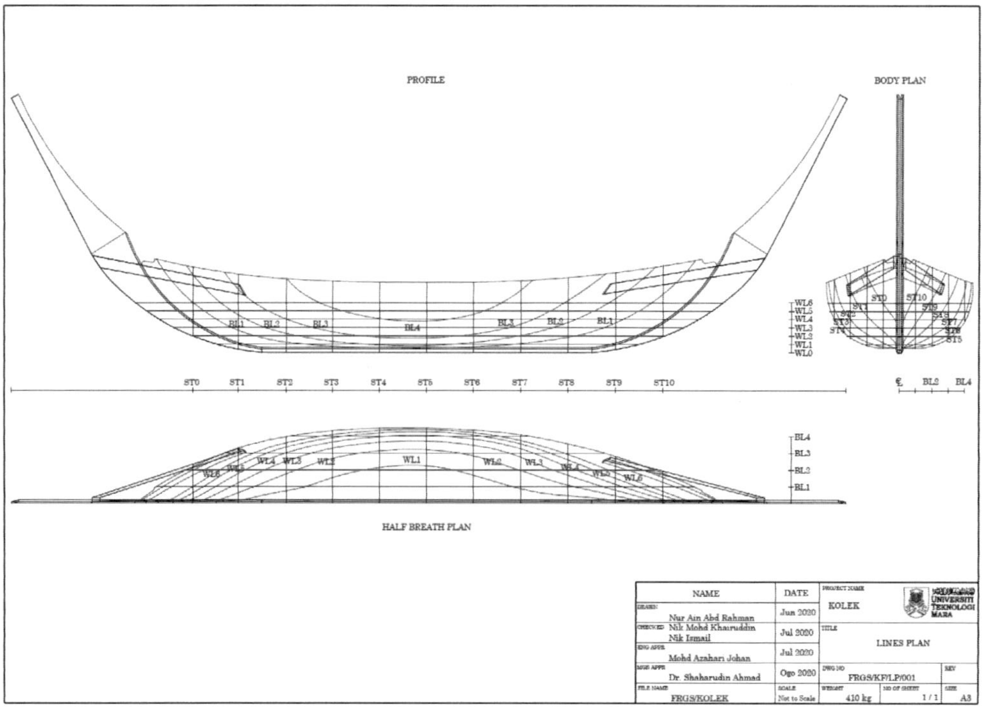

Figure 4: Replica of Kolek Lines Plan

The accuracy of this Lines Plan drawing is essential to ensure that it represents the Kolek that has been measured. Table 1 shows the difference between the measured curves and the curves drawn for the 5 cross sections. It was found that there was a small difference between them where the average was 1.83%. These small differences indicate that the Kolek has been drawn almost identical to the one measured.

Table 1: Cross sections comparison

	1st Cross Section - 2.12 m			2nd Cross Section - 3.14 m			3rd Cross Section - 4.25 m			4th Cross Section - 5.17 m			5th Cross Section - 6.19 m		
Height (m)	Measured (m)	3D model (m)	Diff. %	Measured (m)	3D model (m)	Diff. %	Measured (m)	3D model (m)	Diff. %	Measured (m)	3D model (m)	Diff. %	Measured (m)	3D model (m)	Diff. %
0.15	0.17	0.17	0.02	0.48	0.48	0.03	0.56	0.56	0.02	0.34	0.34	-0.20	0.11	0.11	4.75
0.30	0.38	0.38	0.00	0.68	0.68	-0.01	0.75	0.75	-0.02	0.60	0.60	0.16	0.22	0.24	11.85
0.45	0.54	0.54	0.22	0.78	0.78	0.03	0.84	0.84	0.02	0.72	0.72	0.05	0.33	0.37	11.92
0.60	0.64	0.64	-0.09	0.84	0.84	-0.01	0.88	0.88	-0.01	0.78	0.78	0.03	0.44	0.48	10.05
Deck Edge	0.74	0.74	0.06	0.88	0.88	0.06	0.90	0.90	0.08	0.84	0.84	0.08	0.60	0.65	6.79
		Average	0.04		Average	0.02		Average	0.02		Average	0.02		Average	9.07
														Total Average	1.83

Once the lines plan is produced, the analysis phase can be performed. The first analysis performed was hydrostatics. The hydrostatics analysis is done using an open-source software known as PolyCAD version 10.4. Table 2 shows the results of hydrostatics analysis of the Kolek at various water levels. These hydrostatics data describe the geometric properties of the Kolek. These hydrostatic data are important so that they can be used in stability, and resistance analysis.

Table 2: Hydrostatics data of Kolek at various water levels

T [m]	Disp [Tonnes]	LCB [m]	VCB [m]	KM_T [m]	WPA [m^2]	LCF [m]	TPC [Tonnes/cm]	C_b	C_p	C_{wp}	C_m	BM_T [m]	BM_L [m]	WSA [m^2]	MCT [Tonnes-m]
0.000	0.000	0.000	0.000	0.000	0.15	4.025	0.00	0.000	0.000	0.000	0.000	0.000	0.000	0.000	0.000
0.050	0.016	4.149	0.028	0.043	0.39	4.151	0.00	0.739	0.917	0.959	0.806	0.015	47.202	0.700	0.002
0.100	0.095	4.039	0.072	1.086	2.33	3.999	0.02	0.198	0.482	0.499	0.411	1.014	25.692	2.858	0.005
0.150	0.241	4.016	0.105	1.051	3.33	4.005	0.03	0.251	0.481	0.535	0.522	0.946	16.784	4.035	0.007
0.200	0.435	4.017	0.136	1.008	4.13	4.008	0.04	0.285	0.492	0.554	0.579	0.871	12.816	5.049	0.010
0.250	0.664	4.015	0.167	0.982	4.82	4.012	0.05	0.305	0.495	0.567	0.616	0.814	10.746	5.989	0.012
0.300	0.931	4.018	0.198	0.955	5.43	4.016	0.06	0.322	0.501	0.577	0.642	0.757	9.354	6.888	0.014
0.350	1.223	4.017	0.229	0.933	5.96	4.016	0.06	0.335	0.505	0.585	0.664	0.704	8.388	7.733	0.016
0.400	1.541	4.018	0.259	0.911	6.41	4.018	0.07	0.346	0.507	0.590	0.682	0.652	7.642	8.550	0.018
0.450	1.881	4.019	0.289	0.894	6.82	4.019	0.07	0.356	0.509	0.595	0.698	0.605	7.038	9.365	0.020
0.500	2.243	4.021	0.319	0.880	7.18	4.021	0.07	0.365	0.512	0.600	0.714	0.562	6.538	10.164	0.022
0.550	2.619	4.021	0.349	0.873	7.52	4.022	0.08	0.374	0.513	0.605	0.728	0.524	6.148	10.955	0.024
0.600	3.017	4.023	0.378	0.867	7.82	4.025	0.08	0.382	0.516	0.610	0.741	0.489	5.802	11.757	0.026

Legend :

T	Draft	C_b	Block coefficient
Disp	Mass Displacement	C_p	Prismatic coefficient
LCB	Center of Bouyancy (longitudinal direction)	C_{wp}	Water-plane area coefficient
VCB	Center of Bouyancy (vertical direction)	C_m	Midship section area coefficient
KM_T	Distances from Keel to Transverse Metacentres	BM_T	Distances from Centre of Buoyancy to Transverse Metacentres
WPA	Waterplane Area	BM_L	Distances from Centre of Buoyancy to Longitudinal Metacentres
LCF	Center of floatation (longitudinal direction)	**WSA**	Wetted Surface Area
TPC	Tonne per centimeter immersion	**MCT**	Moment to Change Trim one Centimeter

The measured boat has already been painted with a red underwater coating to indicate the sinking level of the boat. This is done by the boat builder through his many years of experience in the traditional boat building industry. The height of this waterline design when measured is 0.5 meters from the keel as shown in figure 5. Table 3 shows the geometric properties of the Kolek on the waterline design. These data have been taken from table 2 at T=0.5 m.

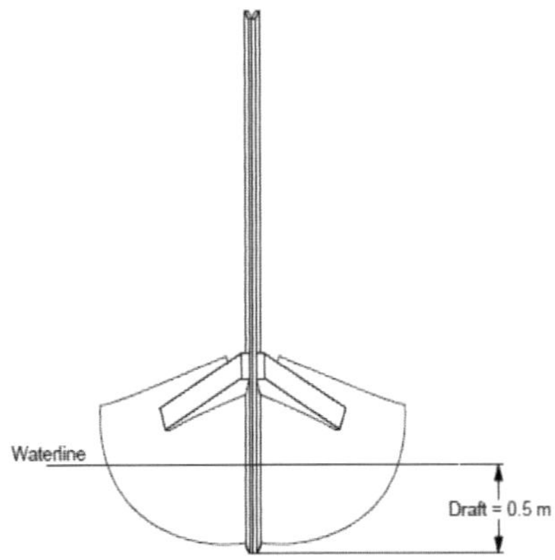

Figure 5: Kolek draft

Table 3: Hydrostatics data of Kolek at design draft

Hydrostatic Particulars	Value	Hydrostatic Particulars	Value
T [m]	0.500	C_b	0.365
Disp [Tonnes]	2.243	C_p	0.512
LCB [m]	4.021	C_{wp}	0.600
VCB [m]	0.319	C_m	0.714
KM_T [m]	0.880	**BM_T [m]**	0.562
WPA [m²]	7.18	**BM_L [m]**	6.538
LCF [m]	4.021	**WSA [m²]**	10.164
TPC [Tonnes/cm]	0.07	**MCT [Tonnes-m]**	0.022

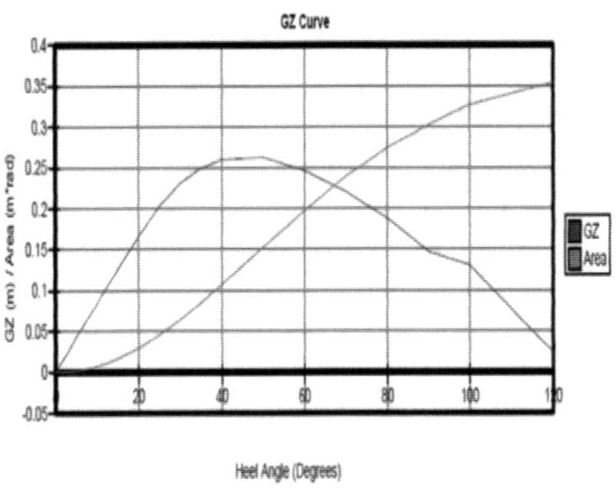

Figure 6: GZ curve of Kolek

Table 4: Compliance of stability criteria

Criteria	Calculated	IMO Requirement	Complience
A_{0-30}	0.0643	> 0.055	comply
A_{0-40}	0.1075	> 0.090	comply
A_{30-40}	0.0433	> 0.030	comply
GZ_{max}	0.2620	> 0.200	comply
Angle of GZ_{max}	50°	> 25°	comply
GM_{awal}	0.464	> 0.150	comply
The boat is Stable			

The last analysis in the analyses phase is resistance analysis. The Kolek resistance analysis was also performed using PolyCAD version 10.4. The resistance curve, as in figure 7, has been successfully produced. This analysis has been done based on the Delft Systematic Yacht Hull Series (DSYHS). From figure 7, it is found that at low Froude numbers (<5 Knots), the frictional resistance is the major component and is dependent on the wetted surface area of the hull (and inherent roughness). While, at higher Froude numbers, the residuary resistance becomes the primary drag component.

The resistance curve can be used to estimate the Effective Power of the Kolek by multiplying R Total [N] and the desired speed. The sail of the Kolek must be able to produce the same number of forces as the Effective Power to keep the boat moving at the desired speed. If you want to modernize the Kolek by installing an engine, the effective power must be multiplied by the loses factor occurs throughout the power train such as loses caused by gearbox, shaft, quasi propulsive coefficient, sea margin and maximum continuous engine rating (MCR) to determine the engine brake horsepower needs to be installed.

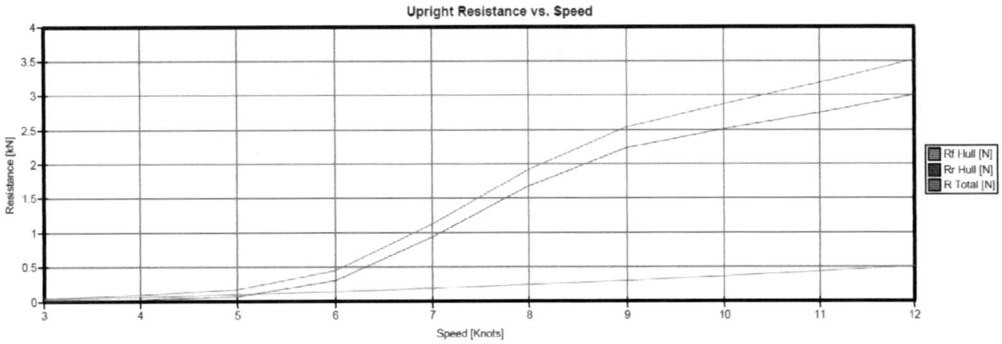

Figure 7: Resistance curve of Kolek

4. Conclusion

Kolek is a significant Malay traditional boat that has been slowly forgotten by Malaysians, mostly the young generation. As a symbol of Malay thought and civilization, she is supposedly preserved. The study has successfully produced a Lines Plan drawing for the Kolek that can be made as a reference for future generations. From the Lines Plan drawing, stability analysis has been done which shows that the Kolek has good stability. The resistance analysis has also been performed to show the power requirements for the Kolek. The results of this resistance analysis can be used to determine the prime mover to be used for the Kolek. Therefore, the objective for constructing a blueprint utilizing modern naval architecture techniques can be achieved to support the preservation of a historical Malay traditional boat.

It is proposed to produce the new construction method of the Kolek using this blueprint. The traditional method of building the Kolek needs to be changed a bit by applying modern methods so that the younger generation can build the new Kolek in the future. This is because, the traditional method of building the Kolek without drawing is quite difficult for the younger generation to master the method.

Acknowledgment

The authors wish to thank Universiti Teknologi Mara, Cawangan Terengganu Kampus Bukit Besi and Haji Abdullah Muda Yard.

References

[1] Wahab M R A, Ramli Z, Ismail N N A, Bakar N A and Azhar W N S W 2018 Symbolism in traditional Malay boat crafting in the East Coast *J. Malaysia Inst. Planners* **16(1)** p 294–302.

[2] Fernandes A, Figueira de Sousa J and Tânia V V 2018 The Preservation of the Tagus Estuary Traditional Boats : New Issues and the preservation of the Tagus Estuary Traditional Boats : new issues and challenges *J. Spat. Organ. Dyn.* **4(2)** p 153–173.

[3] Zamani M and K.Vijaynathan 2000 Cultural Influence on Hull Form Geometrical Features of Malaysian Trawler Fishing Boats *Proc. 7th Int. Mar. Des. Conf.*

[4] Eko S, Triyanto and Mujiyono 2020 Design and Construction of Traditional Fishing Boat in Jepara in the Context of Cultural Ecology : The Implication as Arts Learning Resources *Int. J. Indones. Soc. Cult.* **12(2)** p 209–215..

[5] Wahab M R A, Ramli Z, Samad M A A, Supian N S M, Man M N, Nasir M and Masdey S S 2018 Ancient maritime symbols in Malay traditional boat in the East Coast, Peninsular Malaysia *J. Malaysia Inst. Planners* **16(1)** p 372–380.

[6] Ali I 2009 The Culture of Outrigger Boat in the Malay Archipelago : A Maritime Perspective *Int. J. Hist. Stud.***1(1)** p 57–70.

[7] Haron S J, Koto J and Dalil M 2014 Review on Delivery Issues in Malaysian Traditional Ship Production Process *J. Ocean. Mech. Aerosp.* **13** p 18–21.

[8] Abdul Wahab M R and Ramli Z 2018 Analyzing decreasing factors in the used of Traditional

Malay boats in East Coast *Asian J. Environ. Hist. Herit.* **2** p 279–288.

[9] Yaakob O, Hashim F E, Jalal M R and Mustapa M A 2015 Stability, seakeeping and safety assessment of small fishing boats operating in southern coast of Peninsular Malaysia *J. Sustain. Sci. Manag.* **10(1)** p 50–65.

[10] Di Paola F and Inzerillo L 2018 3D reconstruction-reverse engineering-digital fabrication of the Egyptian Palermo stone using by smartphone and light structured scanner *Int. Arch. Photogramm. Remote Sens. Spat. Inf. Sci.* **42(2)** p 311–318.

[11] Kommula K B, Laudya S T, Eslavath S, Karri S N R and Pandey R 2018 Reverse Engineering for Restoration and Preservation of Old Artifacts and Cultural Heritage *Int. J. Eng. Res. Mech. Civ. Eng.* **3(3)** p 14–18.

[12] Van Damme, Auer J, Ditta M, Grabowski M and Couwenberg M 2020 The 3D annotated scans method: a new approach to ship timber recording *Herit. Sci.* **8(1)** p 1–18.

Effects of Waste Glass Powder Filler on Intumescent Coating for Steel Structures Application

Y M Azmi[1], F Ahmad[1]*, S N Razak[2], M A H A Hadi[1], S Kabir[1] and G H Yeoh[3]

[1] Department of Mechanical Engineering, Universiti Teknologi PETRONAS, Malaysia
[2] Laboratoire Angevin de Mecanique, Procedes et Innovation of Arts et Metiers, France

[3] School of Mechanical & Manufacturing Engineering, University of New South Wales, Australia

*Corresponding author: faizahmad@utp.edu.my

Abstract. Intumescent coating is a passive fire protection system used to prevent and resist the spreading of fire. The coating swells several times of its thickness on exposure to the fire, forming cellular foam and acts as an insulative barrier, thus delaying the heat transfer to the protected substrate. The study aims to analyse waste glass powder filler's effect into intumescent coating formulation limited to one size of waste glass powder. Waste glass is a broken form of leftover by-product from industrial and domestic. Millions of tons of waste glass generated every year, adding with their non-biodegradable nature causing additional environmental pollution. Waste glass powder contains high silica composition known for its high thermal stability, weather resistance, hydrophobicity, low surface tension, and high oxidation resistance. The coating showed improved thermal stability, weather resistance, and adhesion when added into intumescent coating formulation. The lowest recorded backside substrate temperature 154.6 °C compare to coating without the waste glass filler, 333.7 °C. The coating also showed less than 0.1% total weight loss for samples with 4% wt. filler after immersion in seawater for 15 days compared to the control formulation, 0.19%. The char morphology analysis showed the emergence of silica and calcium elements occupying the expandable char's gaps, thus strengthening the char structure.

1. Introduction

Fire losses are among the most disastrous accidents, be it from a human mistake or natural causes. The short time it took to spread and decrease escaping time, resulting in various protection types, was created to either slow it or prevent it. Two significant types of protection available in modern civilisation are active (AFP) and passive fire protection (PFP). Active fire protection requires an action taken to put out a fire, such as a sprinkler system, while passive fire protection prevents or resist the spread of fire. An example of PFP is an intumescent coating (IC). The intumescent coating was applied like paint into the protected substrate and acted to prevent the spreading of fire and delay the heat transfer rate. When exposed to fire, the coating reacts and swell from its original thickness, thus foaming a carbonaceous insulation barrier obstructing the heat transfer to the protected substrate[1]. According to Mariappan (2016), the intumescent coating mechanism comprises five steps. It starts with the decomposition of catalyst material into mineral acid. The acid reacts with carbonic by dehydration to form an ester, thus decomposing and forming a carbonaceous char and releasing the acid. The resin material then melted, forming a skin over the char and the blowing agent releasing an inert gas, resulting in the expansion of char that act as efficient insulation[2]. The modernisation of intumescent coating has shown that the researchers eagerly interested in adding nanofillers for various purposes, such as smoke suppressant, corrosion resistance, thermal conductivity, etc. Filler generally incorporated into the formulation in the range of less than 40% of IC mass.

Glass is a unique multipurpose substance with more comprehensive applications naturally occurring when rock high in silicates melt at high temperature and cool before forming a crystalline structure. Waste glass is a broken form of leftover by-product from industrial and domestic. Millions of tons of waste glass generated every year, adding with their non-biodegradable nature causing additional environmental pollution. In Malaysia, the lack of glass recycling of glass product making a rising effect on the environment. The common types of glass found in domestic are soda-lime glass, lead-alkali-silicate glass and borosilicate glass. These glasses commonly manufactured to form a window, a bottle, glassware and laboratory equipment used in the house and other domestic industries, making the amount of waste glass rising every year.

Research nowadays used waste glass powder (WGP) in the cement and concrete industry. The pozzolanic reaction from the waste glass powder is highly sought due to its advantage of increasing the end product's durability and providing compressive strengths. Decreasing the particle size of glass powder encouragingly increases pozzolanic reactivity [3]–[6]. Glass as a filler also reported by a researcher to help increase the thermal stability and dynamic thermal-mechanical properties and reduce the coefficient of thermal expansion[7]. Few studies on using the waste glass powder in the epoxy and ceramic coating reported, but little to none in the field of intumescent coating formulation studied[8]–[10]. This paper discussed the effect of waste glass powder as filler into intumescent coating limited to one size of waste glass powder. The effect of different size of WGP was not analysed.

2. Experimental Materials and Procedure

2.1. Materials

Bisphenol A diglycidyl ether (DGEBA) with Equivalent Weight (E.E.W) 188, expandable graphite (EG), and zinc borate (ZB) purchased from Mc-Growth Chemical Sdn. Bhd. Boric Acid (BA), melamine (MEL) bought from Sigma-Aldrich (M) Sdn. Bhd. Malaysia. Ammonium polyphosphate (APP) supplied by Clariant (Malaysia) Sdn. Bhd and waste glass obtained from factory glass & plastic packaging (Ipoh), Malaysia.

2.2. Experimental Procedures

All the materials weighed according to their respective weight percentage as per table 1. BA, ZB, Mel, and APP was ground for one minute before mixed with EG and epoxy in a sheer mixer machine at 30 rpm for 15 minutes. The hardener slowly introduced into the mixer, and the process continued for another 10 minutes. The coating then applied to ASTM A36 steel substrate before leaving to cure at room temperature.

Table 1. Intumescent coating with the waste glass powder formulation

Samples	Materials (%)					
	APP	MEL	BA	EG	Epoxy & Hardener	Waste Glass Powder
WG000	11.23	5.5	11.11	5.5	66.66	0
WG001	11.23	5.5	11.11	5.5	65.66	1
WG002	11.23	5.5	11.11	5.5	64.66	2
WG003	11.23	5.5	11.11	5.5	63.66	3
WG004	11.23	5.5	11.11	5.5	62.66	4
WG005	11.23	5.5	11.11	5.5	61.66	5

3. Characterisation and Analysis

3.1. Thermogravimetric Analysis (TGA)
Perkin Elmer Pyris Thermogravimetric Analyser was used for this analysis to determine the residual mass and thermal behaviour of the waste glass powder using a heating rate of 10°C/min under nitrogen flow from room temperature till 900°C.

3.2. Field Emission Scanning Electron Microscope (FESEM) and Scanning Electron Microscope (SEM)
Zeiss SUPRA 55-VP FEGSEM were used to study the morphology structure of char's surface and cross-section. The samples prepared by burning them in the Carbolite furnace for 1 hour at 600°C before subjected to the analysis.

3.3. Fourier Transform Infrared Spectroscopy (FTIR)
The technique based on the functional groups' IR absorption frequencies and spectra was recorded in 450 to 4000 cm^{-1} using the KBR-FTIR method.

3.4. X-ray fluorescence (XRF)
This non-destructive analytical technique used to determine the elemental & oxidation composition of materials. The samples were analyse using Bruker S8 Tiger.

3.5. Direct Fire Test (UL94 Lab scale) and Control (Furnace) Fire Test.
The cured samples exposed to UL94 in a lab-scale test to evaluate the coating's thermal performance. The samples back substrate was attached with three thermocouples and hold with a retort stand. The coating sample then exposed to butane gas fire, and the back substrate's temperature and the physical observation recorded for 1 hour.

A control fire test was performed inside a Carbolite Gero Furnace to examine the morphology and expansion of char in a controlled environment. The cured coating burned for 1 hour at a temperature of 600°C with a 20°C/min rate. The samples thickness before and after burning recorded, and intumescent factor (IF) was calculated based on the following equation:

$$IF = (D_2 - D_0)/(D_1 - D_0) \quad (1)$$

D_2 is the thickness of the char with the substrate, D_1 is the thickness of coating with the substrate, and D_0 is the thickness of the steel substrate.

3.6. Water Immersion Test
The test performs according to ASTM D870-15, Standard Practice for Testing Water Resistance of Coatings Using Water Immersion[11]. All the cured intumescent coating samples were weighted before immerse into seawater obtain from Teluk Batik, Perak. The samples weight was recorded every three days for 15 days, and the last weight recorded after the samples left for drying for 24 hours.

4. Results and Discussions

4.1. Waste Glass Powder Analysis
The waste glass powder was subjected to characterisation analysis to study the material's thermal stability and composition. The graph in figure 1 displayed a low decomposition percentage of 0.78% at the 800°C, even with the sharp slope of -0.0014. This lower value of slope indicates the material is thermally stable and could enhance the intumescent coating's thermal performance. Note the average temperature of the back substrate intumescent coating reported by Ahmad (2015-2017) group was less than 400°C at the end of 1-hour exposure to fire[12]–[14].

The raw samples sent for X-Ray Fluorescence (XRF) to determine the elemental composition of the waste glass powder. The data obtained listed in table 2 and Table 3. As expected, the waste glass powder contains high silicon with 33% of the total elements, with the possibility of the emergence of SiO_2 at 70.60%. Silicon was known for having high thermal stability, weathering resistance, hydrophobicity, low surface tension and high oxidation resistance when added in fire resistance material [15]. In addition, the powder also contains 13% of calcium element with the possibility of the minor components of sodium, aluminium, magnesium, phosphorous, tungsten, iron, potassium, titanium and sulphur.

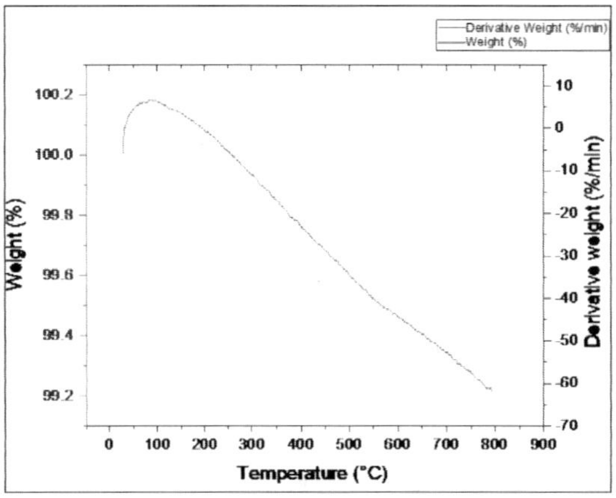

Figure 1: TGA thermogram of raw waste glass powder

Table 2: Waste glass powder composition

Waste Glass powder	Element Composition (%)										
	Si	Ca	Na	Al	Mg	P	W	Fe	K	Ti	S
	33	13.5	3.75	0.778	0.563	0.453	0.438	0.273	0.129	0.119	0.116

Table 3: Waste glass powder oxides composition

Waste Glass powder	Oxides Composition (%)										
	SiO_2	CaO	Na_2O	Al_2O_3	P_2O_5	MgO	WO_3	Fe_2O_3	SO_3	TiO_2	K_2O
	70.60	18.90	5.06	1.47	1.04	0.93	0.55	0.39	0.29	0.20	0.16

4.2. Direct and Control Fire Test

A control fire test performed to evaluate the char expansion with IF values as an indicator. The sample without filler (WG000) showed a brittle like char structure with shrinkage at the substrate's corner. This is not the case for the samples added with waste glass powder. The char expanded fully, and no shrinkage detected between the samples. The leftover char shown in figure 2 still intact with the substrate even after the cross-section cut indicates an excellent adhesive bond successfully created between the substrate and coating ingredient. Chowaniec (2019) studied on adhesive properties of waste glass powder added reported to cause an increase of pull-off strength of the coating. The microstructural analysis of the samples showing waste glass powder penetrates the coating to the substrate and fill the pore, thus strengthening the adhesive[10]. The Intumescent factor used to determine the ratio of the char with the coating. According to Jimenez (2006), char expansion is an essential parameter for evaluating coating performance but not necessary. If a brittle char forms when exposed to fire, it tends to break when subjected to external force (e.g., jet fire), thus decreasing efficiency in the turbulent regime of combustion [16]. This type of protective coating not giving viable security to the steel structure. No significant improvement in the char expansion when waste glass powder added to the formulation.

The average value obtained from samples WG001 to WG005 is 7.55, with WG004 having the slightest highest value of 7.93 compared to WG000 (without WGP) with an IF value of 10.44 in figure 3. Even though calcium reported to increase the char's expansion [17], this is not the case for this research as this might due to less calcium composition in waste glass powder (18.90%) shown in previous table 2. The coating also prepared for direct fire test to evaluate their performance against real fire. A 10 x 10 x 1.5 cm coated substrate used for this test. After an hour, the result tabulated into the graph in figure 4. Sample WG005 recorded the highest temperature obtain after an hour duration with 370.3°C, follow by WG000 with 333.7°C. The other samples containing waste glass powder filler showing temperature less than 300°C with WG002 recorded as the lowest (154.6°C) follow by WG001 (230.4°C), WG003 (257.1°C) and WG004 (259.7°C).

Figure 2: Cross-section Expanded char with and without waste glass powder

The samples excellent performance with waste glass filler might come from the high content of silica with 70.06%. Silica was known for its high thermal stability and the credit gain from the Si-O-Si bond structure[15]. Despite no char expansion improvement when WGP added into the formulation during the control fire test, the direct fire test showing an opposition with an excellent lower temperature sustain for an hour exposure to fire. This showed that the char exhibit a denser structure that delays the heat's penetration to the protected substrate, thus improving the coating's thermal performance.

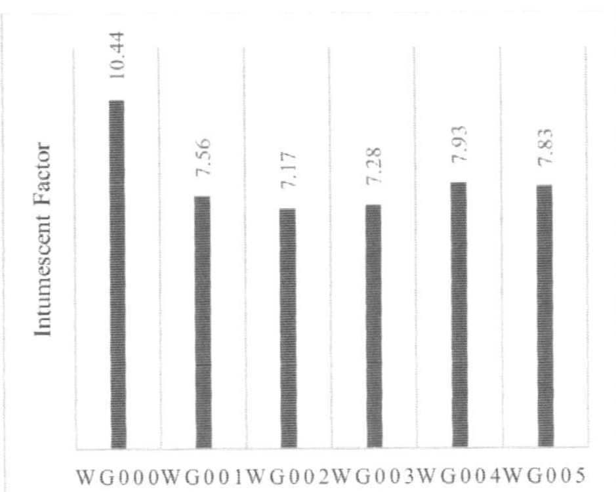

Figure 3: IF values of char expansion with and without WGP.

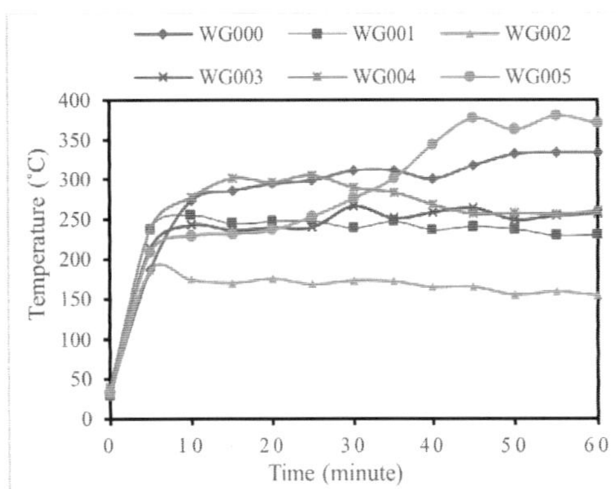

Figure 4: Temperature VS Time graph of intumescent coating with and without WGP.

4.3. Water Absorption of Coating (ASTM D870-15)

The cured intumescent coating fully immersed in seawater for 15 days, and weight loss recorded after drying with a dry cloth every three days. The samples were then left to dry at room temperature 24 hours before weighted again. The bar chart in figure 5 shows a weight-loss trend at the end of 15 days with an additional 24 hours of room temperature dried. The samples with 3% of waste glass powder had the highest final weight loss between samples with filler with 0.128%, while WG004 had the lowest value

IOP Conf. Series: Materials Science and Engineering 1176 (2021) 012029 doi:10.1088/1757-899X/1176/1/012029

with 0.078%, followed by WG002 with 0.089%, and WG001 with 0.103%. WG005 exhibits an increase of weight by 0.035%. Meanwhile the intumescent coating without additional filler exhibits a 0.19% weight loss after dried at room temperature for 24 hours. In the first nine days of immersion, all the coating exhibits permeation and migration simultaneously. In permeation, molecules and ion from seawater penetrate the coating structure leading to an increase of the weight in the coating while hydrophilic fire retardant additives migrate from the coating and are solved in water during the migration process, resulting in a weight loss of the coating [17], [18]. The sample reached equilibrium after nine days of immersions except for WG005, with a significant drop in weight observed. This might occur due to the overloading of waste glass powder added into the formulation, resulting in a higher migration rate, allowing the mineral of salt to accumulate into the pore structure and increasing the samples' final total weight. Overall, the average of less than 0.1% total weight loss and gain for all the samples with waste glass filler proves that the material's water resistance properties improved by adding waste glass filler. However, all the samples demonstrate a suitable water resistance property as it is lower than the maximum requirement weight loss of 3%.

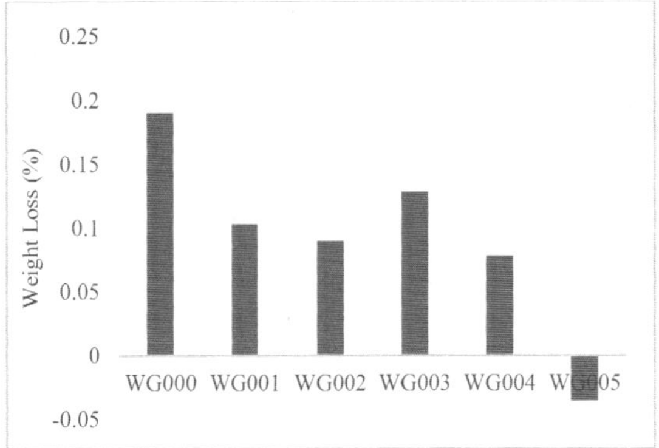

Figure 5: Weight loss after immersed in seawater for 15 days and 24 hours dried

4.4. IR analysis

IR analysis performs to determine the functional groups of the intumescent coating shown in figure 6. A nearly similar trend of spectra peak observes from the figure. A broader and prominent peak was detected between 3100 to 3600cm^{-1} for the sample with waste glass powder compare to WG000. This presence of broadband may due to O-H stretching from strong intramolecular and intermolecular hydrogen bonds[19]. O-H group reported encouraging the hydrogen bonding between the substrate and the coating itself[20].

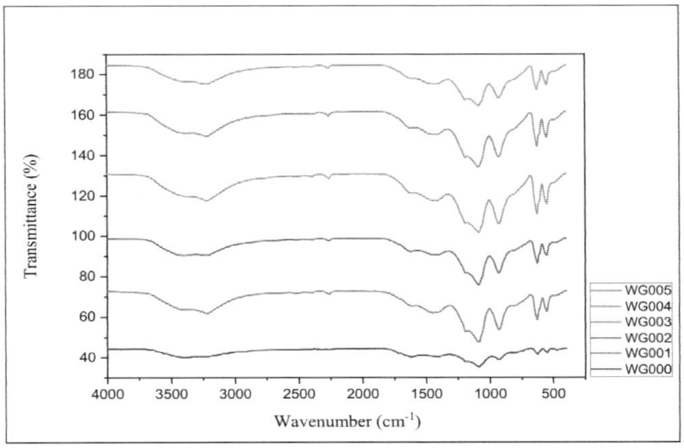

Figure 6: IR spectra of intumescent coating with and without waste glass powder filler

4.5. Morphology of Char

Sample WG002 chosen as the best samples having the lowest back substrate temperature recorded against direct fire with 154.6°C, having the second-lowest for total weight loss on water absorption test and an average of IF 7 value on char expansion. The sample then sent together with the sample without waste glass powder (WG000) for FESEM analysis to further analyse their morphology structure. The interior char morphology of the samples with and without waste glass powder present in figure 7. The sample without the filler showing an empty and open space on the expandable graphite worm-like structure. This situation opposed WG002 with 2% of waste glass powder with an agglomeration of small round shape occupying expandable graphite structure gaps. This shape likely identifies as silica that reported by other researchers[21], [22]. The small flakes like shapes observed in figure 7b, possibly from calcium elements[23].

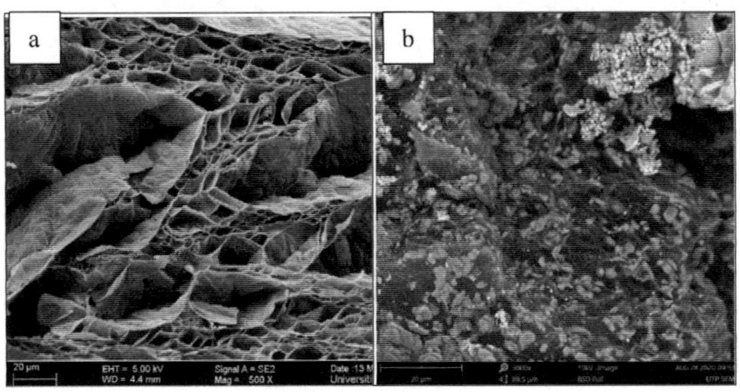

Figure 7: Morphology image of intumescent char samples without filler, WG000 (a) and sample with filler, WG002 (b)

5. Conclusions

Waste glass powder contains as high as 70% silica and has high thermal stability, with a percentage of more than 90% when decompose to 800°C. When added into the intumescent formulation, the coating's thermal stability increases where the lowest back substrate temperature recorded was 154.6°C compare to coating without the waste glass, 333.7°C. The morphology analysis also shows the emergence of silica and calcium elements that occupy and close the char structure gap, thus slowing the heat transfer to the protected substrate. The coating showed a less than 0.1% total weight loss for samples with less than 4% wt filler after immersed in seawater for 15 days compared to the control with 0.19% wt. The char subjected to a cross-section cut, and all the samples still left with char intact to the substrate. This leftover char intact with the substrate indicates an excellent adhesive bond formed between the coating and substrate. The coating's good adhesive bond is shown by the O-H functional group's high intensity promote the hydrogen bonding between those two. Sample WG002 with 2% of waste glass powder was chosen as the best samples with the lowest back substrate temperature recorded against direct fire with 154.6°C, and having the second-lowest for total weight loss on the water absorption test and an average of IF 7 value on char expansion. Hence, this study's findings indicate that waste glass powder's addition to intumescent coating formulation effectively enhances the intumescent coatings' thermal and mechanical properties. It is recommended analysing further the current work, including the char's compactness measurement using a rheometer as the char's measurement has a significant effect on the flow heat to the substrate.

Acknowledgement
The authors would like to acknowledge the Laboratory facilities provided by Advanced Materials Processing, Corrosion Research Centre, Institute of Contaminant Management, Universiti Teknologi PETRONAS, Malaysia, and Ministry of Higher financial support Education (MOHE), Under FRGS grant No. 015MAO-081.

References

[1] Yin J J K, Yew M C, Yew M K, and Saw L H, 2019 *Coatings* **9**.
[2] Mariappan T 2016 *J. Fire Sci.* **34** p 120–163.
[3] Olofinnade O M, Ede A N, and Ndambuki J M 2017 *Int. J. Eng. Technol. Innov.* **7** p 280–291.
[4] Kara P, Csetényi L J, and Borosnyói A 2016 *IOP Conf. Ser. Mater. Sci. Eng.* **123**.
[5] Aliabdo A A, Abd Elmoaty A E M, and Aboshama A Y 2016 *Constr. Build. Mater.* **124** p 866–877.
[6] Islam G M S, Rahman M H and Kazi N 2017 *Int. J. Sustain. Built Environ.* **6** p 37–44.
[7] Teh P L and Jaafar M 2008 Polymers for Advanced Technologies **19** p 308-315.
[8] Chowaniec A and Ostrowski K 2018 *Czas. Tech.* **8** p 99–109.
[9] Haro M de. P, Villafuerte M E, Alcantar B, and Diaz G C 2009 ECERS I*nternational Conference and Exhibition of the European Ceramic Society* **11** p 1–4.
[10] Chowaniec A, Sadowski Ł, and Żak A 2020 *Appl. Surf. Sci.* **504**.
[11] ASTM D870-15 2020 *Standard Practice for Testing Water Resistance of Coatings Using Water Immersion* West Conshohocken, PA: ASTM International.
[12] Arogundade A I, Ahmad F, Bhat A H, and Gillani Q F 2016 *Procedia Eng* **148** p 223–227.
[13] Amir N, Othman W M S W, Ahmad F, and Iskandar B S 2015 *AIP Conference Proceedings* **1669** 020062.
[14] Zia-ul-Mustafa M, Ahmad F, Ullah S, Amir N, and Gillani Q F 2017 *Prog. Org. Coatings* **102** p 201–216.
[15] Yasir M, Ahmad F, Yusoff P S M M, Ullah S, and Jimenez M 2020 *Surf. Eng* **36** p 334–363.
[16] Jimenez M, Duquesne S, and Bourbigot S 2006 *Surf. Coatings Technol.* **201** p 979–987.
[17] Yew M C, Ramli Sulong N H, Yew M K, Amalina M A, and Johan M R 2015 **78** p 59–66.
[18] Amir N et al.,2017 *Key Engineering Materials* **740** p 41-47.
[19] Abd El-Wahab H 2015 *Pigment Resin Technol.* 44 p 101–108.
[20] Md Nasir K, Ramli Sulong N H, Johan M R, and Afifi A M 2018 *Pigment Resin Technol.* **47** p 142–153.
[21] Ahmed A, Saaid I M, Tunio A H, Pilus R M, Mumtaz M, and Ahmad I 2017 *J. Appl. Environ. Biol. Sci* **7** p 56–62.
[22] Azlina H N, Hasnidawani J N, Norita H, and Surip S N 2016 *Acta Phys. Pol. A* **129** p 842–844.
[23] Vu H H T et al 2019 *Sustain.* **11**.

IOP Conf. Series: Materials Science and Engineering 1176 (2021) 012030 doi:10.1088/1757-899X/1176/1/012030

A smart building security system with intelligent face detection and recognition

M H Khairuddin[1], **S Shahbudin**[1*] **and M Kassim**[1]

[1]Faculty of Electrical Engineering, Universiti Teknologi MARA (UiTM),

40450, Shah Alam, Selangor, Malaysia

*Corresponding author: shahranis@uitm.edu.my

Abstract. The security system is the most important application that protects several threats from inside and outside of an organization. The purposes of several security building applications such as the system will know criminals, disgruntled former employees enter the workplace, and enabling building security to work proactively for ensuring that associates and customers are safe from harm. By implementing face detection and recognition of the security system, video surveillance in the building can be improved by upgrading the system of video surveillance with artificial intelligence. The aim of the research is to develop the face detection and recognition in the security system in the building with recognizing human faces and image capture to the cloud storage. The result shows that the video will stream and recognize human faces whether the person is authorized or unauthorized in the system image using Raspberry Pi 3. For analysis purpose, the image captured and stored from the two type of cloud storage namely Cloudinary and Dropbox had been analyzed. Based on the analysis, the image captured and stored into Dropbox cloud storage give the best performance features which gained more 75% of quality images.

1. Introduction

Nowadays, people spend 80 percent of their lives inside buildings [1] such as office buildings, schools, hospitals and homes. The Internet of Things (IoT) offers innovative, smart building solutions to meet these goals for safer and cheaper than the old system uses such closed-circuit television (CCTV) and less power use that can prevent overloaded current power. The creation of this system has grown in developed countries such as the United Kingdom where most every building has an image processing system where the intruder's face will be captured through cameras and the system will go through to the police system.

To enhance the security level of the Internet of Things (IoT) devices, the smart building with intelligence system has been designed for specific purposes such as which system when someone comes into the building and the camera will capture the object's image and then send the object's image in the cloud. The research has been done in home security system [2], face recognition IoT system [3] and an implementation of Raspberry Pi application [4],[5] where the project has been using the Raspberry Pi board in the application and face recognition system based on OpenCV [6]. In [4], live video streaming system using raspberry pie with cloud server is proposed which including video capture module, video compression module and video streaming module using Linux operating system.

Based on issue of fire breaks out at police HQ department of Bukit Aman in 2015 [7], the building has no security system for storing data information permanently because the building used closed circuit television (CCTV) using Network Video Recorder in server room which the data storage may be changed or deleted by someone who is not responsible for protecting criminal data information. In that case, there are no critical criminal investigation documents in the building instead of the criminal investigation department in the other level of the building. By creating the security of face detection and recognition with cloud system, the smart building system can be implemented in front of the door of department for capturing the image of person whether authorized or unauthorized according to references [8] - [24] so the sensor will detect every face for transmitting image processing to the cloud system which the investigator can check the person who entered the building on the cloud.

Therefore, the aim of the research is to develop the smart building security system with intelligent face detection and recognition with the aid of the cloud system in real time using Haar-Cascade method and histogram of oriented gradients. The camera sensor and the raspberry pi board established the Internet of Things (IoT) application system using Dropbox cloud storage implementation. The project used the raspberry pi 3 in model B+ concept to operate the whole system. Furthermore, a connection is needed between Logitech USB webcam with 5 Megapixel and Raspberry pi 3 model B+ using html web page, python programming language and OpenCV based on Linux operating system. For analysis purpose, the image captured and stored from the two type of cloud storage namely Cloudinary and Dropbox will be analyzed.

2. Research Methodology

Figure 1 shows the block diagram of the project. The system is divided into two parts; hardware and software design. In the hardware part, the camera sensor will detect and recognize the presence of human whether the person is authorized or unauthorized in the system using the raspberry pi 3. While in the software design part, the project will be interfaced with html and cloud storage platform on the system using Dropbox.

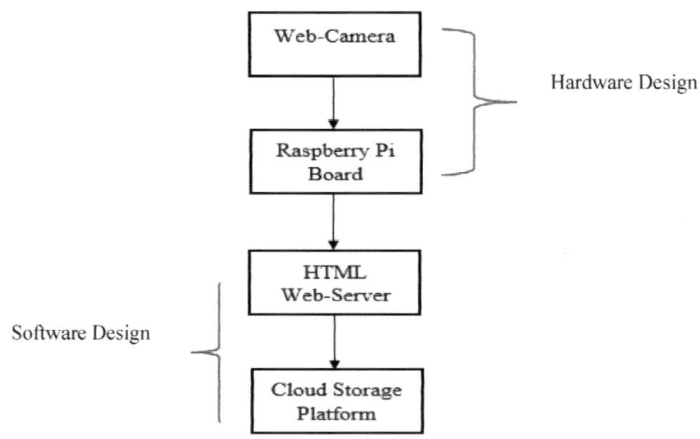

Figure 1. Block Diagram of the System.

The person will be allowed to enter the building with authorization by the system. However, the person who enter the building without authorization by the system can be considered as an intruder. The image captured will store directly to the Dropbox Cloud Storage when the system detects the presence of human.

2.1. Hardware Design

The system has a camera sensor which Logitech model webcam for face detection and recognition connected to the USB port of Raspberry pi 3 model B + interface with html and Dropbox as a data platform. The Raspberry pi connected to the router by using a network cable to configure the network on raspberry pi before raspberry pi can be used as wireless. In addition, Raspberry pi has been using the Raspbian stretch as an operating system that is stored in a 32 GB microSD card. As we can see in Figure 2, it shows the connection between the Logitech webcam, screen monitor and Raspberry pi 3 model B+ have been set up in hardware implementation.

IOP Conf. Series: Materials Science and Engineering 1176 (2021) 012030 doi:10.1088/1757-899X/1176/1/012030

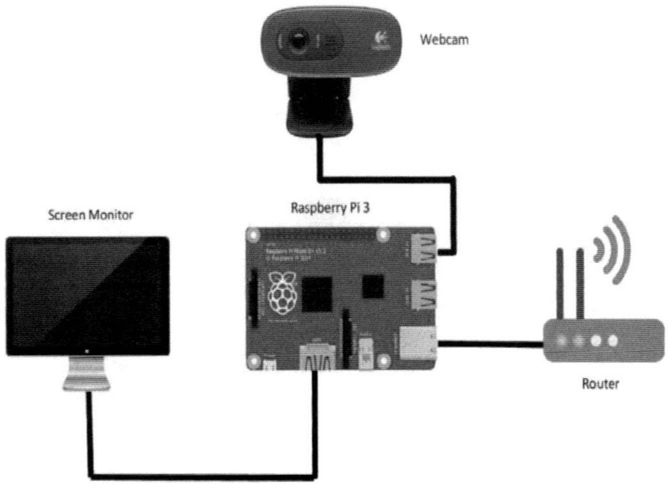

Figure 2. Schematic Diagram of Project.

2.2. Software Design

OpenCV stands for Open Source Computer Vision Library, which has C ++, C, and Python interfaces, running on Windows, Linux, Android and Mac. In this project, the system uses OpenCV in version 3 which up to date compared to the lower version of OpenCV.

Haar Cascade is a face detection algorithm used to identify faces, pedestrians, objects and facial expressions in an image. In Haar-Cascade, the algorithm is provided with several numbers of positive images which are faces of different persons at different backgrounds and negative images which are images that do not face but can be anything else like tree and grass. The feature selection is done along with the classifier training using Adaboost and Integral images. The Adaboost classifier is retrained the algorithm iteratively by choosing the training set based on the accuracy of previous training. There are several stages as related to figure 3 and figure 4 for this project where this project has been using Haar cascade approach method using OpenCV version 3 and python programming language in version 3.5 to process the computer vision and image processing in this project.

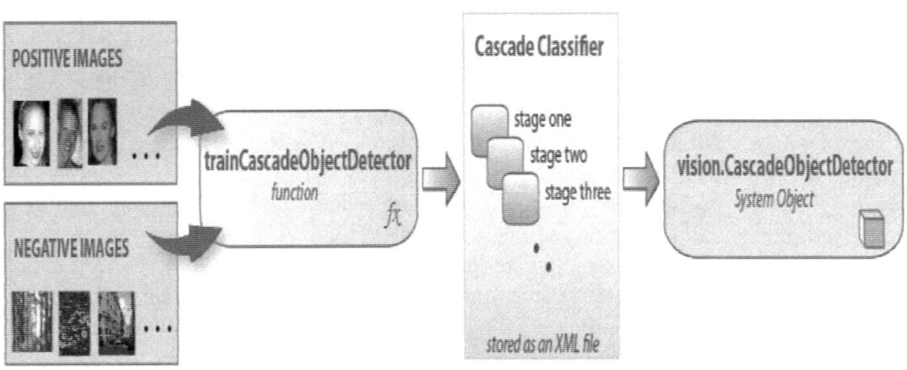

Figure 3. How Haar Cascade Works [25].

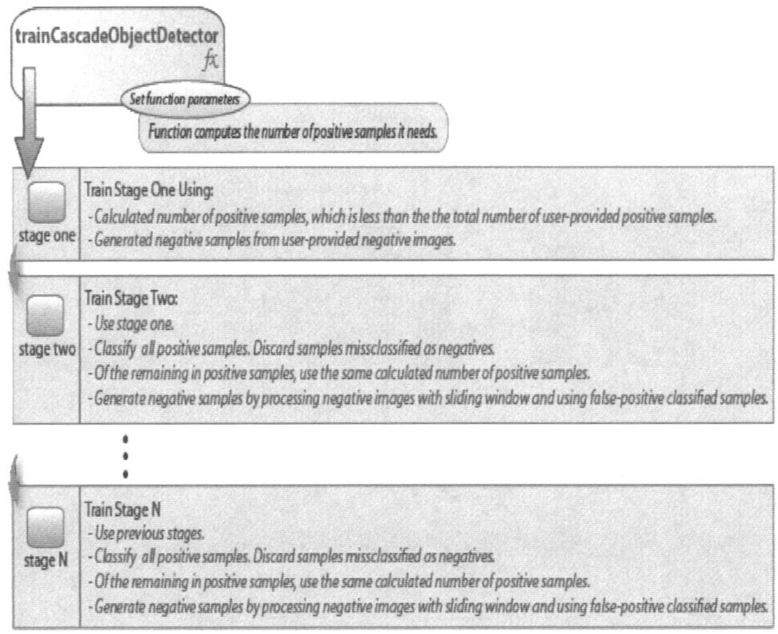

Figure 4. Haar Cascade Training Function [25].

The algorithm used in the face recognition on video streaming is Haar Cascades algorithm. This algorithm is used for pre-trained images on OpenCV platform. Haar Cascade approach method comes with a trainer as well as a detector and already contains man pre-trained classifiers such as a face, eyes and smile. Haar Cascade files are stored in OpenCV folder and basically a classifier which used to detect particular objects from the source. Haar cascade is designed by OpenCV to detect the frontal face and capable of detecting features from the source. Haar Cascade is a machine learning algorithm used to identify objects in image or video. Before pre-trained an image, the system needs to build a dataset first so that the system can recognize faces in the dataset and encode them into 128-d vectors on each image using Histogram of Oriented Gradients (HOG) as a detection method.

Figure 5 shows the flow of proposed system. Based on figure 5, the face recognition system have been implemented using raspberry pi 3 with the Logitech USB camera can stream the video on the website while the system also can store an image of the presence of human directly to Dropbox cloud storage. The system needs WiFi connection by applying a private network in order to run the system. After the system successfully connected to WiFi, the user needs to insert a username and password correctly due to the project has been encrypted for system security.

Next, the user can choose either face recognition on video frame or face detection direct to the Dropbox cloud storage. If the user chooses face recognition, the video stream of face recognition will display on video frame which can recognize the face's staff of the building with the staff names displayed on a video frame. Besides, the system also cannot detect suspicious person by displaying "Unknown" on video frame so that the user can identify a suspicious person in the building. However, if the user chooses face detection with Dropbox, the system will automatically capture the image and send to the Dropbox storage when the system detected the presence of a human. The analysis of cloud storage between Cloudinary and Dropbox has been done in order to obtain the best result resolution of image captured.

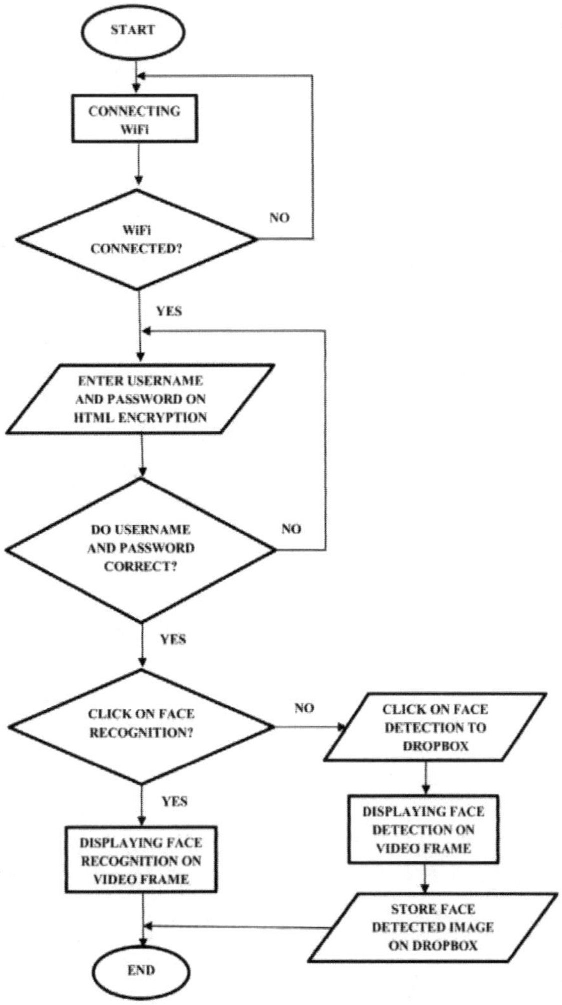

Figure 5. The flow chart of proposed system.

3. Results and Discussion

Figure 6 shows the comparison image from Cloudinary and Dropbox cloud storage. The percentage of an image features from Dropbox and Cloudinary are tabulated in the Table 1. With the same resolution of an image; 640 x 480, the quality image from Dropbox is higher at 81 % of lighting compare to image from Cloudinary that only recorded 79% of lighting. In term of light fixture, image from Cloudinary only display at 74% compared to image from Dropbox that recorded 76% display on light fixture of an image.

IOP Conf. Series: Materials Science and Engineering 1176 (2021) 012030 doi:10.1088/1757-899X/1176/1/012030

(a) (b)

Figure 6. Comparison image captured from Cloud Storage (a) Cloudinary, (b) Dropbox

Table 1. Image Quality Analysis between Cloudinary and Dropbox.

	Cloudinary	Dropbox
Lighting (%)	79	81
Light Fixture (%)	74	76

According to table 1, the analysis showed that the image stored from Dropbox give the best performance features. Therefore, Dropbox cloud storage is applied to the system running on the web server. The main page of the system which encrypted by inserting a username and password for system are shown in Figure 7.

Figure 7. HTML Login Page.

Figure 8. HTML button options on web page.

Figure 8 displays the web page of application system which encrypted by the username and password in order for user allowed to enter the system by clicking the button on web page either face recognition or face detection with Dropbox for displaying a video frame of the application.

The video frame after clicking on the button of face recognition are shown in figure 9. The video frame can recognize human names which registered on dataset using Haar Cascade method and the other person that not authorized, automatically the system will display 'Unknown'. To exit the video frame, the system needs to press 'Q' on keyboard.

Figure 9: Video Frame of Face Recognition.

IOP Conf. Series: Materials Science and Engineering 1176 (2021) 012030 doi:10.1088/1757-899X/1176/1/012030

Figure 10. Video Frame of Face Detection with Dropbox.

Raspberry pi can detect sensitively with motion or the presence of human which the system will capture the presence of human only when the raspberry pi detected as can be seen in figure 10 and the image captured will directly store on Dropbox cloud storage. This project can reduce the usage of storage because the system will only store the image when it has a movement in front of the camera.

Table 2. Accuracy Percentages and Elapsed Time between Cloudinary and Dropbox Image.

	Cloudinary	**Dropbox**
Accuracy (%)	85.3	93.5
Elapsed Time (s)	10.23	5.54

Furthermore, Table 2 shows the comparison of accuracy percentage and elapsed time between Cloudinary and Dropbox image in face detection and recognition. The result shows Dropbox image obtained the higher accuracy with 93.5% compared to Cloudinary image with only recorded 85.3% of accuracy. In term of time, the Dropbox image also recorded the faster time with 5.54 seconds compared to Cloudinary image which only obtained 10.23 seconds. Therefore, it is proved that the Dropbox image gives the best accuracy and fastest time in face detection and recognition.

4. Conclusion

The video streaming based on the web server using Linux Operating System is successfully developed. The system has been implemented with low-cost, portability and easy to maintain and upgrade. The server periodically obtains video recognition from the camera through the private network such as the video streaming transmitted from camera to the server. Therefore, the real time video streaming with intelligent face detection and recognition based on the web server give the better performance and more secure than the video surveillance due to the limitation of video surveillance which is not provide human recognition on the system. For future works, the system will upgrade face recognition system algorithm using machine learning or deep learning method and the WIFI connection will change to public network access.

Acknowledgments

The author would like to thank SRC Grant, (Project File No : 600-RMC/SRC/5/3 (032/2020)) and Research Management Institute, Universiti Teknologi MARA (UiTM), Shah Alam for the support grant in publishing this paper.

References

[1] Gemalto.com. 2018 What is a Smart Building? [online], https://www.gemalto.com/m2m/markets/smart-buildings

[2] Madhesh A, Dr. Kowsigan M, Khishore R and Dr. Balasubramanie P, 2020 IoT based Home Automation and Security System *International Journal of Advanced Science and Technology*, **29** 9 2733 – 2739

[3] Aydin I and Othman N A 2017 A New IoT Combined Face Detection of People by Using Computer Vision for Security Application, *Int. Artif. Intell. Data Process. Sympo (Malatya, Turkey)*

[4] Deepa M, Dharani M, Kalaivani M S and Ms. Parkavi M 2016 Live Video Streaming System Using Raspberry Pi with Cloud Server **2** 11 1075-1077

[5] Manoj R, Biradar, Rekha R, Raju and Sharad A 2020 Smart Home Security System using Iot, Face Recognition and Raspberry Pi. *International Journal of Computer Applications.* **176** 45-47

[6] Zhiguo Z and Yao C 2020 Application of Attitude Tracking Algorithm for Face Recognition Based on OpenCV In The Intelligent Door Lock, *Computer Communications* **154** 390-397

[7] The Star Online 2015 Fire Breaks Out in Bukit Aman. https://www.thestar.com.my/news/nation/2015/07/29/fire-at-bukit-aman/ [Accessed 27 May 2018]

[8] Yang S, Song Y, Ren H, and Huang X 2016 An automated student attendance tracking system based on voiceprint and location *ICCSE 2016 - 11th Int. Conf. Comput. Sci. Educ (Nagoya, Japan)* 214-219

[9] Soniya V, Sri R S ,Titty K S, Ramakrishnan R, and Sivakumar S 2017 Attendance Automation Using Face Recognition Biometric Authentication *2017 Int. Conf. Power Embed. Drive Control (Chennai, India)* 122-127

[10] Islam M M, Hasan M K, Billah M M, and Uddin M M 2017 Development Of Smartphone-Based Student Attendance System," 2017 *IEEE Reg. 10 Humanit. Technol. Conf. (Bangladesh, Dhaka)* 230-233

[11] Soewito B, Gaol F L, Simanjuntak E, and Gunawan F E 2016 Smart Mobile Attendance System Using Voice Recognition and Fingerprint on Smartphone, *Proceeding 2016 Int. Semin. Intell. Technol. Its Appl. ISITIA 2016 Recent Trends Intell. Comput. Technol. Sustain. Energy* 175-180

[12] Lukas S, Mitra A R, Desanti R I, and Krisnadi D 2016 Sistem Kehadiran Pelajar di Bilik Darjah menggunakan teknik pengenalan wajah *2016 Int. Conf. Inf. Komun. Technol. Converg.* 1032-1035

[13] Naidu V, Mudliar K, and Devadkar K 2017 Sistem pengurusan kehadiran WLAN *2017 IEEE Int. Conf. Smart Technol. ICSTM 2017* 3-6

[14] Raghuwanshi A and Swami P 2017 An Automated Classroom Attendance System Using Video Based Face Recognition *2017 2nd IEEE Int. Conf. on Recent Trends in Electronics, Information & Communication Technology (India)* 719-724

[15] Puthea K, Hartanto R, and Hidayat R 2017 A Review Paper on Attendance Marking System based on Face Recognition *5th Int. Conf. on Image Processing, Communication, Engineering and Technology (IPCET-17)* 303-308

[16] Starovoitov V and Samal D 2014 A Geometric Approach to Face Recognition *Proc. of the IEEE-EURASIP Workshop on Nonlinear Signal and Image Processing (NSIP'99)*, Antalya, Turkey 1-4

[17] Ana D and Wati R, 2017 Design of Face Detection and Recognition System for Smart Home Security Application, *2017 2nd Int. Conf. on Information Technology, Information Systems and Electrical Engineering ICITISEE (Yogyakarta, Indonesia)* 341-346

[18] Chilipirea C, Ursache A, Popa D O, and Pop F 2016 Energy Efficiency and Robustness for IOT: Building A Smart Home Security System *2016* IEEE *12th Int. Conf. Intell. Comput. Commun. Process. ICCP 2016 (Cluj-Napoca, Romania)* 43-48

[19] Nayyar C, Valarmathi B, and Santhi K 2017 Home Security and Energy Efficient Home

Automation System Using Arduino *2017 Int. Conf. Commun. Signal Process* Chennai, India 1217-1221
[20] Ali S, Quadri I, and Sathish P 2017 IoT Based Home Automation and Surveillance System *2017 Int. Conf. on Intelligent Computing and Control Systems (ICICCS)* Madhurai, India 861-866
[21] Dhanisha G, Seles M, and Brumancia E, 2015 Android Interface Based Gcm Home Security System Using Object Motion Detection *2015 Int. Conf. on Communications and Signal Processing (ICCSP)* (*Melmaruvathur, India)* 1928-1931.
[22] Daniels M, Muldawer K, Schlessman J, Ozer B and Wolf W 2007 Real-time human motion detection with distributed smart cameras *2007 First ACM/IEEE Int. Conf. on Distributed Smart Cameras* 187-194
[23] Mahler M A, Li Q, and Li A, 2017 SecureHouse: A Home Security System Based on Smartphone *2017 IEEE Int. Conf. Pervasive Comput. Commun. Security* (*Kona, USA)* 11-20
[24] Will Berger 2018 Deep Learning Haar Cascade Explained [online], http://www.willberger.org/cascade-haar- explained/

Design and development of student attendance system using QR-Code for UiTM Cawangan Terengganu

S S Hamzah[1*], U F Bahrin[1], S N S Yasin[1] Z D Eri[1] and H A M A Majid[1]
[1]Faculty of Computer and Mathematical Sciences, Universiti Teknologi MARA (UiTM), 21080 Kuala Terengganu, Terengganu, Malaysia

*Corresponding author: salbiah452@uitm.edu.my

Abstract. This paper describes the development of student attendance using QR-code forUiTM Kuala Terengganu. The existing or manual attendance in Universiti Teknologi Mara Kuala Terengganu Cawangan Terengganu Kampus Kuala Terengganu (UiTMCTKKT) requires lecturer to records student attendance in attendance sheet, calculate the percentage of student attendance manually and give warning letter to the student by hand. There are several problems may occur if lecturer still using manual method to take student attendance such as lost and damaged the attendance sheet, heavy usage of papers, time limitation, and wrong calculation while calculating the attendance percentage. This project work aims to assist the lecturer in taking student attendance, to automatically calculate attendance percentage, generate warning letters to the student and lastly preventing fraudulent activities by the student. This system is developed by using the System Development Life Cycle (SDLC) waterfall model.

1. Introduction

Attendance is the concept of people, individually or as a group, appearing at a location for a previously scheduled event. Measuring attendance could be an important concern for several organizations, which might use such information to determine the effectiveness of their efforts and to arrange for future efforts. Student attendance is used to track and monitor students that attend class. It enables the lecturer to monitor their student attendance and late arrivals.

By using the manual method of taking attendance in class, it will lead to several problems such as time constraint, perjury in signing the attendance sheet and lost the attendance sheet. Therefore, there is a need to propose an attendance system for UiTMCTKKT to manage all the attendance processes that currently be done manually by the lecturer and HEA staff. This proposed system will eliminate the necessity for manual attendance process and cumbersome process. Students can simply log on to the attendance system using the online system and the lecturer can manage all the attendance data in the system. Current process flow are shown in Figure 1 below.

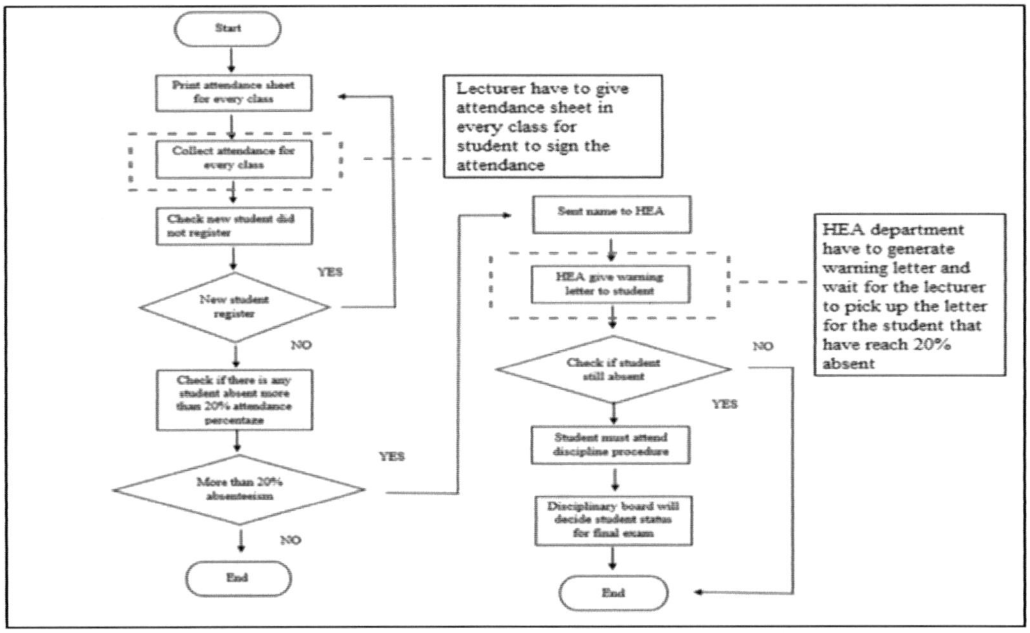

Figure 1. Flowchart of current process flow of Student's Attendance.

1.1. Problem Statement

Based on the process that currently used in UiTMCTKKT, it is difficult for both lecturer and Hal Ehwal Academic department to access the record for student attendance. This is due to attendance in UiTMCTKKT is taken using paper and its take time to find the attendance sheet. Hence, the attendance sheet might lose and may be damage due to human mistake as lecturer have to keep the attendance sheet for 14 weeks for one semester. As a consequence of that, lecturer can no longer trace the student overall attendance record throughout the particular semester.

Currently, the students of UiTMCTKKT sign an attendance paper which was provided by the lecturer's every time they come to a class. The problem to this situation is that students can be dishonest where they can perjury on their sign attendance by asking their friends to sign for them when they did not attend a class. For the lecturer who is handling a big number of students in a lecture hall or tutorial class, they might not notice the students who are not attending the class, but their name was signed as a present student.

Lastly, it is difficult for a lecturer to monitor the student's number of absences. In UiTM, it is a policy for students to attend at least 80% of the class attendance throughout the semester to be able to take the final exam. Therefore, it is indispensable for each lecturer to monitor his/her student's attendances to detect the absenteeism as early as possible. Since the attendance process is done manually, lecturer have to count one by one the attendance percentage for every classes and the tendency to make wrong calculation is very high.

2. Related Works

2.1. Student Attendance System

Attendance system is used to manage people attendance in an organization, it has been used in many institutions and scholarly organizations to control people participation in an organization. Attendance system involves in keeping records, conducting transactions, reporting generations, and consolidating the significant data to be supplied at the multiple management levels [1]. The design of MIS mainly provides management-level control and decision-making characteristics. The implementation of MIS in attendance system helps in making management decisions related to management function mainly especially for attendance system [2].

MIS is a computer-based information system that provides for management-oriented reporting based on transaction processing and business operations of the organization. Management information systems basically concerned with the process of collecting, storing, processing, and transmitting relevant fact to help the management operations. MIS consist of three parts which are management, information and system [3].

2.2. *QR Code*

QR Code is a sign of two dimensions. Denso Wave, one of the predominant Toyota group businesses, created it in 1994 and in June 2000 it was approved as a global ISO standard (ISO / IEC18004). This two-dimensional symbol was originally designed for use in manufacturing control of automotive parts but has become common in other areas such as product tracking, item identification, time tracking, document management, overall marketing and attendance records [8]. A QR code comprises of black squares arranged on a white background in a square grid that can be read by a camera-like imaging device and processed the use of Reed–Solomon error correction until the picture can be properly interpreted. The necessary information is then obtained from patterns current in the image's horizontal as well as vertical aspect.

2.3. *Similar Existing System*

Several similar existing systems implementing the latest technology in taking attendance. There are several latest technologies that can be implemented in taking attendance such as biometric, fingerprint, QR-code, RFID and barcode scanner. All the information about the existing system will be gathering to find the suitable function and benefit for the attendance system, from the research it can help to implement in this system. The example of the existing system is EGBiometric College Poly-Tech Mara (KPTM), E-attendance University Malaya, Student attendance management system and portable Fingerprint-Based Attendance Recording & Monitoring.

Table 1. Comparison of Features between Existing Similar Systems.

Features	EGBiometric (KPTM)	E-attendance University Malaya	SAMS	Fingerprint-Based
Login	✓	✓	✓	✓
User register	X	X	✓	
View attendance	✓	✓	✓	✓
Personal information	✓	✓	✓	✓
Print attendance	X	X	✓	✓
Total attendance percentage	X	X	✓	✓
Print letter	X	X	✓	✓
Logout	✓	✓	✓	✓

3. Methodology

This project goes through six phases which are planning, analysis, design, development and iteration, testing and evaluation and documentation. Table 2 outlines the details methodology followed to accomplish objectives of the project.

Planning is a phase where an interview session was conducted with several lecturers and Hal Ehwal Academic staff in UiTM Kuala Terengganu to gather all information related to the project development. Open-ended is used in the interview session as an open-ended question will encourage the interviewee to respond spontaneously and delivers unstructured responses. Current business flow, as well as problem statements, are the outcome from the interview session.

As for analysis, Interviews with potential users lead to the establishment of refined the problem statement, system and user requirements. It helps to describe the expectation of user on the system. For design, with the guide of knowledge acquired in analysis phase diagrams such as Context Diagram, Data Flow Diagram (DFD), Entity Relationship Diagram (ERD), Site Map and User Interface are produced. Designing those diagrams provide a clear view on how the system operates from beginning to the end.

Development phase objective is to develop a system that has all functionality which follows all the requirements that have been gathered during analysis. Next is testing and evaluation, the system has been tested by 3 expert user and 30 respondents to ensure that the system fulfils all the requirements and is achievable. Documentation is a process where all the project outcome has been documented as a report of the system comprising the whole project activities is deliverable and achievable.

Table 2. Project Development Methodology.

Phases	Activities	Deliverable
Phase 1: Planning	1. Interview with several lecturers, and staff in the HEA department	1. Current process practice, Problem statements and objective has been identified.
Phase 2: Analysis	1. Analysis of the systems development, process and activities	1. Problem statement of the current process, user requirement and System requirement.
Phase 3: Design	1. Design system interface 2. Design Entity Relationship Diagram (ERD) 3. Data Flow Diagram (DFD) 4. Context Diagram 5. Construct Information Table 6. Site Map	1. User interface 2. Entity Relationship Diagram (ERD) 3. Data Flow Diagram (DFD) 4. Context Diagram 5. Construct Information Table 6. Site Map
Phase 4: Development and Iteration	1. Develop the system by writing the programming codes to develop a web-based application	1. Student attendance system
Phase 5: Testing and Evaluation	1. Prepare test plan for the system to test the functionalities of the system 2. Conduct user testing for the system 3. Conduct system testing for the system with expert user	1. Peers and expert feedback 2. User feedback 3. Tested system
Phase 6: Documentation	1. Write a complete report for the system	1. A complete report for the system

4. Results and Discussion

4.1. Student Attendance System

Student attendance system has been developed to be used by three main users which are student, lecturer, and staff. Each user has different access and views in this system. Table 3 outlines system functions accessible for the user.

Table 3. System Functions for User.

User	Functionalities
Student	1. Login into the system
	2. View student personal information.
	3. Take attendance using QR-code.
	4. View attendance percentage.
	5. View date that they did not come to class.
	6. View reminder or warning letter that they received.
Lecturer	1. Login into the system
	2. View student personal information and lecturer information.
	3. View list of student name register for the subject.
	4. View report of student attendance.
	5. View list name of student that get reminder or warning letter.
	6. Create session for class.
	7. Send reminder or warning letter to student.
Staff	1. Login into the system
	2. Manage student and lecturer information.
	3. View reminder or warning letter list.
	4. View report for attendance.

Figure 2 until 4 shows the screenshot of the system which display the process of taking attendance and record the attendance for every class. Lecturer creates session, display the QR-code, view daily attendance report, view overall report for the attendance and student view the attendance percentage and date that student did not come to class.

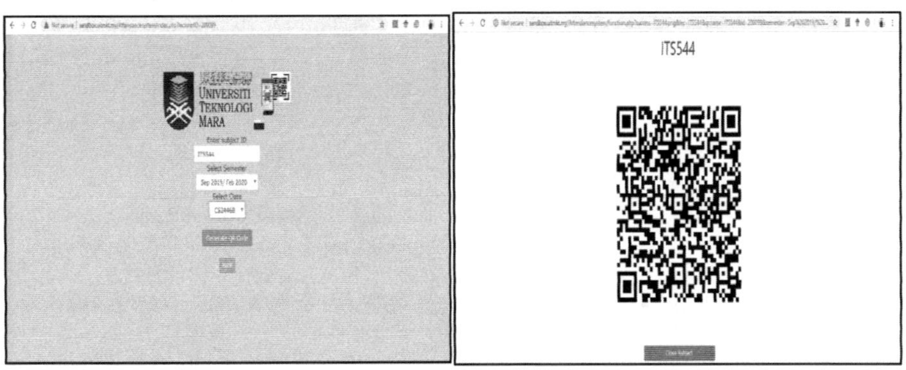

Figure 2. Create session for classes and Display QR-code for student.

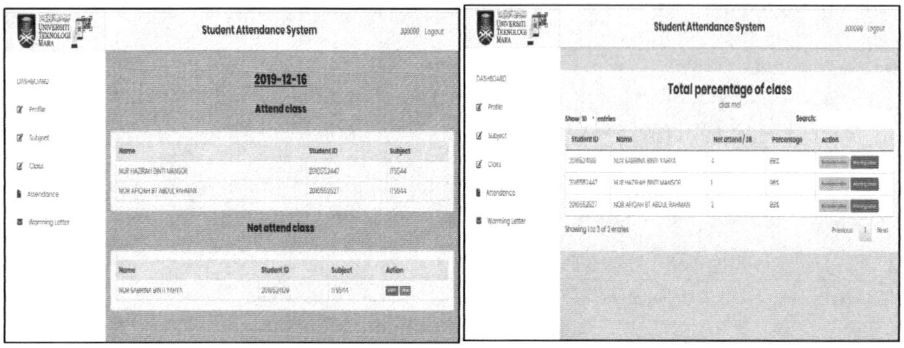

Figure 3. View of daily attendance for classes and view of overall attendance for classes.

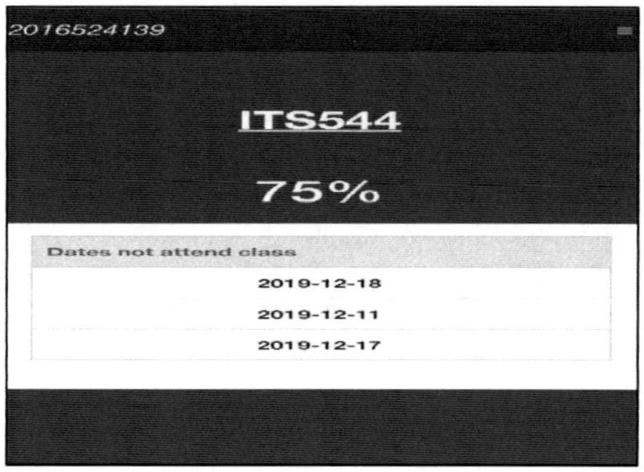

Figure 4. View of attendance percentage and date absent.

4.2. System Testing and Evaluation

System testing is to validate the functionality of the overall system. The test plan was used to record findings of the system. System usability then was evaluated by two type of users namely expert user and other potential users. Three experts have validated the system usability and functionality of the system. Figure 5 shows sample testing result from the experts for two constructs which are the efficiency of the system and the interface.

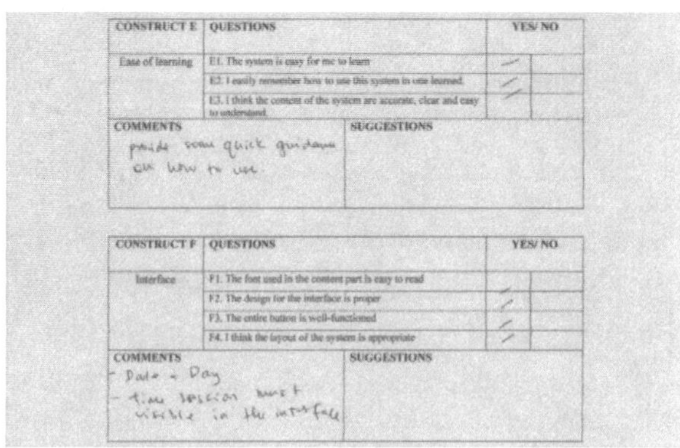

CONSTRUCT E	QUESTIONS	YES/ NO	
Ease of learning	E1. The system is easy for me to learn	✓	
	E2. I easily remember how to use this system in one learned.	✓	
	E3. I think the content of the system are accurate, clear and easy to understand	✓	
COMMENTS: provide some quick guidance on how to use.	SUGGESTIONS		

CONSTRUCT F	QUESTIONS	YES/ NO	
Interface	F1. The font used in the content part is easy to read	✓	
	F2. The design for the interface is proper	✓	
	F3. The entire button is well-functioned	✓	
	F4. I think the layout of the system is appropriate	✓	
COMMENTS: - Date + Day - time session must visible in the interface	SUGGESTIONS		

Figure 5. Expert Evaluation.

For user evaluation, 30 respondents have evaluated the system usability. Findings were recorded using a set of questionnaires. Overall, all means are above 4. It indicates most of the respondents are satisfied with the usability of this system. Figure 6 shows the graph for overall user evaluation for seven constructs which are ease of uses, usefulness, consistency, satisfaction, efficiency, user interface and QR-code.

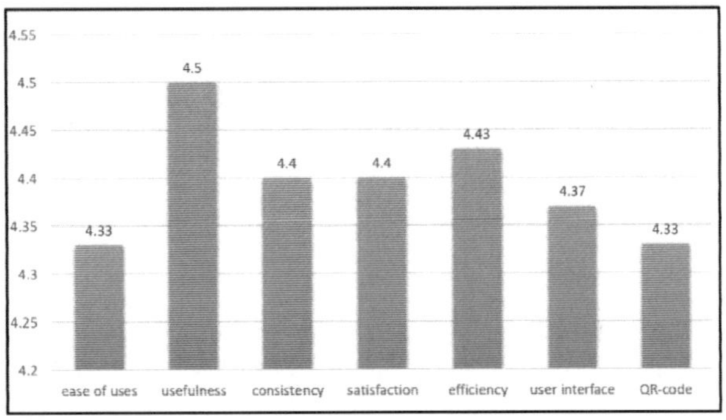

Figure 6. Overall Means for User Evaluation.

5. Conclusion

As a conclusion, Student attendance system using QR-code has improved the current process of recording the attendance in UiTM Kuala Terengganu and solve all the problem occurs while using the manual method of taking the attendance in classes. The objective of the development of this system is to help lecturer and staff to manage the attendance of student and make the process of recording the attendance more systematic and efficient. The evaluation results have provided a positive indicator to enhance the attendance system in the near future.

References

[1] Shoewu O and Idowu O 2012 Development of Attendance Management System using Biometrics *The Pacific Journal of Science and Technology* **13(1)** 300–307.

[2] Kumar P K 2006 Information System-Decision Making IndianMBA Retrieved October 2, 2010 from http://www.indianmba.com/Faculty_Column/FC307/fc307.html.

[3] Nowduri S 2014 Management information systems and business decision making: review, analysis, and recommendations *Journal of Management and Marketing Research* 1-8.

[4] Tobergte D R and Curtis S 2013 A Simulation Model for the Waterfall Software Development Life Cycle *J. Chem. Inf. Model.* **53(9)** 1689-1699.

[5] Balaji S and Sundararajan Murugaiyan M 2012 Waterfall vs v-model vs agile: A comparative study on SDLC *International Journal of Information Technology and Business Management*, **2(1)** 26–30.

[6] Kumar N, Zadgaonkar A S and Shukla A 2013 Evolving a New Software Development Life Cycle Model SDLC-2013 with Client Satisfaction *International Journal of Soft Computing and Engineering* **3(1)** 216-22.

[7] Khan P M and Beg M M S S 2013 Extended decision support matrix for selection of sdlc-models on traditional and agile software development projects *Int. Conf. on Advanced Computing and Communication Technologies (Korea)* 8-15.

[8] Maione C, Turra C, Fernandes E A D N, Bacchi M A, Barbosa F and Barbosa R M 2017 Finding the Most Significant Elements for the Classification of Organic Orange Leaves: A Data Mining Approach *Anal. Lett.* **50(14)** 2292-2307.

IOP Conf. Series: Materials Science and Engineering 1176 (2021) 012032 doi:10.1088/1757-899X/1176/1/012032

An artificial neural network model for forecasting air pollution

W D W A Manan[1*] and N R Abdullah[1]

[1]Faculty of Computer and Mathematical Sciences, Universiti Teknologi MARA (UiTM), 21080 Kuala Terengganu, Terengganu, Malaysia

*Corresponding author: wand@uitm.edu.my

Abstract. Air pollution has caused a lot of problems to people in terms of health and economy, as well as affecting various floras and faunas. Thus, monitoring air quality levels and forecasting the occurrence of air pollution is important so that preventive measures could be taken. In this study, Artificial Neural Network (ANN) was used to forecast the air pollution index (API) in Kuala Terengganu. This study focused on the prediction of API based on 5 years of data of main pollutants' daily concentration taken at the air quality monitoring station in Kuala Terengganu. The aim was to develop an Artificial Neural Network model that can predict the API. A Multilayer Perceptron Neural Network (MLP) engine was implemented in the system prototype and developed by using Keras, a deep learning library in Python. The model's performance was evaluated using the Mean Squared Error (MSE) statistical method and functionality tests were done to ensure the prototype was working correctly. In order to get a good performance model, a hyperparameter tuning process was carried out and the best hyperparameters values were selected. The performance of the model in making predictions was good as the MSE value was 0.0195.

1. Introduction

Clean air is significant for the survival of humans and other living beings. It impacts the well-being and economic development of a country. Air pollution is a grave problem that leads to some crucial concerns in many parts of the world. These worries are mainly about the effects of air pollution on human well-being and health [1], [2], [3]. Air pollution not only endangers people but can also affect crops, forests and animals [1], [4]. It is common knowledge that air pollution also impacts the environment where it causes changes in the climate, creating holes in the ozone layer and increasing the pollution of particulate matter in the air [1], [2]. Therefore, some actions could be taken to reduce its effects if the occurrence of air pollution could be predicted.

This fact motivated this project, which tried to predict air pollution incidence in the city of Kuala Terengganu using an artificial neural network model. The project had three objectives; first, to study algorithms that can be used to predict the occurrence of air pollution. Second, to develop an artificial neural network (ANN) model that can predict air pollution in Kuala Terengganu. Third, to evaluate whether the system prototype built from the ANN model is functioning correctly. At the end of the project, a prototype was built so that users can predict the occurrence of air pollution and find out the status of air quality in Kuala Terengganu.

The user of the prototype is targeted to be someone who wants to predict air pollution in Kuala Terengganu such as workers of the Department of Environment (DOE) and the Meteorological

Department. The prototype could also be used by the general public and other government departments as long as they have the required input data. As the capital of the state of Terengganu, Kuala Terengganu was a suitable focus area for this study. Furthermore, it also has a fairly significant industrial area which is one of the major contributors to air pollution.

All required data for the project were retrieved from the DOE. Five years data of air pollution index (API) and daily concentration of main pollutants such as particulate matters under 10μm (PM10), carbon monoxide (CO), sulphur dioxide (SO_2), ozone (O_3) and nitrogen dioxide (NO_2) in Kuala Terengganu were collected. Figure 1 below shows the conceptual framework of this project.

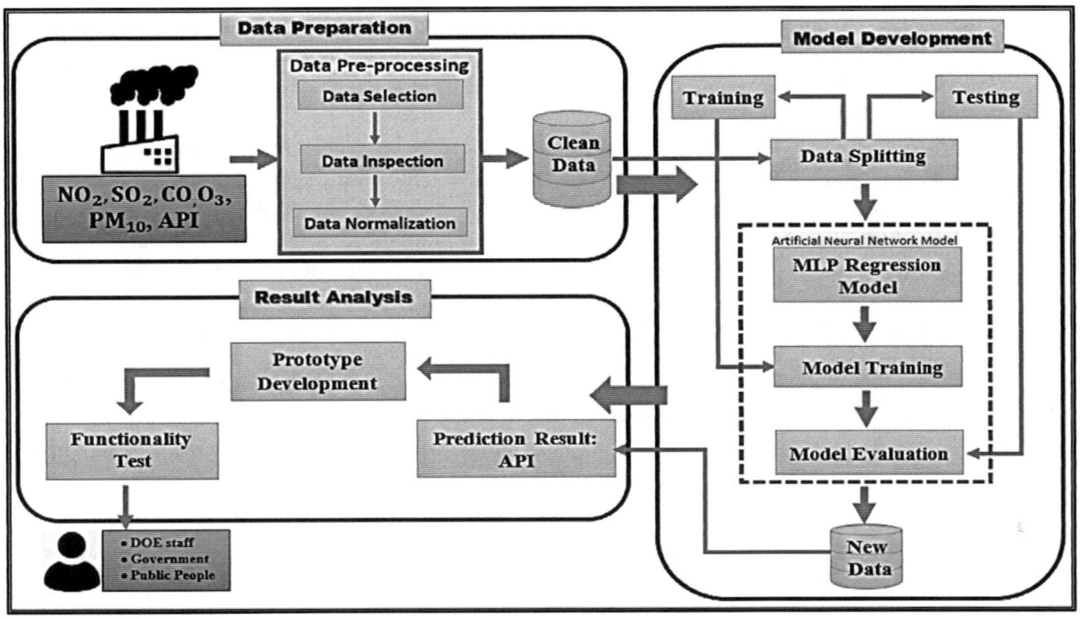

Figure 1. Air Pollution Forecasting in Kuala Terengganu using Artificial Neural Network Project Conceptual Framework.

2. Related Work

2.1. Air Pollution Forecasting

Forecasting of air pollution started in the 1960s, due to rising awareness of the trends and influence of air pollution [4]. The purpose of air pollution forecasting is to predict the composition of air pollution in the atmosphere at a particular time and area. Lately, numerous researchers have conducted various studies to provide accurate air pollution forecasting using ANN. The forecast may result in the prediction of the pollutants' concentration or air quality status. [5] stated that to prevent the rising of air pollution, it is important to forecast the content of pollutants in the air.

Similar to this project, many other kinds of research concentrated on building models to predict the air pollution index. For instance [5] constructed a Feedforward Neural Network (FFNN) model, while [6] used Long Short-Term Memory (LSTM), [7] implemented MLP and [8] used various machine learning techniques in building their prediction models.

2.2. Machine Learning

Most research mentioned above, used machine learning techniques in constructing their prediction models. The Artificial Neural Network technique used in this research is one of the most popular machine learning techniques. [9] defined machine learning as "an artificial intelligence branch that uses intelligent software to enable machines to perform their jobs skilfully". Machine learning gives computers the ability to learn without being explicitly programmed to do so. They stated that methods of statistical learning are the backbone of intelligent software used to develop this intelligence. The learning process starts with observations, direct experience data or feedback which is used to search for patterns of data and make better decisions in the future from the obtained past examples. The main goal of machine learning is to automatically enable computers to learn without human intervention and to take appropriate actions accordingly.

2.3. Applications of Machine Learning and Artificial Neural Network

Machine learning has been applied in areas as diverse as the medical field, the stock market and agriculture. In medical diagnosis, machine learning algorithms were used in diagnosing and detecting several diseases such as heart disease and malaria disease. [10] for example, used Naive Bayes to predict the occurrence of heart disease using machine learning algorithms based on attributes that are related to heart disease.

In time series forecasting, two machine learning algorithms, namely Support Vector Machine (SVM) and K-nearest Neighbour (KNN) were used by [11] in predicting rainfall in the coming year. In the financial market analysis area, research to predict closing stock prices using ANN was conducted by a group of researchers [12]. Machine learning was also used in the agriculture area which can be seen in the research done by [13] that used SVM, ANN and Naive Bayes techniques in the prediction of crop yield.

Artificial Neural Network is a type of machine learning algorithm which is modelled on the human brain. ANN models are designed to mimic the way humans learn. The model is made up of a set of artificial neurons that are interconnected with other neurons that depend on the neural network's weights. The weights are the connections between neurons that will determine the effect of one neuron on another [14]. ANN is a great tool to identify patterns that are too complex for programmers to extract and teach the machine to recognize them automatically.

In many research, ANN has been used in various application areas such as medical diagnosis as well as business and time series forecasting. We can conclude that machine learning algorithms including ANN can be used in many areas due to their ability to improve over time. These algorithms can also provide immediate adaptation without the presence of a human. Thus, these will add intelligence value to the technology or program that uses the machine learning algorithm.

2.4. Similar Application

In predicting the occurrence of air pollution, one of the popular algorithms used is ANN. For example, [5] used two types of artificial neural network models. One of the models was for a short-term prediction that used Feedforward Neural Network (FFNN) and the other one was to model a spatial prediction that used Elman Recurrent Neural Network. In another research on predicting micro particulate matters (PM10), [7] applied MLP algorithm in developing their prediction model. [8] used MLP as one of the machine learning algorithms to develop an air quality prediction model for New Delhi, India. Each research focused on one specific area and several parameters such as main air pollutants, speed of wind and temperature were used.

The researchers also stated that the ANN model can give satisfactory prediction results and better accuracy than other algorithms. For example, the results obtained by [10] indicated that the Feedforward Neural Network model gave more than 70% accuracy rate. Furthermore, the MLP model has proven its great ability in solving complex and nonlinear data. This can be seen from the results of the research done by [7] where the designed MLP model was ready for operational use. Thus, this shows that ANN can be used to forecast air pollution and will be able to produce a good prediction result.

3. Research Methodology

3.1. Project Phases

This research was conducted in six phases namely preliminary study, data collection, system design, implementation, evaluation and documentation. A brief description of each phase is shown in the figure 2 below.

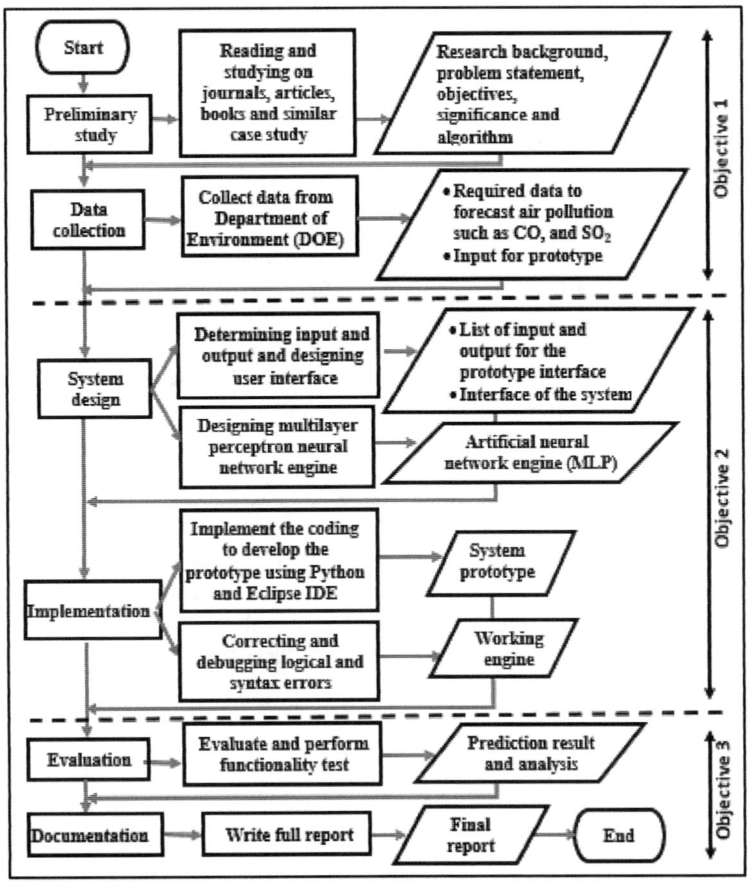

Figure 2. The flow of the research methodology.

3.2. Analysis Phase

3.2.1. Data Collection. All required data were retrieved and collected from the DOE. The data were on the daily concentration of main air pollutants namely particulate matters under 10μm (PM10), carbon monoxide (CO), sulphur dioxide (SO_2), ozone (O_3), nitrogen dioxide (NO_2) and API. Five years of data from 2014 until 2018 were collected from the department for this study.

Since the project was focused on the area of Kuala Terengganu, all data including the API were retrieved from the air quality monitoring station in Kuala Terengganu, located at Sekolah Kebangsaan Chabang Tiga, Kuala Terengganu. Table 1 shows an example of data collected from the DOE.

Table 1. Data from Department of Environment.

Site Id	Site Location	API	SO_2	NO_2	O_3	CO	PM10
CA0034	Sek. Keb. Chabang Tiga, Kuala Terengganu	54	0.003	0.007	0.042	0.63	58.0
CA0034	Sek. Keb. Chabang Tiga, Kuala Terengganu	44	0.004	0.010	0.044	0.63	23.0
CA0034	Sek. Keb. Chabang Tiga, Kuala Terengganu	49	0.004	0.019	0.049	0.81	47.0
CA0034	Sek. Keb. Chabang Tiga, Kuala Terengganu	53	0.007	0.009	0.042	1.08	56.0

3.2.2. Data Pre-processing. Pre-processing of data was done to improve the representation of the collected data and to get a clean dataset. A clean dataset is important as it can affect the accuracy of the machine learning model. Furthermore, a dataset can affect the ability of the model to learn in the training process. Thus, to develop an ANN engine, correct pre-processing data techniques are needed in order to get quality and clean data for the development of the ANN model. Three pre-processing techniques were used in this project, namely data selection, data inspection and data normalization. All these were carried out by using a pre-processing tool provided in the Waikato Environment for Knowledge Analysis (WEKA).

In the data selection process, required data attributes were selected and unnecessary data attributes such as site id and site name were removed from the dataset. Next, further pre-processing was done by inspecting the dataset for missing data. All missing data were replaced by the mean of each data attributes by using the 'replace missing values' filter in WEKA. Then the data were normalized using the 'normalize' filter in WEKA. Normalization was done to make sure the range for all predictors is in the common range of 0 to 1 [15], [16]. After data pre-processing is completed, a clean dataset was produced as pictured in table 2. 80% of this dataset was used to train the MLP engine, while the other 20% was set aside for testing and evaluating the model.

Table 2. Clean Dataset.

SO_2	NO_2	O_3	CO	PM10	API
0.358974	0.145773	0.395349	0.206897	0.053586	54
0.487179	0.233236	0.418605	0.206897	0.014513	44
0.487179	0.495627	0.476744	0.275862	0.041306	49
0.871795	0.204082	0.395349	0.37931	0.051353	53
1	0.145773	0.290698	0.206897	0.03684	43
0.871795	0.204082	0.453488	0.172414	0.023444	47
0.615385	0.437318	0.465116	0.206897	0.034607	48
0.487179	0.233236	0.27907	0.275862	0.032375	39
0.149009	0.204082	0.209302	0.206897	0.020095	28
0.149009	0	0.244186	0.103448	0.011164	29

3.3. Design Phase

In the design phase, determining the input and output of the system and designing the user interface was done as part of building the prototype. Next, a multilayer perceptron neural network engine with ten hidden layers was designed for data training and prediction (see Figure 3). From the available data, the inputs were determined to be the daily concentration of the main pollutants and the output data will be the predicted air pollution index and the air quality status such as "good", "moderate", "unhealthy", "very unhealthy" and "hazardous". The air quality status is set from the standard range of API status as classified by the DOE, shown in table 3. A user interface as shown in figures 7 and figure 8 was also designed in this phase.

Table 3. Standard Air Quality Status.

API	Status
0-50	Good
51-100	Moderate
101-200	Unhealthy
201-300	Very Unhealthy
301-400	Hazardous

3.4. Multilayer Perceptron Neural Network Algorithm

A multilayer perceptron (MLP) is a perceptron that is combined with other perceptrons, lined and arranged in multiple layers which can solve complex and larger problems. An MLP usually consists of a few layers of nodes, namely the input layer as the first layer, the output layer as the last layer and hidden layers [17]. The first layer is the layer that receives information, while the last layer is the output layer where the solution to the problem is produced. The hidden layers meanwhile act as intermediate layers that divide the input and output layers. In this project, the MLP consists of one input layer, ten hidden layers and one output layer. Figure 3 shows the architecture of the MLP that has fully-connected layers.

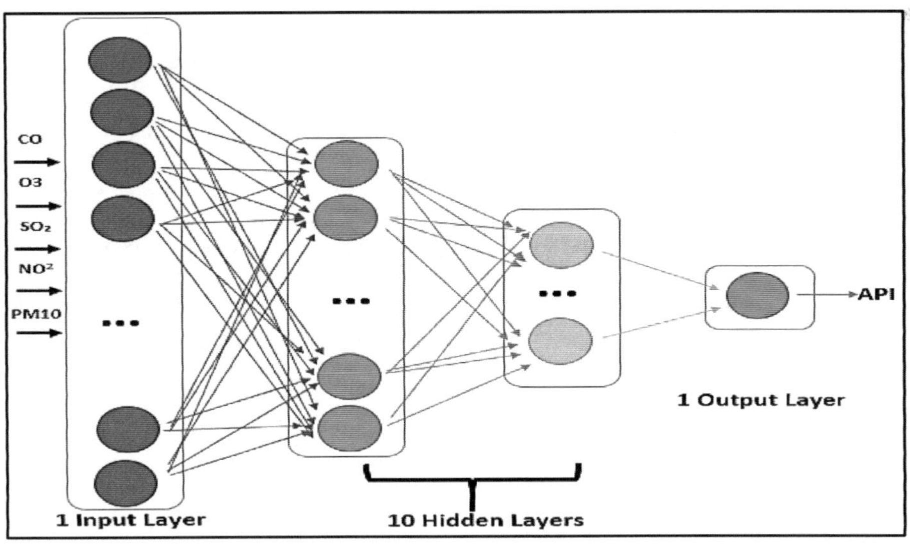

Figure 3. Multilayer Perceptron Architecture Used in This Project.

3.5. Implementation Phase

In this phase, several activities were conducted to develop a prototype containing the MLP training engine for predicting air pollution. The prototype was developed using Python in the Eclipse IDE. All the input and output data were stored in an excel file for users' reference. Testing and debugging were done upon completion of the prototype. In developing the prototype, a personal laptop with 64 bit and

4 GB DDR4 (Double Data Rate 4) memory was used as well as three software, namely Eclipse IDE and Pydev as the code editor and Microsoft Excel for storing data.

3.5.1 Multilayer Perceptron Training Engine. The multilayer perceptron neural network (MLP) engine was developed by using the Keras deep learning library in Python. The Keras library is a high-level API that is written in Python and used for running high-level neural networks [18]. A few steps were needed to build a neural network model using the Keras library [18], [19].

The first step was to define the neural network. In this step, a sequential class which is the container for the layers was defined and created. A Sequential API and the densely-connected layer type were used for creating the structure of the model. The model of this project has twelve densely-connected layers of neurons which consist of one input layer, one output layer and ten hidden layers. This allows the neural network to think widely first before converging to the final prediction. Densely connected layers were used because a multilayer perceptron model must use fully connected layers of neurons. The number of neurons in the input layer and hidden layers was 203 and the dropout rate for each layer was set to 0.0318. The number of inputs or features was specified in the first layer by the *input_dim* attribute which was 5. In the first layer and hidden layers, the Rectified linear (RELU) based activation function was used. These number of hidden layers, neurons per layer, dropout rate and activation function were selected based on an experimental setup described in Section 4.1 as they produced the best score for the mean squared error (MSE). The RELU-based activation function is one of the standard activation functions used today and is defined as $y = max\ (0, x)$ [20]. For the output layer, no activation function was used since this model is a regression model. As we wanted to predict a real-valued or numeric prediction, no activation function was necessary.

The network was compiled after it was defined. The loss function used to evaluate the network and the optimizer used to train the network needed to be specified when compiling the training model. Mean squared error (MSE) loss function and the adaptive moment estimation (Adam) optimization algorithm were used for this model. The Adam optimizer is an efficient stochastic gradient descent algorithm and a widely used optimizer as it can automatically tune itself and produce good results [18]. The learning rate for the Adam optimizer was set to 0.0053. The selection of these two parameters was chosen from an experiment described in the next section.

The third step in developing the MLP engine was to fit the network. Fitting the network means adapting the weights on the training dataset. This is done to ensure that the memory will not be loaded with too many input patterns at the same time. The batch size defines the number of patterns shown by the network before its weight is updated in the epoch, while the epoch defines the number of times the model is exposed to the training dataset. At the start, the number of epochs was set to 50 and the batch size was set to 128 in the *fit()* function. Once fit, the network is ready to be trained.

The next step was to evaluate the network. The performance of the network was evaluated using the test dataset. The MSE value was used to measure the performance of this model. The lower the MSE value, the better the training model. The final step was to make predictions. Once the performance of the model is satisfied, the model can be used to make predictions on new data by using the *predict()* function.

3.5.2 Experiment Setup. An experiment called hyperparameter tuning was conducted to find the best hyperparameters for this model. This is to ensure the best MLP regression model was developed in terms of speed, performance and quality. [18] defined a hyperparameter as a parameter whose value is set before the process of learning starts in training an ANN model. The hyperparameters involved in these experiments were the number of hidden layers, the number of neurons per layer, learning rate, dropout rate, activation function, optimizer and batch size as shown in figure 4.

For this experiment, a total of 25 different trials were conducted. For each trial, the possible combinations of hyperparameter values were evaluated and all output scores of the trials were recorded. The output score was the value of MSE since mean squared error was used as the loss function. The combination of hyperparameter values that produced the best output score among the 25 trials was then selected and used for this model. The best values of the hyperparameters and the score are shown in figure 5.

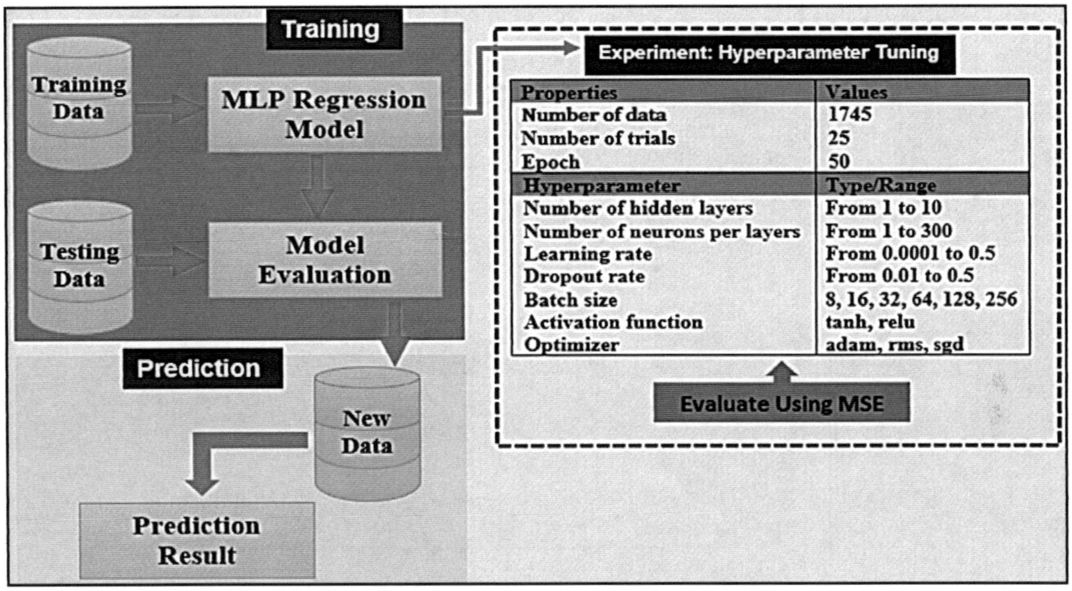

Figure 4. Details of the Experiment Setup.

```
('learning_rate', 0.005328924019769793)
('dropout_rate', 0.0318121261050319)
('num_hidden_layers', 10)
('neurons_per_layer', 203)
('batch_size', 128)
('activation', 'relu')
('optimizer', 'adam')

{'keras_cv': 0.02168685365348093}
```

Figure 5. The Best Values of Hyperparameters and the Output Score.

3.6. Evaluation Phase

The evaluation was done by conducting functionality tests on the prototype. For measuring the performance of the MLP model, a statistical method, namely the mean squared error method (MSE) was used. The MSE method was used to measure the performance of the model in making predictions since it was widely used by many researchers for regression problems. MSE is calculated as the average of squared differences between the predicted API value produced by the prototype and the actual values of the API in the test dataset. The lower the MSE, the better the performance of the model in making predictions as there would be a good match between the actual and predicted dataset [21].

4. Result

4.1 Training Result

The training process was visualized using a loss graph as shown in figure 6. The figure shows that the training and validation values decreased as the number of epochs increased. This means that the loss value, which is the MSE value has reached the minimum value which is zero as shown in figure 6. The learning curve of the loss graph produced from the training process is a good fit or an optimal fit. A good fit learning curve is the goal of the learning algorithm and also shows that the performance of the model on both training and validation dataset were good [22]. The plot of both training and validation loss also decreased to a point of stability. Thus, this means the training model could be implemented in the system prototype and could be used to predict air pollution.

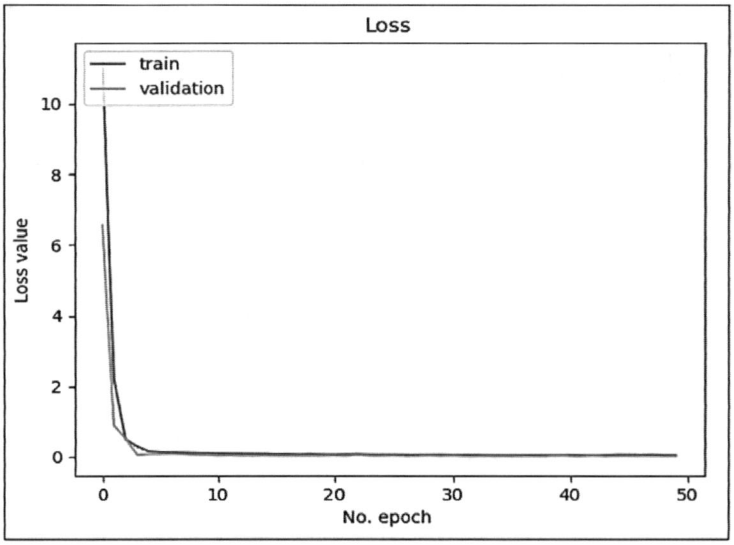

Figure 6. Loss Graph of Training Process.

4.2 System Interface

The system interface is developed to make sure the user can easily understand how to use the system. The system prototype consists of two user interfaces. The first interface contains several input fields, namely date, ozone, sulphur dioxide, carbon monoxide, nitrogen dioxide and particulate matter under 10. The user is required to fill in all the input fields and then clicking on the 'predict button' to make the prediction. The user can clear the input fields using the 'clear button' and close the system prototype using the 'exit button'. Figure 7 shows the first interface of this prototype.

After a user has clicked the predict button, the results of the prediction are shown in the second interface. The second interface consists of two entries which are prediction result and air quality status. The prediction result is in the form of numerical data as it is the value of the predicted API and the air quality status is based on the category of the predicted API. The date of the prediction is displayed at the top of the interface. The user can return to the first interface by clicking the 'back button' and exit the prototype by clicking the 'exit button'. Figure 8 shows the second interface of this prototype.

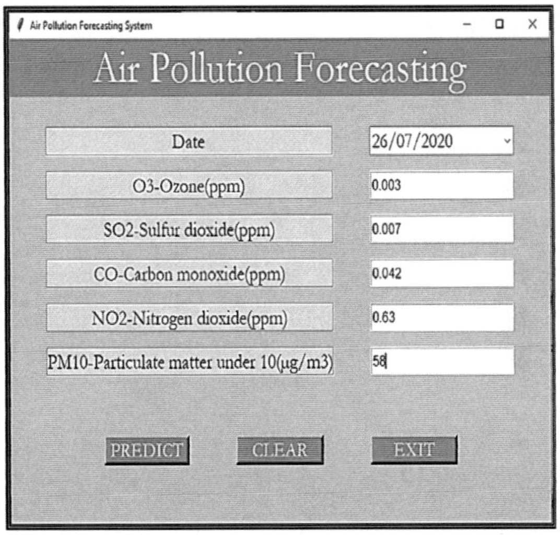

Figure 7. First Interface.

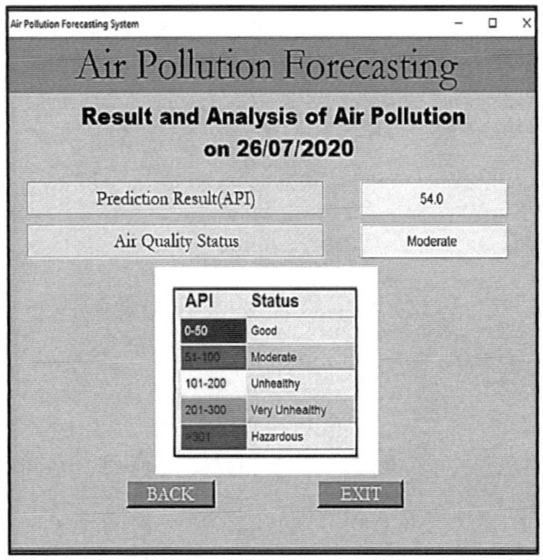

Figure 8. Second Interface.

4.3 Testing and Evaluation

The MLP model was evaluated using test data collected from the DOE. The MSE value during the model evaluation was 0.0195. The lower the MSE, the better the performance of the model in making predictions. Several different research has indicated that the MSE value that is closer to zero is better [21], [23]. Thus, this proved that the performance of the model developed in this project is good and it can be used to predict the air pollution index.

Besides using MSE, predictive and actual results were also compared. The prediction results stored in the excel file were compared with the actual result to determine the accuracy and effectiveness of the system prototype in making the prediction. For this, 140 sets of data were used to evaluate the system prototype. The correct predictions were summed and multiplied by 100 to calculate the percentage of accuracy. The formula used to calculate this is shown in equation (1) below:

$$(\text{Correct predictions}) / (\text{Total predictions}) \times 100 \quad \ldots\ldots\ldots\ldots (1)$$
$$= 110/140 \times 100 = 79\%$$

From this calculation, the accuracy of the system prototype was determined to be 79% where 110 out of 140 predictions were the same as the actual air quality status. This thus indicated that this system prototype was able to successfully predict the air pollution in Kuala Terengganu.

Functionality tests were also done to make sure the system prototype could run properly and was able to make a prediction on air pollution from the data inserted into the system. Several users are selected to test the system and evaluate the system based on their preferences. Several criteria have been selected as the aspects of evaluation based on several observations on a good existing system. The results of the functionality test are summarized in table 4.

Table 4. Evaluation Aspects in Functionality Test.

Aspect	Dissatisfied	Neutral	Satisfied
Users can input the data.			✓
Users can click the buttons.			✓
Users can understand the function of each button.			✓
The prototype gives the desired output.			✓
The input and output data can be saved into an excel file.			✓
User-friendliness.			✓
Prototype efficiency.			✓

5. Conclusion

In conclusion, all three objectives have been achieved after all activities were completed. All the stated methodologies have been implemented in this project. This project has proven that the selected algorithm which is ANN, can be used to solve the problems and is able to give a reasonable prediction result when predicting the air pollution in Kuala Terengganu. This shows that the first objective which is to study algorithms that can be used to forecast air pollution in Kuala Terengganu has been achieved. A system prototype that implemented the ANN algorithm as the engine has also been successfully developed. Hence, the second objective which is to develop an artificial neural network system model that can predict air pollution in Kuala Terengganu was also fulfilled. Evaluation and testing have been done to this system prototype to prove its functionality in predicting the air pollution index. From the evaluation and testing, this system prototype is capable of giving the desired prediction result. Thus, the third objective which is to evaluate the functionality of the system prototype has been achieved. As an overall conclusion, this prototype can be used to forecast air pollution in the city of Kuala Terengganu and is able to give a satisfactory result.

Acknowledgment

Thank you to the Department of Environment Malaysia (DOE) for providing the data for this project.

References

[1] Manisalidis I, Stavropoulou E, Stavropoulos A and Bezirtzoglou E 2020 Environmental and Health Impacts of Air Pollution: A Review *Front. public Heal.* **8** p 14.

[2] Usmani R S A, Saeed A, Abdullahi A M, Pillai T R, Jhanjhi N Z and Hashem I A T 2020 Air pollution and its health impacts in Malaysia: a review *Air Qual. Atmos. Heal.* **13(9)** p 1093-1118.

[3] Mohd Pu'Ad M F, Gunawan T S, Kartiwi M and Janin Z 2019 Development of Air Quality Measurement System using Raspberry Pi *2018 IEEE 5th Int. Conf. Smart Instrumentation, Meas. Appl. ICSIMA (Thailand)* p 28-30.

[4] Bai L, Wang J, Ma X and Lu H, 2018 Air pollution forecasts: An overview *Int. J. Environ. Res. Public Health* **15(4)** p 1-44.

[5] Rahman P A, Panchenko A A and Safarov A M 2017 Using neural networks for prediction of air pollution index in industrial city *IOP Conf. Ser. Earth Environ. Sci.* **87** 4.

[6] Bui T-C, Le V-D and Cha S K 2018 A Deep Learning Approach for Forecasting Air Pollution in South Korea Using LSTM.

[7] Abdullah S, Ismail M and Ahmed A, 2019 Multi-Layer Perceptron Model for Air Quality Prediction *Malaysian J. Math. Sci.* **8(1)** p 15-34.

[8] Srivastava C, Singh S and Singh A P 2019 Estimation of air pollution in Delhi using machine learning techniques *2018 Int. Conf. Comput. Power Commun. Technol. GUCON 2018 (India)*

p 304-309.
[9] Mohammed M, Khan M B and Bashier E B M 2016 *Algorithms and Applications* Boca Raton: Taylor & Francis Group LLC.
[10] N R, Maneesha T, Hafeez S and Krishna H 2018 Prediction of heart disease risk using machine learning *Int. J. Mech. Prod. Eng. Res. Dev.* **8(Special Issue 2)** p 605-614.
[11] Garg A and Pandey H, 2019 Rainfall prediction using machine learning techniques *Int. J. Innov. Technol. Explor. Eng.* **8(7)** p 957-963.
[12] Vijh M, Chandola D, Tikkiwal V A and Kumar A 2020 Stock Closing Price Prediction using Machine Learning Techniques *Procedia Comput. Sci.* **167** p 599-606.
[13] Palanivel K and Surianarayanan C 2019 An Approach for Prediction of Crop Yield Using Machine Learning and Big Data Techniques *Int. J. Comput. Eng. Technol.* **10(3)** p 110-118.
[14] Han S-H, Kim K W, Kim S and Youn Y C 2018 Artificial Neural Network: Understanding the Basic Concepts without Mathematics *Dement. Neurocognitive Disord.* **17(3)** p 83.
[15] Mclean S, Kaiser J and Richard B 2019 Environmental Modelling & Software A review of artifi cial neural network models for ambient air pollution prediction *Environ. Model. Softw.* **119** p 285–304.
[16] Brownlee J, 2020, How to use Data Scaling Improve Deep Learning, *Machine Learning Mastery*. [Online]. Available: https://machinelearningmastery.com/how-to-improve-neural-network-stability-and-modeling-performance-with-data-scaling/. [Accessed: 22-Feb-2021].
[17] Little Z, 2020 Multi-Layer Layer Perceptron (MLP) *Medium*. [Online]. Available: https://xzz201920.medium.com/multi-layer-perceptron-mlp-4e5c020fd28a. [Accessed: 22-Feb-2021].
[18] Brownlee J 2020 Your First Deep Learning Project in Python with Keras Step-By-Step, *Machine Learning Mastery*. [Online]. Available: https://machinelearningmastery.com/tutorial-first-neural-network-python-keras/. [Accessed: 18-May-2020].
[19] Brownlee J 2020 5 Step Life-Cycle for Neural Network Models in Keras - Machine Learning Mastery, *Machine Learning Mastery* [Online]. Available: https://machinelearningmastery.com/5-step-life-cycle-neural-network-models-keras/. [Accessed: 18-May-2020].
[20] Liu D, 2017, A Practical Guide to ReLU, *Medium*. [Online]. Available: https://medium.com/@danqing/a-practical-guide-to-relu-b83ca804f1f7. [Accessed: 11-Jul-2020].
[21] Kazumali E Kr. Biswas S and W Kelsey T 2016 What is the Acceptable MSE value and Coefficient of determination(R2)? *Research Gate*. [Online]. Available: https://www.researchgate.net/post/What_is_the_Acceptable_MSE_value_and_Coefficient_o f_determinationR2. [Accessed: 11-Jul-2020].
[22] Brownlee J 2019 How to use Learning Curves to Diagnose Machine Learning Model Performance, *Machine Learning Mastery*. [Online]. Available: https://machinelearningmastery.com/learning-curves-for-diagnosing-machine-learning-model-performance/. [Accessed: 22-Feb-2021].
[23] Taghavifar H and Mardani A 2014 Application of artificial neural networks for the prediction of traction performance parameters *J. Saudi Soc. Agric. Sci.* **13(1)** p 35-43.

IOP Conf. Series: Materials Science and Engineering 1176 (2021) 012033 doi:10.1088/1757-899X/1176/1/012033

A study of multiple reward function performances for vehicle collision avoidance systems applying the DQN algorithm in reinforcement learning

N J Zakaria[1], M I Shapiai[1] and N Wahid[2*]

[1] Center for Artificial Intelligence (AI) and Robotics Ikohza, Malaysia-Japan International Institute of Technology, Universiti Teknologi Malaysia, Jalan Sultan Yahya Petra, 54100, Kuala Lumpur

[2] Faculty of Electrical Engineering, Universiti Teknologi MARA, Cawangan Terengganu, 23000 Dungun, Terengganu, Malaysia

*Corresponding author: nurbaiti@uitm.edu.my

Abstract. Reinforcement Learning (RL) is an area of Machine Learning (ML) that intends to improve the acts of agents learning from environmental interconnection. The significant concern in RL is to achieve the promising potential of the training process in the model. However, network convergence speed is often sluggish in RL and converges quickly to local optimal solutions. Reward function has been used to deal with these problems as a useful tool to speed up the agent's learning speed. Even though RL convergence properties have been comprehensively explored, there are no specific rules for choosing the reward function. Therefore, searching for efficient potential reward function is still an exciting field of study. This paper discusses the reward function, execute some analysis, and provides the learning agent with the extracted information to increase the speed of learning for collision avoidance task. We provide an experimental study for selecting one reward function in a simulated collision-avoidance environment of an autonomous vehicle by applying the DQN algorithm. It has been conducted on online environments, which is using the CARLA simulator. This experimental study consists of three cases with a various exploration of reward values. Case 1 consists of the range of the penalty value larger than the reward function by 200 times. Case 2 is similar, but with the small range number of the penalty is applied, case 3, which is the reward function and penalty value, is in the same range value. The result shows that case 3 performances outperform case 1 and case 2 with 94% average accuracy; meanwhile, case 1 obtains 70%, and case 2 achieves 85% accuracy. It is may due to the monumental size of the collision penalty in comparison to all else. Hence, the findings obtained show the efficacy of the exploration of the reward function.

1. Introduction

Reinforcement Learning (RL) is one of the Machine Learning (ML) algorithms that acquires information based on environmental observation. Rewards are the consequence of learning in the environment. RL's significant advantage is that, even though the environment is too broad or cannot be defined shortly, it generates the most successful rewards [1]. Nowadays, RL has currently achieved incredible performance in many applications [2]. RL aims to optimize an agent's action based on the environment's reward [1]. The current desirability of environmental states is denoted by reward. The state's value is the final reward that an agent can determine to accumulate for future choices. The decisions are based on the agents' positive and negative outcomes. The agent performs an environmental reward action and adjusts its policy based on the reward. Through this continual interaction with the environment, the agent can learn the optimal policy to achieve the highest cumulative reward. The reward function thus implicitly determines the agent's optimal actions [3]. Q-learning [1] is a common type of RL where the optimal policy is implicitly learned in the form of a Q-function. Therefore, it is an RL method that works by learning an action-value function. The expected utility is obtained by a given action in a given state and following a fixed policy in the future. Q-learning can compare the predicted utility of behavior without having an environmental model. A recent variation in the Q-learning algorithm has shown a valuable enhancement in acquiring positive rewards [1].

However, slowness in convergence [4] is one of the key drawbacks of RL. Therefore, it is also challenging to develop a perfect reward function for a particular situation due to the environment's complexities [2]. Reward shaping is one of an approach to deal with this issue [3]. In this approach, the designer practically creates a virtual reward function and gives it to the learning agent. In the first episodes of learning from this virtual reward function, the agent receives the reward since it does not take any reward from the environment in the early episodes of learning. To optimize its behavior, the agent utilizes this virtual reward. Therefore, the shaped reward could have a vital effect on accelerating the training process. The main idea of reward shaping is to supply additional feedback by the designer, other than that of the environment, for the agent to enhance its convergence rate and training speed [3]. The agent changes its behavior in the following episodes based on the information gained from the environment, and the influence of the formed reward is gradually reduced. Although reward shaping has proven to be a powerful process, it is challenging and sometimes even impossible to determine the values of the shape rewards for extremely large environments.

Therefore, several methods to speed up RL have been proposed. M. J. Mataric [5] suggested a technique for developing reward functions that take advantage of implicit knowledge of the domain. It contains the use of continuous reward functions. E. Wiewiora [5] completed the reward shaping research and proved certain similarities between potential-based shaping and initial Q-values. Besides, with reward shaping, the rewards from the environment are augmented with additional rewards [6]. However, the reward shaping can lead the agent into learning suboptimal policies and thus trap the system. Indeed, selecting the initial Q-values [7] is the most basic approach for biasing training.

Besides, in RL, to adapt to the environment, an agent must interact with the environment, thereby forcing the agent to spend a large amount of time obtaining optimal behavior. In this process, the agent learns how to interact in the environment based on the rewards. To increase this learning process, it is hoped that a reward function can describe the agent's state in a timely and accurate manner. Therefore, the design of a reward function has become an essential aspect of RL. To accelerate learning and training, researchers have proposed numerous methods to design reward functions. Wei *et al.* presented a study of heuristic reward function for RL algorithms [8]. Meanwhile, Yan *et al.* applied a hierarchical RL (HRL) algorithm based on a heuristic reward function to solve a vast state space issue [9]. Next, Maja *et al.* presented a technique for developing reinforcement functions that accelerate learning using implicit domain knowledge [10]. S. Koenig and R. G. Simmons studied different representations of reward functions and the sophistication of Q-learning techniques depending on the preference of RL representation. S. Behnke and M. Bennewitz proposed to give the learning agent access to the Q-values of the experienced agent concerning imitative reinforcement [11]. Mestol *et al.* presented an algorithm that is a variation of RL with the adaptation of reward to the degree of uncertainty of a performed prediction and solved some issues that could not be solved using conventional RL [12]. Next, Wu *et al.* suggested an adaptive network scaling framework to gain an appropriate scale of rewards during training for a promising performance [13]. In a system based on the concept of indirect reciprocity, Xiao *et al.* developed a technique for updating the node reward, which highly improved the system's security [14].

Nevertheless, though RL has been extensively studied and its convergence properties are well known, individuals frequently prefer reward function on one's intuition [4] in practice. This paper presents an analysis of different variations in the reward function of the Q-learning algorithm to obtain better and higher rewards. To propose a generic analysis, we address the effects of RL parameters on the policy. Therefore, we use the Q-learning algorithm to validate our analysis. The key challenge is to shed light on initializing RL parameters correctly to achieve the desired optimal behavior within the framework of a goal-directed task in a minimal time. Therefore, to obtain the optimum Q-value, the critical goal is to increase the efficiency of the Q-learning algorithm and to decrease the number of iterations. We herein concentrate on optimizing the reward function based on the existing reward function to produce better outcomes. By observing the changes in rewards during the RL process, especially when an agent succeeds or fails, we discovered that rewards often change significantly. Throughout this paper, we will focus on the case of three reward functions. We conclude by discussing how the analysis can be generalized to more than two reward functions. In this paper, the reward function consists of the values gained considering the speed of the agent in the environment.

2. Methodology

2.1. Reinforcement learning background

RL has seven crucial keywords: agent, environment, actions, states, episodes, rewards, and policy. RL process/tasks consist of training an agent which associates with its environment. By performing actions, the agent arrives at various scenarios known as states. Next, every action will give on to either positive or negative rewards. The only objective of the agent is to maximize its total reward over an episode. This episode is all events that occur within the environment between the first state and the last or terminal state. The designer reinforces the agent to learn from experience by executing the best actions called policy/strategy. Thus, the increases experience it gains, the highest it's to perform the action.

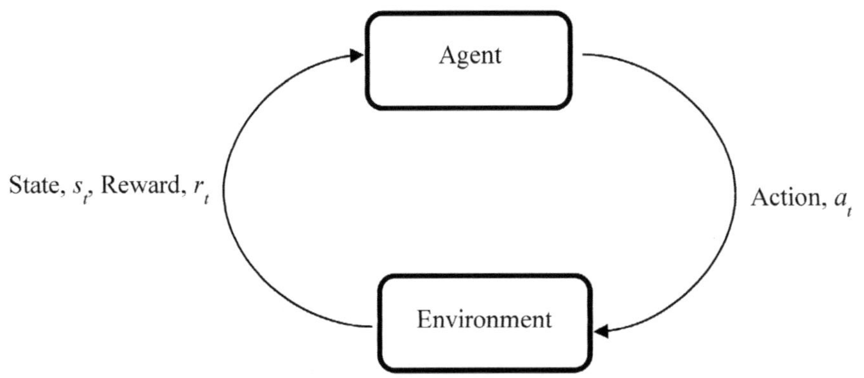

Figure 1. Agent-environment interaction loop.

Figure 1 shows the agent-environment interaction loop that consists of three components which are State (s), Action (a), and Reward (r). State (s) is some instance of the environment at the time (t). While Action (a) is what the agent does in the environment and Reward (r) is concerned about if the action is correct, the reward is high and vice versa. Therefore, the RL also contains two functions: Policy (π) and Value (V). Policy function is a function for choosing actions a while Value function is a function for expected reward (r) given some action (a). The Policy function also decides what actions to take given the input state, while the Value function is the expected return: it notifies the designer how good that action within that state is.

Next, in the RL, the agent is made up of two components: memory and brain, as described in figure 2. Brain encompasses the Neural Network (NN). It is the house of AI systems. Here is the decision-making on what action to takes happens. Meanwhile, the memory stores the data from the previous state (reward + action). The data are used to train in the NN. For every thousand experiences/episodes, the NN gets better at performing specific actions. Eventually, it always performs the actions that always get really good rewards since they will always lead the agent to perform better for the next episodes.

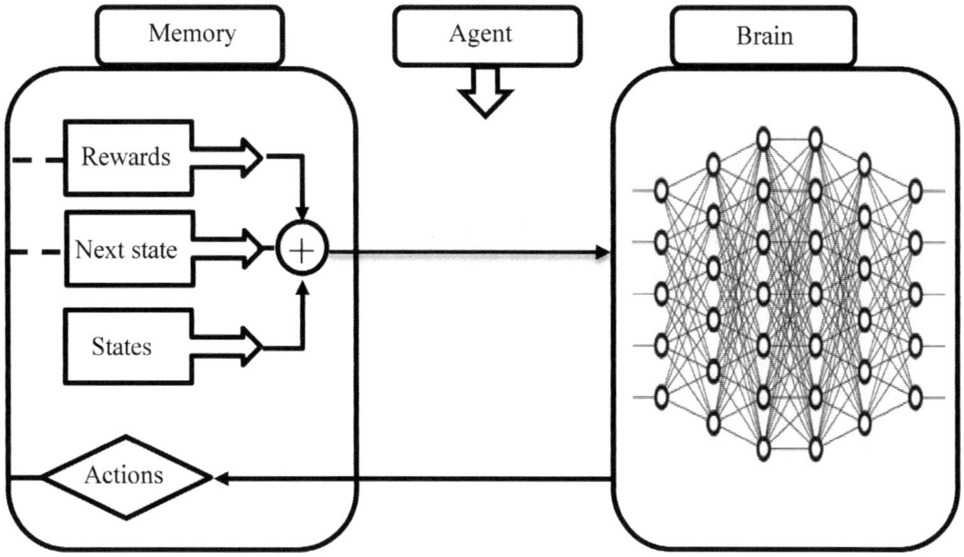

Figure 2. Component of RL.

2.2. Reinforcement learning problem formulation

Learning a policy function ( ) to select the action

$$a_t = \pi(s_t) \tag{1}$$

Select actions that receive maximum reward ( )

$$\pi^* = argmax\ \mathrm{E}[\sum_{t\gg 0} \gamma^t r_t|\pi] \tag{2}$$

Where γ^t is the discount factor, a value between 0 to 1.
$\mathrm{E}[\sum_{t''0} \gamma^t r_t|\pi]$ is known as Expected Return, i.e., V. Thus,

$$\pi^* = argmax\,(V^{\pi}) \tag{3}$$

The policy function denoted by π is a decision-making function that decides what actions to apply. The optimal action is the action with the highest reward. To achieve the optimal action, it is necessary to obtain the optimal policy. So, the equation for the optimal policy, which is denoted by π^* is the maximum of the expected reward given that particular policy at that time. The γ^t is the discount factor. The discount factor is applying because it can't be sure either the reward that will be achieved in the future is as good as a reward earned in the current state. The value of the discount factor is set to 1 and 0. The discount factor controls how an agent regards rewards. The low values of the discount factor encourage myopic behavior where an agent will aim to maximize short-term rewards.

In contrast, high discount factor values cause agents to be more forward-looking and maximize rewards over a longer time frame. The expected value is the average over the whole thing, so that's the optimal policy that the agent can learn. Therefore, the value function means the expected return shows how good was the action for that given state. So, the expected return is equal to the value function. Hence, the optimal policy is simply the maximum value of that value function for that particular policy. The aims are to train the NN to perform the best optimal policy, and it is the task for the RL.

2.3. *Deep Q-Network*

The DQN demonstrates how a Convolutional Neural Network (CNN) can learn successful control policies from just raw data. The network was trained end-to-end and was not provided with any specific information. In DQN-Network, the action is as an input to the NN. It consists of 3 convolutional layers and flattens layer to flatten the output of these layers. It then passes through Fully Connected layers, while the last layer is each node associated with the action. The outcome is a Q value associated with each action, and select the optimal action as the maximum of these Q values. Eventually, given enough experiences that train the NN, the NN will output Q-values associated with all actions given that state. So, it essentially learned the pattern of the possible Q values should be for the experiences that it has gone through for all given actions, and it will select the maximum Q-value, and it will always be the right action. DQN Architecture Network is demonstrated in figure 3.

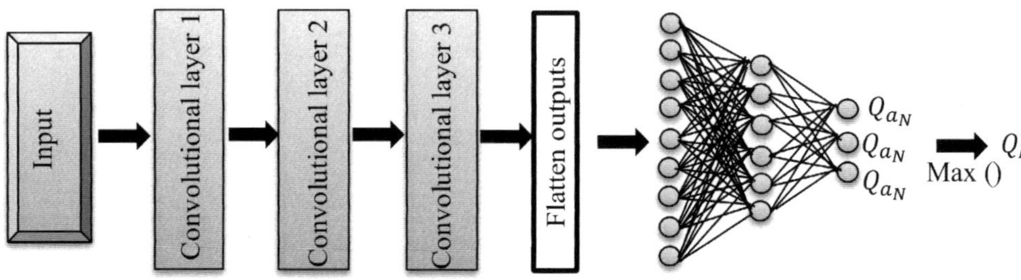

Figure 3. DQN architecture network.

2.4. *DQN problem formulation*

2.4.1. Q Function. Q function is the same as the value function (V), except it takes in state and action pair:

$$Q(s,a) = E[\sum_{t\geq 0} \gamma^t r_t | s, a] \tag{4}$$

The optimal policy is now:

$$\pi^* = argmax\,(Q^*(s,a)) \tag{5}$$

All the value functions obey particular equations, called the Bellman equations. The Bellman equivalent for the Q function:

$$Q^*(s,a) = E_\pi[r_t + \gamma^t max Q^\pi(S_{t+1}, a_{t+1}) | s, a] \tag{6}$$

Q-value function or also known as the action-value function, takes in state and action pair. It also can be defined as the expected return starting at some states but then taking an arbitrary action. Still, after it has achieved some experiences and learn from those experiences, the NN has some policy even though that policy is not strong. So, the Q-value function/action-value function then will be used in the DQN, and this essentially can be a substitute for the state value function in the same way. Hence, the optimal policy function is just a maximum of the optimal action-value function, similar to the optimal policy function being the maximum state value function.

2.4.2. DQN Loss Function. The Bellman Q value is *the* ${Q^*}_{target}$ value. *The output of the model is* $Q_{predicted}$ *value. Loss is simply the Mean Square Error value of the two:*

$$L(w) = E[({Q^*}_{target} - Q_{predicted}(s, a, ; w))^2] \tag{7}$$

IOP Conf. Series: Materials Science and Engineering 1176 (2021) 012033 doi:10.1088/1757-899X/1176/1/012033

The value function obeys the Bellman rule, which is a recursive rule which states that the value of a current state is the reward that gets for that particular state as well as the expected reward for the successor state, so the expected reward for the next state. According to the Bellman equivalent of the action-value function also, the optimal q value or the optimal value for the current state given some action is the reward achieve for that particular time step for that specific state plus the expected value or expected return achieve for the next state and the possible action the agent will take. So, the development equation and basic idea in RL are to estimate this optimal Q value by using this Bellman as an iterative update.

2.5. Policy learning algorithm

The policy learning algorithm for collision avoidance applying the DQN algorithm in RL.

Algorithm 1: Collision Avoidance Policy Learning Algorithm

Input: *Action = A {Move forward, Turn Right, Turn Right}*
State = X {1 …………… N_s}
Output = Q (X, A) optimal State and Action
Let γ [0,1]- Discount factor,
Let α {0,1} – Learning rate = 0.1
Let R – Reward
Initialize of replay memory D to capacity N
Initialize of action-value function network Q with random weights w
Initialize S = Random State, A = Random Action, R = Arbitrary
*Q: S*A – R*
Output: trained w; w - weight
For *epi_idx = 1 to max_episodes* **do**
Randomly initialize the current state S_t;
While *S is not terminal,* **do**
With probability S randomly select an action A with a certain probability;
Otherwise, select action A corresponding to the maximum value of Q through Q-network;
According to the current state S_t and selected action A_t, control the vehicle and interact with the simulation environment in the CARLA;
Calculate Policy (π(x)) - $argmax\ (Q^*(s, a))$
Action – Policy (state)
If *Collision ==True*
reward – R (state, action)
reward = N
else
reward – R (state, action)
update (reward)
end if
s' - T (s,a) //Receive new state
Q (s', a)
s - s'

Calculate the loss function
return
end For

2.6. *Reward function*

To enable the agent to learn the desired strategy, RL algorithms depend on reward functions. The agent is penalized if a collision occurs and gets the rewards for efficiency and safety actions. Here, we adopt an analysis of reward function value R that takes different values using three criteria. In this paper, the shaped reward consists of the values obtained by considering the speed of the agent in the environment. The various evaluations are defined in table 1.

Table 1. Reward function analysis.

Cases	Conditions
1	+1 for each frame driving > 50 KM/H -1 for each frame driving < 50 KM/H -200 for a collision and episode is over
2	+0.005(1/200) for each frame driving > 50 KM/H -0.005(-1/200) for each frame driving < 50 KM/H -1(-200/200) for a collision and episode is over
3	+1 for each frame driving > 50 KM/H -1 for each frame driving < 50 KM/H -1 for a collision and episode is over

2.7. *Traffic scenario*

In a two-lane road scenario, as shown in figure 4, the ego vehicle/agent (yellow car) is run over into the blocking vehicle (red car) at a different speed. The blocking vehicle moves at a low speed compared to the agent, which may collide if the agent maintains at the same speed on the same lane. The ego vehicle in the traffic has two options, either make a lane change or keep the current lanes. If it decides to change lanes, there may be two possible outcomes. If it changes lanes, it may rear collide with the approaching vehicle from another road lane or successfully merge in the new lane and run at its best speed. The blocking vehicle is assumed not to brake abruptly. If the agent chooses to keep the current lane, it has to perform a braking process to slow down the vehicle to match the blocking vehicle's speed. Therefore, the autonomous agent should adapt to different driving scenarios and make a quick and accurate decision.

In this study, the agent is controlled by an RL-base intelligence. It learns how to drive, including longitudinal speed control and lane-changing strategy, from interacting with the environment vehicles. There are also included some assumptions for the agent vehicle to carry out the path planning decision making of collision avoidance as follows:

- The ego vehicle is equipped with sensing capabilities and can measure the relative distances to the blocking vehicle.
- There is no communication between any vehicles.
- Each agent makes its driving decisions according to the driver's decision-making model.

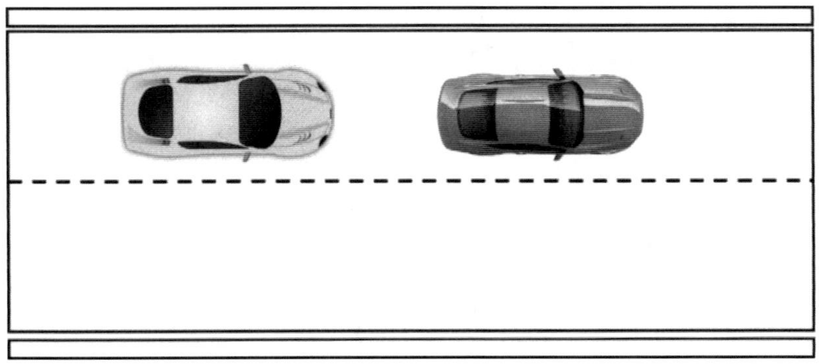

Figure 4. Two-lane traffic scenario.

3. Performance Evaluation

The performance evaluation is evaluated using three criteria which are accuracy, loss, and reward average. The accuracy is used to evaluate each episode's performance whether the autonomous driving behavior is completed the task on the road segment. Successful cases, which are indicated by the accuracy percentages, show no collision; meanwhile, the failure cases show that the agent's accidents occur. Therefore, the excellent loss value is supposed to keep the value of 0 and below to get the model's best performance. For the reward average, the network's promising performance indicates that the reward average will keep increase and gradually stable and optimal after the agent gets enough experience in the environment. For this study, the model is trained for 10000 episodes.

To carry out the study of the multiple reward function performance first, we set the penalty evaluation value to -200, which is 200 more times than the reward evaluation. A penalty value larger than the reward value is suppressed. The penalty value is set to ≥200 times higher because it is the amount to make sure that the agent will be more efficient in the learning process when it has given the higher punishment for the mistake/collision from the action that it takes.

3.1. Analysis performance: case 1.

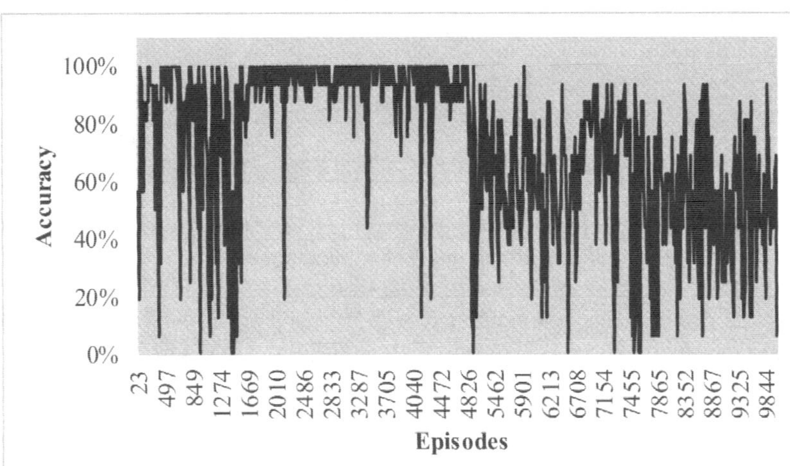

(a) Accuracy

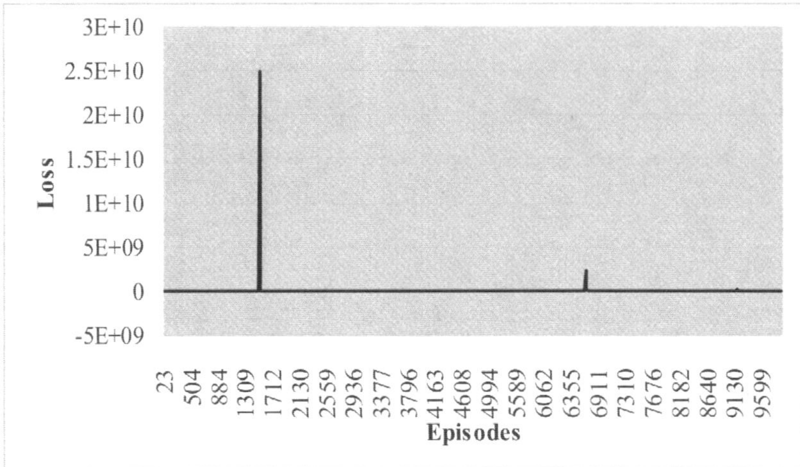

(b) Loss

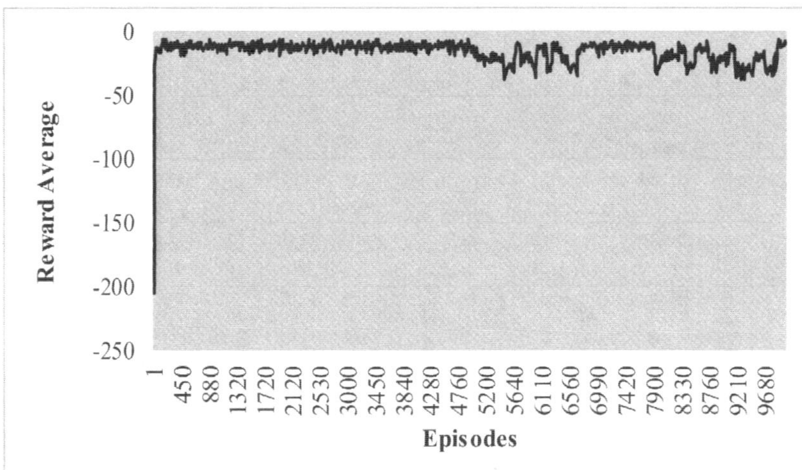

(c) Reward average

Figure 5. Analysis performance for case 1 (a) Accuracy, (b) Loss, and (c) Reward average.

Figure 5 shows the analysis performance for case 1. Figure 5(a) shows the accuracy performance of the network. The line graph of accuracy shows that it fluctuates, which means that the learning process of the model is not satisfied/ reach the aim for the designated task. While figure 5(b) shows the loss value, which is there are the specific time that the performance is exploding which is at 1121 to 1534 and 6280 to 7525 episodes. Next, as shown in Figure 5(c), the average reward obtained is fluctuated in the simulation process, which means that the model is not stable in performing the task. By setting the penalty value range higher than the reward value, we found that this was likely exploding the Q values, which also seemingly, and now obviously, due to the monumental size of the collision penalty in comparison to all else. Also, possibly the bounds being out of range. Therefore, from the analysis, it cannot perform good collision avoidance.

Then, as the Q-value is exploding next, we set the penalty value in the form of a lower integer but still in the same range value as in case 1. So, the value of the reward and penalty are divide by 200 to get the lower integer but the same range value in case 1. Thus, the penalty value is -1, and the reward value is set to +0.005.

3.2. Analysis performance: case 2. The analysis performance case 2 shows that in figure 6(a), we can see that at first, the accuracy is exploded at the beginning to learn from the process. It became stable with a high accuracy value; however, at episode 3814, it becomes fluctuate until the end of the training. Meanwhile, for the loss graph, as shown in figure 6(b), it explodes at the 7177 episode, which is supposedly it will explode at the early learning to learn the experience and gradually become constant at 0 value. Meanwhile, figure 6(c) show that the reward average improved pretty consistently. At the moment, it has slipped, but not for necessarily longer than it historically did a few times. Overall, the model did improve compared to case 1.

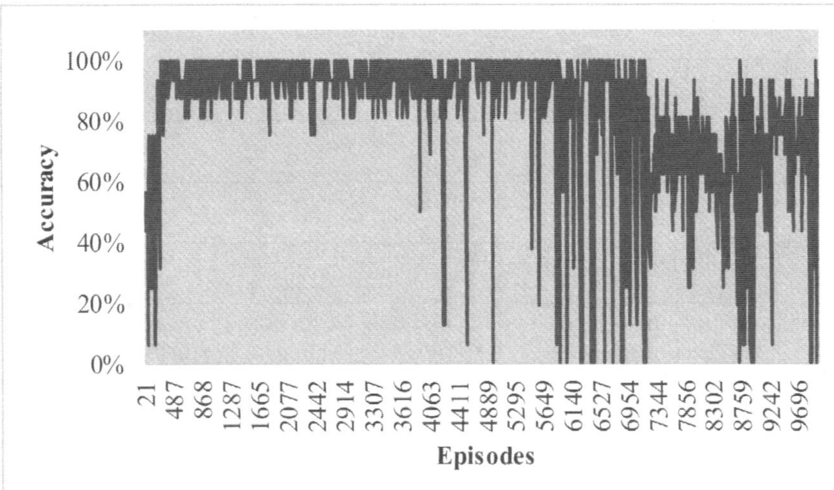

(a) Accuracy

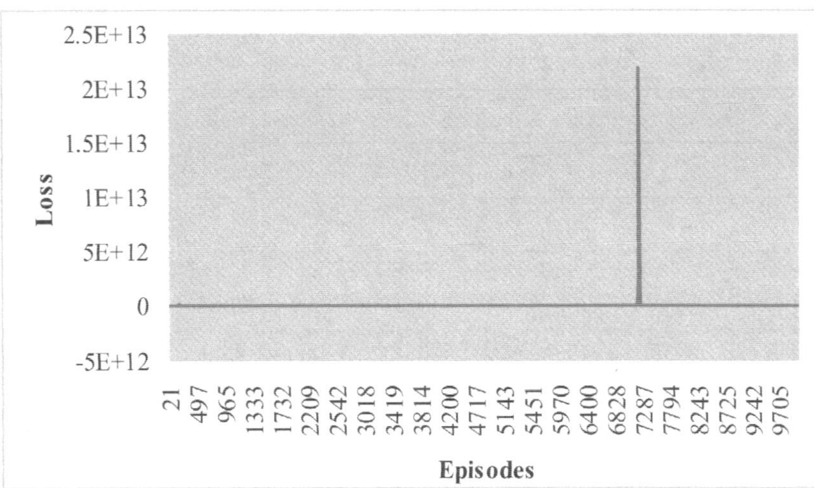

(b) Loss

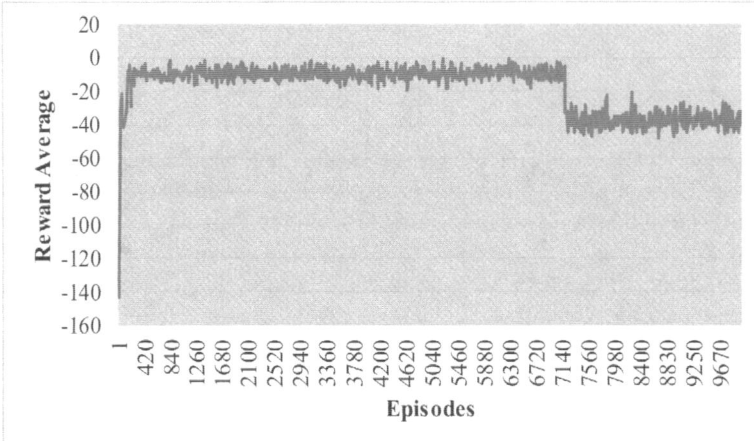

(c) Reward average

Figure 6. Analysis performance for case 2 (a) Accuracy, (b) Loss, and (c) Reward average.

3.3. Analysis performance: case 3. Next, in Case 3, the penalty value is slightly increased than the reward. Which is there is no much gap of values ranges between penalty and reward. The agent will get the penalty (-1) if it is driving speed is <50 KM/H, which is in deaccelerate state, and (-2) if during the decelerations it collides with the obstacle. Analysis performance in case 3 shows that the model gives more stable performances. Figure 7(a) shows the fluctuate graph until episodes 789 become much more constant with high accuracy (>60%). From 0-789 episodes, the agent is still learning from the experience it gets in that environment. After 789 episodes, it is learning to avoid the collision, and the efficiency of the model is increased. As the highest percentage accuracy is achieved, the highest chance of the agent to complete the task assign by the designer while avoiding all of the obstacles. In terms of loss, as indicated in figure 7(b), the loss graph, which is at the beginning is exploded, and after that, it became constant at 0. This means that during the explode period, the agent is still getting the learning process experience then, after that, it can perform well, which is a constant 0 of the error value. Next, figure 7(c) shows the promising result of the average reward value as the line keeps increasing and gradually stable and optimal.

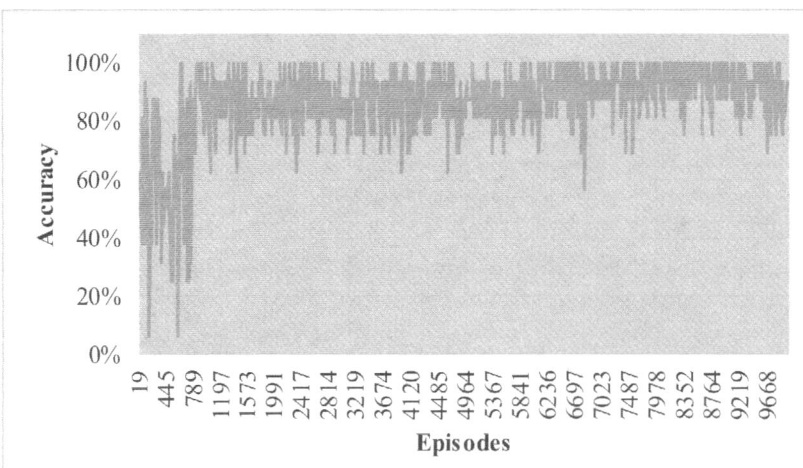

(a) Accuracy

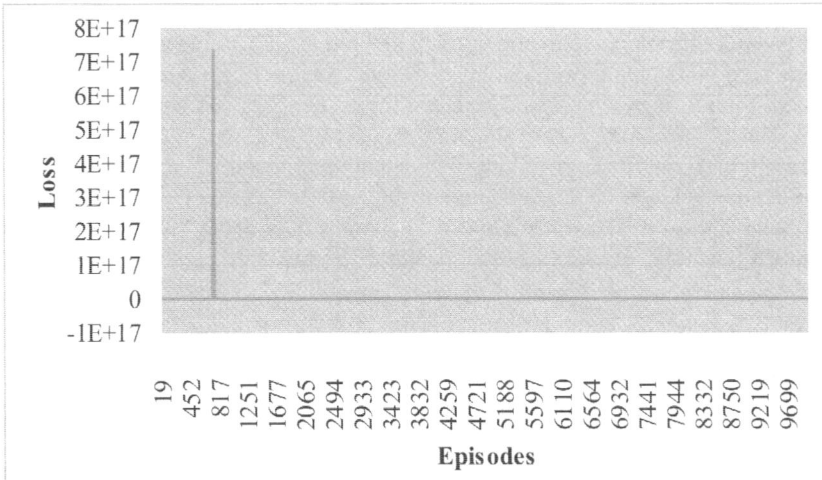

(b) Loss

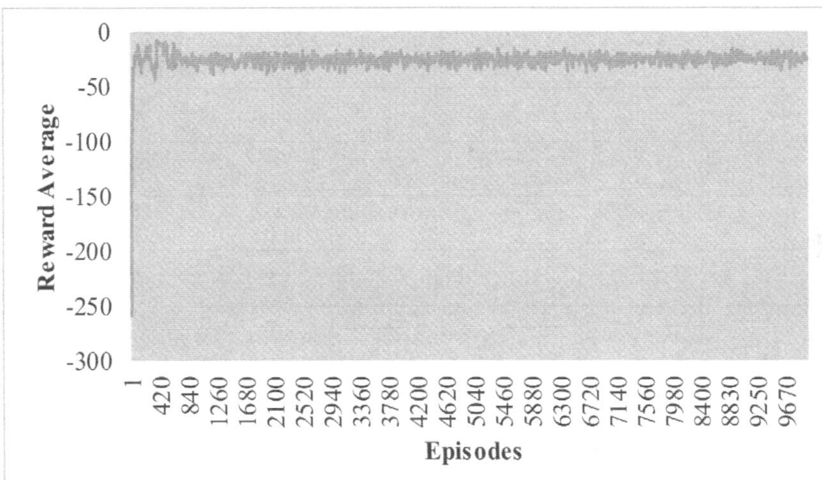

(c) Reward Average

Figure 7. Analysis performance for case 3 (a) Accuracy, (b) Loss, and (c) Reward average.

The average accuracy of the three cases is demonstrated in table 2. Setting the parameter of reward function for the DQN Algorithm, from the analysis, shows that there is a disadvantage in setting the penalty value higher than the reward value as the Q-value may explode. It is necessary to give a slightly different value between the reward and penalty as the model can perform/learn much better.

Table 2. Average accuracy.

Cases	Conditions
1	70%
2	85%
3	94%

4. Conclusion

The slow convergences and the urge to slip into the optimal local solution are improvements in RL. In this paper, the reward function performance for collision avoidance applying the DQN algorithm in RL has been studied. Significantly, the study provided the three cases for selecting the reward function model. We use the time-weighted reward function, which is speed, with the 50kmH threshold. Setting the reward function parameter from the study shows that there are advantages to set the penalty value higher than the reward value as the Q-value may explode. It is may due to the monumental size of the collision in comparison to all else. Therefore, by adopting different types of analysis using various appropriate reward functions, we were able to obtain the higher positive rewards and the optimum Q-values.

Acknowledgement

This work was supported in part by the CRG 8.1: AI Predictive Model for HVAC Operation under Grant Q.K130000.2443.04G73, and CRG 8: Development Smart Embedded-Analytics Sensors, Cloud-Based Management Systems for Heating, Ventilation, and Air Conditioning (HVAC) System under Grant Q.K130000.2443.04G64 and in part by the Malaysia Ministry of Education under Fundamental Research Grant Scheme FRGS/1/2019/TK08/UITM/03/1 File No. 600-IRMI/FRGS 5/3 (466/2019)

References

[1] Manju M S 2011 An analysis of Q-learning algorithms with strategies of reward function *Int. J. Comput. Sci. Eng.* **3** 814–20
[2] Hu Z, Wan K, Gao X and Zhai Y 2019 A dynamic adjusting reward function method for deep reinforcement learning with adjustable parameters *Math. Probl. Eng.* **1** 1–10
[3] Marashi M, Khalilian A and Shiri M E 2012 Automatic reward shaping in reinforcement learning using graph analysis *2012 2nd Int. eConf. Comput. Knowl. Eng. ICCKE 2012 (Mashhad, Iran)* 111–6
[4] Matignon L, Laurent G J and Le Fort-Piat N 2006 Reward function and initial values: Better choices for accelerated goal-directed reinforcement learning *Lect. Notes Comput. Sci. (including Subser. Lect. Notes Artif. Intell. Lect. Notes Bioinformatics)* 840–9
[5] Shaping P, Initialization Q and Wiewiora E 2003 Potential-based shaping and Q-Value initialization are equivalent *J. Artif. Intell. Res.* **19** 205–8
[6] Ng, Andrew Y. Daishi Harada S R 2554 Policy invariance under reward transformations: Theory and application to reward shaping *Int. Counc. Mach. Lubr.* **3**
[7] Hailu G and Sommer G 1999 On amount and quality of bias in reinforcement learning *Proc. IEEE Int. Conf. Syst. Man Cybern.* **2** 1491–5
[8] Wei Y, Zhao M, Zhang F and Hu Y 2004 Research of a heuristic reward function for reinforcement learning algorithms *Proc. World Congr. Intell. Control Autom.* **3** 2676–80
[9] Yan Q, Liu Q and Hu D 2010 A hierarchical reinforcement learning algorithm based on heuristic reward function *IEEE Int. Conf. Adv. Comput. Control*
[10] Mataric M J 1992 Reward functions for accelerated learning *Int. Conf. Mach. Learn.* 181–9
[11] Behnke S and Bennewitz M 2005 Learning to play soccer using imitative reinforcement *ICRA 2005 Work. "Social Asp. Robot Program. through Demonstr.*
[12] Mesot B, Sanchez E, Pena C A and Perez-Uribe A 2002 SOS++: Finding smart behaviors using learning and evolution *8th Int. Conf. Simul. Synth. Living Syst. Alife (*Sydney, Australia) 264–73
[13] Wu Y H, Sun F Y, Chang Y Y and Lin S De 2018 ANS: Adaptive network scaling for deep rectifier reinforcement learning models *arXiv:1809.02112v3 [cs.LG]*
[14] Xiao L, Chen Y, Lin W S and Liu K J R 2012 Indirect reciprocity security game for large-scale wireless networks *IEEE Trans. Inf. Forensics Secur.* **7** 1368–80

IOP Conf. Series: Materials Science and Engineering 1176 (2021) 012034 doi:10.1088/1757-899X/1176/1/012034

Semiconductor defect classification by using cascaded convolutional neural network

P N A Tajudin[1*] and M I Shapiai[1]
[1]Center of Artificial Intelligent and Robotics, Malaysia-Japan International Institute of Technology, Universiti Teknologi Malaysia, 54100 Kuala Lumpur, Malaysia.

*Corresponding author: putrinorazila21@gmail.com

Abstract. The semiconductor industry is one of the rapidly growing industries globally. To be on par in both quality and quantity, most of the semiconductor industry has an inspection procedure in line with every production process. The dataset defect classes' main features can be group into two main groups base on their pattern similarity. Previously, most of the studies did not consider the pattern similarity and fed all the input into one classifier, increasing the risk of misclassification into more classes. One classifier does not feed all. Therefore, this study proposed a cascaded convolutional neural network classifier framework for semiconductor defects to enhance the defect pattern analysis. The cascaded convolutional neural network consists of multiple models trained based on various pre-process inputs. The data pre-processing includes a median filter method by using the OpenCV library. The input has to be pre-processed to distinguish the main feature, and the model can classify it based on those features. Comparing the proposed method with the single classifier show a slight increase in accuracy from 86.7% to 87.7%. In the future, further investigation will be carried out, especially on certain defect wafer compromise of multiple defect classes and how does it reduce the capability of the cascaded convolutional neural network model.

1. Introduction

The semiconductor industry is one of the booming industries that overgrow from over $1 billion in 1964 to $335.8 billion in 2014. In 2017, semiconductors worldwide sales exceeded the US $400 billion for the first time[1]. Higher demand impacts the manufacturing process, and to maintain yield, the manufacturer opted for the inspection system throughout the production floor process. Silicon wafer is a primary component of the majority of semiconductor devices. Silicon wafer undergoes numerous fabrication processes that commonly fall into four general processes: deposition, removal, patterning, and modification of electrical properties[2]. Throughout this process, the wafer will undergo multiple inspection processes depending on the companies standard operating procedure. A 100 percent yield is impossible because it will go through various methods, whether it is a fully automated process or even a process involving human workers. The inspection process maintains yield by informing the engineer's abnormalities for rapid maintenance response; it also maintains product quality to the consumer.

Up until recently, wafer defects were analyzed by trained process engineers using high-resolution microscopes by monitoring process criteria of the wafer map defect, such as location, size, and color. These abnormalities are categorized into Type A, Type B, and Type C[3]. Type A defect is considered as evenly random with stable mean density. This type of defect has generated randomly; no particular clustering pattern is perceptible, as shown in Figure 1(a). Type B defect is considered a systematic and repeatable defect, as shown in Figure 1(b). This type of defect has a perceptible clustering pattern and usually caused by the production process. Type C defect is a combination of Type A and Type B defect shown in figure 1(c).

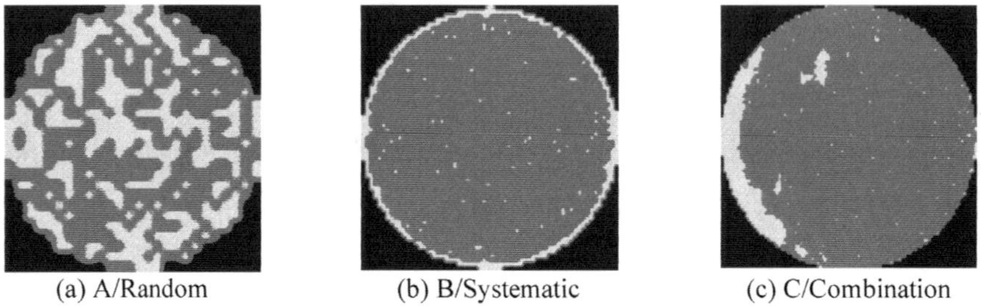

(a) A/Random (b) B/Systematic (c) C/Combination

Figure 1. Wafer Defect Type (a) A/Random, (b) B/Systematic, and (c) C/Combination.

A public dataset of the semiconductor wafer is hard to come by, primarily to obtain the defect type that might tarnish the quality of the factory processes. However, there is a silicon wafer dataset provide by Multimedia Information Retrieval (MIR) Laboratory, which provides the WM-811K silicon wafer dataset for public use. Collected data came from real semiconductor industries. This particular dataset contains nine defect classes that the process engineer has uniquely identified. The defect class name is *Center, Donut, Local, Edge-Local, Edge-Ring, Random, Scratch, Near-Full,* and *None*[4].

On this particular set of the dataset, there have been various studies that employ the usage of artificial intelligence, such as Support Vector Machine(SVM)[4], Artificial Neural Network (ANN)[5], Generative Adversarial Network (GAN)[6,7], and Convolutional Neural Network (CNN). SVM and ANN are a part of machine learning, and these kinds of methods required a lot of feature extraction that required manual intervention. On the other hand, CNN is a part of deep learning, and it does not require much manual intervention[8]. GAN is one of the methods introduced in deep learning, but it is costly due to the complex computational operation. Maksim *et al.* [9] proposed using a single Resnet50 for defect classification using only six classes out of nine defect classes. Batool *et al.*[10] propose a similar method but by using more basic CNN architecture. However, those research disregard some of the classes, and even with discarding some classes, the average accuracy of Maksim *et al.* paper is less than our proposed method.

This study proposed a cascaded convolutional neural network classifier framework for semiconductor defect problems to enhance the defect pattern analysis. The defect classes' main features can be categorized into two major groups based on their pattern similarity. The cascaded convolutional neural network consists of multiple classifiers trained based on different pre-process inputs that enhance the defect's primary feature. These studies also include all nine defect classes.

2. Research Methodology

This segment will briefly present the dataset, demonstrate the augmentation technique, and depict the CNN model's essential structure.

2.1. Dataset

WM-811K is a public dataset provided by MIR Lab. It contains 811 457 wafer map fabrication data collected in the actual fabrication process[4]. There are approximately 20% of the labeled dataset, and most of it comes from the *"None"* pattern class. The dataset contains nine defect types. The utilization of the label *"None"* class is just a portion of the entire label *"None"* class.

On the other hand, other classes are utilized fully. The distribution for training: validation: testing is 6:1.5:2.5. Figure 2 shows the distribution of data.

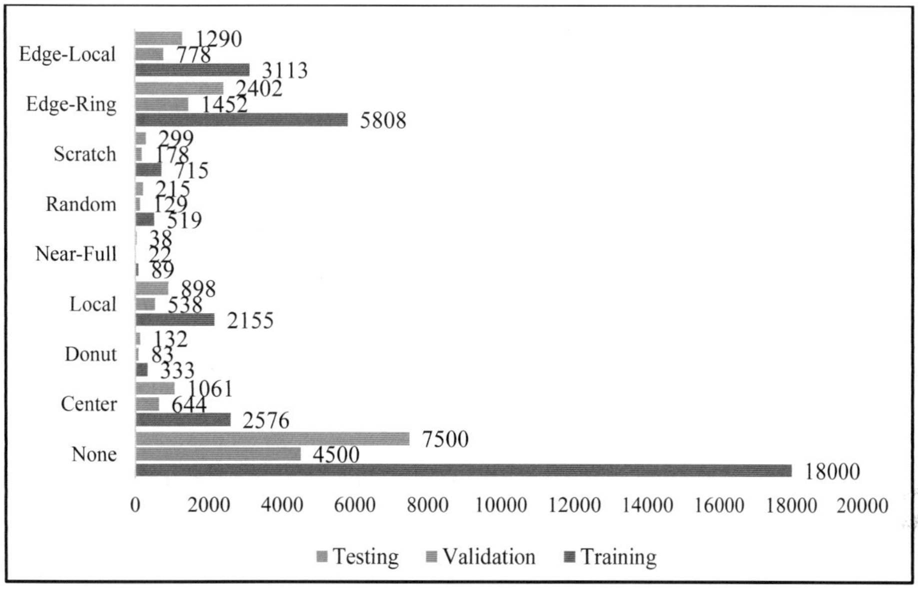

Figure 2. Dataset Distribution.

2.2. Data Pre-processing

The WM-811K wafer data contain different sizes of the die from 6×21 to 300×202. Die size of 24x24 and 224x224 selected for the classifier training and testing. The dataset was then filtered out to remove out the noise without removing the defect's primary characteristic. This experiment utilizes a median filter that supports by OpenCV. The median filter replaced the current location's pixel value with the median of all neighboring pixels[11]. The kernel for the median filter varies between the classifier. Classifier two received median filter input with kernel size five, while classifier three received median filter input with kernel size 3. The random rotation transformation method was applied to the input before feeding the three classifiers' input for training. Random rotation of a maximum of 180 degrees was applied using an "on the fly" augmentation technique.

2.3. Proposed Cascaded Classifier

The proposed classifier contains three separate classifiers, as shown in figure 3. Classifier A acts as a first filter that differentiates the input from classifier B or classifier C. The result of classifier one will send a file name either to classifier two or classifier three. Classifier B and classifier C will then retrieve input from their respective median filter kernel size folder. Input class that goes through classifier B is ***Edge-Ring, Edge-Local, Local***, and ***None***.

In contrast, the other four classes, ***Donut, Center, Random***, and ***Scratch***, will go through classifier C. Classifier B are grouped due to the similarities of their defect pattern. Those defects pattern mainly have higher noise to remaining classes. Thus, it needs a higher value kernel size to smooth out the input images.

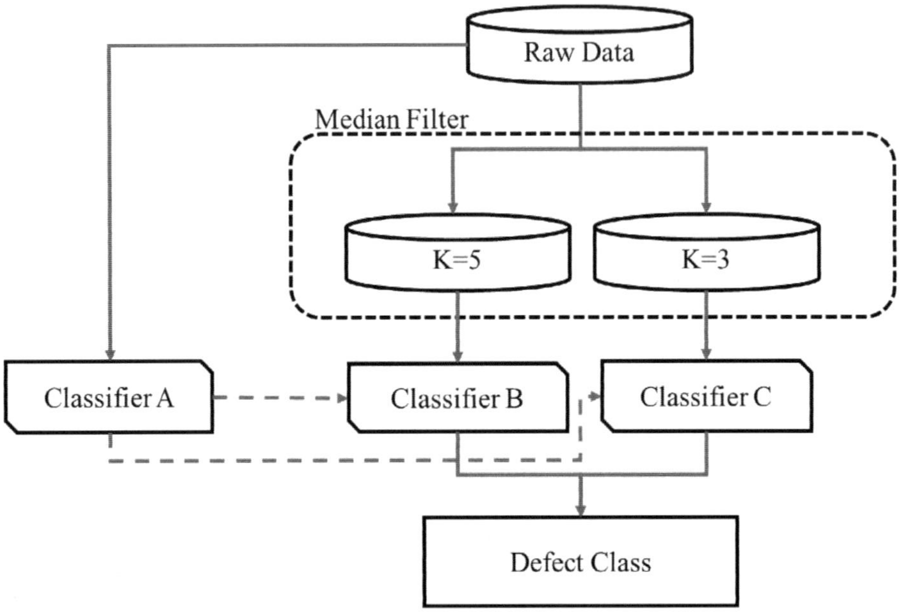

Figure 3. Flowchart of Cascaded CNN.

2.4 Convolutional Neural Network

Classifier A employs VGG-16 architecture.VGG-16 is an architecture contain thirteen convolutional layers and three fully connected layers with a Rectified Linear Unit (ReLU) activation. Along with these, there are five max-pooling layers implement in between the convolutional layers. The kernel for the convolutional layer is three, and the input size is 224x224[12].

The classifier B and C are a custom CNN, consisting of a combination of 2 convolutional layers, two max-pooling layers, and one dropout layer followed by a two-dense layer and the output layer in the end. The kernel size for the convolutional layer is five, while for the max-pooling layer and the kernel size is two. Classifiers B and C received input images of size 24x24.

The optimizer for classifier A and B is SGD, while for classifier C the optimizer is Adam. The different optimizer types because during the hyperparameter optimization, the respective optimizer works well with the respective classifier. Table 1 shows the network layer of the classifier and tabulated hyperparameter are in table 2. Visualization of the network configuration is in figure 4.

Table 1. The network layer of classifier B and classifier C.

Layer	Type	Feature Map	Output Size	Activation
IN	Input	3	24x24	-
C1	Convolutional	16	20x20	ReLU
P1	Max-Pooling	16	10x10	-
C2	Convolutional	32	6x6	ReLU
P2	Max-Pooling	32	3x3	-
L1	Fully-Connected	256	-	ReLU
L2	Fully-Connected	84	-	ReLU
OUT	Output	4@5	-	LogSoftmax

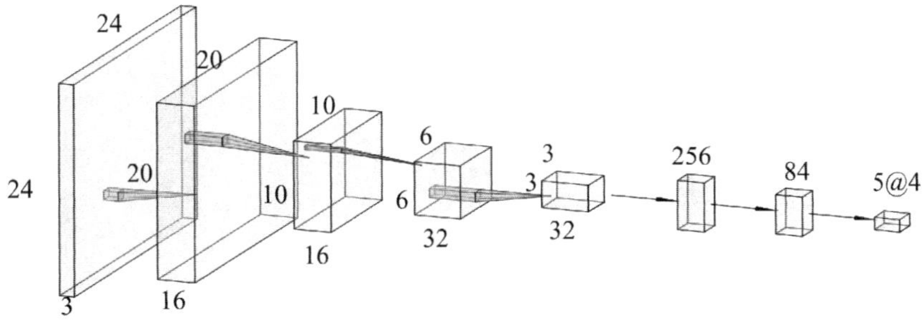

Figure 4. Convolutional neural network for classifier B and classifier C.

Table 2. Training hyperparameter for all three networks.

Parameter	Classifier A	Classifier B	Classifier C
Optimizer	SGD	SGD	Adam
Learning rate	0.001	0.02	0.0001
Weight decay	$6e^{-5}$	$5e^{-5}$	$5e^{-5}$
Transfer learning	Yes	No	No
Epoch	100	100	100

3. Result and Discussion

This section will briefly present the performance, result, and discussion of the proposed cascaded CNN.

3.1. *Performance*

The batch size for the training is 32, and the number of epochs is set to 100 epochs for each classifier. During the training phase, all classifiers' input is randomly rotated with the "on-the-fly" augmentation method support by the Pytorch library. Training loss for classifier A, classifier B, and classifier C is approximately 0.025,0.1, and 0.1, respectively. Validation loss for all classifiers is approaching zero. Compared with training loss, the lower validation loss is due to the augmentation technique used during the training phase, leading to validation data being much easier to predict. Besides that, all classifiers contain a dropout layer, which only activates during the training phase. The summarization of the losses shown in figure 5.

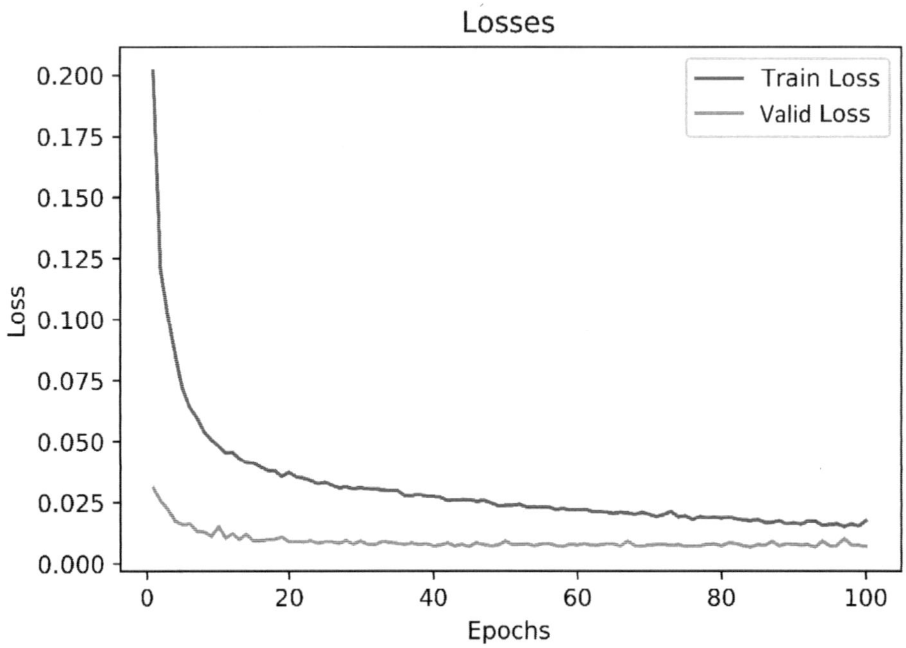

(a) Classifier A

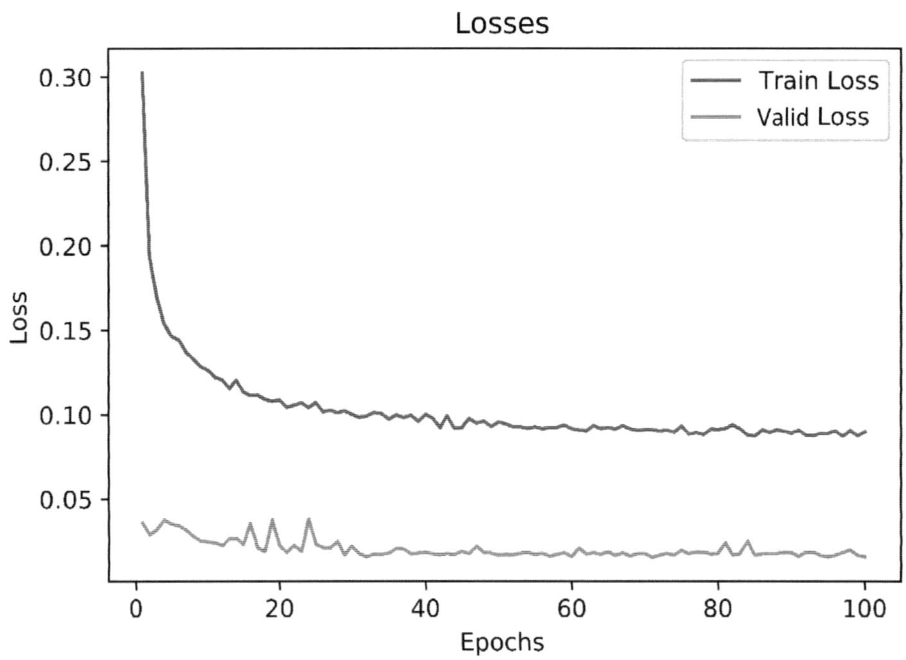

(b) Classifier B

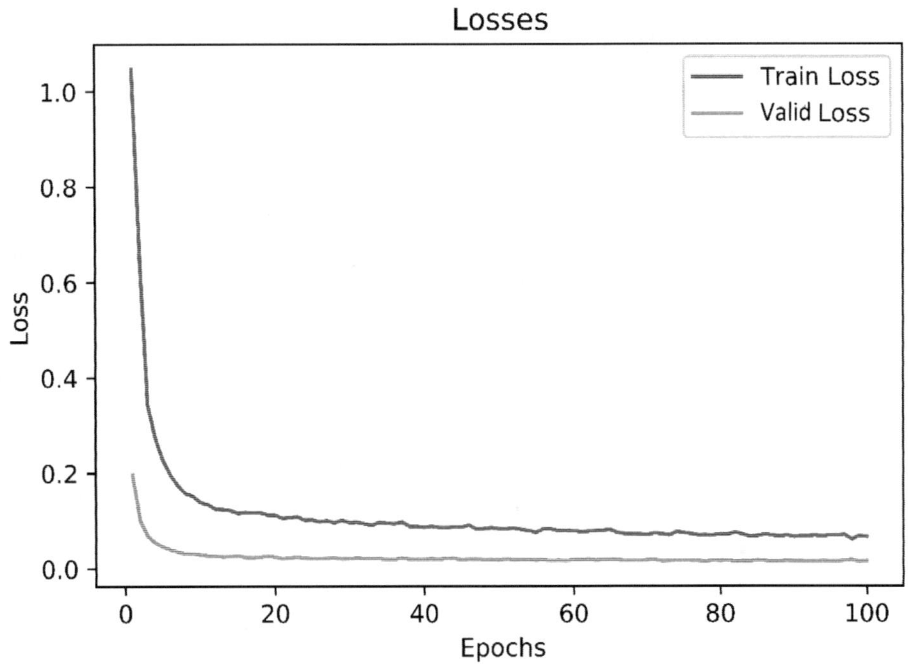

(c) Classifier C

Figure 5. Losses (a) Classifier A, (b) Classifier B and, (c) Classifier C.

The confusion matrix in figure 6 demonstrates the model's prediction accuracy on the test data in terms of the number of samples and the percentage. The model test a total of 13835 images, which comprise 25% of the label class. It can be seen from the confusion matrix that *None, Center, Donut, and Random* class perform well with accuracy above 90%. The misclassification between Donut and Random class is not high even though those two classes are among the lowest percentage of data in terms of the total data use for training. The reason is that those two classes have a very distinguishable pattern defect. *Edge-Local and Edge-Ring* show the worst performance, which got confused with classes within the B classifier grouping. The reason is due to the similarity between the defect pattern in grouping B itself.

The cascaded classifier shows that it can correctly handle the *None* class base on the performance matric in table 3. This is important to avoid false alarms on the production floor. The Donut, Local, and Random class performance matric show that the class is well detected, but the classifier does include another class point.

Table 3. Cascaded classifier performance matric.

Class	Precision	Recall	F1-Score
Center	0.96	0.91	0.94
Donut	0.81	0.94	0.87
Edge-Local	0.71	0.78	0.74
Edge-Ring	0.99	0.79	0.88
Local	0.80	0.89	0.84
Random	0.74	0.96	0.83
Scratch	0.79	0.82	0.81
Near-Full	0.70	0.82	0.76
None	0.95	0.97	0.96

True Label \ Prediction	Edge-Loc	Edge-Ring	Local	None	Center	Donut	Random	Scratch	Near-Full
Edge-Loc	1,006 (0.78)	24 (0.02)	68 (0.05)	127 (0.10)	9 (0.01)	1 (0.00)	25 (0.02)	30 (0.02)	0 (0.00)
Edge-Ring	269 (0.11)	1,907 (0.79)	1 (0.00)	209 (0.09)	3 (0.00)	0 (0.00)	3 (0.00)	10 (0.00)	0 (0.00)
Local	59 (0.07)	0 (0.00)	801 (0.89)	12 (0.01)	0 (0.00)	20 (0.02)	4 (0.00)	0 (0.00)	2 (0.00)
None	78 (0.01)	3 (0.00)	95 (0.01)	7,282 (0.97)	11 (0.00)	0 (0.00)	14 (0.00)	17 (0.00)	0 (0.00)
Center	0 (0.00)	0 (0.00)	40 (0.04)	18 (0.02)	964 (0.91)	8 (0.01)	20 (0.02)	0 (0.00)	11 (0.01)
Donut	0 (0.00)	0 (0.00)	1 (0.01)	0 (0.00)	0 (0.00)	130 (0.94)	0 (0.00)	1 (0.00)	0 (0.00)
Random	1 (0.00)	0 (0.00)	1 (0.00)	4 (0.02)	0 (0.00)	1 (0.00)	208 (0.96)	0 (0.00)	0 (0.00)
Scratch	3 (0.01)	1 (0.00)	2 (0.01)	30 (0.10)	14 (0.05)	1 (0.00)	1 (0.00)	247 (0.83)	0 (0.00)
Near-Full	0 (0.00)	0 (0.00)	0 (0.00)	0 (0.00)	0 (0.00)	0 (0.00)	7 (0.18)	0 (0.00)	31 (0.82)

Figure 6. Confusion matrix defect classes of the proposed method.

3.2. Comparison Between Cascaded CNN And Single VGG-16

The comparison between using a single VGG-16 and the proposed classifier achieves a slight increase in the average accuracy. Overall test accuracy of 87.7% against 86.7% of single VGG-16 has been observed. Both *Edge-Local* and *Edge-Ring* have the most badly predicted class by VGG-16 show a slight increase by the proposed classifier. This justifies the proposed method to categories the group into two main groups.

The primary characteristic of group B classes can hardly be distinguished with naked eyes such as *Edge-Local, None, and Local.* The proposed method tries to eliminate the risk of misclassification between group B and group C while emphasizing classifier B to learn to distinguish the features of four classes in group B more thrivingly.

In the case of classifier C, the worst performed class is *Near-Full* compared with single VGG-16. Near-Full class misclassification with *Random* class caused by the median filter process that erases the Near-Full class's primary feature. Overall, the classifier shows an increase in accuracy for six classes while underperforming in three other classes. A comparison of the testing accuracy results is in table 4.

Table 4. Comparison of testing accuracy.

Class	Method	
	Cascaded Classifier	Single-VGG-16
Center	**90.9**	81.4
Donut	**94.2**	94.02
Edge-Local	**77.9**	77.6
Edge-Ring	**79.4**	77.8
Local	89.2	**89.8**
Random	**96.3**	89.8
Scratch	**82.3**	81.3
Near-Full	81.6	**89.5**
None	97.2	**98.6**
Average	**87.7**	86.7

The simulation of this experiment utilizes Nvidia GeForce GTX 1080 GPU. The time taken to evaluate one image is 9.4ms for a cascaded classifier and 6.8ms for a single VGG-19. The cascaded classifier needs a slightly higher time due to three classifiers than the VGG-16 that only uses one classifier.

4. Conclusion

This study proposed a cascaded convolutional neural network for defect classification that thoroughly analyses the defect's characteristics. The comparison between using a single VGG-16 and the proposed classifier achieves a slight increase in the average accuracy. The proposed network show an average test accuracy of 87.7% against 87.7% of single VGG-16. Even though the accuracy did not significantly increase, it shows that with more proper characteristic analysis and much more enhance CNN for classifier B and C, the proposed method can improve significantly. The cascaded classifier shows the more practical use of CNN in solving real problems because raw data is usually corrupt with noises, and these noises vary between classes. The usage of multiple classifiers can distinguish the characteristics more thoroughly. There is a problem with the imbalance of data training with the "on the fly" augmentation methods. In the future, further investigation will be carried out, especially on certain defect wafer that compromises of multiple defect classes, and how does it reduce the capability of the cascaded convolutional neural network model.

Acknowledgment

The authors would like to thank Universiti Teknologi Malaysia for Industrial Defect Detection for Wafer Semiconductor using Artificial Intelligence Technology grant (Q.K130000.3043.01M84) and Imbalanced Strategy for Wafer Defect Classification Using Fully Convolutional Neural Network grant (Q.K130000.2543.21H17).

References

[1] Semiconductor Industry Association Annual Semiconductor Sales Increase 21.6 Percent, Top $400 Billion for First Time - Semiconductor Industry Association.

[2] Bera B 2019 Silicon Wafer Cleaning: A Fundamental and Critical Step in Semiconductor Fabrication Process *Int. J. Appl. Nanotechnol.* 8–13.

[3] Chien J C, Wu M T and Lee J Der 2020 Inspection and classification of semiconductor wafer surface defects using CNN deep learning networks *Appl. Sci.* **10** 1–13.

[4] Wu M J, Jang J S R and Chen J L 2015 Wafer map failure pattern recognition and similarity ranking for large-scale data sets *IEEE Trans. Semicond. Manuf.* **28** 1–12.

[5] Saqlain M, Jargalsaikhan B and Lee J Y 2019 A voting ensemble classifier for wafer map defect patterns identification in semiconductor manufacturing *IEEE Trans. Semicond. Manuf.* **32** 171–82.

[6] Ring E Using GAN to Improve CNN Performance of Wafer Map Defect Type Classification.

[7] Wang J, Yang Z, Zhang J, Zhang Q and Chien W T K 2019 AdaBalGAN: An Improved Generative Adversarial Network with Imbalanced Learning for Wafer Defective Pattern Recognition *IEEE Trans. Semicond. Manuf.* **32** 310–9.

[8] Schmidhuber J 2015 Deep Learning in neural networks: An overview *Neural Networks* **61** 85–117.

[9] Maksim K, Kirill B, Eduard Z, Nikita G, Aleksandr B, Arina L, Vladislav S, Daniil M and Nikolay K 2019 Classification of wafer maps defect based on deep learning methods with small amount of data *2019 Int. Conf. Eng. Telecommun. EnT 2019* 1–5.

[10] Batool U, Shapiai M I, Fauzi H and Fong J X 2020 Convolutional Neural Network for Imbalanced Data Classification of Silicon Wafer Defects *2020 16th IEEE Int. Colloq. Signal Process. its Appl. CSPA 2020 (Langkawi, Malaysia)* 230–5.

[11] Culjak I, Abram D, Pribanic T, Dzapo H and Cifrek M 2012 A brief introduction to OpenCV *MIPRO 2012 - 35th Int. Conv. Inf. Commun. Technol. Electron. Microelectron. (Crotia)* 1725–30.

[12] Simonyan K and Zisserman A 2015 Very deep convolutional networks for large-scale image recognition *3rd Int. Conf. Learn. Represent. ICLR 2015 (San Diego, USA)* 1–14.

IOP Conf. Series: Materials Science and Engineering 1176 (2021) 012035 doi:10.1088/1757-899X/1176/1/012035

Predicting vehicle parking space availability using multilayer perceptron neural network

M H Ismail[1], T R Razak[1], R A J M Gining[1], S S M Fauzi[1] and A Abdul-Aziz[1]

[1]Faculty of Computer and Mathematical Sciences, Universiti Teknologi MARA (UiTM), 02600 Arau, Perlis, Malaysia.

*Corresponding author: mohammadhafiz@uitm.edu.my

Abstract. In this study, we have investigated potential use of Multilayer Perceptron (MLP) to predict parking space availability for use within Field Programmable Gate Array (FPGA) accelerated embedded devices. While previous studies have explored the use of MLP for classification problem in FPGA, very little studies concentrated on the potential use of MLP in regression problem, especially in parking space forecasting. Therefore we formulated five Multi-Layer Perceptron (MLP) models with varying hidden units to perform single-step prediction to forecast parking space availability within the next 15 minutes based on the previous one-hour parking occupancy. The proposed models were trained on the historical data of Kuala Lumpur Convention Center dataset and evaluated against baseline ARIMA models. The results have shown that our proposed MLP model performed relatively well against baseline model with the root mean square error between (RMSE) 78.25 to 78.41 and mean absolute error (MAE) between 37.02 to 39.17.

1. Introduction

For city dwellers, finding a parking space is critical especially in a densely populated area such as the city centre. Therefore, modern commercial buildings have installed vehicle parking sensors as part of their Parking Guidance and Information System (PGIS) to detect parking space availability and utilisation. The availability of parking spaces is communicated to drivers through electronic variable message signboard (VMS) which is controlled with a series of embedded computing devices. The information provided by PGIS assists drivers in planning their route by reducing the time spent on searching and waiting for the parking space to become available. This is supported by [1], which states drivers who have prior knowledge on space availability are more successful in finding a parking spot for their vehicle compared to those who do not.

Consequently, drivers waste an average between 3.5 to 14 minutes just to find a vacant parking spot [2], which also inadvertently waste fuel and causes unnecessary pollution [3]. Reducing search time leads to less vehicle movements in the parking area, which subsequently reduces traffic congestion and noise pollution, especially near parking entrances [4][5].

PGIS detects availability of parking space through a series of embedded computer sensors placed on the parking lot. The sensors detect parking lot vacancy and send the information to a centralised computer. The parking information is stored in the database and then displayed on the variable message signboard (VMS) via computer networks at various location around the vehicle parking complex. Some PGIS posts parking information on the website or display it on mobile application to further assist drivers in finding available parking spots.

However, through our findings, most PGIS only report current parking spot availability in real-time or periodically. While this information is helpful for drivers who are already in the vicinity of parking complex, it is less useful driver who are still planning for a trip to the city centre. Thus, the parking space availability forecast is more useful as it could help drivers in the latter category to estimate the space availability in advance before they reached the parking complex. Moreover, parking space forecast can help drivers avoid congestion by assisting them in planning their route.

The advent of edge computing enables data to be processed directly on the edge with minimal dependency on remote computers. This is crucial as on-site forecasting can eliminate network delay problems associated with remote processing. Additionally, an implementation of PGIS which can perform on-site forecasting is preferable because it can operate independently in event of network disruption.

Recent development in electronics have enable the use of Field Programmable Gate Array (FPGA) in embedded computers. The introduction of FPGA allows embedded devices to perform complex mathematical tasks such as classifications or regression on raw data without depending on remote processing from cloud computers [6]. According to [7][8], Artificial Neural Network (ANN) is one the most used machine learning algorithms for classification tasks in FPGA. This is due to the parallel processing capabilities of FPGA which allows ANN models to be executed more efficiently [8]. In addition, Multilayer Perceptron Network (MLP) is a type of feedforward ANN which is commonly implemented on FPGA. Due to its flexibility, MLP is also used in classification problems as well as regression tasks.

While previous studies have explored the potential of implementing Multilayer Perceptron Network (MLP) in FPGA for various near real-time classification problems [9][10][11][12][13], very little research has been done in the area of regression, especially in the area related to parking space availability forecasting. Moreover, other studies on parking space forecasting implementations do not specifically target FPGA-based platforms. Thus, this study intends to focus on parking space forecasting problems which can be implemented on FPGA.

Therefore, we propose short-term parking space prediction model with MLP. Our goal is to determine the forecasting performance of MLP before implementing the network on embedded computer powered by FPGA. The model should be able to perform single-step prediction based on the previously seen one-hour parking occupancy records.

2. Literature Survey

Current research in the field has delved into predictive methods for short-term parking space availability forecasting intended for drivers. Yang et al. [15] suggested the use of neural network on historical data time series for short term parking space prediction as one of the methods to improve the efficiency of PGIS. The authors suggested a three-layer neural network model which consist of of traffic flow, weather, events, and parking garage availability as multivariate input. The authors noted that this approach depends on a large historical data sample size which may be difficult to obtain and the difficulty in determining the suitable number of neurons to get accurate prediction.

Later work by Yanjie, Dounan, Blythe, Weihong and Wei [16] uses Wavelet neural network to predict short-term parking space availability. The authors trained the Wavelet Neural Network on time-series data with 1-minute interval and compared its performance to Lyapunov exponents method. The authors found out that Wavelet Neural Networks is more precise in one-step forecasting (1-minute look ahead) in regular days.

Yanxu et al. [3] compared Regression Tree, Support Vector Regression, and Neural Network Model performance in predicting parking occupancy and found the Regression Tree performed better in prediction performance than Neural Network and Support Vector Regression model. The authors also found that the computational complexity of the Neural Network model is between the Regression Tree and the Vector Regression Support Model.

The use of nonlinear auto-regressive Neural Network (NARNN) model has been proposed by [17] for predicting parking space utilization within a large time window of more than half of the day. The implementation uses a recurrent feedback connection within the network unit. However, the results are inconclusive as the dataset used in the study is limited.

Fengquan et al. [18] proposed ARIMA model to predict unoccupied parking space in real-time. This study was performed on a dataset collected from a mall parking lot. The dataset contains 2,880 datapoints representing unoccupied parking space in 15-minutes interval. The study found that the ARIMA (2,1,3) performs slightly better than the Neural Network model with a Root Mean Square Error (RMSE) of 4.47 compared to 5.45.

The use of Long Short-Term Memory (LSTM) networks and Recurrent Neural Network in parking space availability prediction has been explored by [19]. The study uses parking occupancy dataset consists of twelve million parking events recorded by various sensors in a year. The parking events data is converted into a time-series. The model can predict parking space occupancy within 1 minute, 5 minutes, 15 minutes, and 30 minutes ahead of time.

A Deep LSTM was proposed by [20] as part of smart car parking availability architecture. The Deep LSTM consist of multiple stacked LSTM, in which the output of a layer is passed to the next layer. The dataset consists of data collected from thirty parking locations within a period of 8 months. The model was trained on a time series with 1-hour interval, and 1-hour look ahead prediction time.

Provoost et al. [21] compared the ANN, Convolutional Neural Network (CNN) and Random Forest (RF) performance in predicting the parking space occupancy from Web of Things application. The optimum ANN number of hidden layers and neurons was determined using grid search technique. The Convolutional Neural Network uses the same number of hidden layers and neurons as the ANN with a 4x4 kernel used as look-back window. Grid search is also used to determine the maximum features and optimum depth of the RF model. The study was performed on a dataset consisting of parking occupancy information, traffic flow, date, time of the day and weather conditions. However, the authors found out that time of the day and weather conditions has smaller impact on prediction accuracy.

Barreto et al. [22] has found that Field Programmable Gate Array (FPGA) enables edge computing approach of processing sensor data with Convolutional Neural Network (CNN) on the embedded computer itself. Furthermore, [23] have found that FPGA is more energy efficient compared to GPU in executing matrix multiplication and workload typical for Neural Network computing within edge computing use-case. Therefore, it is accepted that Multi-Layer Perceptron neural network (MLP) is most suitable to be implemented on edge computing device that are integrated with FPGA.

From the literature survey being done, it can be concluded that prior studies have proposed different approach for predicting parking space availability. While the previous studies' focus is centred around forecasting performance of parking spots through various machine learning models, our proposed approach concentrates more on evaluating MLP forecasting model that are more suitable to be implemented on edge computing device.

3. Methodology

3.1. KLCC Parking Occupancy Dataset

This research uses Kuala Lumpur Convention Center (KLCC) parking availability data obtained from [24]. The original dataset comprises of 47,603 parking occupancy records from June 2016 to November 2017. The dataset attributes consist of sequence number, parking availability, and timestamp. The parking availability data is observed within 15-minutes interval.

For this study, we truncate the dataset to only include observation between June 2016 and July 2016, thus we only concentrated on the first 3,264 rows. Table 1 presents the sample data taken from the dataset.

Table 1. KLCC Parking Occupancy Dataset.

Sequence	Space	Timestamp
1	1642	6/1/2016 10:12
2	1609	6/1/2016 10:15
3	1458	6/1/2016 10:30
...	...	...
3197	1870	7/4/2016 17:45
3198	2013	7/4/2016 18:00

The "space" column denotes the number of available parking space at observation time. KLCC has 5,500 total parking space. However, the KLCC PGIS will output "FULL" if there are no more parking space available and "OPEN" if there is a problem with the sensor reading. For this study, amu instance of "OPEN" will be treated as missing value.

3.2 Dataset Preprocessing and Transformation

The dataset is then preprocessed where missing values (labeled as "OPEN") are removed and non-numeric string value of "FULL" is replaced by 0 to denote no parking space available. This results in 145 rows removed from the dataset, resulting in a timeseries with 3,119 data.

The Augmented Dickey-Fuller unit root test was performed to determine the timeseries dataset stationarity. The Null hypothesis (H_0) of the test is the timeseries dataset is non-stationary and will require differentiation. The Alternative hypothesis (H_1) is the timeseries is stationary. The p-value of ADF test is 0.001 which less than 0.05, and thus the H_0 is rejected and the timeseries is deemed to be stationary. The parking availability data is then normalized so that the value is within 0 to 1. Normalization is performed to make the time-series data suitable to be trained in MLP.

3.3 Multi-Layer Perceptron Model Design

A Multilayer Perceptron (MLP) is a type of Artificial Neural Network (ANN) which comprise of layers of neurons and their connections. Generally, the MLP layers are divided into input layers, hidden layers, and output layers.

The number of neurons in input and output layers within MLP model depends on the specific problem domain or tasks that we are trying to solve. Since the parking records observation are recorded in 15-minutes interval time, we have decided to design a single-step forecasting model which are able to predict the number of available parking space in the next 15-minutes based on the previously seen one hour (60 minutes) observation of parking occupancy data. Thus, the input unit is set to four (60 minutes = 15 minutes interval x 4). The model only has one output because we are constructing single-step prediction model.

Therefore, the input layer for the MLP model is set to four input and one output. The optimum number of hidden units is depending on the number of input units (which is four) and are constrained by the limited memory resources of embedded computing environment with FPGA [25].

Based on this consideration, we have decided to build five MLP models 8, 16, 24,32 and 40 hidden units respectively and evaluate their performance. Figure 1 illustrates the MLP model used in our study.

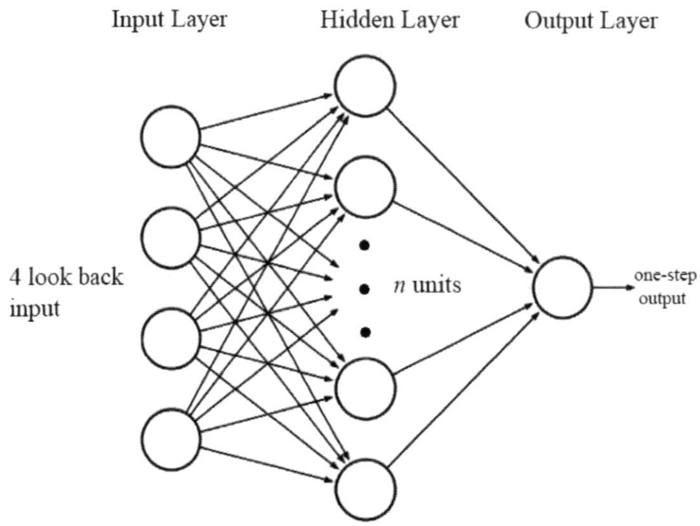

Figure 1. Multilayer Perceptron Model.

The MLP models uses the rectified linear unit (ReLU) activation function as it is faster than sigmoid activation function [26], [27]. ReLU also perform well on regression task in MLP [28].

3.4 Model Training and Evaluation

The timeseries data is split training (2,089 data points) and validation set (1030 data points) with 0.67 ratio. The timeseries data is then trained on each MLP models using mean squared error loss function and Adam optimizer. Adam optimizer is chosen because it is computationally efficient and require small memory to implement [29] which is an important factor to consider when implementing the model on resource-constrained edge computing device.

The training is facilitated using [21] and [22] as backend. Due to the stochastic nature of neural network, each training and validation run would produce a slightly different result. Thus, we use a seed value of 12 to ensure reproducibility and consistency.

We use the root square error (RMSE) and mean absolute error (MAE) as an evaluation metrics to compare the models' performance against our chosen baseline models, ARIMA (4,0,0) and ARIMA (4,0,1). ARIMA was chosen as baseline as it is a commonly used regression model in forecasting task for stationary time-series. Moreover, ARIMA conveniently includes integrated autoregressive functionality and moving average component which are useful features for creating a baseline meant for performance comparison.

4. Results and Discussions

The result of MLP training and model evaluation is outlined in table 2. The MLP model performance is compared against RMSE and MAE metrics, where the lowest value signifies the mean of predicted parking space availability is closest to actual data.

Table 2. Result and MLP Model Performance Comparison.

Model	MSE	RMSE	MAE
ARIMA (4,0,0)	7127.226	84.4229	42.9670
ARIMA (4,0,1)	7131.414	84.4477	43.7597
MLP (8 hidden units)	8046.09	89.7000	47.3162
MLP (16 hidden units) **	**6123.876**	**78.2552**	**37.0242**
MLP (24 hidden units)	6197.468	78.7240	38.0600
MLP (32 hidden units)	6149.038	78.4158	39.1756
MLP (40 hidden units)	6416.01	80.1000	42.1660

From the result, it can be seen that the MLP models performed well compared to the baseline ARIMA models. The MLP with 16 hidden units has the best forecasting performance followed by MLP (24 hidden units) and MLP (32 hidden units), the MLP performance seems to degrade with MLP (40 hidden units). We do not investigate the value beyond 40 hidden units as our aim is to implement the MLP within embedded computing environment with the least number of hidden units as possible. Increasing the number of hidden units will increase the MLP complexities and consume a lot of resources on a constrained FPGA-powered embedded computing environment.

5. Conclusion and Future Works

We have proposed and formulated several Multilayer Perceptron (MLP) models for predicting parking space availability. The proposed models are intended to be implemented in resource constrained FPGA devices. Thus, the MLP models should be simple and small without sacrificing forecasting accuracy. We have found that MLP performed comparatively well against ARIMA model with the most accurate MLP model (MLP - 16 hidden units) is able to predict the parking space availability with a mean absolute value of 37.0242 and root mean square error of 78.2552 for the KLCC parking space availability dataset. Although other models performed consistently close to each other, the MLP with 16 hidden units is preferable to be implemented in FPGA as it uses the least number of neurons and is friendlier to the resource-constrained embedded computing device.

For future work, the next natural step is to implement the best performing MLP model into FPGA-based embedded computers and tests the hardware-accelerated implementation against real-time live data stream of parking space data obtained from PGIS.

References

[1] Caicedo F, Robuste F and Lopez-Pita A 2006 Parking management and modeling of car park patron behavior in underground facilities *Transp. Res. Rec.* **1956(1)** pp. 60–67.

[2] Vlahogianni I, Kepaptsoglou K, Tsetsos V and Karlaftis M 2016 A real-time parking prediction system for smart cities *J. Intell. Transp. Syst.* **20(2)** pp. 192–204.

[3] Zheng Y, Rajasegarar S, and Leckie C 2015 Parking availability prediction for sensor-enabled car parks in smart cities in *IEEE 10th Int. Conf. on Intelligent Sensors, Sensor Networks and information Processing (Singapore)* pp. 1–6.

[4] Caicedo F, Blazquez C, and Miranda P 2012 Prediction of parking space availability in real time *Expert Syst. Appl.* **39(8)** pp. 7281–7290.

[5] Kotb A, Shen Y, and Huang Y 2017 Smart parking guidance, monitoring and reservations: a review *IEEE Intell. Transp. Syst. Mag.* **9(2)** pp. 6–16, 2017.

[6] Gaikwad N, Tiwari V, Keskar A, and Shivaprakash NC 2019 Efficient FPGA implementation of multilayer perceptron for real-time human activity classification *IEEE Access* **7** pp. 26696–26706.

[7] Shawahna A, Sait S, and El-Maleh A 2018 FPGA-based accelerators of deep learning networks for learning and classification: A review *IEEE Access* **7** pp. 7823–7859.

[8] Tisan A and Chin J 2016 An end-user platform for FPGA-based design and rapid prototyping of feedforward artificial neural networks with on-chip backpropagation learning *IEEE Trans. Ind. Informatics* **12(3)** pp. 1124–1133.
[9] Gschwend D, 2020 Zynqnet: An fpga-accelerated embedded convolutional neural network *arXiv Prepr. arXiv2005.06892.*
[10] Sarić R, Jokić D, Beganović N, Pokvić L, and Badnjević A 2020 FPGA-based real-time epileptic seizure classification using Artificial Neural Network *Biomed. Signal Process. Control* **62** p. 102-106.
[11] Zairi H, Talha M, Meddah K, and Slimane S, 2019 FPGA-based system for artificial neural network arrhythmia classification *Neural Comput. Appl* ***32*** pp. 1–16.
[12] Sahin S, Becerikli Y, and Yazici S 2006 Neural network implementation in hardware using FPGAs *Int. Conf. on Neural Information Processing (Hong Kong)* pp. 1105–1112.
[13] Qiu J, Wang J, Yao S, Guo K, Li B, Zhao E, Yu J and Tang T 2016 Going deeper with embedded FPGA platform for convolutional neural network *Proc. of the 2016 ACM/SIGDA Int. Symp on Field-Programmable Gate Arrays (Monterey California)* pp. 26–35.
[14] Libano F, Rech P, Tambara T Tonfat J and Kastensmidt F 2017 On the reliability of linear regression and pattern recognition feedforward artificial neural networks in FPGAs *IEEE Trans. Nucl. Sci.* **65(1)** pp. 288–295.
[15] Yang Y, Liu H, and Wang X 2003 The research on the key technologies for improving efficiency of parking guidance system *Proc. of the 2003 IEEE Int. Conf. on Intelligent Transportation Systems (Shanghai, China)* **2** pp. 1177–1182.
[16] Ji Y, Tang D, Blythe P, Guo W, and Wang W 2014 Short-term forecasting of available parking space using wavelet neural network model *IET Intell. Transp. Syst.* **9(2).**
[17] Zhao P, Guan H, and Zhang K 2019 A Parking Utilization Prediction Approach Based on NARNN Model *Int. Conf. of Transportation CICTP (Nanjing China)* pp. 4437–4448.
[18] Yu F, Guo J, Zhu X, and Shi G 2015 Real time prediction of unoccupied parking space using time series model *Int. Conf. on Transportation Information and Safety (Wuhan China)* pp. 370–374.
[19] Shao W, Zhang Y, Guo B, Qin K, Chan J and Salim F 2018 Parking availability prediction with long short term memory model *Int. Conf. on Green, Pervasive, and Cloud Computing (Hangzhou)* pp. 124–137.
[20] Ali G, Ali T, Irfan M, Draz U, Sohail M, Glowacz A, Sulowicz M and Martis C 2020 IoT Based Smart Parking System Using Deep Long Short Memory Network *Electronics* **9(10)**.
[21] Provoost J, Kamilaris A, Wismans L, Drift S and Keulen M 2020 Predicting Parking Occupancy via Machine Learning in the Web of Things *Internet of Things* **12.**
[22] Barreto R, Lobo J, and Menezes P 2019 Edge Computing: A neural network implementation on an IoT device *5th Experiment Int. Conf. (Madeira Island Portugal)* pp. 244–246.
[23] Biookaghazadeh S, Zhao M, and Ren F 2018 Are FPGAs suitable for edge computing? *1st USENIX Workshop on Hot Topics in Edge Computing (Boston USA).*
[24] Ismail M H 2017 KLCC Parking Occupancy 2016 *Kaggle* doi:10.34740/kaggle/dsv/8000
[25] Ortigosa E M, Cañas A, Rodriguez R, Diaz J and Mota S 2006 Towards an Optimal Implementation of MLP in FPGA *Int. Workshop on Applied Reconfigurable Computing* Delft *(Netherlands)* pp. 46–51.
[26] Rynkiewicz J 2019 Asymptotic statistics for multilayer perceptron with ReLU hidden unit *Neurocomputing* **342** pp. 16–23.
[27] Dahl G E, Sainath T N, and Hinton G E, Improving deep neural networks for LVCSR using rectified linear units and dropout *2013 IEEE Int. Conf. on Acoustics, Speech And Signal Processing (Vancouver, Canada)* pp. 8609–8613.
[28] Schmidt-Hieber J 2020 Nonparametric regression using deep neural networks with ReLU activation function *Ann. Stat.* **48 (4)** pp. 1875–1897.
[29] Kingma D P and Ba J 2014 Adam: A method for stochastic optimization *arXiv Prepr. arXiv:1412.6980 arXiv1412.6980*, 2014.
[30] Chollet F 2015 Keras *Github*

[31] Abadi M, Barham P, Chen J, Chen Z, Davis, A, Dean J, Devin M, Ghermat S, Irving G, Isard M, Kudlur M, Levenberg J, Monga R 2016 Tensorflow: A system for large-scale machine learning *12th Symp. on Operating Systems Design and Implementation (Berkeley, California)* pp. 265–283.

Technology acceptance model (TAM) for analysing cloud computing acceptance in higher education institution (HEI)

M T Amron[1]* and N H M Noh[1]

[1]Faculty of Computer and Mathematical Sciences, Universiti Teknologi MARA (UiTM), 21080 Kuala Terengganu, Terengganu, Malaysia

*Corresponding author: talmizie@uitm.edu.my

Abstract. Cloud computing is a fast-growing and up-and-coming technology. Cloud computing allows educational users a more flexible, accessible, and wide range of education resources. In this study, the technology acceptance model (TAM) was used as the foundation to explore the effects of perceived usefulness, perceived ease of use with two proposed additional variable: security and learning environment towards using cloud computing. The questionnaire was distributed using convenient sampling and a total of 664 higher education institution members consisting of lecturers, students and staffs participated in this research. The response was analysed using the Partial Least Squares Structural Equation modeling (PLS-SEM), which provides evidence of the reliability and validity of the Technology Acceptance Model (TAM) in this study. The result shows that perceived ease of use and perceived usefulness positively and significantly influence the intention to use cloud computing. It is also seen that security and learning environment also have a positive and significant impact on the intention to use cloud computing.

1. Introduction

The evolution of computing today has given consumers more options to be more frugal and use more user-friendly technological facilities. Cloud computing offers users more convenience when they no longer need to install software on their device [1][2]. Even more economical, they do not have to bear the cost of purchasing or expensive software licenses. Cloud computing is a platform where data is stored in a network-connected via the internet, and applications and software are shared across servers [3]. Various cloud computing service providers are available such as Google Cloud, Amazon Web Server, IBM Cloud, Alibaba Cloud and Microsoft Azure.

As a result of the COVID-19 pandemic, many sectors were severely affected and even faced problems and challenges in surviving operations. The education sector was also affected when schools and universities had to be closed, and online learning had to be done from home. Therefore, reliance on online resources becomes very important to continue learning sessions from home. The use of online applications such as Google Classroom, OpenLearning.com, Teachable and other learning management systems (LMS) is beneficial [4]. Almost all these applications use the cloud computing platform where information resources, applications, and others are shared. For example, Google Cloud provides Google Classroom, Google Drive, Google Form, Google Docs, Google Sheets, and many other applications that help with the teaching and learning process.

A study by [5] revealed that only 29% of the study respondents said they use cloud-based applications, while 90% of respondents used it. This situation shows that many users are unaware that they have been exposed to and using cloud-based applications. They are probably using cloud application already yet do not know it. Many studies have been conducted in higher education to evaluate cloud computing technology adoption in organisations, especially to measure organisational readiness, ICT infrastructure compatibility, service quality, cost, security, and risk. However, very few studies evaluate individuals such as students, lecturers, and administrative staff who use this technology, especially in developing countries.

This study will understand the determinants and the underlying relationship of cloud computing acceptance. A practical acceptance of cloud computing can be facilitated in the education field, particularly in the new norm of teaching and learning environment. Therefore, this study investigates the factors and attempts to develop and test the proposed modified TAM model that makes the model more relevant to cloud computing in HEI. Furthermore, this study attempts to fill a research gap by addressing the effects on security and learning environment with intentions towards the use of cloud computing.

2. Theoretical background

There have been many previous studies that have adapted various theories of acceptance of new technologies, such as the theory of planned behaviour (TPB), the theory of reason action (TRA), the technology acceptance model (TAM) and the unified theory of acceptance and use of technology (UTAUT). These theories are widely used for individual units of analysis. For this study, the author uses TAM theory to accept cloud computing technology among higher education institution students, lecturers, and staff in Malaysia.

2.1. Technology acceptance model

TAM is a theory developed by [6] that models how users accept and use technology. In general, TAM focuses on individual acceptance of technology by emphasising two main factors: perceived usefulness (PU) and perceived ease of use (PEOU). PU is defined as the degree to which a person believes that using technology would enhance their job performance. While PEOU refers to the degree to which a person believes that using a particular system would be free from effort.

Literature has demonstrated TAM's capacity to clarify the factors that affect the decision to implement new technology. As proposed by [7–9], it is suitable for analysing new technology acceptance. However, the finding of past studies also concluded that adopting new technology is limited to TAM's original factors. This study uses TAM because this theory is one of the earliest fundamental theories of acceptance that allows external variables to be tested together with the two factors of the theory [10]. Besides, TAM is also suitable for predicting factors that influence the acceptance of technology [11].

2.2. Security and Learning Environment, and acceptance of technology

Security aspects are critical in ensuring data protection and information stored in cloud computing [12]. Therefore, a standard security mechanism needs to be applied and implemented in the cloud environment platform. The security aspect is also a question and doubts for many users to start accepting cloud computing technology. Many users are unclear about the security and privacy of information in the cloud, causing them to avoid using cloud-based applications. Therefore, this study will directly assess the security factors to the influence of intentional use.

The environment greatly influences the acceptance of technology. In this study, the learning environment either helped or became a barrier to individuals' acceptance of cloud computing. According to [13], elements such as self-determination, learning content, support system and knowledge sharing in the learning environment will influence technology acceptance in education. Therefore, the learning environment is seen as a vital factor to be tested in this study, especially in online and distance learning that requires the use of technologies such as cloud computing.

External security factors and learning environment are associated with the use intention to test the relationship between them. Past studies also did the same method of the direct relationship of external factors to the use intention, and it showed mixed results on the studies [14–16]. Figure 1 shows the study model based on the theory of TAM.

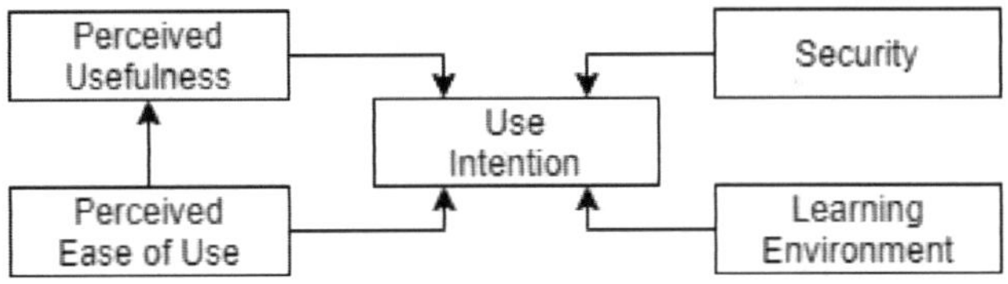

Figure 1. Research model.

2.3. Hypothesis development

The research model consists of five factors: perceived usefulness (PU), perceived ease of use (PEOU), security (SEC), learning environment (LE), and use intention (UI). PU means whether someone perceives that the technology to be useful for what they want to do. A study by [17] proved that PU influenced university students in Britain to accept new technologies. Hence, the study proposes that:
H1: PU is positively related to the intention to use cloud computing among HEIs.

The PEOU refers to the individuals feeling that technology is easy to use and is less difficult. This sense of fun accelerates one's use of new technology. A study by [14] concludes that PEOU is a key factor that allows many students to accept the use of e-Book in learning. Hence, the study proposes:
H2: PEOU is positively related to the intention to use cloud computing among HEIs.
H3: PEOU is positively related to the PU to use cloud computing among HEIs.

The SEC is a factor that is difficult to separate from new technologies, especially when it involves data and information. In cloud computing technology, the SEC is often the subject of controversy, especially in security and trust matters [18]. Many past studies have concluded that this factor greatly influences the adoption of new technologies such as cloud computing [19][20]. Hence, the study proposes:
H4: SEC is positively related to the intention to use cloud computing among HEIs.

LE refers to the current state of the learning and teaching (T&L) process that drives a person to adopt cloud computing technology. The current situation opens space for the T&L process to be more dynamic and no longer revolve around conventional methods. According to [21], the environment greatly influences individuals to adopt new technologies. A study by [22] agrees that in implementing e-Learning, the learning environment is very important in encouraging students to accept new technologies. Hence, the study proposes that:
H5: LE is positively related to the intention to use cloud computing among HEIs.

3. Research Method

An individual is the unit of analysis in this study. Students, lecturers, and staff at HEI in Malaysia are the respondents in this research. The instrument was developed based on five variables of the research model. All the items were adopted from the previous study. There were five items of Perceived Usefulness adopted from [6] and [23]; while five items of Perceived Ease of Use adopted from [23]; four items of Use Intention adopted from [24] and [23]. The seven items of security adopted from [25] and five Learning Environment items are self-developed due to no previous studies on these variables.

For data collection, a convenience sampling method was used as supported by [26] and [27] stated that convenience sampling allows researchers to get responses cost-effectively. The G*Power software has been used to determine the sample size of the respondents. The minimum number of respondents needed is 85. An online survey questionnaire was distributed to the sample of this study via email and WhatsApp group application. A total of 664 completed responses were received, which is above the scholar's recommendation [28].

4. Results and Discussion

4.1. Data

The survey questions are divided into two major sections: the respondent profile and the technology acceptance model's application. The items and measurements used in this questionnaire were mainly adapted from previous TAM related studies published by renowned scholars. Some questions were changed to match the context of the research. Respondents were asked about adopting cloud computing in HEI using a five-point Likert scale ranging from Strongly Disagree to Strongly Agree for each statement.

To investigate the research model, partial least squares (PLS) analysis was conducted using the SmartPls Software [29]. The measurement model (validity and reliability of the measures) was tested followed by an examination of the structural model (testing the hypothesised relationship) [30–32] by following the recommended two-stage analytical procedures developed by [33]. A resample of 3000 was used to find out the significance of the path coefficients and the loadings.

When data is collected through self-reported questionnaires, common method variance needs to be examined, particularly when both the predictor and criterion variables are obtained from the same individual [34]. To test response bias, Harman's single factor test was assimilated [35]. The maximum co-variance explained by a single factor was only 42.565% which is less than 50%, indicating that common method bias is not a likely issue in this study [36][37].

Data normality was assessed via multivariate skewness and kurtosis analysis as recommended by [30] and [38] at https://webpower.psychstat.org/models/kurtosis/results.php?url=dcd49fd96ae7dca77016c3031fe29dab. The Mardia's multivariate skewness ($\beta = 1.1858$, $p<0.01$) is above the cut of +1 and Mardia's multivariate kurtosis ($\beta = 47.86$, $p<0.01$) above the cut of +20 computed showed that the data collected is not multivariate normal. Thus, the use of PLS-SEM in this analysis is justified.

4.2. Measurement model

The construct validity for each latent variable in the model was first checked prior to proceeding to the hypothesis testing. Five latent constructs comprise the measurement model used in this research. Before proceeding to hypothesis testing, the construct validity for each latent variable in the model was first verified. In this analysis, the measurement model used consists of five latent constructs: Use Intention, Perceived Ease of Use, Perceived Usefulness, Learning Environment and Security. Table 1 presents the output assessment of reliability and validity in the data of the study. Composite reliability (CR) of more than 0.7 values suggest that these constructs have a reasonable degree of internal consistency [39]. The average variance extracted (AVE) values above the minimum threshold value of 0.5 indicate good convergent validity [40]. As shown in table 1, no items were dropped since all the loadings are more than 0.5. The convergent validity is then verified and now can assess the data's discriminant validity.

Table 1. Convergent validity.

Items	Outer Loadings	CR	AVE
UI1	0.7463	0.908	0.714
UI2	0.8573		
UI3	0.8797		
UI4	0.8883		
PE1	0.8657	0.938	0.753
PE2	0.8515		
PE3	0.8974		
PE4	0.8874		
PE5	0.8351		
PU1	0.7948	0.927	0.717
PU2	0.8585		
PU3	0.8623		
PU4	0.8697		
PU5	0.8457		
LE1	0.7595	0.851	0.533
LE2	0.7579		
LE3	0.6756		
LE4	0.7360		
LE5	0.7181		
SEC1	0.7828	0.934	0.669
SEC2	0.8404		
SEC3	0.8415		
SEC4	0.8367		
SEC5	0.7655		
SEC6	0.8621		
SEC7	0.7908		

The discriminant validity of all latent variables in the model was determined by using heterotrait-monotrait (HTMT) ratio of correlations criterion [41]. Since the correlation values corresponding to the respective constructs did not exceed the threshold of the HTMT 0.90 criterion as shown in table 2, it is safe to say that the discriminant validity is established in the measurement model. Consequently, structural measurement research is now sufficient to proceed.

Table 2. Discriminant validity.

	UI	PEOU	PU	LE	SEC
UI					
PEOU	0.724				
PU	0.629	0.666			
LE	0.559	0.623	0.542		
SEC	0.422	0.439	0.457	0.408	

4.3. Structural model

The bootstrapping procedure is used to test the hypotheses to produce results for each path relationship in the model, as shown in table 3. To allow the procedure to estimate the model for each sub-sample, bootstrap sub-samples with 3000 cases were computed. All path relationships are found to be significant at 99 and 95 % confidence level (PU -> UI, $\beta = 0.2182$, $p < 0.01$; PEOU -> UI, $\beta = 0.4771$, $p < 0.01$; PEOU -> PU, $\beta = 0.6658$, $p < 0.01$; SEC -> UI, $\beta = 0.0653$, $p < 0.05$; LE -> UI, $\beta = 0.1169$, $p < 0.01$). Thus, all five hypotheses in this study are supported. Using the suggested threshold value of 3.3 [42] for the criteria of variance inflation factor (VIF), the multicollinearity index among the variables computed in the model did not cause concern.

Table 3. Path coefficient and hypothesis testing.

H	Beta	SE	T values	P values	LL	UL	VIF	Decision
1	0.2182	0.0493	4.4275	0.0000	0.1383	0.3003	1.9763	**Supported**
2	0.4771	0.0397	12.0143	0.0000	0.4098	0.5425	2.2135	**Supported**
3	0.6658	0.027	24.6392	0.0000	0.6189	0.7093	1.0000	**Supported**
4	0.0653	0.0316	2.0648	0.0195	0.0133	0.1182	1.3472	**Supported**
5	0.1169	0.0384	3.0433	0.0012	0.0531	0.1819	1.7547	**Supported**

SmartPLS 3 software can also generate quality outputs for the coefficient of determination (R^2) and the effect size (f^2) of all independent variables on the dependent variable, as shown in table 4. The values for R^2 of 0.5757 suggests that the independent variables in this study explain 57.57 % of variances in UI, while R^2 of 0.4425 indicate 44.25% variance in PU can be attributed to UI, respectively.

Additionally, the F^2 effect size values exhibit the importance of each independent variables to the dependent variable. A scholar [43] defined values near 0.02 as small, near 0.15 as a medium, and above 0.35 as large. It is shown that the effect size of PEOU on UI ($F^2 = 0.2431$) is medium to large followed by PU ($F^2 = 0.0565$), LE ($F^2 = 0.0184$) and SEC ($F^2 = 0.0074$) which has small effective size respectively. It is observed that PEOU has the largest effect size on PU ($F^2 = 0.7963$).

Table 4. Effect size and prediction summary.

Constructs	R^2	F^2		Q^2
		UI	PU	
UI	0.5757			0.5433
PEOU		0.2431	0.7963	
PU	0.4425	0.0565		0.4404
LE		0.0184		
SEC		0.0074		

To test for predictive efficiency, PLS Predict analysis was carried out. Predictive validity implies that a given set of measures can predict a given outcome variable for a specific construct [44][45]. The corresponding Q^2 from the PLS-Predict analysis was UI (0.5433) and PU (0.4404), which are all greater than 0, suggesting sufficient predictive relevance.

5. Conclusion

The study revealed that the positive perception of ease of use of cloud computing is a stronger predictor of intention than perceived usefulness. Perceived ease of use is defined as the degree to which the user thinks that using cloud computing will be effortless, and perceived usefulness is defined as the degree to which the user thinks that using cloud computing will be useful in completing tasks. The result implies that the intention to use cloud computing is largely influenced by the perception that cloud computing is effortless and easy to use more than its benefit. At the same time, perceived ease of use is a strong contributor to perceived usefulness, suggesting that if the user thinks it's easy to use and operate cloud computing, they'll think it's going to beneficial to make the job done.

This study also offers evidence that both the learning environment and security are possible factors influencing cloud computing use intention among part of the lecturers, students, and staff among HEIs. The learning environment had slowly shifted from face-to-face to blended learning, embracing several innovations. Some of these involve the use of technology through blended learning. Given the covid-19 pandemic situation, students and lecturers are forced to implement distance learning. Following these conditions, students, lecturers and staffs must use online tools, and one of them is the cloud computing application.

On the other hand, security refers to how much the users feel safe to use Cloud Computing application. It is quite clear that if users feel safe and secure upon using Cloud Computing, they will be generally open to using it and learning it to perform their tasks. This study reveals that both the learning environment and security are a positive predictor that contributes to the use intention of cloud computing among part of lecturers, students, and staff among HEIs. This indicates that the requirement of a learning environment nowadays had a positive impact on their intention to use cloud computing and feel safe and secure using it.

The development in cloud computing has produced innovative functions towards helping the education and learning endeavours of teachers and students. Numerous educational frameworks and functions, particularly cloud computing, have been introduced in recent years. Thus, understanding academicians' intention, students and HEIs staff towards using cloud computing applications are crucial. This study validated the proposed theoretical model for predicting the intention to use cloud computing among HEIs in Malaysian. The convenience sampling technique reduces the reliability of the findings to HEIs in Malaysia, so the findings should be interpreted with caution.

References

[1] Maresova P and Kacetl J 2015 Cloud Computing in the Public Sector–Case Study in Educational Institution *Procedia - Soc. Behav. Sci.* **182** 341-348.

[2] Mokwena S, Hlebela C 2018 Factors affecting the adoption of Software as a Service in South African Small Medium Enterprises *Open Innov Conf.* 1–6.

[3] Amron M T, Ibrahim R, Abu Bakar N A, Chuprat S 2019 Acceptance of cloud computing in the Malaysian public sector: a proposed model *Int. J. Eng. Bus. Manag.* **11**.

[4] Dhawan S 2020 Online Learning: A Panacea in the Time of COVID-19 Crisis *J. Educ. Technol. Syst.* **49(1)** 5–22.
[5] Dixon C 2014 New research shows cloud storage awareness, usage high [internet] nScreenMedia [cited 2020 Dec 11] Available from: https://nscreenmedia.com/new-research-shows-cloud-storage-awareness-usage-high/
[6] Davis F D 1989 Perceived Usefulness, Perceived Ease of Use, and User Acceptance of Information Technology *MIS Q.* **13(3)** 319–40.
[7] Ashtari S Eydgahi A Student perceptions of cloud applications effectiveness in higher education. *J. Comput. Sci.* **23** p173-180.
[8] Aharony N 2015 An exploratory study on factors affecting the adoption of cloud computing by information professionals *Electron. Libr.* **33(2)** 308–23.
[9] Taib S M, De Coster R, Nyamu J 2016 Innovation diffusion of wearable mobile computing: Pervasive computing perspective *Int. Conf. on Information Society (Dublin, Ireland)* p 97–101.
[10] Hong S H and Yu J H 2018 Identification of external variables for the Technology Acceptance Model(TAM) in the assessment of BIM application for mobile devices *IOP Conf. Series: Materials Science and Engineering.*
[11] Sharma S K, Al-Badi A H, Govindaluri S M and Al-Kharusi M H 2016 Predicting motivators of cloud computing adoption: A developing country perspective *Comput. Human. Behav.* 62 p 61-69.
[12] Amron M T, Ibrahim R, Bakar N A A 2020 Cloud computing acceptance among public sector employees *Telkomnika (Telecommunication Comput. Electron. Control.* **19(1)** 124–33.
[13] Hew T S and Kadir S L S A 2016 Predicting instructional effectiveness of cloud-based virtual learning environment *Ind. Mana.g Data Syst.* **116(8)** 1557–84.
[14] Ahmed M M, Houssein E H, Hassanien A E, Taha A and Hassanien E 2018 Adoption of E-Book for University Students *Int. Conf. on Advanced Intelligent Systems and Informatics (Cairo, Egypt)* p 481–94.
[15] Gangwar H, Date H and Ramaswamy R 2015 Understanding determinants of cloud computing adoption using an integrated TAM-TOE model *J. Enterp. Inf. Manag.* **28(1)** 107–30.
[16] Farooq M S, Salam M, Jaafar N, Fayolle A, Ayupp K, Radovic-Markovic M, et al 2017 Acceptance and use of lecture capture system (LCS) in executive business studies: Extending UTAUT2 *Interact. Technol. Smart. Educ.* **14(4)** 329–48.
[17] Hajli N, Wang Y, Tajvidi M and Hajli M S 2017 People, Technologies, and Organisations Interactions in a Social Commerce Era *IEEE Trans. Eng. Manag.* **64(4)** 594–604.
[18] Amron M T M, Ibrahim R and Chuprat S 2017 A review on cloud computing acceptance factors. *Proc. Computer Science* p 639–46.
[19] Rahi S and Abd. Ghani M 2018 The role of UTAUT, DOI, perceived technology security and game elements in internet banking adoption *World J. Sci. Technol. Sustain. Dev.* **15(4).**
[20] Zhang H, Tang Z and Jayakar K 2018 A socio-technical analysis of China's cybersecurity policy: Towards delivering trusted e-government services *Telecomm. Policy* **42(5)** 409–20.
[21] Mahesh D D, Vijayapala S and Dasanayaka S W S B S B 2018 Factors affecting the intention to adopt big data technology : A study based on financial services industry of Sri Lanka *4th Int. Multidisciplinary Moratuwa Engineering Research Conf. (Sri Lanka)* p 420–5.
[22] Nyeko S and Moya M 2017 Determinants of eLearning adoption among instructors in Ugandan public universities *IST-Africa Week Conf.* 1–10.
[23] Ibrahim R, Leng N S, Yusoff R C M, Samy G N, Masrom S and Rizman Z I 2017 E-Learning Acceptance Based on Technology Acceptance Mode (TAM) *J. Fundam. Appl.* **9(4S)** 871–89.
[24] Lin C-H, Shih H-Y and Sher P J 2007 Integrating technology readiness into technology acceptance: The TRAM model *Psychol. Mark.* [Internet] **24(7)** 641–57.
[25] Amron M T, Ibrahim R, Bakar N A, Chuprat S, Abu Bakar N A and Chuprat S 2019 Development and Validation of a Questionnaire to Measure the Acceptance of Cloud Computing in Public Sectors. *Open Int. J. Informatics.* **7(2)** 85–95.
[26] Hair J F, Matthews L M, Matthews R L and Sarstedt M 2017 PLS-SEM or CB-SEM: updated

guidelines on which method to use *Int. J. Multivar. Data Anal.* **1(2)** 107.
[27] Ngah A H, Ramayah T, Ali M H and Khan M I 2019 Halal transportation adoption among pharmaceuticals and comestics manufacturers *J. Islam Mark.*
[28] Baruch Y 1999 Response Rate in Academic Studies *Human Relations.* **52** p 421–38.
[29] Ringle, Christian M, Wende S and Becker J-M 2015 *SmartPLS 3* [Internet]. Bönningstedt Available from: http://www.smartpls.com
[30] Hair J F, Hult G T M, Ringle C M and Sarstedt M A 2017 *Primer on Partial Least Squares Structural Equation Modeling (PLS-SEM). 2nd ed.* SAGE Publication.
[31] Ramayah T, Yeap J A L and Ignatius J 2013 An Empirical Inquiry on Knowledge Sharing Among Academicians in Higher Learning Institutions *Minerva* **51(2)** 131–54.
[32] Ramayah T and Lee J W C 2011 JBC *Network collaboration and performance in the tourism sector. Serv Bus.* **5(4)** 411–28.
[33] Anderson J C and Gerbing D W 1988 Structural Equation Modeling in Practice: A Review and Recommended Two-Step Approach *Psychol Bull.* **103(3)** 411–23.
[34] Podsakoff P M, MacKenzie S B, Lee J Y and Podsakoff N P 2003 Common Method Biases in Behavioral Research: A Critical Review of the Literature and Recommended Remedies *J. Appl. Psychol.* **88(5)** 879–903.
[35] Podsakoff P M, MacKenzie S B and Podsakoff N P 2012 Sources of Method Bias in Social Science Research and Recommendations on How to Control It *Annu. Rev. Psychol.* **63(1)** 539–69.
[36] Eichhorn B R 2014 Common Method Variance Techniques *Midwest SAS Users Gr Conf. (Chicago)* 1–11
[37] Dupuis M, Khadeer S and Huang J 2017 I Got the Job: An exploratory study examining the psychological factors related to status updates on facebook. *Comput. Human Behav.* **73** 132–40.
[38] Cain M K, Zhang Z and Yuan K H 2017 Univariate and multivariate skewness and kurtosis for measuring nonnormality: Prevalence, influence and estimation *Behav. Res. Methods.* **49(5)** 1716–35.
[39] Gefen D, Straub D and Boudreau M-C 2000 Structural Equation Modeling and Regression: Guidelines for Research Practice *Commun. Assoc. Inf. Syst.* **4(October)**.
[40] Bagozzi R P and Yi Y 1988 On the evaluation of structural equation models. *J. Acad. Mark. Sci.* **16(1)** 74–94.
[41] Henseler J, Ringle C M and Sarstedt M 2015 A new criterion for assessing discriminant validity in variance-based structural equation modeling *J. Acad. Mark. Sci.* **43(1)** 115–35.
[42] Hair J F, Black W C, Babin B J and Anderson R E 2014 *Multivariate data analysis. 7th ed.* Vol. 16 Statistica Neerlandica. England: Pearson Education Limited.
[43] Cohen J 1988 Statistical Power Analysis for the Behavioral Sciences. *2nd ed. Lawrence Erlbaum Associates* 567 p.
[44] Shmueli G, Ray S, Velasquez Estrada J M and Chatla S B 2016 The elephant in the room: Predictive performance of PLS models. *J. Bus. Res.* **69(10)** 4552–64.
[45] Felipe C M, Roldán J L and Leal-Rodríguez A L 2017 Impact of organisational culture values on organisational agility *Sustain.* **9(12)**.

Extending technology acceptance model (TAM) to measure the students' acceptance of using digital tools during open and distance learning (ODL)

N H M Noh[1*], R Raju[1], Z D Eri[1] and S N H Ishak[1]
[1]Faculty of Computer and Mathematical Sciences, Universiti Teknologi MARA (UiTM), 21080 Kuala Terengganu, Terengganu, Malaysia

*Corresponding author: nurhidayah0738@uitm.edu.my

Abstract. Ever since the Covid-19 pandemic outbreak, we are taught to live in the new norm. Higher learning institutions are forced to adopt a new norm of open and distance learning (ODL) immediately. Since that, students are facing issues, especially in courses that need a conventional setting. During ODL, students and lecturers rely on digital tools, social media, pre-recorded video, and video conference to assist the learning process. However, there are unexplored topics on the students' intention to use digital tools during ODL. A sample of 367 students who responded from different diploma and degree programs of Universiti Teknologi MARA (UiTM) Terengganu was collected via an online survey. Data collected are analysed on its correlation and coefficient determinant, regression, and analysis of actual use. Technology Acceptance Model (TAM) is expanded to include external factors of technology access and online learning skills to be measured with perceived usefulness, perceived ease of use, attitude, behaviour intention to use and actual system use. Findings on the study will help uncover significant understanding of students' acceptance of ODL in using digital tools for a better decision-making and students' support assistance. Hence, this study benefits us to foresee the possible future need for courses in conducting ODL successfully.

1. Introduction

The pandemic coronavirus 2019 (COVID-19) has become an international concern and affects all fields, including the education industry, as UNESCO stated [1]. On 30th January 2020, the World Health Organization (WHO) declared the permissibility of the COVID-19 virus as a general health risk and a worldwide pandemic on 11th March 2020. [2]. As in the education line, COVID-19 has caused severe problems for students, instructors, and educational institutions worldwide. [3]. Due to the pandemic, the Malaysian ministry of higher education had issued all academic activities to fully transform online [4]. Many higher learning courses are facing issues, especially the courses that need physical contact. According to UNESCO, this pandemic has impacted over 87% of the world's student population in more than 160 countries [5].

As learning processes must continue, educational institutions' services must continue to run even with the various situations and challenges amid the outbreak of a pandemic. UiTM had enforced open and distance learning (ODL) for online facilitation and other activities supporting education [6]. This situation illustrates how online learning delivery has allowed instructors and learners to incorporate flexibility into their lesson plans. However, there are unexplored topics on the students' intention to use digital tools during ODL. Thus, this study investigates students' intent to use digital tools during ODL based on the Technology Acceptance Model theory (TAM). A random sample of 367 students who responded from different diploma and degree programs from UiTM Terengganu were collected via an online survey. Data collected are analysed on its reliability, correlation and coefficient determinant, regression, and analysis of actual use. Findings on the study will help understand students' intention to use digital tools during ODL to improve decision-making and support assistance for students' needs.

1.1. Technology Acceptance Model

The Technology Acceptance Model (TAM) is introduced by Davis in 1989 [7] to explain how user accepts and use technology. This model proposes that perceived usefulness and perceived ease of use of technology predict user attitude towards using the technology, subsequent behavioural intentions, and actual usage [8]. In TAM, perceived usefulness refers to the perception that using the technology will improve the user's work performance. Perceived ease of use refers to how seamless the user perceives it to be. The user's attitude towards using technology and the user's preference for using the technology influence their intention to use it [9]. TAM has been applied in various studies, and many refer to TAM for technology acceptance [10]. Figure 1 illustrates the theoretical framework for this study, adapted based on TAM's original version [7]. TAM is expanded to include external factors, which are technology access (TA) and online learning skills (OLS) to be measured with perceived usefulness (PU), perceived ease of use (PEOU), attitude (ATT), behaviour intention to use (UI) and actual use. Technology access is defined as the program infrastructure, including the internet and wi-fi. The use of computers and the internet form a significant component of online learning. Online learning skill encompasses basic computer skills and techniques to engage in online education. Previous findings reveal that computer experience, skills and personal knowledge of Internet technology contribute to the students' achievement in an online computer environment [11].

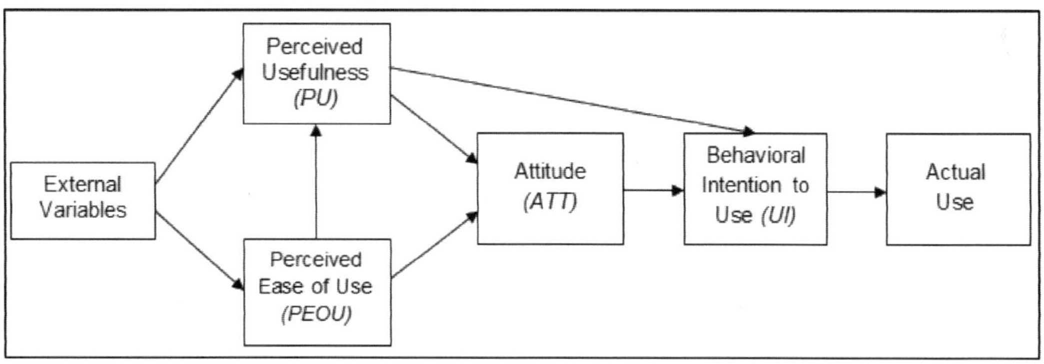

Figure 1. Technology Acceptance Model (TAM) as a theoretical framework.

1.2. Open and distance learning

The Malaysian Ministry of Higher Education (MoHE) decided to halt all face-to-face lectures as part of the covid-19 pandemic control measures and urged all higher education providers to prepare for open and distance learning (ODL). The majority of teaching is administered by someone remote and separated from their learner during ODL. The goal seeks to enable students to have a more comprehensive degree of flexibility and accommodations and its programme or other structure elements[12]. Distance learning also can be defined as a method of teaching and learning in different geographical location, either near or far, with the use of electronic media and devices as tools for enhancing the availability of training, communication, and interaction, and that helps in accepting novel ways of comprehending and the interaction and knowledge in establishing learning [5].

Distance education provides affordable educational options for those who would like to learn. Still, their circumstance cannot afford a formal university education since educators and students can replace physical interaction with digital media [13]. In this time of deadly pandemics, online social media platforms that allow video conferencing, online discussions, and online lecture access are necessary. Internet connections must be good, enabling instant feedback on students' progress from remote locations and can be used as an alternative to face-to-face classes [14]. The development of reliable and sufficient technology resources, such as digital learning tools and platforms and online courses, are essential for this situation [15].

1.3. Digital Tools

The use of distance learning in higher education institutions has expanded globally [16]. During distance learning, students and lecturers rely on technology to assist the learning process. Orlando and Attard stated that teaching with technology does not apply to everyone. Its implementation will depend on the technological resources available and the curricular needs of the students [17]. The study in Hong Kong University shows some of the aspects that affect technology usage for learning to have been students' skill with computers, their attitude towards technology, learning styles, as well as teachers and peers' support [18].

Distance learning using digital tools can be defined as "the use of electronic technology to deliver, support and enhance both learning and teaching and involves communication between learners and teachers utilising online content" [19]. Distance learning has the potential to facilitate the transition from a teacher-centred approach, where the student might be a passive observer at the hands of the teacher, to a student-centred approach where a student might be a vocal participant in their own education as well as with other students. Utilising digital tools in distance learning may help cope with the variability in learning environments and may provide individuals with self-paced learning and content [20], [21]. The digital educational tools are divided into eight major categories: Learning Management System (LMS), Massive Online Open Courses (MOOCs), Mind mapping, Interactive tools, Web conferencing, Infographics, Research visibility tools and Web-based Content Management Systems. A recent study among nursing students suggests that only one-third of them preferred distance learning using digital tools [22]. Moreover, it was found that students' increased willingness to complete research work effectively was associated with digital platforms for education and learning [23]. Digital technology is advantageous to lesson study and learning, which has enabled rigorous collaboration, synchronous observations, ensuring better data management, leading to a more profound understanding of information [24].

2. Research Method

The study-in-progress investigated TAM as a basis for hypothesising the effects of proposed external variables in ODL, which are technology access and online learning skills. An attempt is made whether attitudinal belief such as perceived ease of use and perceived usefulness have a relationship and impact towards the ODL adoption and the suitability of the technology acceptance model in the study's context. Convenience sampling is frequently used in quantitative research, where the researcher selects subjects that are more readily accessible. The questionnaires were distributed through an online survey among UiTM Terengganu students at three campuses located in Dungun, Bukit Besi and Kuala Terengganu. 370 students from various background participated in this study. The questionnaire developed was based on 6 constructs which consist of 30 items. The items in the construct of Perceived Ease of Use (PEOU), Perceived Usefulness (PU), Attitude (ATT) and Use Intention (UI) were adopted from Shittu, Gambari and Sule, [25] while the items in the construct of Technology Access (TA) and Online Learning Skills (OLS) are adopted from Puspitasari and Listyarini [26].

The research model consists of seven (7) factors where four (4) factors are adopted from TAM and three (3) other factors identified from the literature are technology access (TA), online learning skills (OLS) and attitude (ATT). These three (3) additional factors are identified as the external variables. The hypotheses of this research are:

H1: PU is positively related to the intention to use (UI) digital tools among students for online learning.
H2: PU is positively related to the attitude (ATT) among students for online learning.

H3: PEOU is positively related to the PU to use digital tools among students for online learning.
H4: PEOU is positively related to the attitude (ATT) among students for online learning.
H5: ATT is positively related to the UI to use digital tools among students for online learning.
H6: TA is positively related to the PU to use digital tools among students for online learning.
H7: OLS is positively related to the PU to use digital tools among students for online learning.
H8: TA is positively related to the PEOU to use digital tools among students for online learning.
H9: OLS is positively related to the PEOU to use digital tools among students for online learning.

3. Results and discussions

3.1. Descriptive Analysis

A total of 370 participated in this study, where 36% of them are bachelor students and 64% are diploma students. 74.32% of the students are female, whereas 25.68% are male as sown in figure 2 and figure 3.

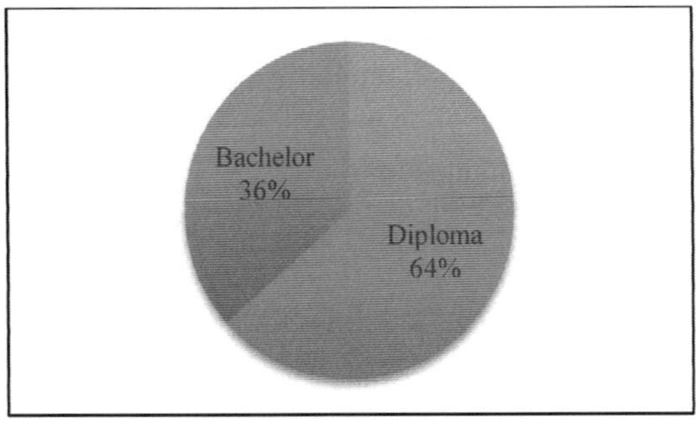

Figure 2. Number of respondents by the level of education.

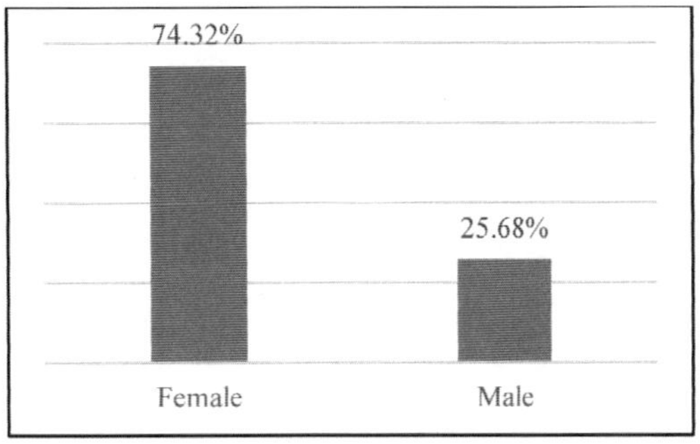

Figure 3. Number of respondents by gender.

3.2. Reliability Analysis

Table 1 represents the reliability analysis from this study.

Table 1. Reliability analysis.

Construct	Number of Items	Cronbach's Alpha
Technology Access	3	0.782
Online Learning Skills	8	0.875
Attitude	3	0.925
Perceived Ease of Use	6	0.940
Perceived Usefulness	6	0.955
Use Intention	4	0.931

Cronbach Alpha is a measure used to evaluate the reliability or the internal consistency of a scale or test item set. In other words, the reliability of any given measurement refers to the degree of consistency of an assumption measurement. One way to measure the strength of that consistency is Cronbach Alpha. According to Sekaran and Bougie [27], Cronbach Alpha of 0.6 is categorised as poor, 0.7 is good, 0.8 is considered very good and 0.9 is indicated as excellent. As can be seen in Table 1, the value for Cronbach Alpha for technology access (0.782) indicates good reliability, online learning skills (0.875) indicates very reliable whereas attitude (0.925), perceive ease of use (0.94), perceived usefulness (0.955) and use intention (0.931) shows excellent internal consistency.

3.3. Correlation and Coefficient of Determination Analysis
Correlation coefficient measures the strength of a relationship between two variables. Values ≤ 0.35 are generally considered as weak correlations, 0.36 to 0.67 represents moderate correlations and 0.68 to 1.0 shows strong correlations by Taylors [28]. The result in table 2 shows that the relationship between PU -> UI (0.807), ATT -> UI (0.698), PEOU -> ATT (0.680), PU -> ATT (0.697) and PEOU -> PU (0.819) are strong positive relationship. OLS -> PU (0.636) and OLS -> PEOU (0.618) exhibit moderate positive relationship. On the contrary, TA -> PU (0.465) and TA -> PEOU (0.433) shows weak positive relationship. All of the relationships are significant (p-value <0.05).

Table 2. Correlation and coefficient of determination analysis.

Hypothesis	Correlation	Sig.	Strength	R^2
Perceived Usefulness -> Use Intention	.807**	0	Strong Positive	0.660
Attitude ->Use Intention	.698**	0	Strong Positive	0.467
Perceived Ease of Use -> Attitude	.680**	0	Strong Positive	0.477
Perceived Usefulness -> Attitude	.697**	0	Strong Positive	0.474
Perceived Ease of Use -> Perceived Usefulness	.819**	0	Strong Positive	0.678
Technology Access -> Perceived Usefulness	.465**	0	Weak Positive	0.208
Online Learning Skills -> Perceived Usefulness	.636**	0	Moderate Positive	0.380
Technology Access -> Perceived Ease of Use	.433**	0	Weak Positive	0.183
Online Learning Skills -> Perceived Ease of Use	.618**	0	Moderate Positive	0.380

R^2 determines the coefficient of determination. This value measures how much variation the independent variable explained the dependant variable. Table 2 shows that the variation in UI is explained by PU (66.0%) and ATT (46.7%). The variation in ATT is explained by PEOU (64.7%) and PU (47.4%). The variation in PU is explained by PEOU (67.8%), TA (20.8%) and OLS (38.0%), whereas the variation in PEOU is explained by TA (18.3%) and OLS (38%).

3.4. Regression Analysis
Regression analysis is used to estimate the relationships between a dependent variable and one or more independent variables. It is an analysis to see the impact of an independent variable on the dependent variable. Since all the relationship, as shown in Table 2, is significant (p-value <0.05), we can carry on the regression analysis.

As shown in aable 3, the impact of PU and ATT on UI are both significant (p-value <0.05). The B coefficient means that when PU and ATT increased by 1 unit, the UI will increase by 0.813 and 0.683, respectively.

Table 3. Regression analysis of independent variable vs use intention.

Model		Unstandardised Coefficients		Standardised Coefficients	t	Sig.
Dependent Variable	Independent Variable	B	Std. Error	Beta		
	(Constant)	0	0.031		-0.006	0.995
	Perceived Usefulness	0.813	0.031	0.813	26.643	0

Use Intention	(Constant)	0	0.038		-0.007	0.994
	Attitude	0.683	0.038	0.683	17.885	0

Table 4 shows that the impact of PU and PEOU towards ATT are both significant (p-value <0.05). The B coefficient means that when PU and PEOU increased by 1 unit, the ATT will increase by 0.688 and 0.691, respectively.

Table 4. Regression analysis of independent variables vs attitude.

Model		Unstandardised Coefficients		Standardised Coefficients	t	Sig.
Dependent Variable	**Independent Variable**	**B**	**Std. Error**	**Beta**		
Attitude	(Constant)	8.28E-05	0.038		0.002	0.998
	Perceived Usefulness	0.688	0.038	0.688	18.127	0
	(Constant)	-3.65E-06	0.038		0	1
	Perceived Ease of Use	0.691	0.038	0.691	18.248	0

Table 5 shows that the impact of PEOU, TA and OLS towards PU are all significant (p-value <0.05). The B coefficient means that when PEOU, TA and OLS increased by 1 unit, the PU will increase by 0.823, 0.456 and 0.617, respectively.

Table 5. Regression analysis of independent variables vs perceived usefulness.

Model		Unstandardised Coefficients		Standardised Coefficients	t	Sig.
Dependent Variable	**Independent Variable**	**B**	**Std. Error**	**Beta**		
Perceived Usefulness	(Constant)	0	0.03		-0.004	0.997
	Perceived Ease of Use	0.823	0.03	0.823	27.723	0
	(Constant)	-5.89E-05	0.047		-0.001	0.999
	Technology Access	0.456	0.047	0.456	9.797	0
	(Constant)	-4.25E-05	0.041		-0.001	0.999
	Online Learning Skills	0.617	0.041	0.617	14.962	0

Table 6 shows that the impact of TA and OLS towards PEOU are both significant (p-value <0.05). The B coefficient means that when TA and OLS increased by 1 unit, the PEOU will increase by 0.428 and 0.616, respectively.

Table 6. Regression analysis of independent variables vs perceived ease of use.

Model		Unstandardised Coefficients		Standardised Coefficients	t	Sig.
Dependent Variable	Independent Variable	B	Std. Error	Beta		
Perceived Ease of Use	(Constant)	9.07E-05	0.047		0.002	0.998
	Technology Access	0.428	0.047	0.428	9.057	0
	(Constant)	0	0.041		0.003	0.998
	Online Learning Skills	0.616	0.041	0.616	14.943	0

3.5. Descriptive Analysis of Actual Use

From figure 4, it is safe to say that majority of the respondents (62.105) have been using digital tools between 2-3 years and most of the respondents (49.9%), according to figure 5, had been using digital tools daily. From figure 7, most of the respondents have experience using word processing documents (82.1%) followed by tablet and cell phone applications (75.1%), slide presentation software (72.9%), educational applications that can be found on the internet (69.6%) and electronic reference materials (55.6%). Summing up the regression analysis, the model established in this study as in figure 7.

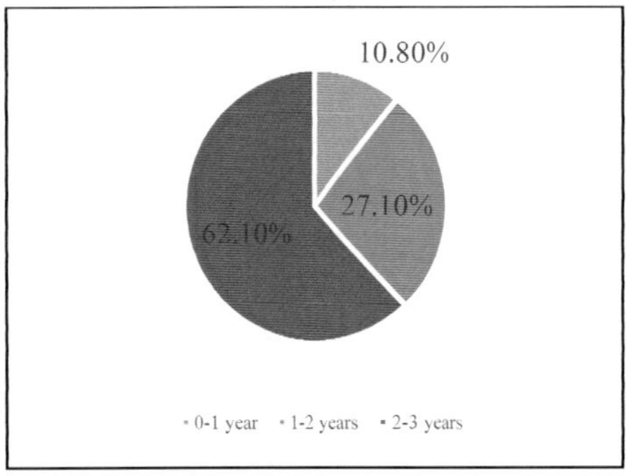

Figure 4. Experience using digital tools.

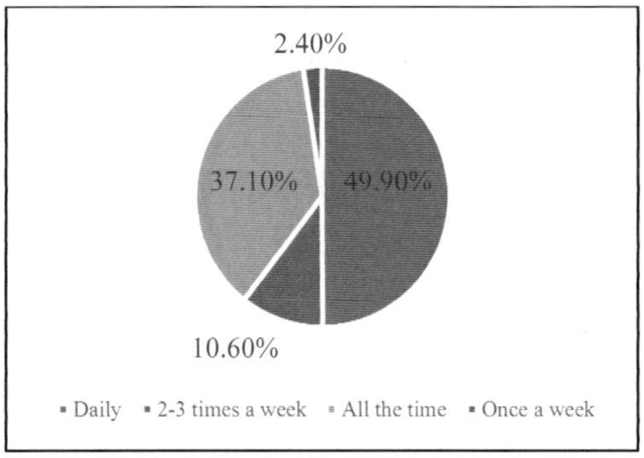

Figure 5. Frequency using digital tools.

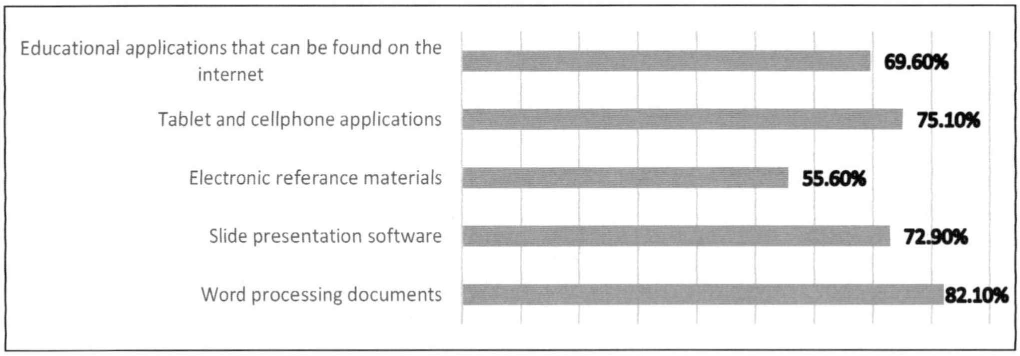

Figure 6. Digital tools used.

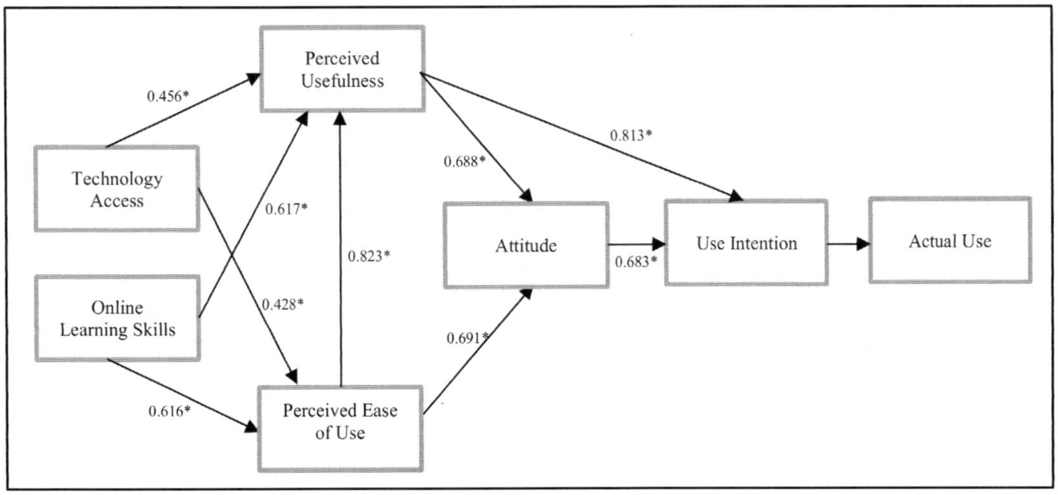

Figure 7. Model summary (*: coefficient significant at $p < 0.05$).

4. Conclusions

The goal of this research, which is based on the TAM model, is to add new variables, namely technology access and online learning skills, as the external variables to the model and explore students' intention to use digital technology in learning during ODL. The TAM model was extended in this study to investigate the university students' digital tools usage and the factors influencing intention relating to its usage. Online learning skills means how well the student can operate a computer and their ability to use online tools, whereas technology access refers to the condition of the computer that they have, internet access and adequate software. The result proves that online learning skills exert a more substantial influence on perceived usefulness and perceived ease of use than technology access. Therefore, it is advisable to take measures to reinforce students' technology access in terms of their computer condition, internet access and adequate software provided in assisting online and distance learning.

This study provided information that perceived ease of use and perceived usefulness towards attitude to use digital tools are encouraging. It implies that if the student thinks that using the digital tools will be easy, clear and understandable and that digital tools may enhance their performance, they will have a positive attitude towards using digital tools during ODL. This study also reveals that digital tools' use intention was more affected by perceived usefulness than attitude, though both are significant contributor. It indicates that if the students feel that using digital tools would improve their efficacy and enhance their learning, they are more likely to use digital tools during ODL.

To conclude, this research showed that part of UiTM Terengganu students has a positive intention to use digital tools in their learning activities during ODL. Hence, it is safe to say that students are ready to use digital tools in online learning. However, there are some limitations to this study. The use of the convenience sampling technique limits the generalizability of the findings to all UiTM Terengganu students; therefore, the results should be interpreted with caution.

References

[1] UNESCO U. COVID-19 Educational Disruption and Response [Internet]. 2020 [cited 2021 Feb 28]. Available from: https://en.unesco.org/themes/education-emergencies/coronavirus-school-closures.

[2] Cucinotta D and Parmensis M V 2020 WHO Declares COVID-19 a Pandemic *Acta Bio Medica Atenei Parm.* **91(1)** 157.

[3] Almanthari A, Maulina S and Bruce S 2020 Secondary school mathematics teachers' views on E-learning impleme ntation barriers during the COVID-19 pandemic: the case of Indonesia.

Eurasia J. Math. Sci. Technol. Educ. **16(7)** em1860.
[4] Sani R Overseas Dream Put on Hold [Internet] *New Straits Times* 2020 [cited 2021 Feb 28]. Available from: https://www.nst.com.my/education/2020/06/599586/overseas-dream-put-hold
[5] Krishnan K S T and Hussin H 2017 E-learning readiness on Bumiputera SME's intention for Adoption of Online Entrepreneurship Training in Malaysia *Management* **7(1)** 35–9.
[6] Karim M A UiTM Moves to Online Learning Mode [Internet] *New Straits Times* 2020 [cited 2021 Feb 28]. Available from: https://scholar.google.com/scholar?hl=en&as_sdt=0%2C5&q=%5B5%5D%09K.+S.+T.+Krishnan+and+H.+Hussin%2C+%282017%29+"“E-learning+readiness+on+Bumiputera+SME's+intention+for+adoption+of+online+entrepreneurship+training+in+Malaysia%2C”'+Management%2C+vol.+7%2C+
[7] Davis F D 1989 Perceived Usefulness, Perceived Ease of Use, and User Acceptance of Information Technology *MIS Q.* **13(3)** 319–40.
[8] Weng F, Yang R-J, Ho H-J and Su H-M 2018 A TAM-Based Study of the Attitude towards Use Intention of Multimedia among School Teachers *Appl. Syst. Innov.* **1(3)** 36.
[9] Masrom M 2007 The Technology Acceptance Model and The E-learning *Technology* **21(24)** 81.
[10] Ching-Ter C, Hajiyev J and Su C R 2017 Examining the students' behavioral intention to use e-learning in Azerbaijan? The General Extended Technology Acceptance Model for E-learning approach *Comput. Educ.* **111** 128–43.
[11] Bouhnik D and Carmi G 2014 Computer skills and prior experience with e-learning courses as factors that affect development of intellectual thinking dispositions in online learning *Multicultural Awareness and Technology in Higher Education: Global Perspectives, IGI Global* p 302–23.
[12] UNESCO U 2002 Open and Distance Learning: Trends, Policy and Strategy Considerations
[13] Faith K 1988 *Toward new horizons for women in distance education: International perspectives. 1st Ed.*
[14] Basilaia G, Dgebuadze M, Kantaria M and Chokhonelidze G 2020 Replacing the classic learning form at universities as an immediate response to the COVID-19 virus infection in Georgia *Int. J. Res. Appl. Sci. Eng. Technol.* **8(3)** 101–8.
[15] Huang R H, Liu D J, Tlili A, Yang and Wang J F 2020 Handbook on Facilitating Flexible Learning During Educational Disruption: The Chinese Experience in Maintaining Undisrupted Learning in COVID-19 Outbreak Rights and Permissions *Handbook on Facilitating Flexible Learning During Educational Disruption.*
[16] Poon J 2013 Blended Learning: An Institutional Approach for Enhancing Students' Learning Experiences *J. Online Learn. Teach.* **9**.
[17] Orlando J and Attard C 2016 Digital natives come of age: the reality of today's early career teachers using mobile devices to teach mathematics *Math. Educ. Res. J.* **28(1)** 107–21.
[18] Lee C, Yeung A S and Ip T 2016 Use of computer technology for English language learning: do learning styles, gender, and age matter? *Comput. Assist. Lang. Learn.* **29(5)** 1033–49.
[19] Durmaz A, Aklime D, Emre C and Sen C 1998 Effect of screen-based computer simulation on knowledge and skill in nursing students' learning of preoperative and postoperative care management: a randomised *CIN Comput. Informatics, Nurs.* **30(4)** 196–203.
[20] Bowden J A and Marton F 1998 *The University of Learning. Psychology Press.*
[21] Biggs J and Tang C 2007 *Teaching for quality learning at university Maidenhea*d Berkshire, UK: McGraw-Hill Education.
[22] Langegård U, Kiani K, Nielsen S J and Svensson P A 2021Nursing students' experiences of a pedagogical transition from campus learning to distance learning using digital tools. *BMC Nurs.* **20(1)** 23.
[23] Likhachev S, Likhacheva T, Silchenkova L, Krivorotova E and Plesnik L 2020 Controlling Research Activity of Students by Digital Tools *SHS Web Conf.* 79:01010.

[24] Weaver J C, Matney G, Goedde A M, Nadler J R and Patterson N 2021 Digital Tools to Promote Remote Lesson Study *Int. J. Lesson Learn Stud.*
[25] Shittu A T, Gambari A I, Sule A O 2013 Students' Attitude and Behavioural Intention on Adoption of Internet for Learning among Al-Hikmah University Students in Nigeria: A Test of Technology Acceptance Model *Malaysian J. Distance Educ.* **15(2)** 89–107.
[26] Puspitasari A K and Listyarini S 2018 E-Learning readiness of distance learners *ASEAN J. Open Distance Learn.* **10(2)** 62–71.
[27] Sekaran U and Bougie R 2013 R*esearch methods for business: A skill-building approach. 6th ed.* West Sussex, UK: John Wiley & Sons Ltd.
[28] Taylor R 1990 Interpretation of the Correlation Coefficient: A Basic Review *J Diagnostic Med Sonogr.* **6(1)** 35–9.

IOP Conf. Series: Materials Science and Engineering 1176 (2021) 012038 doi:10.1088/1757-899X/1176/1/012038

The development of a preschool event management system using usability heuristic theory

N S M Yusoff[1], N A A A Bakar[1*] and N M N Daud[1]

[1]Faculty of Computer and Mathematical Sciences, Universiti Teknologi MARA (UiTM), 21080 Kuala Terengganu, Terengganu, Malaysia

*Corresponding author: azila268@uitm.edu.my

Abstract. Preschool Event Management System for Genius Aulad (PEMS) is a computerized prototype system, developed to facilitate preschool event management process handling. Genius Aulad Kuala Terengganu has been selected as a case study. Adapting a more systematic data management approach is critical to elevate problems inherent from the current manual way used in managing the event management related processes. The issues that lead to the proposal of the new system are difficult to inform the upcoming events to the parents, missing confirmation messages due to unstructured conversation, paper-based registration form used cause inefficient data storage and retrieval, the payment for event program always overdue and limited type of report can be generated. The System Development Life Cycle (SDLC) using Adapted Waterfall Model serves as a development methodology. Furthermore, the Usability Heuristic for User Interface Design is adapted in developing the system interface. The PEMS is tested and evaluated in terms of functionality and usability by the developer, and five target users. The test case and scenario are used to evaluate the functionality of the system. Meanwhile, the usability metric is applied to evaluate the usability of the system. The SUS score for PEMS is 82.5, which means the users are satisfied with the overall usability of the system. All the recommendations received during the testing session can be used as a reference for future work. With the existence of PEMS it is hoped that it will bring benefits to the organization when using this system.

1. Introduction

A Management Information System (MIS) is an information system used in an organization for decision-making and for information communication, monitoring, analysis and visualization. MIS contributes to systematic information management within a company or organization. Numerous types of events usually been conducted in many organizations. Handling event management process involves among others data related to participant registration, event details, fee payments and event related reminders or notifications to the participants. Managing these data requires major efforts to ensure smooth event execution, thus having an information system that facilitates event related data store, processing and information generation is significant.

A study conducted in an organization providing pre-school education namely Genius Aulad Kuala Terengganu (GAKT) has been done to gather understanding on event process handling data management. Currently, manual method of handling the event related process is still in place. All event data are recorded manually using regular forms and is placed in files. Upon receiving information of events need to be conducted from the headquarters, the staff of GAKT will distribute memos to the parents via invitation letter. The parents will respond to the invitation by filling in student's details in the invitation form and register student's participation through chat messages. To confirm the registration, the parents then need to be presented to GAKT and complete the payment. Reminder for payment will be generated through chatting application if it is not received beyond the due date.

Growing number of events involving GAKT students imply directly to number of information communication to the parents. Communication through existing platforms have several drawbacks. The staff need to prepare an invitation letter and registration form for the parents every time a new event is initiated. Reminder through chat messages sometimes will need to be generated through several cycles. Upon receiving registration, the data written in the form need to be summarized manually to get overall picture on the registered participants, events registered, payment made and many more. Furthermore, communication made through chat messages caused some messages to be missed. Missing important messages gives undesirable perception between GAKT and the parents. Occasionally, the parents forgot to make the payments for event fees. Overdue payments may affect the event management process. Therefore, the staff need to keep reminding the parents to make payments before the days of the events. In long term performance, chat messages will contribute to the two-way communication problem. In addition, as of now the only information generated from manual data management method is list of the students' names who participate in the events. Producing reports from manual kept records requires more effort from the staff thus hindering reports needed by the management to be generated at the right time.

Hence, a computerized system is seen as solution to facilitate the work process related to event management. The Preschool Event Management System (PEMS) is targeted to manage events related data in a more proper way by keeping data about the participants, parents, activity details, registration and payment status in a systematic manner thus helping the system to later generate accurate information on the events. The parents can check the upcoming event easily and register for the event via the system. Moreover, the system enables to generate payment reminder to the parents which can reduce payment overdue. Indeed, the system facilitates in generating various types of report related to the events such as yearly event report, participant detail report as well as payment related report. Table 1 summarizes the scope of PEMS.

Table 1. Summary of the Scope for PEMS.

Process	**Data**	**User**	**Functionality**
Notification of upcoming events, registration and payment	Event details, registration and payment	Parents	View notification of the upcoming events View the upcoming event details Register for the event Make payments for event fees Receive notification payment reminder (if any)
		Staff	Notify the upcoming events Add/Delete parents View payment status Generate payment reminder Generate event related report
		Advisor/ System Administrator	Add/Delete staff Add/Update information about the event Update registration status View payment collection

2. Literature Study

2.1. Management Information System

Furduescu [1] stated that MIS is a combination of human and computer resources focus on collecting, storing, organizing, and retrieving data into information that managers use in performing their tasks to achieve effective management. The main purpose of MIS as further cited from Alcami et al. [2] is to provide organizations with any information they needed to make the decision-making process and solve any vital issues in the organization. Moreover, according to an article from Shrivash [3], MIS has six characteristics which are management-oriented, integrated, future-oriented, common data flows, long-term planning and central database. Generally, the MIS plays a vital role in the management, administration and operations of an organization. Table 2 indicates a summary of the Management Information System.

Table 2. Summary of Management Information System.

Definition of MIS	Purpose of MIS	Characteristic of MIS
MIS is a planned system of collecting, storing and transform the data into useful information through a process to carry out the functions of management.	The purpose of MIS is to provide the necessary information to the users by helping them to manage the functions of organising, planning, control and decision making.	▪ Management-oriented ▪ Integrated ▪ Future-oriented ▪ Common data flows ▪ Long-term planning ▪ Central database

2.2. Event Management System features

An event management system is widely used to manage the event process by allowing users to register and pay the event fees directly through the system. The event management application is designed to encounter the problems of event planning in order to make it more organized as indicated by Saipriya et al. [4]. Based on research conducted by Mishra et al. [5], they also stated that event management system has five characteristics which are multiple event options, real-time, event notification, feedback section and filter for fees features to make it easier for the customer to make fees comparison for each event. According to Sharadha et al. [6], event management provides a basic flow of information for managerial decision-making and helpful for managing the event process. Furthermore, event management system also must include four characteristics which are instant notifications, instant real-time update, status fees and feedback section to evaluate the events based on their experience of the events Sharadha et al. [6]. Table 3 shows the characteristics of event management system from two authors.

Table 3. Characteristic of Event Management System.

Characteristic \ Author	Mishra et al. [5]	Sharadha et al. [6]
Status fees	X	√
Multiple event option	√	X
Filter fees	√	X
Real-time registration	√	√
Notifications	√	√
Feedback section	√	√

2.3. Usability Heuristic Theory

In general, Nielsen's heuristic is one of the most commonly used sets of heuristics to evaluate systems or applications as stated by Quiñones et al. [7]. The Usability Heuristic theory is extensively used for web-based systems and is intended for user interface design. Based on Miller et al. [8], the context of heuristics design represents broad rules of thumb to attain optimal design. Usability Heuristic has ten general principles that can be implemented in the development process for user interface design state Nielsen et al. [9]. The principles of Usability Heuristic are the visibility of system status, match between system and the real world, user control and freedom, consistency and standard, error prevention, recognition rather than call, flexibility and efficiency of use, aesthetic and minimalist design, help users recognize, diagnose and recover from errors and help and documentation.

According to Nielsen [10], the results of the heuristic evaluation showed different characteristics identified different numbers and types of usability problems. However, Nielsen suggested that between three and five characteristics are sufficient because when the number of characteristics used increases, the number of problems identified increases in turn.

3. Methodology

This section discusses the project methodology used in developing Preschool Event Management System which is the adapted Waterfall Model. Each phase in the methodology includes different activities to be carried out during the development process. The major phases are system planning, system development, and system documentation. Table 4 outlines project development methodology for the project execution.

Table 4. PEMS Project Development Methodology.

PHASE	ACTIVITY	TECHNIQUE	DELIVERABLE
PHASE 1: SYSTEM PLANNING	• Consult with the supervisor the title of the proposed system • Conduct interview with the advisor of Genius Aulad • Identify the flow of the current business process • Identify the problem based on the business process	• Brainstorming • Face-to-Face Interview	• Title of the proposed system • Flow of current business process • Problem Statement • Chapter 1
PHASE 2: SYSTEM DEVELOPMENT		**ANALYSIS**	
	• Define user requirement and system requirement for the proposed system • Identify suitable system development model and theory for the proposed system	• Interview • Research • Analysis and fact-finding from scholarly materials such as articles, journals and books	• User requirement • System requirement • Adapted Waterfall Model • Ten Usability Heuristics Theory • Chapter 2

	DESIGN		
	• Design Context Diagram • Design Data Flow Diagram • Design Entity Relationship Diagram • Design User Interface • Design Site Map • Design Test Plan • Design Questionnaire	• Draw.io • Microsoft PowerPoint	• Context Diagram • Data Flow Diagram (DFD) • Entity Relationship Diagram (ERD) • User Interface Design • Site Map • Test plan and questionnaire • Chapter 3
	DEVELOPMENT		
	• Develop Preschool Event Management System for Genius Aulad	• Wamp Server • Notepad++ • Google Chrome	• A functional Preschool Event Management System (PEMS)
	TESTING & EVALUATION		
	• Conduct system testing to test on system functionalities • Demonstrate system to users • Conduct expert and user testing to test on system usability • Distribute questionnaire to the target users	• Test Case Scenario • Questionnaire	• Usability and functionality evaluation result of the proposed system
PHASE 3: SYSTEM DOCUMENTATION	• Compiling and refining a complete report	• Microsoft Word	• A complete project report

3.1. System Planning

In the system planning phase, the objectives are to gain related information and identify the problems faced by Genius Aulad. The process begins by discussing the title of the proposed system with the supervisor. After that, the interview is conducted with the advisor of the organization, Mrs. Edora Leezia at Genius Aulad Kuala Terengganu. During the interview session, there are various important information and business flow has been gathered. The information that has been collected, such as the issues faced by Genius Aulad and the current event registration are documented. Next, the current event management handling process flow is gathered for further analysis in the next stage. Figure 1 shows the information collected during the interview session with the organization.

INTERVIEW CONDUCTED WITH GENIUS AULAD

1. **What is the problem with the current event registration process?**
   - Staff need to prepare invitation letter to the parent regarding to the upcoming events
   - Sometimes the invitation letter is misplaced or lost
   - Parents need come to the school to make register and make payment
   - Staff need to keep reminding the parents to make payment through chat messages
2. **What is the suggestion about the system function that will be developed?**
   - Parents can register and make payment online
   - Parents receive notification about the upcoming events through the websites
   - Staff can manage parents and student details
   - Advisor can add or delete event information and view monthly report related to the events

Figure 1. Information Collected During the Interview Session.

3.2. System Development

In the first process, an analysis activity involves identifying the functional and non-functional requirements and user requirements. These requirements have been gathered by conducting an interview session with the advisor of Genius Aulad. The analysis activities consist of comparing the similar characteristics of the existing system, identifying a suitable system development model, and researching suitable theory for the proposed system. Next is the design process. In this process, Context Diagram, Data Flow Diagram, Entity Relationship Diagram, Site Map, and User Interface Design have been designed. This process aims to provide a clear perspective on how the proposed project will operate at the end of the project.

The development process takes place when all the design are completed. In this process, the code is produced, and the focus for the developer is to produce a well-functioning system based on the collected requirements and the chosen theory. Table 5 summarizes six principles of the theory that had been implemented on PEMS. All six principles are adapted into the system development due to its suitability with PEMS.

Table 5. Implementation of Usability Heuristics Theory to PEMS.

Principle	Description
Visibility of system status	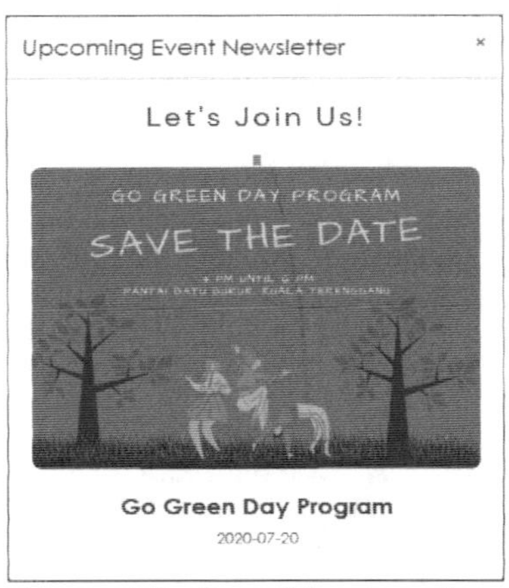 ▪ The system displays notification of the upcoming event to notify the user of the latest events of the week.

	 ▪ The system informed the users that they have login successfully into the system. ▪ Keep users informed regarding system responses or actions with feedback.
Match between system and the real world	 ▪ The system uses familiar terms that are simple and easy to understand by the users. ▪ The recycle bin icon is similar to a real bin in the real world.
Consistency and standard	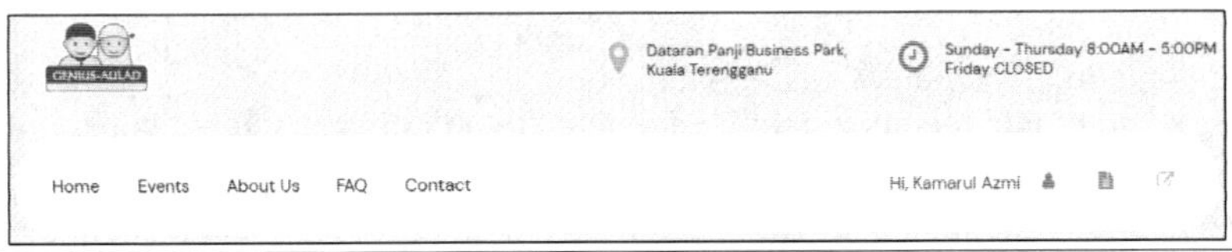 ▪ The layout of the system is standardized in one specific style and display the same content structure for each of the pages in the system.
Error prevention	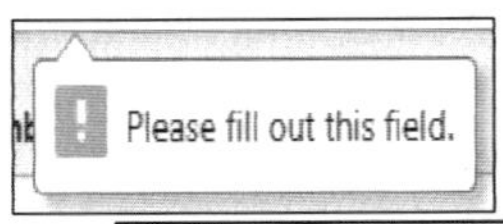 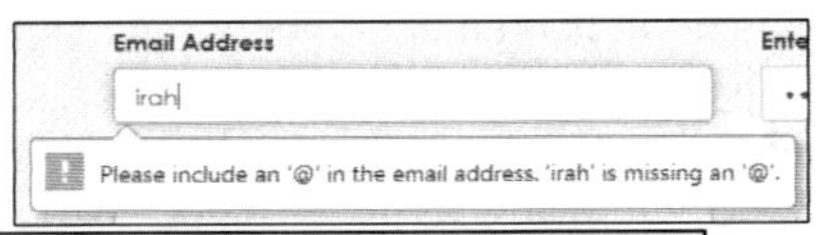 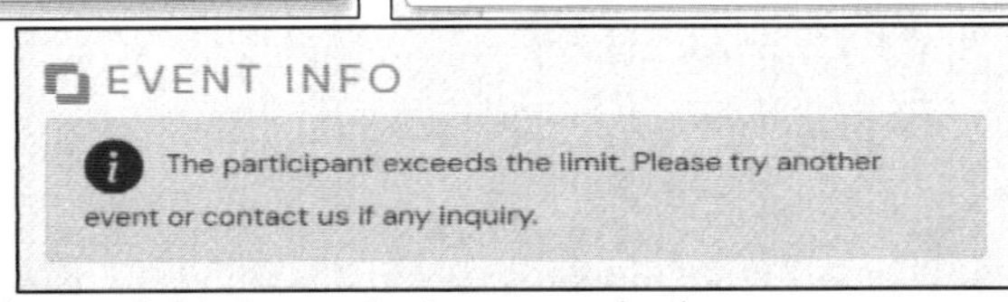 ▪ The input field is marked as required to ensure user fill in all information as requested. If not, the system will display an alert to the user. ▪ The system will display an alert whenever the user fills in the wrong data/input such as email. ▪ An alert will be displayed whenever the registration exceeds the limit of registration.
Aesthetic and minimalist design	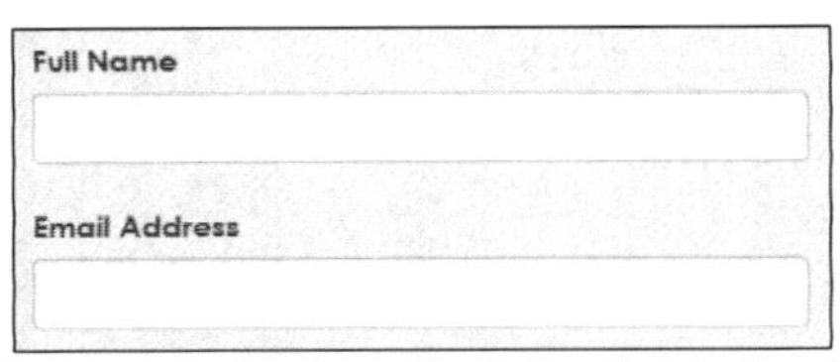 ▪ The system focuses on simple and minimalist design that helps the users to understand the flow of the system better without confusing the users.
Help and documentation	

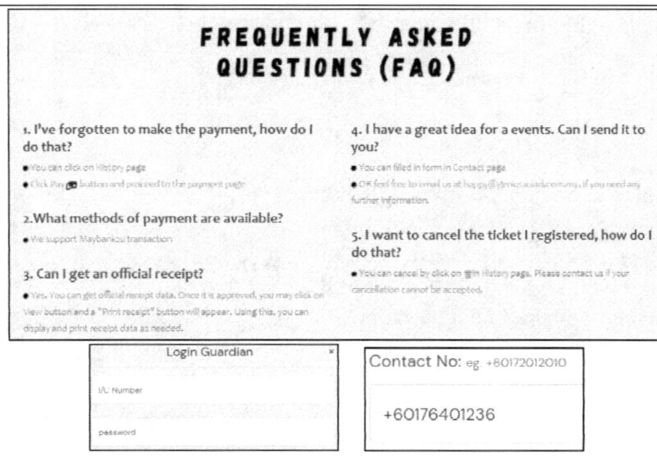

- The system provides help to users of the functionality and possible issues related to the system.
- The input field consists of placeholder to ensure the user fill in the correct input and provides an example of an input to avoid incorrect input.

Lastly is testing and evaluation activity which involves testing the functionality and usability of the system. The developer initially tests the system using three created testing scenarios which are Normal, Alternative and Exception Scenarios. The results are recorded in the test case table as shown in table 6. Usability test is done by demonstrating the system to five target users. User evaluation is one of the methods to collect data regarding the usability of the system. The overall perceived usability was measured using the Software Usability Scale (SUS) questionnaire. It was chosen because it is a reliable and valid measure of perceived usability and is widely applied. A set of questionnaires is distributed to the respondents to evaluate the usability of the system. Likert scale of 1-5 where 1 is for strongly disagree and 5 for strongly agree is used as the measure. Figure 2 shows the score for the satisfaction of the PEMS is 82.5, which means the users are satisfied with the overall usability of the system.

Table 6. Example of Scenario Segment and Test Case Question for Functionality Test.

Normal Scenario

Noraishah Mohd is one of the parents in Genius Aulad who would like to register the event for their children using the system-based application. When open the system homepage, the notification of upcoming events will be displayed,and show the latest events in the week. Next, she must login by fill in the I/C number and password to make the registration. Once login is successful, she can click on "more about events" in the event list. Next, it will display the details of the particular events and she must click on "register now" to register. After click on the "register now" button, the system will display form of the registration where she needs to select their children's name, fill in the medical declaration and read the terms and regulations. If she agrees with the terms and regulations, she must click on the checkbox to continue to the next section otherwise the checkbox will show an alert message "Please check this box if you want to proceed"....

Test Case ID	#A01		**Date Tested**	__/__/2020
Reviewed By			**Version**	
Created By	Nur Syakireen Bt Mat Yusoff		**Test Case Result**	Pass / Fail
Test Case Description	Test the parents to login the website			
Pre-requisite	1. Access to browser chrome			
	2. The user account must exist			
Test Data	1. guardian_ic = "890504145230"			
	2. guardian_password = "nor1234"			

No	Test Case	Expected Results	Actual Results	Pass / Fail / Not executed / Suspended
1.	Navigate to http://localhost:88/EMS/dashboard/websites/index.php	Site should be open		
2.	Click on "Login" icons	The system will show the login form		
3.	User enter the guardian IC number and password	Input field can be entered		
4.	Click on "Login" button	Successfully logged in		
Remark				

	System Usability Scale	Strongly Disagree 1	2	3	4	Strongly Agree 5
I1	I think that I would like to use this feature frequently.	0	0	0	2	**3**
I2	I found the system unnecessarily complex.	0	0	0	**3**	2
I3	I thought the system was easy to use.	0	0	0	**3**	2
I4	I think that I would need the support of a technical person to be able to use this system.	**3**	1	1	0	0
I5	I found the various functions in this system were well integrated.	0	0	1	**3**	1
I6	I thought there was too much inconsistency in this system.	2	**3**	0	0	0
I7	I would imagine that most people would learn to use this system very quickly.	0	0	0	2	**3**
I8	I found the system very cumbersome to use.	**3**	2	0	0	0
I9	I felt very confident using the system.	0	0	1	1	**3**
I10	I needed to learn a lot of things before I could get going with this system.	2	**3**	0	0	0

Figure 2. SUS Score for PEMS.

3.3. System Documentation

The last phase is where all the information in the system is combined into one report after the project is completed. System documentation provides the flow of the project and serves as a reference to the user about the PEMS project. Therefore, the documentation needs to be created for an explicit purpose and can be understood by future users.

4. Result

At the end of the project, the developer has successfully produced a system that fulfils the characteristics of an event management system has been reviewed in table 3. Screenshots for registration form and multiple event options are as depicted in Table 7. The Usability Heuristics principles has also been integrated in the system development and are shown in Table 5 in section 3. Furthermore, testing and evaluation has been conducted and the result for testing on five target users has shown satisfaction on system usability with the SUS score of 82.5. According to Thomas [11], the average system usability scale score is 68 points, and 80.3 or higher is the highest score point for the usability scale score to measure the system's satisfaction.

Table 7. Screenshots of the System.

Real-time registration	**Multiple event option**
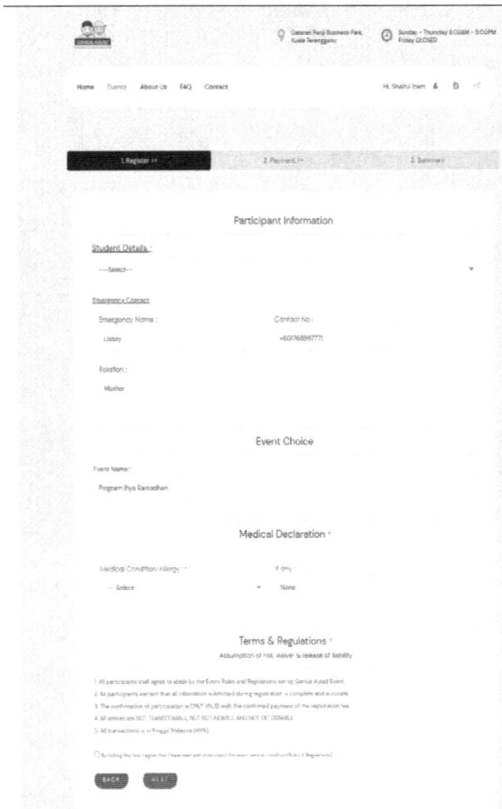	

IOP Conf. Series: Materials Science and Engineering 1176 (2021) 012038 doi:10.1088/1757-899X/1176/1/012038

5. Conclusion

Preschool Event Management System for Genius Aulad (PEMS) has been successfully tested by the target users. Based on the positive response revealed through analysis, it seems that the system has helped to alleviate problems inherited from the current business process in handling event related data. The development of PEMS is to facilitate communication between parents and the organization, payment process, notification process and report generation. The problems faced due to the unstructured way to communicate information regarding the upcoming events and event related data management have been countered. Evaluation results obtained will be used as a reference to come out with solutions on how to enhance PEMS in the future.

References

[1] Furduescu B A 2017 Management information systems *Holistica* **8(3)** p 60-71.
[2] Alcami R L and Caranana C D 2012 *Introduction to management information systems* Publicacions de la Universitat Jaume.
[3] Shrivash B K 2018 *Characteristics of management information system (MIS)* Vikas ® Publishing House Pvt. Ltd
[4] Saipriya S, Shruti S, Sanchana H, and Bushra S 2016 Event Management an android application *Int. J. Innov. Res. Sci. Eng. Technol.* **5(3)** p 3255-3259.
[5] Mishra V, Dubey M, Banarjee P, Jumle A, Raipure P and Wankhede P 2016 Event management system *International Journal of Trend in Research and Development* **3(6)** p 2394-9333
[6] Sharadha N L, Sushmitha H, Sangeetha S, Patil S, and Rashmi C 2018 Event management information system *International Journal of Advanced Research in Computer Science* **9 (Special Issue 3)** p 55
[7] Quiñones D and Rusu C 2019 Applying a methodology to develop user experience heuristics *Comput. Stand. Interfaces* **6** 103345
[8] Miller K, Capan M, Weldon D, Noaiseh Y, Kowalski R Kraft R, and Arnold R 2018 The design of decisions: matching clinical decision support recommendations to Nielsen's design heuristics *International Journal of Medical Informatics* **117** p 19-25.
[9] Nielsen J, and Molich R L (Eds.) 1994 *Usability Inspection Methods* **1** New York: Wiley
[10] Nielsen J 1992 Finding usability problems through heuristic evaluation *Proceedings of the SIGCHI Conference on Human Factors in Computing Systems* p 373-380.

Designing Roller Coaster Loop's By Using Extended Uniform Cubic B-Spline

N M Sukri[1]*, S M Nor-Al-Din[1], N K Razali[1] and M A A Mazwin[1]
[1]Faculty of Computer and Mathematical Sciences, Universiti Teknologi MARA (UiTM), 21080 Kuala Terengganu, Terengganu, Malaysia

*Corresponding author: nursyazni@uitm.edu.my

Abstract. Roller coaster is one of the attractions to the public. The design of roller coaster plays an important role for the smoothness and safety of each roller coaster rail. This research is about designing a smooth roller coaster rail loop using extended cubic uniform B-spline method as one of the alternative in designing a smooth roller coaster rail. Two-dimensional design of roller coaster rail loop is form by using extended cubic uniform B-spline with degree 3, 4 and 5 and shape parameter, λ= 0.5, and 1. Then, three-dimensional cubic B-spline is formed by using sweep surface which is translation method. The loop is analysed by the distance between curve and control polygon. Moreover, the G-felt with the different radius of the loop has been calculated. The value of G-felt for each method used will be compared to the G-felt value of the original roller coaster. The result from this research showed that by using extended cubic B-spline method degree 4, λ=0.5 is the best way to design roller coaster rail because it has the best G-force value compared to others designs.

1. Introduction

Roller coaster is one of the thrill rides in every theme park. It is an elevated railway with steep inclines and descents that carries a train of passengers through sharp curves and sudden changes of speed and direction for a brief thrill ride [1] - [3]. Most of the actual design and layout of a roller coaster is done on a computer. In the roller coasters with all those drops, twists, turns and changes of the loop have gravitational force and acceleration force that acting on it. G-force is a force of gravity on an object and it was measured in g's where 1 g is equal to the gravitational force at the Earth's surface which is $9.8ms^{-2}$ [1]. The engineers and the designers of the roller coasters will make sure those forces interplays for the passengers realy feel it acting on their body. Furthermore, inertia also happens in the roller coaster ride. Inertia is the resistance of an object to a change in its state of motion. The force that the passengers feel is the opposite of the force that acting on the passengers itself [2], [4]. Therefore, the safety of the roller coaster is determined by the design of the loops.

Extended cubic B-spline is one of the techniques that can be applied in system of design or in creating new objects. Extended cubic B-spline is an extension of B-spline [5]. One free parameter, λ, is introduced within the basis function where this parameter can be used to alter the shape of the generated curve. The value of λ can be varied to obtain different numerical results [6]. Extended cubic B-spline curves have the same properties of cubic uniform B-spline which is symmetry, geometric invariability, and convex hull [5].

The main objective of this paper is to design two and three dimensional roller coaster loop by using extended cubic uniform B-spline curves degree 4,5 and 6 with $\lambda = 0.5$ and 1. Then, G-force for each two-dimensional roller coaster loop design is analyzed to determine the best method to be used in roller coaster loop design.

2. Methodology

2.1. Extended Cubic B-Spline Curve

Extended Cubic Uniform B-Spline is used to design curve and generate the object. The basis functions are as follows [5].

Basis function for degree 4

$$\begin{cases} b_0^4(t)=\frac{1}{24}(4-\lambda-3\lambda t)(1-t)^3 \\ b_1^4(t)=\frac{1}{24}\left[16+2\lambda-12(2+\lambda)t^2+12(1+\lambda)t^3-3\lambda t^4\right] \\ b_2^4(t)=\frac{1}{24}\left[4-\lambda+12t+6(2+\lambda)t^2-12t^3-3\lambda t^4\right] \\ b_3^4(t)=\frac{1}{24}\left[4(1-\lambda)+3\lambda t\right]t^3 \end{cases} \tag{1}$$

The basis functions of extended cubic uniform B-spline degree 5 for $t \in [0,1]$ as follows:

$$\begin{cases} b_0^5(t)=\frac{1}{40}\left[(5-\lambda-4\lambda t)(1-t)^4\right] \\ b_1^5(t)=\frac{1}{40}\left[30+2\lambda-20(3+\lambda)t^2+40(1+\lambda)t^3-5(1+7\lambda)t^4+12\lambda t^5\right] \\ b_2^5(t)=\frac{1}{40}\left[5-\lambda+20t+10(3+\lambda)t^2-20(1+\lambda)t^3-5(1-5\lambda)t^4-12\lambda t^5\right] \\ b_3^5(t)=\frac{1}{40}\left[5(1-\lambda)+4\lambda t\right]t^4 \end{cases} \tag{2}$$

The basis functions of extended cubic uniform B-spline degree 6 for $t \in [0,1]$ as follows:

$$\begin{cases} b_0^6(t)=\frac{1}{60}\left[(6-\lambda-5\lambda t)(1-t)^5\right] \\ b_1^6(t)=\frac{1}{60}\left[48+2\lambda-30(4+\lambda)t^2+40(3+2\lambda)t^3-30(2+3\lambda)t^4+6(3+7\lambda)t^5+5\lambda t^6\right] \\ b_2^5(t)=\frac{1}{60}\left[6-\lambda+30t+15(4+\lambda)t^2-20(3+2\lambda)t^3+15(2+3\lambda)t^4-6(3+2\lambda)t^5-5\lambda t^6\right] \\ b_3^5(t)=\frac{1}{60}\left[6(1-\lambda)+5\lambda t\right]t^5 \end{cases} \tag{3}$$

Theorem 1: The basis functions $b_i^k(t)$, where k=4,5,6, and $i = 0,1,2,3$, satisfy

i. $\sum_{i=0}^{3} b_i^k(t) = 1$

ii. $b_i^k(t) = b_{3-i}^k(1-t)$

iii. When $-k(k-2) \le \lambda \le 1$, $b_i^k(t) \ge 0$, $t \in [0,1]$.

The polynomial curve segments for $u \in [u_i, u_{i+1}]$, $i = 3,4,\ldots,n$ are defined as follows [5].

$$C_{j,k}(\lambda : t) = \sum_{i=0}^{3} b_i^k(t) P_{j+i-3}, \quad k = 4,5,6 \tag{4}$$

Where, $P_i \in R^d$ $(d = 2,3,\ i = 0,1,2,\ldots n)$ is control point and knots is $u_1 < u_2 < \ldots < u_{n+1}$.
The polynomial curve is defined as follows [5].

$$C_k(\lambda; u) = C_{i,k}\left(\lambda; \frac{u - u_i}{h_i}\right), \quad u \in [u_i, u_{i+1}] \tag{5}$$

Where,

$$t = \frac{u - u_i}{h_i}, h_i = u_{i+1} - u_i.$$

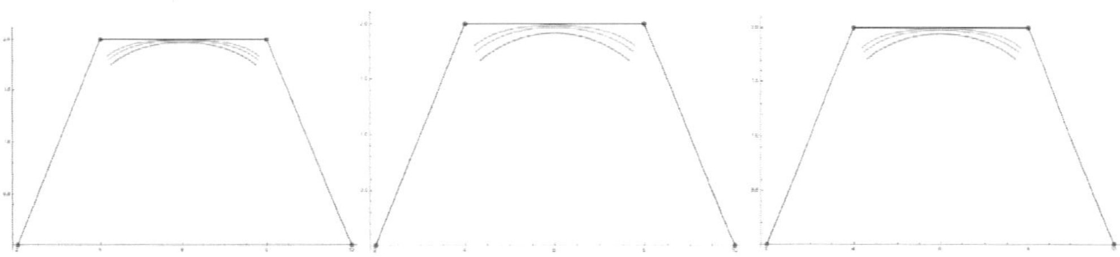

$\lambda = 0$ $\lambda = 0.5$ $\lambda = 1$

Figure 1. Extended cubic B-spline curves degree 4, 5 and 6 with the different shape of parameter, λ

Figure 1 above show that the extended cubic B-spline curves degree 4, 5 and 6 with the different shape of parameter $\lambda = 0$, 0.5 and 1 in the same control polygon. The result shows that all the curves obtained does not touch the start and end point. It is called as close curve. The distant between close curve and control polygon is changed based on the values of shape parameter, λ and the degree of Extended Cubic Uniform B-spline curve. The curves degree 4 with $\lambda = 0$, 0.5 and 1 is far from the control polygon while extended cubic B-spline curves degree 6 with $\lambda = 0$, 0.5 and 1 is the closest curve from control polygon for all values of shape parameter. Curve with $\lambda = 0$ is closest to the control polygon compared to curve with $\lambda = 0.5$ and 1.

2.2 Translation Technique

Translation of sweep surface method is known as transition of sweeping where the contour moves from one end of the line segment to another line segment perpendicularly. The transformation in three dimensional spaces which is moves in any direction is also called as translation [7] - [11]. Figure 2 shows the transformation of two dimensional curve to three dimensional object by using the equations of translation technique below [12], [13];

$$g(u) = [g_1(u), g_2(u)]^T, \ r(v) = [r_x(v), r_y(v), r_z(v)]^T \tag{6}$$

$$T(v)=\frac{r'(v)}{\|r'(v)\|},\ N(v)=\frac{T'(v)}{\|T'(v)\|},\ \text{and}\quad B(v)=T(v)+N(v) \tag{7}$$

$$S(u,v)=r(v)+g_1(u)N(v)+g_2(u)B(v) \tag{8}$$

where, $g(u)$= contour blended, r(v) = trajectory curve, N(v) = normal vector , T(v) = the unit tangent vector of the trajectory, and B(v) = binormal vector.

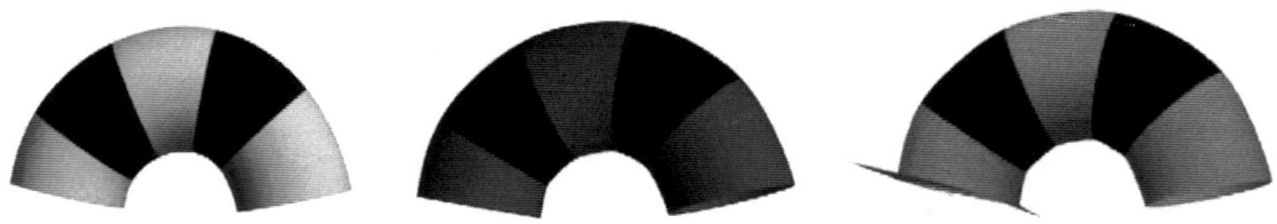

Figure 2. Three dimensional curves formed by translation

2.3 G's felt

A roller coaster is considered for amusement parks if it falls into the Goldilocks zone. The G-forces in the range 0.5g to 2g are suitable for children rides and for family rides is in range to 0g to 3 g. Thrill rides often involve 4 g or more, as well as negative g's, where the body lifts from the seat and must be held in place by the restraint system. The roller coaster will be too dangerous if it has many g's and too boring if it has few g's [3], [14]. Therefore, this is important to the designer keeping these within safe values.

For gravity force, the number represents how many times the force of gravity is felt at a particular point. In everyday life, everybody experiences 1G regular force of gravity when in still position. If a theme park says a roller coaster rider will experience 3G then the rider will briefly experience three times the force of gravity [1].

Positive G-Force occurs at the bottom of a drop when the roller coaster wants to continue moving in the same direction, but the track forces it in a different direction. The sharpness of this change of direction determines the positive G. Positive G forces typically occur when the train pulls up a hill after going down a drop. Passengers will experience G's at the loops. Formula from circular motion will be used calculate the G's felt. The centripetal acceleration must be calculated inside the loop [1].

$$a_{centripetal}=\frac{V^2}{R} \tag{9}$$

After the centripetal acceleration is obtained, the G's felt can be calculated

$$\text{G's force}=\frac{a_{centripetals}}{9.8\frac{m}{s^2}} \tag{10}$$

Lastly, the value of G-force will be used to obtain the g-felt.

$$\text{G's felt}=\text{G's force}-1 \tag{11}$$

3. Results and Discussion

3.1. Roller coaster's loop design

Extended cubic B-spline curve degree 4, 5 and 6 can be used to design a roller coaster's loop in two and three dimensional by using shape parameter λ= 0.5 and 1. The degree of curve and value shape parameter used will determine the smoothness of roller coaster loop. Figure 3, figure 4, figure 5 and figure 6 show the design obtained.

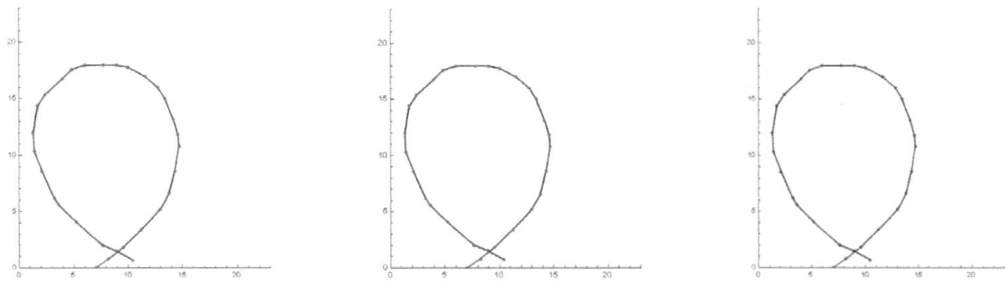

(a) Degree 4 (b) Degree 5 (c) Degree 6

Figure 3. Two dimensional curves by using shape parameter λ= 0.5

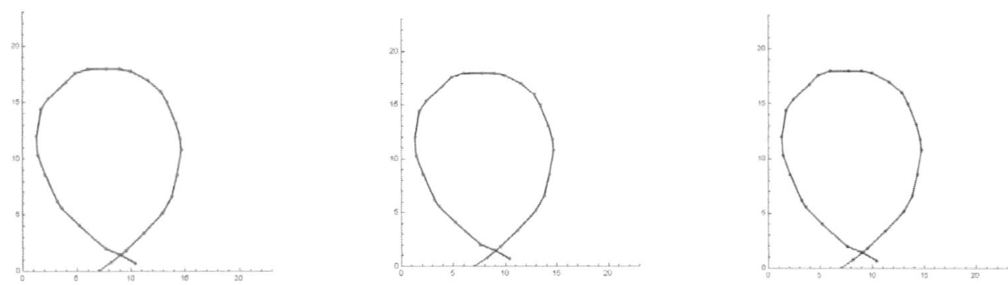

(a) Degree 4 (b) Degree 5 (c) Degree 6

Figure 4. Two dimensional curves by using shape parameter λ= 1

(a) Degree 4 (b) Degree 5 (c) Degree 6

Figure 5. Three dimensional curves by using shape parameter λ= 0.5

IOP Conf. Series: Materials Science and Engineering 1176 (2021) 012039 doi:10.1088/1757-899X/1176/1/012039

(a) Degree 4 (b) Degree 5 (c) Degree 6

Figure 6. Three dimensional curves by using shape parameter λ= 1

3.2. G-force value

Table 1 represents the result for G-felt in the loop of extended cubic B-spline degree 4,5 and 6 with shape parameter λ=0.5 and 1.

Table 1. Result for G-felt in the loop of extended cubic B-spline degree 4,5 and 6 with shape parameter λ=0.5 and 1.

Position	On the top	On the right	On the left
Real roller coaster	-1	0.58	0.53
Degree 4, $\lambda = 0.5$	-1	0.00	0.04
Degree 4, $\lambda = 1$	-1	0.00	-0.12
Degree 5, $\lambda = 0.5$	-1	0.16	0.14
Degree 5, $\lambda = 1$	-1	-0.12	-0.11
Degree 6, $\lambda = 0.5$	-1	0.19	0.00
Degree 6, $\lambda = 1$	-1	0.04	-0.08

The result obtained is compared to the G-force value of the real roller coaster's loop. It is clearly shows that extended cubic B-spline method degree 5 with λ=0.5 is the best way to design roller coaster loop because it has the best G-force value and most accurate to the G-force value of the real roller coaster loop.

4. Conclusion

In conclusion, extended uniform B-spline can be used to form two-dimensional roller-coaster loop's design. Then, it can be transformed into three dimensional loop by using translation method. The smoothness of loop's design is depending on the values of shape parameter used. G-felt values for each loops are calculated in order to determine which degree and value shape parameter is the best use to design the roller coaster loops. After that, the G-force for each of the roller coaster loop are calculated. G-force value is really important for the safety and comfort of the roller coaster's passengers. As the result, the best method in designing the loop of roller coaster is by using extended cubic B-spline method degree 5 with λ=0.5. This is because the G-force value is almost the same with the standard G-force value of the real roller coaster's loop. Nevertheless, using extended cubic B-spline method degree 5 with λ=0.5 produce the smooth roller coaster loop.

The extended cubic B-spline loops can be used as one of the alternative in design the roller coaster. However, this research can be more interesting and improved if further studies is applied in getting a better curve of the roller coaster rail. Furthermore, the next research can apply B-spline method in designing the whole roller coaster rail from the start point until its end point.

Acknowledgements
The authors would like to acknowledge UiTM Terengganu Branch Kuala Terengganu Campus for its support. All the objects were generated using Mathematica 11 software.

References:

[1] Pendrill A M 2005 Rollercoaster loop shapes *Phys. Educ* 517-521.
[2] Pendrill A M 2013 Student investigations of the forces in a roller coaster loop *European Journal of Physics* 1379-1389.
[3] Pendrill A M and Eager D 2020 Velocity, acceleration, jerk, snap and vibration: forces in our bodies during a roller coaster ride *Phys. Educ.* **55** 13.
[4] Nordmark A B and Essen H 2010 The comfortable roller coaster – on the shape of tracks with constant normal force *European Journal of Physics* **31** 1307.
[5] Gang X and Zhao W G 2008 Extended Cubic Uniform B-Spline and α -B-Spline *Acta Automatica Sinica* **34** 8.
[6] Hamid A N N, Majid A A and Ismail A I 2010 Extended Cubic B-spline Interpolation Method Applied to Linear Two-Point Boundary Value Problems *Int. J. Comp. Sci.* **38.**
[7] Pocock L and Rosebush J 2002 *The Computer Animator's Technical Handbook* (Burlington, USA: Morgan Kauffman Publisher).
[8] Azernikov S 2008 Sweeping Solids on Manifolds *Proc. of the 2008 ACM Symp. on Solid and Physical Modeling (Stony Brook New York)* 249-256.
[9] Marhl M, Guid N, Oblonsek C and Horvat M 1996 Extensions of Sweep Surface Constructions *Computer & Graphics* **20** 893-903
[10] Tai L and Loe K F 1996 Surface design via deformation of periodically swept surfaces *The Visual Computer* **12** 10 475-483
[11] Elber G 1997 Global Error Bounds and Amelioration of Sweep Surfaces *Computer-Aided Design* **29** 441-447
[12] Ali M J 2005 Permukaan Sapuan Translasi dan Putaran Lengkung Kuartik Serupa Bezier *Simposium Sains Kebangsaan Matematik ke XIII (Johor, Malaysia)* 495-499
[13] Salomon D 2007 *Curves and surfaces for computer graphics* (Springer Science & Business Media)
[14] Eriksson U and Pendrill A M 2019 Up and down, light and heavy, fast and slow-but where? *Phys. Educ.* **54**

Application of TOPSIS Method for Decision Making in Selecting the Best New Car in Malaysia

S M Nor-Al-Din[1]*, N K Razali[1], N M Sukri[1] and M A Rosli[1]

[1]Faculty of Computer and Mathematical Sciences, University Teknologi MARA (UiTM), 21080 Kuala Terengganu, Terengganu, Malaysia

*Corresponding author: sitim907@uitm.edu.my

Abstract. Purchasing and selecting of vehicles especially cars in the market is a tough decision to make due to the variety of the operational and technical parameter specifications such as design, body shape, technology used, and many more. Therefore, selection procedure techniques are required in order to overcome this problem. In the decision-making model, Technique for Order of Preference by Similarity to Ideal Solution (TOPSIS) is one of the selection procedure techniques to be embraced into this study. This technique provides a base for decision-making processes where there are limited numbers of choices but each has a large number of attributes. In this project, the selected vehicles were ranked based on their segments. The best car for sedan is Proton Persona, Proton Iriz was ranked as the best hatchback, Perodua Alza is the best MPV and Perodua Aruz was ranked as the best SUV. This result was based on the main criteria which are cost and total displacement, then follow by seat capacity, boot capacity, and top speed of the vehicle.

1. Introduction

The selection of automobile is crucial for the purchaser due to the complexity created by fake publicity and advertisement from the dealers. Choosing the best one among the choices given becomes a critical decision-making problem. The biggest constraint faced by most purchasers are the possible budget on which car to buy. Other important criteria in selecting the best car includes: fuel economy, comfort, and convenience features, life span, suspension, style, cost, reliability, safety, technology, capacity, convenience, etc. The outcome of this decision-making process will be significant and beneficial for both customers and manufacturers. In this paper, some cars are considered with different attributes and select the best car using TOPSIS technique.

In this study, it will be focusing on the ranking selected car among 1.5 and 1.6 engine capacity (cc) in Malaysia's market, which released from 2017 until latest 2019. This project will mainly focus on the following objectives:

- To identify the important criteria's that can be a factor which influenced the consumer in buying a new car.
- To use the TOPSIS method in evaluating the best cars based on selected criteria's.
- To provide the rank for each selected cars based on the car segment in Malaysia.

IOP Conf. Series: Materials Science and Engineering 1176 (2021) 012040 doi:10.1088/1757-899X/1176/1/012040

2. Literature review

The TOPSIS method is a multiple attribute decision-making method. This paper presents the principles and calculation steps of TOPSIS method in detail and applies it to the quantitative evaluation in selecting the best car based on the customer's priority. Through the analysis of practical application, we found that the TOPSIS method has relatively high rationality and applicability when it is used to evaluate the comprehensively competitive ability during the car selection [1]. TOPSIS method is called an ideal solution and also an effective multiple attribute decision-making method. This method is to construct the ideal solutions and minus idea solutions to the problems of multiple attributes and uses the two benchmarks of being close to the ideal solutions and being far away from the minus ideal solutions as the criteria for evaluating the feasible projects [2]. Based on experiments conducted on 60 datasets for employee placement with AHP, SAW, TOPSIS, and PROMENTHEE methods, obtained different accuracy levels of each method in terms of ranking employees. In terms of accuracy, TOPSIS method has the highest accuracy of 95% followed by PROMENTHEE of 93.34% and SAW 81.67%, and the last AHP of 50%. Regardless of the level of accuracy, TOPSIS is better to work on [3].

3. Methodology

- Step 1: Setting Criteria and Alternative

  This research considers nine criteria that can influence the ranking of the car. Table 1 state about criteria which is considered to complete the ranking of the car by their segments. The alternatives involved vehicles from 14 brands such as Perodua, Proton, Honda, Kia, Toyota, Nissan and etc, and make 44 alternatives in total.

Table 1. List of Criteria

Main Criteria	Main Code
Cost	C1
Seat Capacity	C2
Engine Power (horse power)	C3
Fuel Capacity (liter)	C4
Total Displacement (cc)	C5
Weight	C6
Boot Capacity	C7
Top Speed	C8
Maximum Engine Power (rpm)	C9

- Step 2: Establish the decision matrix

  The first step of the TOPSIS method involves the construction of a Decision Matrix (DM).

$$DM = \begin{array}{c} \\ L_1 \\ L_2 \\ \vdots \\ L_m \end{array} \begin{array}{c} \begin{array}{cccc} C_1 & C_2 & \cdots & C_n \end{array} \\ \begin{bmatrix} X_{11} & X_{12} & \cdots & X_{1n} \\ X_{21} & X_{22} & \cdots & X_{2n} \\ \vdots & \vdots & \ddots & \vdots \\ X_{m1} & X_{m2} & \cdots & X_{mn} \end{bmatrix} \end{array}$$

where i = criterion index ($i = 1 \ldots n$)
n = number of potential sites
j = alternative index ($j = 1 \ldots m$).
$C_1, C_2 \ldots, C_n$ = criteria
$L_1, L_2 \ldots, L_m$ = alternative locations

- Step 3: Calculate a normalized decision matrix
  The normalized values denote the Normalized Decision Matrix (NDM) which represents the relative performance of the generated design alternatives.

$$NDM = R_{ij} = \frac{X_{ij}}{\sqrt{\sum_{i=1}^{m} X^2_{ij}}}$$

- Step 4: Determine the weighted decision matrix
  Not all the criteria may be equal importance and hence weighting were introduced from pairwise comparison technique to quantify the relative importance of the different selection criteria. The weightage preferences are shown in table 2 while the weightage preference for criteria selected was presented in table 3.

Table 2. Weightage preference

Weight	Weight Preference
Not at all important	1
Slightly important	2
Important	3
Fairly Important	4
Very Important	5

Table 3. Weightage preference for criteria

Criteria	Weight preference
Cost	5
Seat Capacity	4
Engine Power (hp)	3
Fuel Capacity (liter)	3
Total Displacement (cc)	5
Weight	3
Boot Capacity	4
Top Speed (km/h)	4
Max Engine Power (rpm)	3

- Step 5: Identify positive and negative ideal solution
  The positive ideal (A+) and the negative ideal (A-) solutions are defined according to the weighted decision matrix.

$$PIS = A^{+} = \{V_1^{+}, V_2^{+}, \ldots, V_n^{+}\},\ \text{where: } V_j^{+} = \{(\max i (V_{ij}) \text{ if } j \in J);\ (\min i\, V_{ij} \text{ if } j \in J')\}$$

$$NIS = A^{-} = \{V_1^{-}, V_2^{-}, \ldots, V_n^{-}\},\ \text{where: } V_j^{-} = \{(\max i (V_{ij}) \text{ if } j \in J);\ (\min i\, V_{ij} \text{ if } j \in J')\}$$

- Step 6: Calculate the separation distance of each competitive alternative from the ideal and non-ideal solution.

$$S^{+} = \sqrt{\sum_{j=1}^{n} \left(V^{+}_{j} - V_{ij}\right)^{2}} \qquad i = 1, \ldots, m \qquad\qquad S^{-} = \sqrt{\sum_{j=1}^{n} \left(V^{-}_{j} - V_{ij}\right)^{2}} \qquad i = 1, \ldots, m$$

where i = criterion index
j = alternative index

- Step 7: Measure the relative closeness of each location to the ideal solution.

- Step 8: Rank the preference order.

The flow chart in figure 1 represents the research steps that have been used in this project:

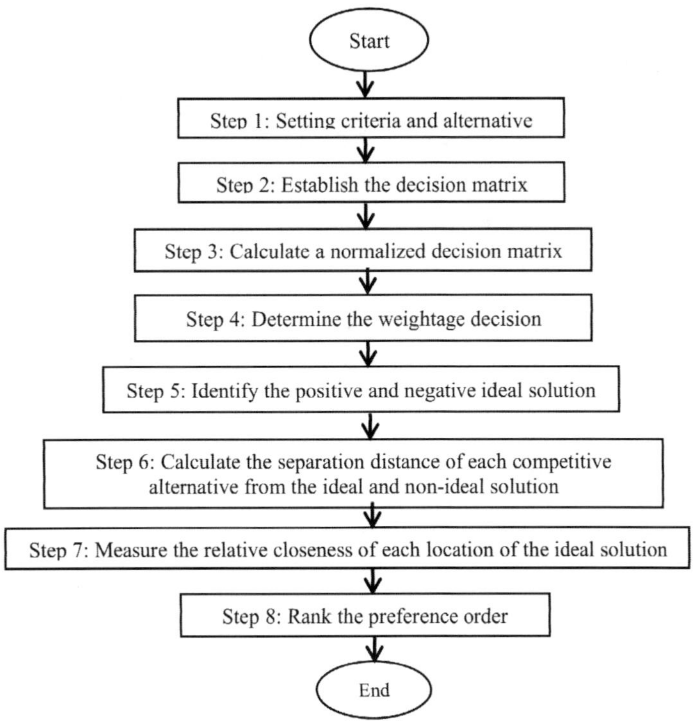

Figure 1: Flow chart of research steps

4. Results and discussion

Pair-wise comparison has been used to obtain the weightage for every criterion. Table 4 shown the weightage of the main criteria:

Table 4. Weightage of the main criteria

Criteria	Weightage %
Cost	17.3913043
Seat Capacity	13.0434783
Engine Power (hp)	6.52173913
Fuel Capacity (litres)	6.52173913
Total Displacement (cc)	17.3913043
Weight (kg)	6.52173913
Boot capacity (litres)	13.0434783
Top Speed (km/h)	13.0434783
Max Engine Power (rpm)	6.52173913

According on the table above, the most weight criteria are cost of the car and total displacement that are same weight and followed by three criteria with same level which is seat capacity, boot capacity, and top speed. Next, engine power, fuel capacity, weight of the car and max engine power in revolution per minutes got the last place among all the criteria where it is not really important to concern about when looking for car. The overall ranking was established by using TOPSIS model but the car ranking based on their segment will be more beneficial. Table 5 shows car ranking by their segment.

Table 5. Car Ranking by Segment

Segment	Rank	Car	Total final weightage
Sedan	1	Persona	0.997416
	2	Preve	0.997276
	3	Almera	0.997081
	4	Mazda 2	0.997023
	5	City	0.996976
	6	Vios	0.996837
	7	Vento Comfortline	0.996495
	8	Cerato	0.995281
	9	Civic	0.994944
	10	Focus Sedan	0.992666
	11	408 e-THP	0.991799
	12	Mazda 3	0.991785
	13	Veloster Turbo / Turbo Sport	0.989966
	14	BMW 3 Series	0.983223
	15	C-Class 200 Avantgarde	0.970992
Hatchback	1	Iriz	0.997379
	2	Myvi	0.99737
	3	Jazz	0.997067
	4	Yaris	0.99705
	5	Polo MPI Comfortline Vienna	0.996895
	6	Fiesta	0.996546
	7	Ioniq HEV / HEV Plus	0.995917
	8	208 Gti	0.992675
MPV	1	Alza	0.997305
	2	Exora	0.997262
	3	Avanza	0.996550
	4	5008	0.989196

SUV	1	Aruz	0.996964
	2	BR-V	0.99667
	3	Rush	0.996016
	4	HR-V	0.994888
	5	Tucson T-GDI	0.992055
	6	CR-V	0.990469
	7	3008 THP Plus	0.990435
	8	5008 THP Plus	0.988275
	9	GLA 200 Style	0.978605
	10	GLA 200 Night Edition	0.974304
	11	Countryman ALL4	0.973002
Coupe	1	CLA 200 Night Edition	0.974732
	2	C-Class Coupe AMG Line	0.947835
	3	i8 Coupe eDrive	0.128265
Wagon	1	508 SW THP	0.987217
Sports	1	i8 Roadster eDrive	0.0001745

5. Conclusion

Choosing the best and right option becomes a critical decision-making problem. To overcome this problem, Technique for Order of Preference by Similarity to Ideal Solution (TOPSIS) method has been used. TOPSIS method was used in evaluating the best cars based on numerous criteria's. This study also significant in providing the rank for each selected cars based on segment in Malaysia which is hatchback, sedan, SUV, MPV, coupe and sports. This outcome can be beneficial to the consumer who are intend to choose the best car based on the criteria selected. According to the survey, the criteria that be the best option to be consider are the cost matter, follow by total displacement (cc), and follow with seat and boot capacity and also top speed. Thus, TOPSIS has been successfully applied in to accomplish the objective of this study. Other than this method, Fuzzy TOPSIS, AHP, MAUT also can be used to find the best car in the market. The results shown that there might be a slight different if it used the other multi-criteria decision-making method. In the further studies, other models of multi-criteria decision-making method can be used for this problem and the results of this study and further studies can be compared. In conclusion, there are many methods that can be used for various type of problems.

References

[1] Zhongyou X 2012 Study on the Application of TOPSIS Method to the Introduction of Foreign Players in CBA Games *Phys. Procedia* **33** 2034 – 2039.

[2] Wang G 2006 Theory and Method of Decision-making [M]. *Anhui: Press of University of Science and Technology of China* 47-49.

[3] Widianta M M D, Rizaldi T, Setyohadi D P S et al 2018 Comparison of multi-criteria decision support methods (AHP, TOPSIS, SAW & PROMENTHEE) for employee placement *J. Phys. Conf. Ser.* vol 1 **953** 012116.

[4] Neil S 2010 Pairwise Comparisons *Encyclopaedia of Research Design* 2012

[5] Vahdani B, Mousavi S and Tavakkoli-Moghaddam R 2011 Group decision making based on novel fuzzy modified TOPSIS method *Appl. Math. Model.* vol 9 **35** 4257-4269

[6] Yue Z 2011 A method for group decision-making based on determining weights of decision makers using TOPSIS *Appl. Math. Model.* vol 4 **35** 1926-1936.

[7] Bulgurcu B 2012 Application of TOPSIS Technique for Financial Performance Evaluation of Technology Firms in Istanbul Stock Exchange Market *Procedia Soc. Behav. Sci.* **62** 1033-1040.

[8] Shuyu X 2012 Ecology Sports Studies Carried Out in School Sports *Phys. Procedia* **33** 2027-2033.

[9] Cengiz K, Sezi C O and Basar O 2015 Fuzzy Multicriteria Decision-Making: A Literature Review *International Journal of Computational Intelligence Systems* vol 4 **8** 637-666

[10] Jayant, Arvind and Sharma J 2018 A comprehensive literature review of MCDM techniques electre, promethee, vikor and topsis applications in business competitive environment *International Journal of Current Research* **10** 65461-65477.

[11] BMW Group Malaysia 2019 All Model.

IOP Conf. Series: Materials Science and Engineering 1176 (2021) 012040 doi:10.1088/1757-899X/1176/1/012040

[12] Perodua 2019 Our Models.
[13] Kia Motors Malaysia | Sedans, Hatchbacks, SUVs & MPVs 2019
[14] Malaysia Fantastic Rewards 2019.
[15] Jan J C, Pablo R and Brito P 2019 Multi-criteria decision making for the choice problem in mining and mineral processing: Applications and trends. *Expert Syst. Appl.* **121** 393-417
[16] Zigwheels New Cars Prices, *Read Car Reviews: Zigwheels Malaysia (n.d).*
[17] Rana M Z, et al 2020 Selection of medical clinic for disease diagnosis by using TOPSIS Method *Int. J. Pharm. Sci. Rev. Res.* vol 1 **61** 22-27.

Potential and Opportunities of Building Information Modeling (BIM) For Housing Maintenance

A M Sarman[1*], C Jaison[1], E M Mazlan[2] and N N Nazeri[1]

[1] Civil Engineering, Faculty of Engineering, Universiti Malaysia Sabah, Jalan UMS, 88400 Kota Kinabalu, Sabah, Malaysia.

[2] Development and Maintenance Department, Universiti Malaysia Sabah, Jalan UMS, 88400 Kota Kinabalu, Sabah, Malaysia.

*Corresponding author: sr.asmawan@gmail.com

Abstract. Building Information Modelling (BIM) represents an approach that aims to form and manage information over the building's whole life cycle. BIM has been mainly adopted for commercial or large scales building projects while being mostly neglected by small or residential building. Therefore, the objectives of this study are to evaluate the readiness of BIM implementation for housing maintenance in Sabah, to access the types of priority maintenance in house maintenance, and to determine the potential benefits to apply BIM for housing maintenance. This research is focused on the participants of the construction industry in Sabah to obtain their opinions towards the opportunities and strategy of Building Information Modeling (BIM) adoption in housing maintenance. Two method was done which was a pilot survey that was analyse using SPSS and Revit modelling. The result shows that most of the industry sectors had a moderate knowledge about BIM but they agree that it should be implemented by the authority and the percentage of willingness to change from the conventional maintenance to BIM was high. The 3 highest ranked of priority maintenance is the pipes linkage, electrical faults and sanitary appliance failure. Furthermore, a modeling is created using Revit application and access of information which called the Level of Detail (LOD) regarding the items in houses is easily obtain using schedule list for the purpose of maintenance.

1. Introduction

One of the difficulties in projects is always the need to have sufficient information on products readily available for any maintenance operation, such as: specifications, previous maintenance work, list of specialist professionals to conduct work, etc [1].

BIM comprises ICT frameworks and tools that can support stakeholders' collaboration over projects life-cycle. While BIM is thought to transform the way of the built environment is working, studies related to BIM have often been focusing on storing and sharing technical information. BIM capabilities can be enhanced by knowledge-based techniques, such as Case-Based Reasoning (CBR), to enable both information and knowledge-sharing that will benefit stakeholders. By incorporating Knowledge Management principles embedded in CBR systems with Information Management principles embedded in BIM systems, the transformation from 'Building Information Modelling' to 'Building Knowledge Modelling' can be better recognized [1].

On the other hand, Building Information Modelling (BIM) represents an approach that aims to form and manage information over the building's whole life cycle [2]. BIM could potentially provide more flexible and efficient approach where customers are informed about their property during and after the delivery process [3].

The main benefit of BIM is towards facility management. Opportunities for utilising BIM for facility management have also been investigated with the focus on supporting maintenance planning and identifying useful information in a BIM model needed by facility managers [4]. This is possible with

BIM as BIM provides a niche of software that all parties involved in the project could access and share information, amend adjustment and develop the design simultaneously within the design team [4]. Nowadays, there is a high demand of Building Information Modeling (BIM) as much international organization and government taking the steps in promoting BIM over the industry life cycle. The life cycle in BIM is primary sets it apart from preceding digital technologies, which were designed in specific phases of the building life cycle for specific sectors of the building industry, such as design, construction and Facility Management (FM) [5].

Most information created during the design and construction process that is of value to facility managers can only be found elsewhere and in scattered sources: in written construction specifications, warranty certificates and operations and maintenance manuals. However, the current practice BIM in scope of FM, stakeholders are not entirely implementing BIM in FM industry and in current FM operations that applied BIM, most functions still done manually even facility manager knowing by adopting BIM during operational building can decrease chance of errors and increase efficiency [5]. Managing facilities failure will raise the issues that occur repeatedly including assets are not registered, assets are not labelled as government asset, weaknesses on disposal and maintenance, asset does not be used or being wasted and asset misappropriation [6], [7]. Indeed, BIM has vast implications for the FM phase of buildings [5] with regard the owner and facility managers are in need of computerized supports that will improve how they operate and maintain their facility [8]. Here we can see that how important is the facilities management in construction industries especially on maintaining and operating. Based on an analysis, O&M activities can be divided into 5 groups which are maintenance and repair, energy management, emergency management, change or relocate management and securities [9].

Therefore, BIM have not been implemented in FM and have not been used for house maintenance. There is a gap of the BIM for house maintenance especially in Sabah. This is probably due to the reduction in workflow productivity of interactions between both professionals and organization, ineffective information delivery, and poor communication channels, reworks, and recurrent changes, among others [10]. The objectives of this research are to evaluate the readiness of BIM implementation for housing maintenance in Sabah. In addition, to access the types of priority maintenance in house maintenance as well as to determine the potential and opportunities to apply BIM for housing maintenance.

1.1. The Important Aspects in House Maintenance

Building Information Modeling (BIM) has constituted a technological and team coordination breakthrough in the architectural, engineering, and construction (AEC) industry, it has been mainly adopted for commercial and/or large scales building projects while being mostly neglected by small or residential building project teams [11]. The main reasons include:

- Home-builders' unawareness of BIM, or their misunderstanding of BIM as a mere innovative and more sophisticated software producing fancy 3D models, thus ignoring the new multi-disciplinary collaboration processes among project parties for building design [12];
- The high first investment for new software, training, and/or new hires to implement BIM [13];
- Home-builders' perception that residential projects are simple in comparison to, e.g., commercial projects, and that BIM practices significantly improve project outcomes (e.g., time, cost, and quality) only if used in complex projects [14], [15]; and
- The severe crisis in the residential building market during the last decade [16] impacting business owners' ability to invest in innovative methods and technologies, along with the higher probability or bankruptcy of small businesses if profits are not quickly produced in such market [17]

Furthermore, BIM for residential projects is gaining popularity. The National Association of Home Builders (NAHB) conducted a survey where 15% of home-builders claimed to be very familiar with BIM, and another 30% would like to use it in the future [12]. BIM is attracting home-builders' attention promising a great qualitative leap in residential project outcomes including time, cost, quality, and sustainability [15], [18]. Therefore, since there are some attentions for BIM implementation in residential project, it can be said that there is a potential and opportunities for BIM to be used in house maintenance.

1.2. The potential and opportunities of BIM in House Maintenance

The advantages of BIM technology are either to reduce cost, materials usage or indirectly through efficiency gains throughout the three major phases in the building lifecycle: design, construction and management.

Moreover, BIM can provide the necessary information needed for Facilities Management (FM) that is important for maintenance and repair, management of energy, and the commissioning [19].
Burgess G, Muir K and Jones M [20] said BIM is a way that will reduce costs, save time and energy, and summarizes the main benefits of BIM as:

- Reduction of design errors, by making the errors easier to be identify before a site commencement that help in cost and time saving by eliminating work.
- Avoiding conflicts and design clashes which affect in construction changes. Therefore, this saves on construction costs and time.
- Construction of sustainable buildings, by using environment-friendly materials which reduces energy use.
- Efficiency improvements in maintenance and operation companies. Maintenance time can be saved up to 15% and the costly site visit can be removed by having the historical design data instantly available.

2. Methodology

In order to achieve the aim and objectives of the research, questionnaire was sent out to the people in the construction industry by link of Google Form via WhatsApp. This method was widely used in conducting survey because it was the most economical method which can offer relatively high validity of results. As for Modelling, an AutoCAD drawing of houses was used which was the project of SPNB terrace housing in Ranau, Sabah.

However, there is some limitation when dealing with questionnaire survey. One of which, there is no guarantee that the people who complete the survey are the right person that you stated in the questionnaire form. Other than that, respondent might answer the survey generally and also based on their knowledge or heard from others rather based on their understanding towards the current industry.

The first test was the frequency analysis is used to represent the summary of the respondent"s profile which will then be tabulated out. Basically, the respondent's profile was from the construction industry. By knowing the distribution of the respondents' group, the analysis can be done easily based on the ratio of it.

Next was the Crosstab or Cross Tabulation is used to aggregate and jointly display the distribution of two or more variables by tabulating their results one against the other in 2-dimensional grids. It uses a process of creating contingency tables from the multivariate frequency distribution of variables, presented in a matrix format. Crosstab is widely used in survey results to find out interrelationships and interactions between variables.

The Chi-square test is intended to test how likely it is that an observed distribution is due to chance. It is also called a "goodness of fit" statistic, because it measures how well the observed distribution of data fits with the distribution that is expected if the variables are independent. A Chi-square test is designed to analyze categorical data. That means that the data has been counted and divided into categories.

Another method that was use was creating a modelling. A modelling was created using Autodesk Revit application. A housing model was shown to have a clear picture of the benefits obtain. One of which the goal is to produce a schedule list.

In a nutshell, after the data analysis has been conducted, the feasibility of this research can be proven and the opinion towards the particular issues can be summarized out and analyzed in more details in the next chapter. As for the modelling, it will be as the supporting material for the objective number 3 which was to show the benefits of BIM.

3. Results and discussion

There are 51 sets of questionnaires answered by various profile of respondents. Based on table 1, most of the respondent are from Contractor firm (41.2%) followed by Consultancy firm (39.2%). Furthermore, most of the respondent's profession are Civil Engineer (78.4%). Most of the questionnaire are replied by officer (60.8%). In addition, most of the respondent are less than 30 years old (88.3%) and most of them have an experience in their own profession of less than 5 years (86.3%). Additionally, their educational background are mostly Degree holders (62.7%).

Table 1. Respondent Profile

Descriptive	Frequency	Percentage (%)	Descriptive	Frequency	Percentage (%)
Industry Sector Organization			**Profession**		
Contractor	21	41.2	Civil Engineer	40	78.4
Consultant	20	39.2	Electrical Engineer	4	7.8
Government	7	13.7	Quantity Surveyor	4	7.8
Developer	3	5.9	Mechanical Engineer	3	5.9
Total	51	100	Total	51	100
Age			**Years of Experience in Profession**		
<=30	45	88.2	<= 5	44	86.3
31 - 40	5	9.8	<= 10	6	11.8
41 - 50	1	2	<= 15	1	2
Total	51	100	Total	51	100
Qualification			**Position in current organization**		
Degree	32	62.7	Officer	31	60.8
Diploma	14	27.5	Supervisor	14	27.5
Master	4	7.8	Manager	4	7.8
Certificate	1	2	Board of Director	2	3.9
Total	51	100	Total	51	100

From table 2, the Government (100%) had the highest percentage that willing to change from conventional to BIM in house maintenance. Then, the Consultant (75%), Contractor (38.1%), and Developer (33.3%). As an overall, most of them agreed by 60.8% and the response maybe by 37.3% and the rest answer no by 3.9%.

Table 2. Cross Tabulation of Industry Sector Organization and Agree Implement of BIM in House Maintenance by Authority

		Agree Implement BIM House Maintenance			Total
		yes	maybe	no	
Consultant	Count	15	5	0	20
	% within Industry Sector	75.0%	25.0%	0.0%	100.0%
Contractor	Count	8	11	2	21
	% within Industry Sector	38.1%	52.4%	9.5%	100.0%
Developer	Count	1	2	0	3
	% within Industry Sector	33.3%	66.7%	0.0%	100.0%
Government	Count	7	0	0	7
	% within Industry Sector	100.0%	0.0%	0.0%	100.0%
Total	Count	31	18	2	51
	% within Industry Sector	60.8%	35.3%	3.9%	100.0%

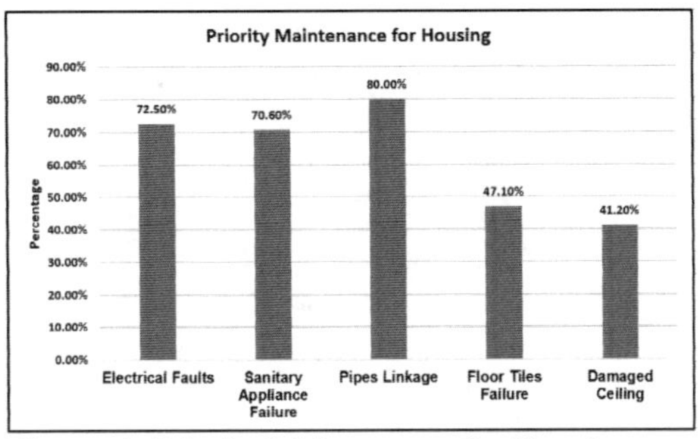

Figure 1. Priority Maintenance for Housing

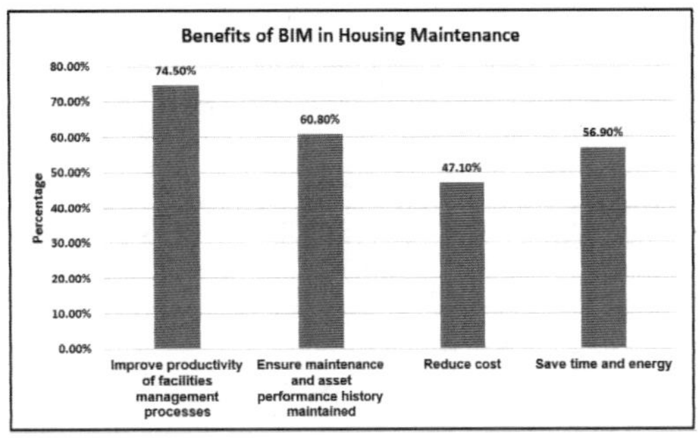

Figure 2. Benefits of BIM for Housing Maintenance

A question was asked about their opinion on which part of the house they think was the priority maintenance. According to Figure 1, the top three highest percentage was the pipes linkage (80.4%), follows by electrical faults (72.5%) and sanitary appliance failure (70.6%). Figure 2 shows that most of the respondent think that BIM will benefits in improve productivity of facilities management process (74.5%). Follows by the ensure maintenance and asset performance history maintained (60.80%), save time and energy (56.90%) and reduce cost (47.10%).

4. Revit modelling

A revit is an application built for Building Information Modeling with several features. It is not a BIM, but it and other applications made for BIM help designers design, simulate, visualize and collaborate in order to capitalize on the advantages of the interconnected data within a BIM model.

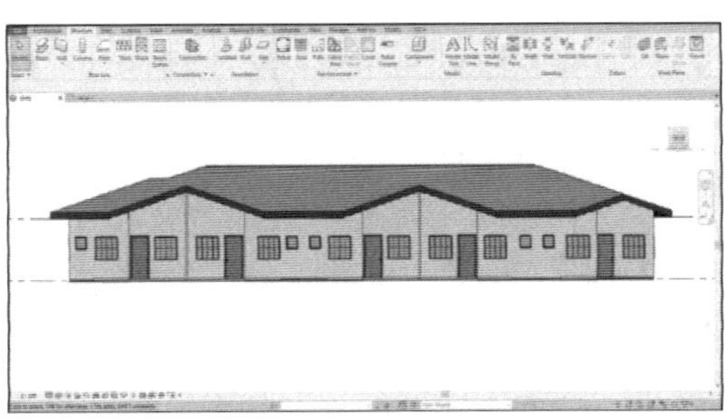

Figure 3. Front View of the Terrace House Model

Figure 3 shows the front view of a model of the SPNB project which is located in Taman Saujana, Ranau, Sabah, that is created in Revit. It is a terrace house project and the model consists of 5 single storey house.

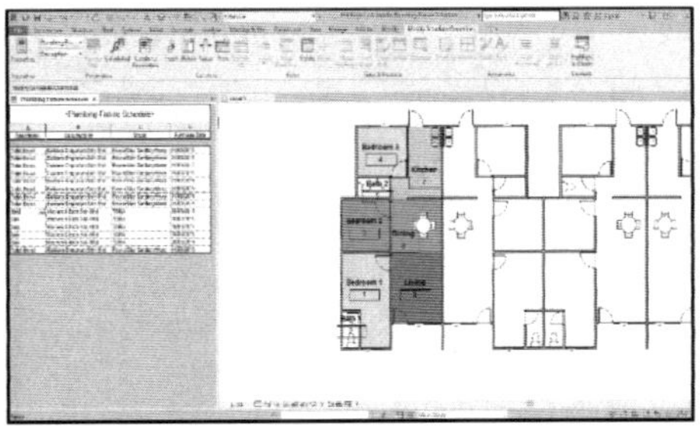

Figure 4. Plumbing Fixture

<Plumbing Fixture Schedule>			
A	B	C	D
Description	Manufacturer	Model	Purchase Date
Toilet Bowl	Builders Emporium Sdn Bhd	PowerStar SanitaryWare	01/03/2015
Toilet Bowl	Builders Emporium Sdn Bhd	PowerStar SanitaryWare	01/03/2015
Toilet Bowl	Builders Emporium Sdn Bhd	PowerStar SanitaryWare	01/03/2015
Toilet Bowl	Builders Emporium Sdn Bhd	PowerStar SanitaryWare	01/03/2015
Toilet Bowl	Builders Emporium Sdn Bhd	PowerStar SanitaryWare	01/03/2015
Toilet Bowl	Builders Emporium Sdn Bhd	PowerStar SanitaryWare	01/03/2015
Toilet Bowl	Builders Emporium Sdn Bhd	PowerStar SanitaryWare	01/03/2015
Toilet Bowl	Builders Emporium Sdn Bhd	PowerStar SanitaryWare	01/03/2015
Sink	Kitchen & Bath Sdn Bhd	TORA	20/03/2015
Sink	Kitchen & Bath Sdn Bhd	TORA	20/03/2015
Sink	Kitchen & Bath Sdn Bhd	TORA	20/03/2015
Sink	Kitchen & Bath Sdn Bhd	TORA	20/03/2015
Sink	Kitchen & Bath Sdn Bhd	TORA	20/03/2015
Toilet Bowl	Builders Emporium Sdn Bhd	PowerStar SanitaryWare	01/03/2015

Figure 5. List of Plumbing Fixture

Figure 4 and figure 5 shows the plumbing fixtures schedule list that shows the description which is the specific name of the items, manufacturer, the model or brand, and purchase date of an item. The information needs to only be key in once and it is automatically applied to all same items in the project. Furthermore, revit will update any changes made to the schedule. Therefore, if any problem occurs that required maintenance, this list of schedules can be refered to get extra information about the items. For instance, if the toilet bowl needed to be replaced and the owner wants the same as the old types of toilet bowl or only the manufacturer had the parts that need to be replaced, it can be easily get since the name of manufacturer is recorded. In addition, whenever a maintenance is done, the date can be updated. It is a suitable criterias and one of convenient methods for planned maintenance.

5. Conclusion and recommendation

This research found that the industry of construction agreed with the implementation of BIM for the purpose of housing maintenance (Objective 1). In addition, there are many items that needed to be maintain in a single house and some findings related to the priority maintenance was found (Objective 2) and several potential benefits that agree by the construction industry (Objective 3).

It is crucial to identify that all industry sectors are mostly ready if the authority implemented BIM for housing maintenance and are willing to turn from the conventional method. As stated in chapter 2 that the authority plays an important role institutes a mandate for BIM implementation to be followed by its affiliated agencies. Therefore, the BIM implementation in Sabah can be achieved.

Moreover, the top three ranked priority maintenance for houses are pipes linkage it may due to leakage problem, second priority is the electrical faults which can be dangerous to maintain ourselves and last is the sanitary appliance failure which can be complicated to maintain.

Generally, the potential benefits of BIM are analyse using bar chart and the benefits stated which is improve productivity of facilities management process which is highly agree by the respondent. However, several benefits can not be neglected since there are more than half of the respondent agree to it. The rest are ensuring maintenance and asset performance history maintained, save cost and save time and energy.

As mentioned on the objectives 3 which is the benefits when apply BIM for housing maintenance. Once a building is complete, it enters into the operation and maintenance phase of its lifecycle. The Revit modelling support the statement where it can show the schedule list of items which will display the information regarding the items clicked.

The subject on Building Information Modeling (BIM) is continuously under study. Based on the study, research of an alternative based on the identified barriers in promoting adoption of BIM can be possible which beneficial to the industry. Moreover, some development on suggestion to the authority of an appropriate system can be done for BIM implementation.

References

[1] Motawa I and Almarshad A 2013 A knowledge-based BIM system for building maintenance *Autom. Constr.* **29** 173–182.
[2] Liu R and Issa R 2013 Issues in BIM for facility management from industry practitioners'perspectives *Computing in Civil Engineering* 411-418.
[3] Parn E, Mayouf M, Laycock P and Edwards D J 2015 Building information modelling (BIM) for uk housing: exploring potential drivers and opportunities to deliver better customer experience *Mindanao Journal of Science and Technology* **13** 51–64.
[4] M Sarman A, Mazlan E M, Jompiuh A, Lim C H and M Hayati M F 2020 *The 6th Int. Online Conf. on Science, Technology, and Interdisciplinary Research (Universiti Teknologi Malaysia).*
[5] Aziz N D, Nawawi A H, and Ariff N R M 2016 *Procedia Soc. Behav. Sci.* **234** 353-362
[6] Berahim N, Jaafar M N, and Razali M N 2013 *Proc. of the Global Conf. on Business, Economics and Social Sciences* pp 262-273.
[7] Mahadi A and Hussin H 2007 Pengurusan aset kerajaan berkomputer: Satu anjakan paradigma
[8] Akcamete A, Akinci B, Garrett J H and Jr 2010 *Proceedings of the International Conference on Computing in Civil and Building Engineering (UK: Nottingham University Press)* p 151
[9] Gao X and Pishdad-Bozorgi P 2019 BIM-enabled facilities operation and maintenance: A review. *Advanced Engineering Informatics* **39** 227-247.
[10] Muñoz La Rivera F, Vielma J and Herrera Rodrigo F and Carvallo J 2019 *Advances in Civil Engineering* 1-16
[11] NAHB 2016 *Residential builders adopt commercial building tools to increase sales* (Washington, D.C: NAHB- Now, National Association of Home Builders)
[12] NAHB 2014 *Building information modeling (BIM) survey* (Washington, D.C: National Association of Home Builders)
[13] Nellis S 2012 *Master Thesis submitted to the Construction Management Program at East Carolina University*
[14] Excelize 2015 *Advantages of BIM for residential designs* (Chesterfield, New Jersey: Excelize)
[15] Fabris P 2010 *Professional Builder Magazine*
[16] McGraw Hill Construction 2014 *Green multifamily and single-family homes: growth in a recovering market* (New York City, NY: SmartMarket Report, McGraw Hill Financial)
[17] Robinson C, O'Leary B and Rincon A 2006 *Business start-ups, closures and economic churn: A review of the literature* (BERR, UK: Enterprise Directorate)
[18] Poirier E A, Staub-French S and Forgues D 2015 *Autom. Constr.* **58** 74-84.
[19] Eastman Charles M 2008 *BIM Handbook: A Guide to Building Information Modeling for Owners, Managers, Designers, Engineers and Contractors*
[20] Burgess G, Muir K and Jones M 2018 *BIM in the UK house building industry: opportunities and barriers to adoption* (UK: Cambridge Centre for Housing & Planning Research)

IOP Conf. Series: Materials Science and Engineering 1176 (2021) 012042 doi:10.1088/1757-899X/1176/1/012042

Construction of a Tributyrin Calibration Curve using High Performance Liquid Chromatography

F Atan[1], A H Mohamed[1], N N Dzulkifli[1], Z M Zain[2] and N M Ahmad[1*]

[1]School Of Chemistry and Environment, Universiti Teknologi MARA Cawangan Negeri Sembilan, Kampus Kuala Pilah, 72000 Negeri Sembilan, Malaysia

[2] Electrochemical Material and Sensor Research Group, Faculty of Applied Sciences, Universiti Teknologi MARA, Shah Alam, Selangor. Malaysia

*Corresponding author: normonica@uitm.edu.my

Abstract. Most of the explored research works on Triglycerides or any of the Triglycerides families was based on the profiling identification than a quantification. To quantify them, a calibration curve needs to be constructed at a certain range of concentration. In this study, the calibration curve for the shortest chain of Triglycerides namely Tributyrin was accomplished using high performance liquid ultraviolet detector with two variables parameters. The composition of mobile phase for tributyrin was found optimum using 50:50 acetonitrile: acetone while the optimum flowrate for tributyrin was 0.55 mL/min using a 20 μL injection, constant condition at 35°C and 215 nm wavelength. The calibration curve of standard tributyrin was found linear from 0.0025 - 0.0375 mM with the correlation coefficient (R2) of 0.986. The developed calibration curve could serve as a future platform for the quantification of TB in real samples with various matrixes such as food and serum.

1. Introduction

Triglycerides (TGs) are important in human metabolism as an energy source for transportation of dietary fat. TG is an ester derived from glycerols that contain three hydroxyl (OH) groups and three molecules of fatty acid with a carboxyl group (COOH). It includes other individual TGs, which are tributyrin (1,2,3-tributyrylglycerol), tricaprylin (1,2,3-trioctanoylglycerol), tripalmitin (1,2,3-trihexadecanoylglycerol), tristearin (1,2,3-trioctadecanoylglycerol), triolein (1,2,3-tri-[c-9]-octadecenoylglycerol), and trilinolein (1,2,3-tri-[c9,c12]-octadecadienoylglycerol), depending on the number of carbon associated with the number of fatty acid attached to one molecule of glycerol. However, an excess of TGs in food and blood may cause coronary heart disease with a permitted concentration of 40 to 160 mg dL−1 in men and 35–135 mg dL−1 in women [1]. Among the TGs, tributyrin (TB) which is the shortest medium-chain TGs and hydrophobic in nature [2, 3], has been used as a model of TGs substrate in a biosensor development [4,5]. The major advantage of quantification of TG analysis is it allows food-related authorities and professionals to establish information in providing sufficient knowledge to the public, therefore, many efforts have been done on TGs using HPLC.

The selection of the detector in HPLC is a crucial factor in TG profiling analysis. Refractive detector (RI), ultraviolet detector (UV), mass detector (MS), and evaporative light scattering detector (ELSD) are the common detector in HPLC which owe their advantages and disadvantageous to TGs detection. For example, RI was found unsuitable for a gradient elution system which requires longer time for stabilization. Even though it is not affected by the polarity of the mobile phase nature, it is found as a suitable detector for the fractionation and quantification of the TGs peaks [6]. Another approach in TGs chromatography analysis involved HPLC-ELSD due to its stability which is not strongly affected by the polarity of the mobile phase and is available for gradient elution [2,7,8]. Even though this detector is likely to be a universal detector for TGs detection, it is not suitable for the quantitative analysis and the fractionation of the peaks because the signal for calibration curves produced by this detector could not produce a linear equation [6]. Attention has been focused on HPLC-MS system as another universal detector and a new method for the TGs molecular species analysis due to its ability to identify the structure of each peak from the fragmentation pattern along with quantification. However, this instrument is known to be expensive which requires highly-skilled personnel and time investment for personnel training [9]. HPLC-UV is influenced by the nature of the mobile phase and is able to analyze compounds containing chromophore through absorption at selective wavelengths such as 210, 215, and 255 nm. However, a study by Naviglio et al. [10] on TGs using HPLC-UV found that detection still does not involve quantification analyte but requires transesterification of triglycerides as phenethyl esters before being examined through this detector to create a chromophore that allows detection using UV detection methods. This is well supported by Lee et al. [7] who agreed that the UV detector has limitations on species with the absence of many double bonds such as TGs and other lipid species.

The preparation of samples containing fat or oil required a simple method as reported by Naviglio et al. [10]. In the work, anhydrous fat was dissolved in hexane, and 1 μL of this mixture was injected prior to profiling analysis. However, the analysis on quantification was not reported, therefore developing an urgency to study the conditions for producing a calibration curve before quantitation of TG can be performed. TGs can be effectively eluted by HPLC using octadecyl groups (RP-18) as a stationary phase as has been reported in many research. To observe the elution, good criteria of mobile phase are highly needed which would enhance the polarity of the mobile phase, increasing peak sensitivity and reducing retention times. This can be achieved with the incorporation of acetonitrile with an organic modifier such as acetone that improves the solubility of TGs [11]. Although there were reported uses of other mobile phases such as diethyl ether, alcohol, tetrahydrofuran, and hexane to enhance the solubility of TGs, the improvement on the sensitivity is almost absent in HPLC-UV detection due to the absence of the chromophore in the mobile phase. Retention time is the time required for a compound to travel in the stationary phase before being detected by a detector. Reversed-phase high-performance liquid chromatography (RP-HPLC) that utilized the polar mobile phase and nonpolar stationary phase has been widely explored for the analysis of the TG mixtures where the retention time is mostly affected by the total length of the acyl chains together with the degree of unsaturation [12]. Retention time is influenced by the polarity and affinity of a compound against the mobile phase and stationary phase. A long period of analysis involving HPLC will usually increase the cost of the consumption of the mobile phase. In addition, at the end of a long elution, a blank run should be performed using a mobile phase for at least 30 minutes to ensure optimal column cleaning between samples as agreed by King et al. [13]. Due to the many compounds in TGs, elution involving a typical integration for 15 families of triglycerides normally require almost half an hour to complete one elution [6, 14, 15]. It is found to be impractical if a researcher only needs one compound from TGs especially if it involves components in earlier elution such as TB. Most of the research works on TGs are on profiling identification instead of quantification that has been explored using ELSD [2] MS [7] UV [10] and RI [11]. Quantification had been successfully performed with a combination of two detectors which are charged aerosol detection (CAD) in line with the UV detection and MS that permits the detection and quantification of both short-chain volatile molecules and non-volatile complex TGs [13]. However, the calibration curve for individual standards of each triglyceride to assess linearity was not employed in those studies. In this study, the development of the calibration curve of individual TGs containing the shortest carbon chain, which is TB as a model, was performed through a variable affecting the TB significant peaks. These changes are a composition of the mobile phase and mobile phase flow rate that contribute to the significant TB in HPLC peak.

2. Research Method

2.1. Chemicals and instrumentation
N-hexane, acetone, and acetonitrile with 99.9% purity were used as a mobile phase (Merck, Germany). The standard solution of tributyrin with 99.9% purity was purchased from Sigma, Germany. The solvents and prepared standards were degassed in an ultrasonic bath prior to HPLC analysis and filtered through a vacuum filter to remove possible particulate matter and prevent clogging. A HPLC-UV instrument was used in this study. This instrument consists of carbon 18 column (150 mm × 4.6 mm, 5 μm) equipped with a quaternary pump with a 20 μL loop injector and UV detector (Agilent Technologies 7820A). The temperature of the column and the detector was set at 35 °C respectively. The mobile phase is acetonitrile/acetone at a certain ratio and a flow rate was used for optimization. The wavelength was set at 215 nm.

2.2 Preparation of stock, standard solution and calibration curve of Tributyrin
The stock standard solution was prepared in 1617 μL hexane containing 382 μL of tributyrin standard. To prepare the serial standard solution, the stock was diluted with a certain volume of hexane to obtain 0.3 mM, 0.15 mM, 0.075 mM, 0.0375 mM, 0.018 mM, 0.009 mM, 0.0045 mM, and 0.00225 mM. Using the optimized condition, 20 μL of each concentration was injected three times into the injector loop. The calibration curve was plotted using the peak area as the y-axis and concentration as the x-axis. The correlation coefficient and equation on calibration curves were constructed using the formula from Microsoft Excel.

2.3 Effect of mobile phase composition and flowrate
To further the experiment, five different mobile-phase compositions were optimized which are 70:30, 60:40, 50:50, 40:60, and 30:70 of acetonitrile: acetone. The study was done by using 0.018 mM standard tributyrin at 0.75 mL/min flowrate and 215 nm wavelength. Next, studies on various flow rates were performed to identify the optimum flow rate of acetonitrile and acetone. In this analysis, five flow rates which were 0.35, 0.55, 0.75, 0.95, and 1.00 mL/min have been used throughout the experiments. The wavelength was set at 215 nm. The optimization process was done by using 0.018 mM standard tributyrin at 50:50 acetonitrile: acetone mobile phase composition.

3. Result and Discussion

3.1. Effect of mobile phase composition
Five compositions of mobile phase were used in the analysis of tributyrin starting from the most polar to the least polar solvent which is acetonitrile: acetone (70:30), (60:40), (50:50), (40:60), and (30:70). The effect of the mobile phase composition of acetonitrile: acetone to the peak area of tributyrin using 0.018 mM and wavelength of 215 nm is shown in figure 1. From the graph, the highest peak area was found using 50:50 acetonitrile: acetone and started to decrease at 40:60 acetonitrile: acetone and 30:70 acetonitrile: acetone. Therefore, the composition of 50:50 acetonitrile: acetone was selected for subsequent optimization. TB is a polar compound with the presence of ester linkage. It is more polar compared to other TG due to its shortest chain of fatty acid. The equal combination of acetonitrile and acetone composition is suitable possibly due to the characteristic of the medium polarity solvent for both solvents with a polarity index of 5.1 and 5.8, respectively. Acetonitrile is one of the known solvents that has successfully improved the solubility of TG, peak selectivity, retention time, and selectivity of TG compound with the use of another organic modifiers such as acetone [11].

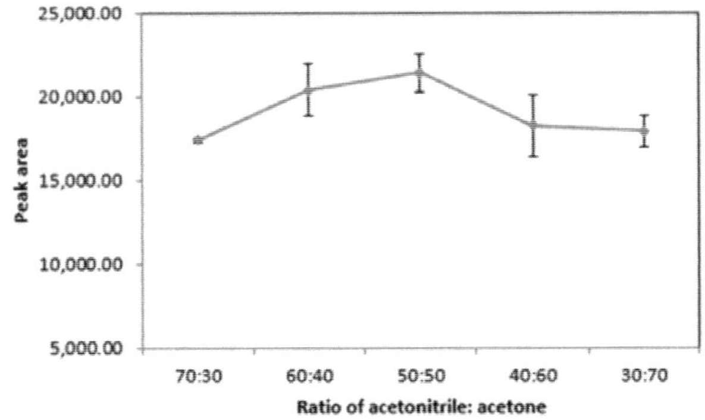

Figure 1. Effect of mobile phase composition of acetonitrile: acetone to the peak area of tributyrin using 0.018 mM and wavelength of 215 nm.

3.2. *Effect of flowrate*

Various flow rates reduce retention time as a higher flow rate will sweep the elution faster. Figure 2 shows the effect of flow rates on the peak area of tributyrin. The peak area was found to increase when the flow rate was increased from 0.35 mL/min to 0.55 mL/min. However, the peak area was found to start decreasing at flow rates of 0.75, 0.95, and 1.0 mL/min. Therefore, the flow rate of 0.55 mL/min was chosen for the subsequent experiments due to the optimum peak area compared to the other flow rates. This finding was almost similar to Zeb and Murkovic's [16] result who obtained a flow rate of 0.6 mL/min using the MS detector. However, this contradicts the mentioned flowrates of 1.4 mL/min [11] and 1.5 mL/min [17].

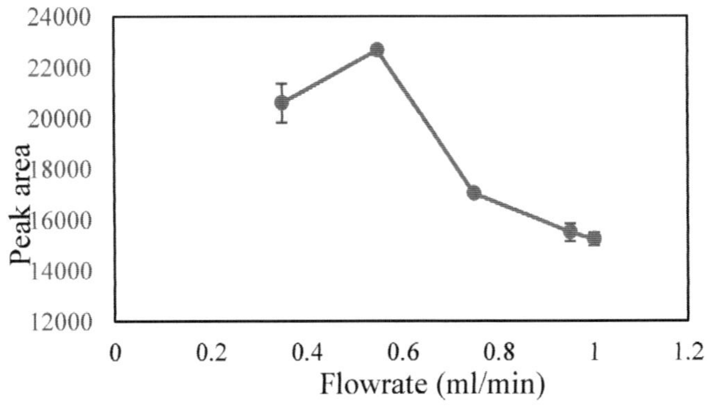

Figure 2. Effect of flowrate to the peak area of 0.018 mM Tributyrin at wavelength of 215 nm.

3.3. *Calibration curve*

To investigate the linearity of the HPLC-UV towards the TB standard, several concentrations of TB were injected to the injector which were 0.3 mM, 0.15 mM, 0.075 mM, 0.0375 mM, 0.018 mM, 0.009 mM, 0.0045 mM, 0.00225 mM, and blank. Figure 3 shows the chromatogram of the TB at different concentrations. A sharp peak that was successfully obtained indicates that the optimized parameters are suitable for the elution of tributyrin. Figure 4 shows the calibration curve of the TB standard at 0.55 mL/min and 50:50 mobile phase composition with a correlation coefficient of 0.986. Based on the graph shown, the peak area was found proportionally linear to the concentration of standard tributyrin at a range of 0.025–0.0375 mM. Concentrations higher than 0.0375 mM were not included in the calibration curve since it reaches a saturated point in the chromatogram. A similar study has been explored by Rombout et al. [8] but using gradient elution with acetone, acetonitrile, and dichloromethane as the mobile phase which requires a higher cost than the present study. This present study was also better than

IOP Conf. Series: Materials Science and Engineering 1176 (2021) 012042 doi:10.1088/1757-899X/1176/1/012042

reported by Aoki et al. [6] who obtained difficulties to get a linear calibration of TGs curve through gradient elution using HPLC-UV. So far, the study to quantify TGs is a huge challenge as quantification of complex TGs is difficult due to the large possible variation in the carbon number of the fatty acids at each position on the glycerol backbone and the presence of the configuration of cis and trans to the double bonds on each fatty acid (Wei et al., 2021). Therefore, an exploration of the fundamentals of the calibration curve is highly demanded prior to quantification. A calibration curve is made possible in the present study due to the reverse phase chromatography which used carbon 18 as a column. A similar column was reported by Chen et al. (2020) on TG profiling of species in human milk. In the future, the approach of quantifying serum samples using the developed calibration curve of TGs could be a benchmark to support the past available profiling study. This is very important since the heart and cardiovascular disease become a worldwide concern that may be contributed by elevated TGs in human serum (Ritchey et al., 2019). To express the importance of quantification, to date, another advanced technique has been explored by Bin Peng et al. (2021) using the enzymatic esterification of oil samples to elucidate the effects of different TGs on lipid accumulation in oleic acid-induced human hepatic cells culture.

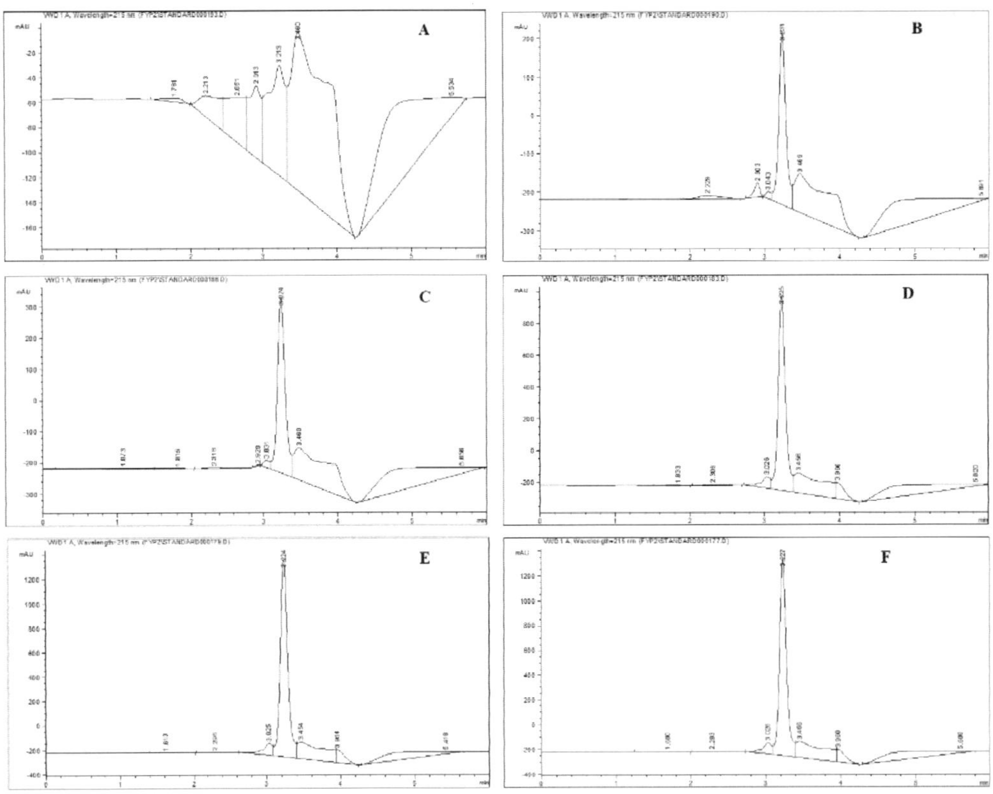

Figure 3. HPLC chromatogram of A) blank, B) 0.00225 mM, C) 0.0045 mM, D) 0.009 mM, E) 0.018 mM and F) 0.0375 mM. Condition: Wavelength was set at 215 nm.

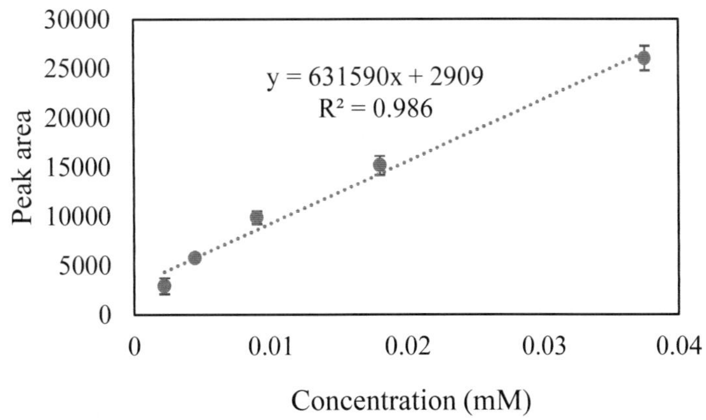

Figure 4. Calibration curve of standard tributyrin at 0.55 mL/min and 50:50 acetone: acetonitrile.

4. Conclusion

The composition of 50:50 acetonitrile: acetone and flow rate of 0.55 mL/min were used to develop the calibration curve of tributyrin. The calibration curve was successfully constructed at 0.0025–0.0375 mM based on the best condition of the variable on the mobile phase composition and flow rate. This finding can be a future platform for the quantification of TB in real samples with various matrixes such as food and serum.

Acknowledgements

The authors gratefully acknowledge UiTM Cawangan Negeri Sembilan, Kampus Kuala Pilah for giving permission to access high performances liquid chromatography-ultraviolet (HPLC-UV) throughout the study.

References

[1] Bhardwaj S K, Chauhan R., Yadav P, Ghosh S, Mahapatro A K, Singh J and Basu T 2019 Bi-enzyme functionalized electro-chemically reduced transparent graphene oxide platform for triglyceride detection *RSC Chem. Biol.* p 1-9.

[2] Lee J-H, Jones K C, Lee K-T, Kim M-R and Foglia T A 2003 High-Performance Liquid Chromatographic Separation of Structured Lipids Produced by Interesterification of Macadamia Oil with Tributyrin and Tricaprylin *Chromatographia* **58** p 653-658.

[3] Pundir C S and Narwal V 2018 Biosensing methods for determination of triglycerides: A review in *Biosens. Bioelectron.* **100** p 214–227.

[4] Solanki P R, Dhand C, Kaushik A, Ansari A A, Sood K N and Malhotra B D 2009 Nanostructured cerium oxide film for triglyceride sensor *Sens. Actuators B Chem.* **141(2)** p 551-556.

[5] Solanki S, Pandey C M, Soni A, Sumana G and Biradar A M 2016 An amperometric bienzymatic biosensor for the triglyceride tributyrin using an indium tin oxide electrode coated with electrophoretically deposited chitosan-wrapped nanozirconia *Mikrochim. Acta* **183** p 167–176.

[6] Aoki T, Otake I, Gotoh N, Noguchi N and Wada S 2004 Quantification Method for Triglyceride Molecular Species in Fish Oil with High Performance Liquid Chromatography-Ultraviolet Detector *J. Oleo Sci.* **53** p 285-294.

[7] Lee J H, Jones K C, Foglia T A, Lee C-R, Akoh C C and Lee K-T 2008 High performance liquid chromatographic separation of interesterified, palm oil with tributyrin *Food Science and Technology* p **41(8)** 1446–1451.

[8] Rombaut R, Clercq N D, Foubert I and Dewettinck K 2009 Triacylglycerol Analysis of Fats and Oils by Evaporative Light Scattering Detection *J. Am. Oil Chem. Soc* **86** p 19–25.

[9] Manohar M and Marzinke M A 2016 Application of Chromatography Combined With Mass Spectrometry in Therapeutic Drug Monitoring in Clinical Challenges *Ther. Drug Monit.* p 45-70.
[10] Naviglio D, Dellagreca M, Ruffo F, Andolfi A and Gall M 2017 Rapid Analysis Procedures for Triglycerides and Fatty Acids as Pentyl and Phenethyl Esters for the Detection of Butter Adulteration Using Chromatographic Techniques *J. Food Qual.* **2017** p 1-11.
[11] Moreda W, Pérez-Camino M C and Cert A 2003 Improved method for the determination of triacylglycerols in olive oils by high performance liquid chromatography *Grasas y. Aceites* **54** p 175-179.
[12] Marini F 2016 Triacylglycerols: Characterization and Determination *Reference Module in Food Science Encyclopedia of Food and Health* p 345-350.
[13] King H M, Cosslett A G, Thomas C P and Price-Davies R 2018 A HPLC method to monitor the occurrence of lipid peroxidation in intravenous lipid emulsions used in parenteral nutrition using in-line UV and charged aerosol detection *Clin. Nutr. ESPEN* **28** p 1-7.
[14] Hmida D, Abderrabba M, Tchapla A, Heron S and Mouss F 2015 Comparison of iso-eluotropic mobile phases at different temperaturesfor the separation of triacylglycerols in Non-Aqueous Reversed Phase Liquid Chromatography *J. Chromatogr. B* **990** p 45–51.
[15] Wulff-Pérez M, Barrajón-Catalán E, Micol V, Martín-Rodríguez A, Vicente J D and Gálvez-Ruíz, M J 2014 In vitro duodenal lipolysis of lipid-based drug delivery systems studied by HPLC–UV and HPLC–MS *Int. J. Pharm* **465** p 396–404.
[16] Zeb A and Murkovic M 2010 Analysis of triacylglycerols in refined edible oils by isocratic HPLC-ESI-MS *Eur. J. Lipid Sci. Technol.* p 844–851.
[17] Andrikopoulos N K, Giannakis I G and Tzamtzis V 2001 Analysis of Olive Oil and Seed Oil Triglycerides by Capillary Gas Chromatography as a Tool for the Detection of the Adulteration of Olive Oil *J. Chromatogr. Sci.* **39** p 137-145.
[18] Wei N, Wang M, Adams S J, Yua P, Avula B, Wang Y-H, Pan K, Wang Y. and Khan I A 2021 Comparative study and quality evaluation regarding morphology characters, volatile constituents, and triglycerides in seeds of five species used in traditional Chinese medicine *J. Pharm. Biomed. Anal.* **194** 113801.
[19] Chen Y J, Zhou X H, Han B, Yu Z, Yi H X, Jiang S L, Li Y Y, Pan J C and Zhang L W 2020 Regioisomeric and enantiomeric analysis of primary triglycerides in human milk by silver ion and chiral HPLC APCI-MS *J. Dairy Sci.* **103(9)** Article in Press.
[20] Ritchey M D, Wall H K, George M G and Wright J S 2019 US trends in premature heart disease mortality over the past 50 years: Where do we go from here? *Trends Cardiovasc. Med.* Article in Press
[21] Peng B, Luo T, Chen F, Wang M, Zheng L-F, Li J and Deng Z-Y 2021 Medium- and long-chain triglycerides attenuate lipid accumulation and regulate the expression of proteins related to lipid metabolism in oleic acid-induced lipid deposition in human hepatic LO2 cells *J. Funct. Foods* **78** 104354.

Detection of Myristic and Palmitic Acid in Butter By Hydrolysis And Acid Methylation

N S Tapak[1], S Senin[1], A H Mohamed[1], N N Dzulkifli[1], Z M Zain[2] and N M Ahmad[1*]

[1]School Of Chemistry and Environment, Universiti Teknologi MARA Cawangan Negeri Sembilan, Kampus Kuala Pilah, 72000 Negeri Sembilan, Malaysia

[2] Electrochemical Material and Sensor Research Group, Faculty of Applied Sciences, Universiti Teknologi MARA, Shah Alam, Selangor. Malaysia

*Corresponding author: normonica@uitm.edu.my

Abstract. Myristic acid and palmitic acid in butter was successfully detected by Gas Chromatography Flame Ionization Detector (GC-FID) after hydrolysis and methylated to fatty acid methyl esters (FAMEs). Optimization of initial oven temperature and carrier gas flow rate were determined earlier, and it was found that 50°C and 2.0 mL/min were chosen as optimum initial oven temperature and carrier gas flow rate respectively. Two different calibration curves were constructed using methyl myristate and methyl palmitate. Both were found linear at 20 - 100 ppm with correlation coefficient (R^2) 0.9972 and 0.9933 respectively. Methyl stearate were used as an internal standard (IS). Limit of detection (LOD) where the lowest concentration that can be detected by the instrument was found at 16 ppm. In order to determine the concentration of FAMEs in butter, fatty acids were hydrolyzed using lipase enzymes and were converted to FAMEs by acid catalyst methylation method. Butter with Brand A, B and C was found to contain 24 ppm, 21 ppm and 30 ppm Methyl Myristate. Methyl Palmitate was only found in butter Brand B and not detected for another two brands.

1. Introduction

Butter is a general food of the public regardless of age as it is often used as a seasoning for many dishes such as stews, pasta, and rice. This is due to the desirable flavours possessed by butter itself that almost judged to be "natural" by public. If stored at low temperatures, it is suitable as a smear on other foods such as bread, biscuit, and pastry [1]. Therefore, it is considered as an important and quick add-on menu by a family because it can be used directly as one of the breakfast components and oil substitutes for various menu dishes [2]. Nevertheless, consuming foods that high in Fatty acid (FA) such as butter affect long-term health including the risk of cardiovascular and coronary heart diseases [3]. Therefore, reliable data and methods need to be established for the detection of FA in butter for the authorities related to food, dieticians and consumers. High level of unsaturated fatty acids (FAs) such as monounsaturated fatty acids (MUFA), polyunsaturated fatty acid (PUFA) and cis or trans leads to cardiovascular diseases, colon cancer, and various diseases such as obesity, hypertension and arthritis [4,5]. Therefore, accurately quantification method must be applied to measure the concentration of fatty acids (FAs) with high sensitivity and precision.

Gas chromatography (GC) equipped with flame ionization detector (FID) was found as the most common techniques for determination of FA composition [6,7,8]. To make it compatible with GC requirement, samples need to be volatiles prior to analysis which normally involve derivatization methylation process. A common approach involved the use of several catalysts such as in acid catalyst, base catalyst or acid-base catalyst [8, 9,10] to obtain a fatty acid methyl ester (FAMEs) which is more volatilize and are compatible to GC-FID. However, it was found that base catalyst method produces by-products and impurities that will lead to arise interferences to GC analysis [10] whilst acid-base catalyst is not able to fully methylates the FAs under mild condition [11]. To overcome the matters, acid catalyst method provides a complete methylation for all types of commonly used fatty acids [10]. To quantify FAMEs, GC couple with flame ionization detector (GC-FID) was found as alternative instruments to analysed FAMEs due to a volatility and compatibility of compound [12]. Moreover, this instrument able to detect all classes of FAMEs with higher resolution [13].

Since butter contains triacylglycerol in addition to fatty acids, a considerable hydrolysis step using the lipase enzyme should be considered to obtain the actual FA content prior to methylation process. In this research, three butter samples were hydrolysed using lipase enzyme before methylated using acid catalyst method to form a respective fatty acid methyl ester (FAMEs). Using a chromatogram of GC-FID, the observed retention time was compared with the standard FA to confirm their presence in the butter followed by quantification through an internal calibration curve.

2. Research Method

2.1. Samples, reagent and chemical

Three butter samples were purchased from a supermarket in Kuala Pilah, Malaysia. Analytical grade of methyl stearate (C18:0), methyl palmitate (C16:0) and methyl myristate (C14:0) were purchased from Sigma chemicals. HPLC grade of Diethyl ether and methanol (CH_3OH) were purchased from Fisher Scientific. Arabic gum, bovine serum albumin, lipase enzyme, hydrochloric acid (HCl), sodium hydrogen carbonate ($NaHCO_3$), and anhydrous sodium sulphate (Na_2SO_4) for hydrolysis and methylation proses were purchased from Merck (analytical grade). Deionised water from Purite was used throughout the analysis.

2.2. Instrumentation

Data management and analysis were performed using gas chromatography (GC) equipped with Flame Ionization Detector (FID) and HP5 capillary column (30 m x 250 μm x 0.25 μm) brand Agilent Technologies 7820A. The condition of GC-FID was followed [7, 14] with slight modification. In brief, nitrogen and hydrogen were used as a carrier gas with the initial oven temperature was set at 50 °C for 8 minutes at 2.0 mL/min gas flow rate. The injection port of GC-FID was 250 °C with 40:1 split ratio at 250 °C detector temperature. The retention time of methyl myristate, methyl palmitate and methyl stearate are expected to appear at 15.060, 19.613 and 24.102 minute, respectively.

2.3. Preparation of stock solution and working solution.

Methyl myristate (C14:0) and methyl palmitate (C16:0) was used as a standard solution. Accurate amount of 0.1 g of individual fatty acid methyl ester (FAME) was dissolved with 100 mL of diethyl ether to prepare 1000 ppm of individual FAME. Then, mixture of working solution containing methyl myristate (C14:0) and methyl palmitate (C16:0) were diluted to 20 ppm, 40 ppm, 60 ppm, 80 ppm, and 100 ppm for calibration purposes. Methyl stearate (C18:0) at 60 ppm was used as internal standard (IS).

2.4. Hydrolysis of triglycerides

The hydrolysis process was carried out using method from [15] with slight modification. In brief, 5 mL of butter emulsion was prepared by mixing 0.5 mL of butter sample with 4.5 mL Arabic gum at 10 % (w/v). The hydrolysis process was performed for 30 minutes at 400 rpm in a beaker with constant temperature of 37 °C and pH 8.0. Subsequently, in the same beaker that containing 5 mL of butter emulsion, 10 mg/mL lipase enzyme, 10 mL distilled water, and 50 µL of bovine serum albumin 12.5 % (w/v) were added into the reaction beaker.

2.5. Methylation

Methylation process were carried out using method from [10] with slight modification. Briefly, 5 mL of hydrolysed samples was measured approximately. Then, the samples were transferred to a 10 mL test tube containing 1 mL of 4% (v/v) HCl/MeOH that was prepared earlier by diluting 4 mL of concentrated HCl with methanol in 100 mL volumetric flask. The mixture was placed in water bath and gradually heated for 30 minutes at 60-70 °C. Next, 3.0 mL of 2% (w/v) $NaHCO_3$ was used to neutralise the reaction mixture. Then, the sample was extracted with 5 mL diethyl ether. The upper layer which is the organic layer of diethyl ether was dried using anhydrous Na_2SO_4 then it was concentrated under nitrogen gas. 1 mL of internal standard (IS) of methyl stearate at 60 ppm were added to the sample before the injection.

2.6. Identification and quantification

To identify the FAMEs samples, GC-FID equipped with HP5 capillary column (30 m x 250 µm x 0.25 µm) brand Agilent Technologies 7820A will be used. Nitrogen and hydrogen were used as a carrier gas. Initial oven temperature was set at 50 °C for 8 minutes. Then, 5-fold of serial dilution of working mixture of methyl myristate and methyl palmitate were injected and measured. Then, 5-points of standard mixture will be used to plot a calibration curve [6]. Graph of peak ratio (peak area of interest/peak area for IS) against concentration was plotted as a standard calibration curve using fives working mixture consist of 20 ppm, 40 ppm, 60 ppm, 80 ppm, and 100 ppm.

The linearity of peak ratio obtained from the standard curve was used to fit in regression equation of $y = mx + c$ where m is the slope and c are the intercept [16]. Limit of detection (LOD) which is the lowest concentration detected by the instrument were determined by injecting the different concentration that consist of 12 ppm, 14 ppm, 16 ppm, and 18 ppm mixture of methyl myristate, methyl palmitate and methyl stearate at same GC condition. The concentration of the methyl myristate and methyl palmitate in three butter sample were determined in triplicate by linear regression equation of $y = mx + c$ obtained from the calibration curve where y is the peak ratio and x is the concentration of analyte [16].

3. Result and Discussion

3.1. Peak identification of methyl myristate, methyl palmitate, and methyl stearate

Sixty (60) ppm methyl myristate, methyl palmitate, and methyl stearate were used for identification of mixture of FAMEs using retention time. Figure 1 illustrates the chromatogram of individual FAMEs which are methyl myristate, methyl palmitate, and methyl stearate respectively which is summarized in table 1. Based on this study, the retention time of methyl myristate, methyl palmitate and methyl stearate were 5.419 minutes, 5.938 minutes and 6.433 minutes respectively. Regarding to the retention time of the three FAMEs, methyl myristate was eluted first because it has the lowest boiling point compared to methyl palmitate and methyl stearate. This was contrary from the previous study where the retention time of methyl myristate, methyl palmitate, and methyl stearate were found at 34.846 minutes, 39.571 minutes and 43.873 minutes respectively [17]. This difference may be due to the different used of solvent to prepare the stock solution. In the previous study the solvent used was heptane that have a boiling point of 98 °C while the solvent used in this study was diethyl ether (DEE) that have a boiling point of 34.4 °C [18,19]. For the comparison purpose, the individual standard was mixed as a mixture standard FAMEs and injected to GC-FID. Retention time was compared to a previous individual FAMEs and was summarized table 2. It was observed that the retention time of FAMEs mixture was found similar between mixture and individual retention time indicates that none of the interferences occurred to a mixture standard.

Retention time is depending on to a boiling point of a compound. The lower of vapour pressure and volatility is contradict to the retention time. In this study, diethyl ether which have lower boiling point and more volatile lead to a shorter retention time [19]. As methyl myristate, methyl palmitate, and methyl stearate have different values of boiling point which are 296.85 °C, 325.85 °C, and 351.85 °C respectively, the compound will elute with different retention time [20]. Methyl myristate was eluted first because it has the lower boiling point which means that it is more volatile than methyl palmitate and methyl stearate. Then, followed by methyl palmitate and methyl stearate due to its increasing boiling point and decreasing volatility. Another factor may cause by the polarity of analyte. As methyl myristate is less polar than methyl palmitate and methyl stearate, it will retain shorter in the column because it has lower affinity and less interaction in the non-polar HP5 capillary column. Therefore, methyl myristate was eluted first followed by methyl palmitate and methyl stearate.

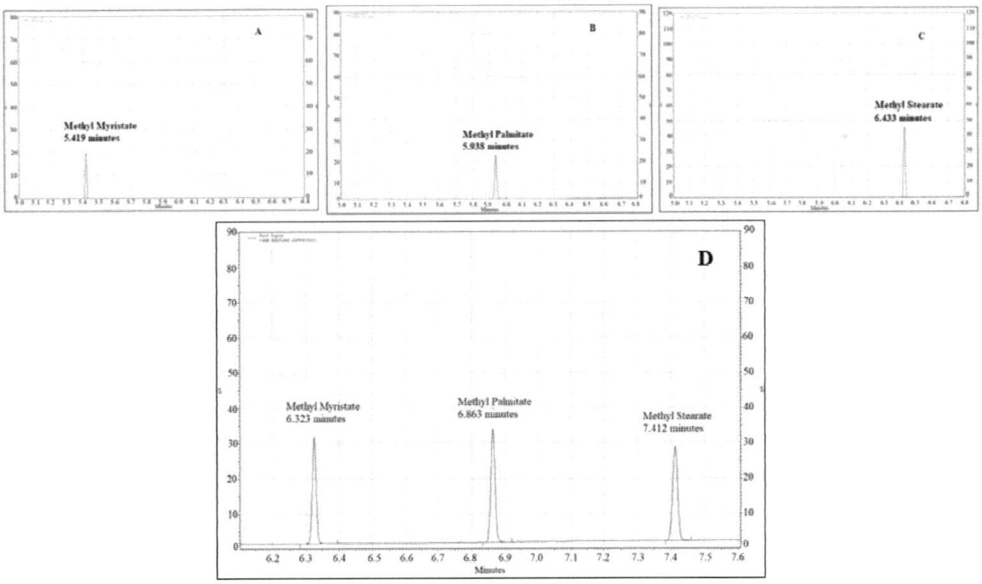

Figure 1. GC-FID chromatogram of A) methyl myristate, B) methyl palmitate and c) methyl stearate and D) mixture of methyl myristate, methyl palmitate and methyl stearate. Flow rate: 2.0 mL/min, initial oven temperature: 50 °C.

Table 1. Retention time and peak area for individual FAMEs (60 ppm)

FAMEs	Boiling point (°C)	Retention time (min)	Peak area
Methyl Myristate	296.85	5.419	179690
Methyl Palmitate	325.85	5.938	166253
Methyl Stearate	351.85	6.433	202508

Table 2. Retention time and peak area of FAMEs mixture (60 ppm)

Peak	Retention time (min)	Peak area
Methyl Myristate	5.407	167411
Methyl Palmitate	5.933	200797
Methyl Stearate	6.430	207239

3.2. Effect of initial oven temperature
Initial oven temperature is an importance parameter to measure because it will give an effect on the retention time of the solute. In this study, the separation based on initial oven temperature of 50 °C, 100 °C, 150 °C, and 200 °C were optimized. Figure 2A, 2B, 2C and 2D show the chromatograms for different initial oven temperature. Based on previous study, the higher the initial oven temperature resulting to shorter retention time but at too high temperature will trigger to co-elution of a compound and peak tailing [21]. An optimum initial oven temperature must be chosen to avoid the peak tailing or overlapping of an analyte.

Figure 2 demonstrates the graph of peak area against initial oven temperature. It was found the peak area of each FAMEs decreases from 50 °C to 200 °C. The optimum initial oven temperature at 50 °C will be used for subsequent analysis. This finding was almost similar to previous study which used 60 °C as an initial oven temperature even though different column size (30m x 0.53mm x 0.50 μm) and solvent (heptane) were reported [22].

IOP Conf. Series: Materials Science and Engineering 1176 (2021) 012043 doi:10.1088/1757-899X/1176/1/012043

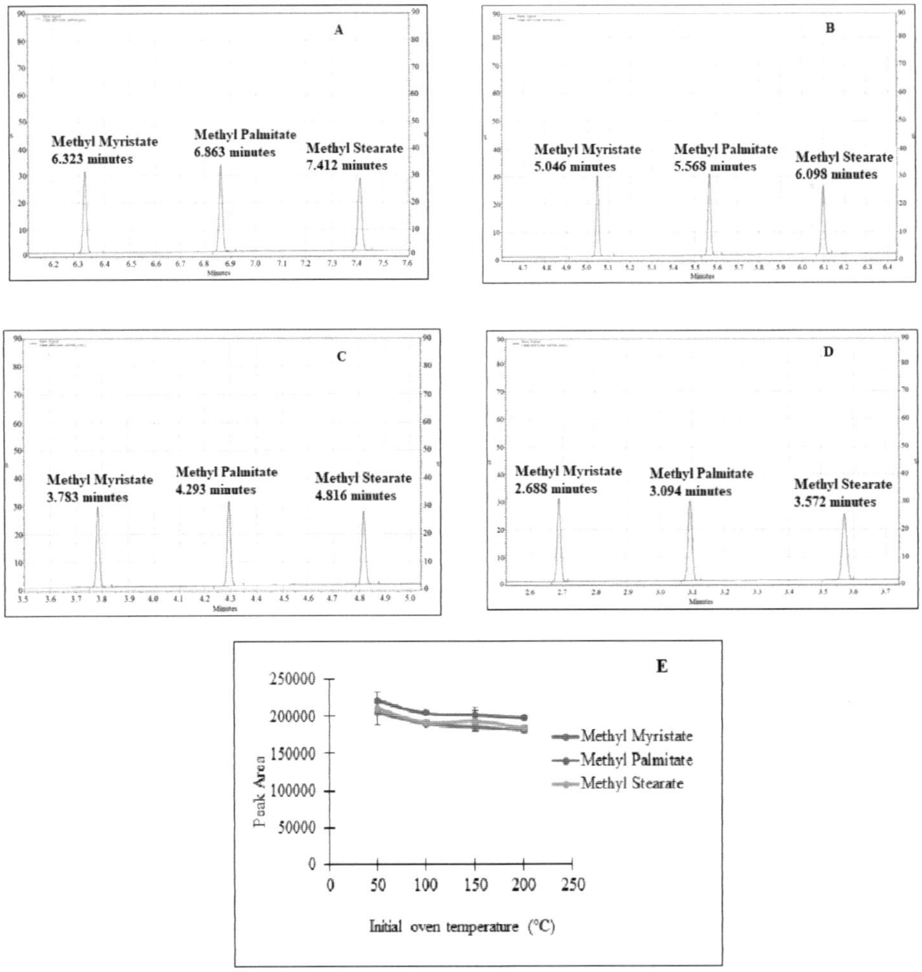

Figure 2. GC-FID chromatogram for mixture of methyl myristate, methyl palmitate, and methyl stearate. Initial oven temperature: A) 50 °C, B) 100 °C, C) 150 °C, D) 200 °C and E) optimization graph for initial oven temperature.

3.3. Effect of flow rate

Flow rate is a parameter that determine the velocity of the mobile phase normally given in volume/time and have relationship with the quantification results [23,24]. Through this study, the carrier gas flow rate was set at 0.5 mL/min, 1.0 mL/min, 1.5 mL/min, 2.0 mL/min and 2.5 mL/min. Figure 3A, 3B, 3C, 3D and 3E shows the chromatogram for different flow rate at 50 °C initial oven temperature. Based on the analysis, the retention time are inversely proportional to the flow rate. This may be due to the interaction of the analyte with the stationary phase. In this study, 0.5 mL/min flow rate shows the longest retention time while 2.5 mL/min flow rate shows shortest retention time. An optimum flow rate must be selected to ensure a better separation and reduce the risk of peak broadening. Based on figure 3F, the optimum flow rate (2.0 mL/min) will be used in the subsequent experiments as the peak area at this flow rate is very significant and will likely give a high response to the construction of the calibration curve. This outcome was contrary to that of Gao *et al.* (2015) [25], who found that 2.2 mL/min was used as optimum flow rate. This discrepancy could be attributed to the uses of single carrier gas which was hydrogen instead of mixture of hydrogen and nitrogen gas while this study also used different types of capillary column which was BP70 capillary column.

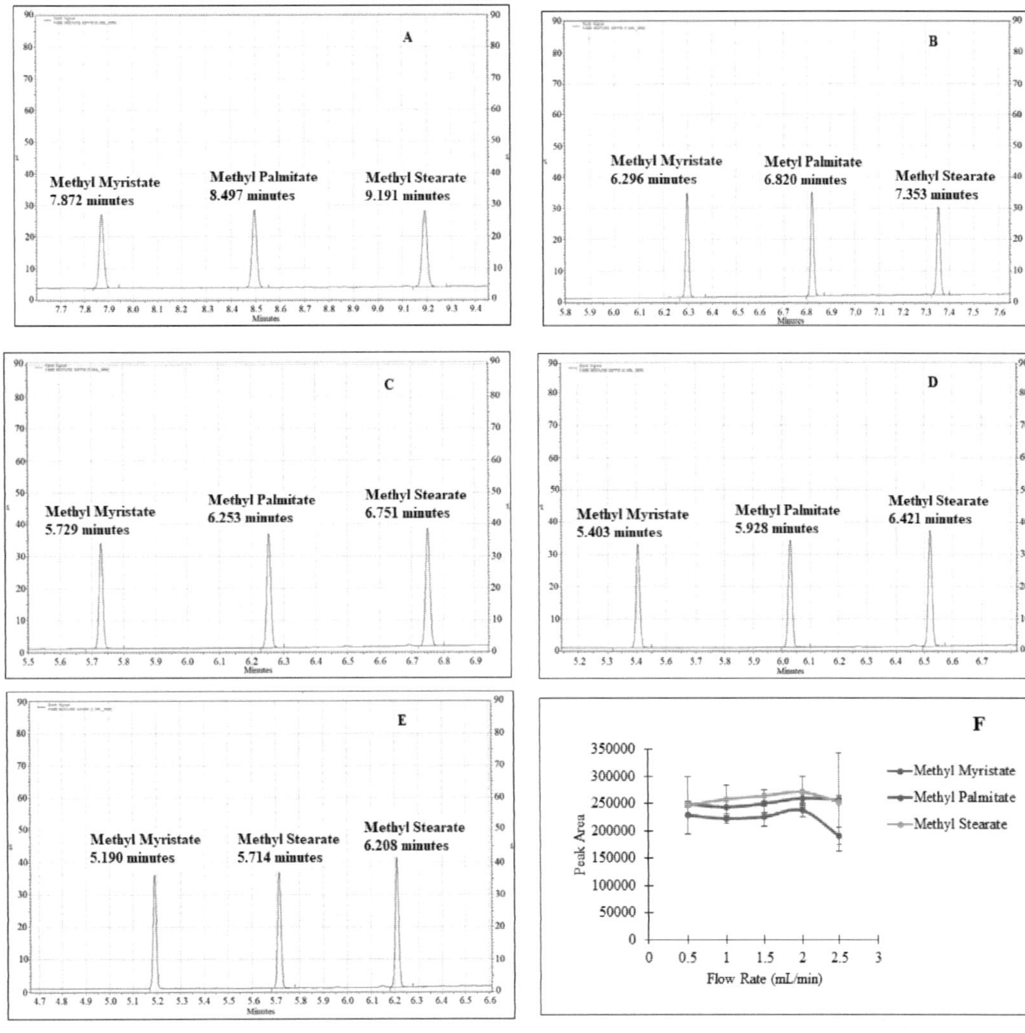

Figure 3. GC-FID chromatogram for mixture of methyl myristate, methyl palmitate, and methyl stearate. Flow rate: A) 0.5 mL/min, B) 1.0 mL/min, C) 1.5 mL/min, D) 2.0 mL/min, E) 2.5 mL/min at initial oven temperature: 50 °C and F) optimization graph of flow rate (mL/min)

IOP Conf. Series: Materials Science and Engineering 1176 (2021) 012043 doi:10.1088/1757-899X/1176/1/012043

3.4. Calibration curve for standards FAMEs

Calibration curves measured the relationship between peak ratio (peak area of interest/peak area for IS) and the concentration of the components [13]. Graph of peak ratio against concentration were plotted as a calibration curve where peak area ratio as y-axis were calculated using formula below where R is peak ratio, A_1 is FAMEs peak area and A_2 is internal standard peak area.

$$R = \frac{A_1}{A_2} \quad (1)$$

Previous study has noted the importance of internal standard as it will rectify the error occurred throughout sample preparation or even during the sample injection [26]. In this study, methyl stearate was used as IS since it has the significant peak area compared methyl myristate and methyl palmitate. This is well agreed by [27] who reported methyl stearate as an internal standard in their study. Figure 4 (A) and (B) show the calibration curve of methyl myristate and methyl palmitate with correlation coefficient (R^2) of 0.9972 and 0.9933, respectively. It was found linear at 20 – 100 ppm using methyl stearate as an IS. This finding is contrary with previous study which reported a different linearity at 20 to 700 ppm with correlation coefficient of 0.9998 and 0.9999 respectively [28]. This is because, concentration greater than 100 ppm was not injected throughout this study to avoid column degradation for the good of the lifespan of the column.

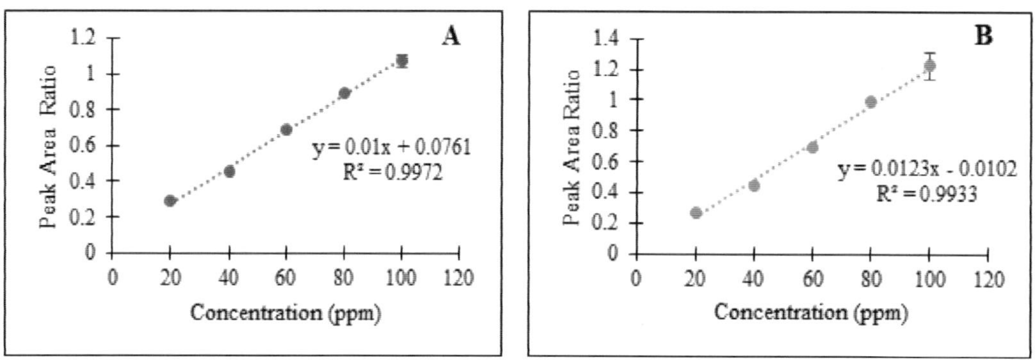

Figure 4. Calibration curve of methyl myristate (A) and methyl palmitate (B) obtained from GC-FID.

3.5. Limit of detection (LOD)

Limit of detection (LOD) is a qualitative measurement which reported the smallest concentration of an analyte that can be detected by the method [13,23]. There were no signal or peak detected by the instrument at 12 and 14 ppm because they were too low to be detected by this GC-FID. However, the lowest concentration that can be detected by the instrument was found at 16 ppm. It was differed from previous study conducted by [28] that obtained lower LOD at 5 ppm. The result was contrary due to different column used in the study which was CP FFAP CB which has high polarity rather than HP5 capillary column that is non-polar. Besides that, the thickness of the column (0.32 μm and 0.25 μm) for CP FFAP CB and HP5 capillary column respectively are different with the present study.

3.6. Determination of methyl myristate and methyl palmitate in real sample

For the analysis of butter samples, the developed GC-FID method was profitably applied to simultaneous determination of methyl myristate and methyl palmitate. The concentration of the methyl myristate and methyl palmitate were determined by linear regression equation $y = mx + c$ obtained from the calibration curve where x is the concentration of analyte. Figure 5 show the chromatogram of sample in Brand A, B and C. Based on the study, the concentration of methyl myristate in Brand A, B and C was found to be 24 ppm, 21 ppm and 30 ppm respectively. However, only Brand B contained methyl palmitate with concentration of 20 ppm. Some problems may arise from the derivatization process which typically involved incomplete conversion of FA to FAMEs, alteration of FA composition, overlap of methyl ester peaks and damage to the chromatographic column due to traces of esterifying reagents [29]. Most of the problems has been avoided in the present research which may be due to the hydrolysis step that has been taken before methylation. However, limitation was seen at the figure 6(A), (B) and (C) where there are significant unidentified and unquantified peaks in each sample that could belongs to the unconverted fatty acids. The presence of such unconverted components may reduce the actual concentration of calculated myristic acid and palmitic acid which should be resolved in future study. To date, another lipid samples such as flaxseed oil [30] rat and human faeces along with fermentation fluids [31] and human milk [32] have been reported so far, quantified using GC-FID method.

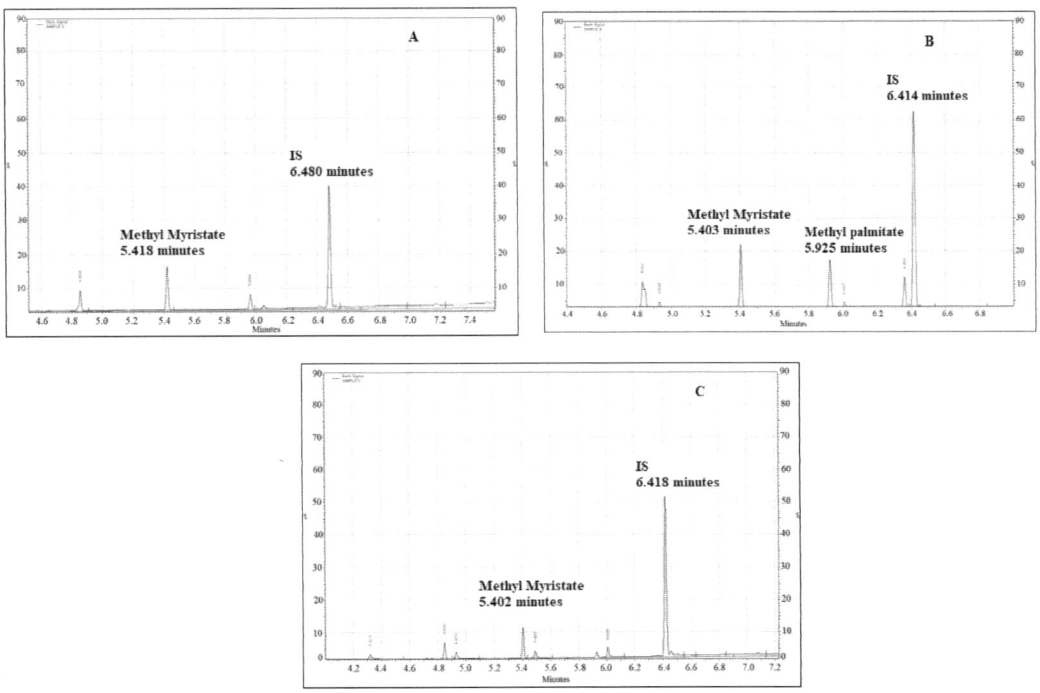

Figure 5. chromatogram of A) sample Brand A, B) Sample B and C) Sample C

4. Conclusion

A GC-FID method was successfully designed to determine the concentration of myristic acid and palmitic acid in butter samples by hydrolysis process using lipase enzyme and acid catalyst methylation method to methylate the fatty acids into fatty acids methyl esters (FAMEs). Initial oven temperature and flow rate were found optimized at 50 °C and 2.0 mL/min respectively which were used to construct the calibration curves. Calibration curve of methyl myristate and methyl palmitate was found linear from 20 to 100 ppm where the correlation coefficient (R2) was 0.9972 and 0.9933 respectively with LOD at 16 ppm. Quantification using calibration curve reveal the concentration of Methyl Myristate and Methyl Palmitate in the butter samples. Therefore, hydrolysis, acid methylation process and condition of GC-FID of this study will be a great platform to analyse all FAMEs in butter samples. In future, FAMEs profiling and quantification should be carried out by mass spectroscopy detector for a comparison study.

References

[1] Bobe G, Zimmerman S, Hammond E G, Freeman A E, Porter P A, Luhman C M, and Beitz D C 2007 Butter composition and texture from cows with different milk fatty acid compositions fed fish oil or roasted soybean in *J. Dairy Sci.* **90(6)** p 2596–2603.

[2] Fíndík O and Andiç S 2017 Some chemical and microbiological properties of the butter and the butter oil produced from the same raw material *LWT-Food Science and Technology* **86** p 233–239.

[3] Morenga L T and Montez J M 2017 Health effects of saturated and trans-fatty acid intake in children and adolescents: Systematic review and meta-analysis *Plos One* **12(11)** p 1-20.

[4] Bystrická Z, and Ďuračková Z 2016 Gas chromatography determination of fatty acids in the human erythrocyte membranes – A review *Prostaglandins Leukotrienes and Essential Fatty Acids* **115** p 35–40.

[5] Ganguly R, and Pierce G N 2015 The toxicity of dietary trans fats *Food. Chem. Toxicol.* **78** p 170–176.

[6] Adjepong M, Valentini K, Pickens C A, Li W, Appaw W and Fenton J 2017 Quantification of fatty cid and mineral levels of selected seeds, nuts, and oils *Ghana Journal of Food Composition and Analysis* **55(11)** p 43–49.

[7] Salimon J, Omar T A and Salih N 2017 An accurate and reliable method for identification and quantification of fatty acids and trans fatty acids in food fats samples using gas chromatography *Arabian Journal of Chemistry* **10** p 1875–1882.

[8] Abugri D A, McElhenney W H and Willian K R 2012 Comparison of transesterification methods for fatty acid analysis in higher fungi: application to mushrooms *Food Anal. Methods.* **5** p 1159–1166.

[9] Carvalho A P and Malcata F X 2005 Preparation of Fatty Acid Methyl Esters for Gas-Chromatographic Analysis of Marine Lipids: Insight Studies *J. Agric. Food Chem.* **53(13)** p 5049–5059.

[10] Igarashi M, Tsuzuki T, Kambe T and Miyazawa T 2004 Recommended Methods of Fatty Acid Methylester Preparation for Conjugated Dienes and Trienes in Food and Biological Samples *J Nutr. Sci. Vitaminol* **50(2)** p 121–128.

[11] Salimon J, Omar T A, and Salih N 2014 Comparison of two derivatization methods for the analysis of fatty acids and trans fatty acids in bakery products using gas chromatography *The Scientific World Journal* p 1-10.

[12] Jones A 2016 GC / FID Analysis of Fatty Acid Methyl Esters without Correction Factors Using the Polyarc Reactor Application *Activated Research Company* p 1–6.

[13] Aparicio-Ruiz R, García-González D L, Morales M T, Lobo-Prieto A and Romero I 2018 Comparison of two analytical methods validated for the determination of volatile compounds in virgin olive oil: GC-FID vs GC-MS *Talanta* **187** p 133–141.

[14] Costa N, Cruz R, Graça P, Breda J and Casal S 2016 Trans fatty acids in the Portuguese food market *Food Control* **64** p 128–134.

[15] Ghattas N, Abidi F, Galai S, Marzouki M N, and Salah A Ben 2014 Monoolein production by triglycerides hydrolysis using immobilized Rhizopus oryzae lipase *Int. J. Biol. Macromol.* p **68** 1–6.

[16] Ponphaiboon J, Limmatvapirat S, Chaidedgumjorn A, and Limmatvapirat C 2018 Optimization and comparison of GC-FID and HPLC-ELSD methods for determination of lauric acid, mono-, di-, and trilaurins in modified coconut oil *J. Chromatogr. B Analyt. Technol. Biomed. Life Sci.* **1099** p 110–116.

[17] Dixit S and Das M 2012 Fatty acid composition including trans -fatty acids in edible oils and fats: probable intake in indian population *J. Food Sci.* p 188-199.

[18] Mohon M, Associate R, Calder J, Student G and Wang W 2016 Cold start idles emissions from a modern Tier-4 turbo-charged diesel engine fueled with diesel-biodiesel, diesel-biodiesel-ethanol, and diesel-biodiesel-diethyl ether blends *Appl. Energy* **80** p 52–65.

[19] Hage D S and Carr J D 2011 *Analytical Chemistry and Quantitative Analysis. (1st ed.).* United States: Daniel Kaveney p 507-536.
[20] Hashimoto N, Nomura H, Suzuki M, Matsumoto T and Nishida H 2015 Evaporation characteristics of a palm methyl ester droplet at high ambient temperatures *Fuel* **143** p 202–210.
[21] Gómez-Cortés P, Rodríguez-Pino V, Juárez M, and Fuente M A De 2017 Optimization of milk odd and branched-chain fatty acids analysis by gas chromatography using an extremely polar stationary phase *Food Chemi.* **231** p 11–18.
[22] Yurchenko S, Sats A, Poikalainen V, and Karus A 2016 Method for determination of fatty acids in bovine colostrum using GC-FID *Food Chem.* **212** p 117–122.
[23] Vékey K, Telekes A, and Vertes A 2008 *Medical Applications of Mass Spectrometry. (1 st ed.). (Amsterdam)* Elsevier B.V p 61-89.
[24] Zuas O and Budiman H 2016 Estimating precision and accuracy of GC-TCD method for carbon dioxide, propane and carbon monoxide determination at different flow rate of carrier gas *Hem. Ind* **70(4)** p 451–459.
[25] Gao F, Yang S, and Birch J 2015 Physicochemical characteristics, fatty acid positional distribution and triglyceride composition in oil extracted from carrot seeds using supercritical CO_2 *J. Food Compost. Anal.* **45** p 1-26.
[26] Laura A, Sobrado A, and Mariella L F 2016 Comparison of gas chromatography-combustion-mass spectrometry and gas chromatography-flame ionization detector for the determination of fatty acid methyl esters in biodiesel *J. Chromatogr. A* **1457** p 1-27.
[27] Potier S, Bernard M M, Schikorski D, Buatois B, Duriez O, Gabirot M, Leclaire S and Bonadonna F 2018 Preen oil chemical composition encodes individuality, seasonal variation and kinship in black kites Milvus migrans *Journal of Avion Biology* **49(7).**
[28] Mannion D T, Furey A, and Kilcawley K N 2016 Comparison and validation of 2 analytical methods for the determination of free fatty acids in dairy products by gas chromatography with flame ionization detection *J. Dairy Sci.* **99(7)** p 1–17.
[29] Simionato J I, Garcia J C, Santos G T D, Oliveira C C, Visentainer J V and Souza N E D 2010 Validation of the Determination of Fatty Acids in Milk by Gas Chromatography *J. Braz. Chem.* **21(3)** p 520-524.
[30] Danish M and Nizami M 2019 Complete fatty acid analysis data of flaxseed oil using GC-FID method *Data in Brief* **23** p 1-6.
[31] Scortichini S, Boarelli M C, Silvi S and Fiorini D 2020 Development and validation of a GC-FID method for the analysis of short chain fatty acids in rat and human faeces and in fermentation fluid *J. Chromatogr. B* **1143** 121972
[32] Rydlewski A A, Manin L P, Pizzo J S, Silva P D, Silveira R, Tavares C B G, Paula M D, Pereira O, Santos O O and Visentaine J V 2021 Lipid profile by direct infusion ESI-MS and fatty acid composition by GC-FID in human milk: association with nutritional status of donors *J. Food Compost. Anal.* Journal pre proof.

IOP Conf. Series: Materials Science and Engineering 1176 (2021) 012044 doi:10.1088/1757-899X/1176/1/012044

Production and Characterisation of Cheddar Cheese-like from *Cocos Nucifera L.*

S N Ismail[1], M S A Latip[2*] and M A Mohamad[3]

Faculty of Hotel and Tourism Management, Universiti Teknologi MARA Cawangan Terengganu Kampus Dungun, Jalan Sura Hujung 2300 Dungun, Terengganu, Malaysia

*Corresponding author: safuanlatip@uitm.edu.my

Abstract. Traditionally, cheese is made from animal milk and had been introduced a long time ago. The increasing demand for cheese has caused a substantial commercialise production that led to animal exploitation and environmental pollution. Henceforth, this study aims to produce and characterise Cheddar Cheese-like attributes made of coconut cream from *Cocos Nucifera L.* The shelf life analysis, colour analysis used Chroma Meter CR-410, pH value analysis, meltability analysis, and sensory evaluation were conducted through the meltability test, descriptive test, paired comparison test and a consumer preference test. The results illustrated that shelf life is extended below danger zone temperature, and there is a distinct colour of Cheddar Cheese-like colour compared to regular cheese. The Cheddar Cheese-like managed to mimic the effect of melting cheese and was accepted by most of the respondents based on the appearance (93.3 per cent), aroma (86.7 per cent), taste (93.3 per cent), texture (73.3 per cent), and consistency (86.7 per cent) characteristic. Subsequently, preferences analysis shows 56.7 per cent of the respondents prefer Cheddar Cheese-like over the regular cheese. Therefore, based on the result, this study concludes that coconut milk can be an alternative to producing a cheese-like product that will benefit the consumer of different diet and allergies and at the same time preserve the environment for sustainability reason.

1. Introduction

Cheese had been introduced a long time ago and is traditionally made from animal milk. Milk itself is the main raw material in the dairy sector. Indeed, approximately 6 billion people worldwide consume milk and dairy products. The global report indicates 81% of world milk production is cow's milk, 15% buffalo milk, and 4% is a combination of goat, sheep and camel milk [1]. The statistic in 2018 showed an increase of 1.6% in world milk production which is predicted to grow exponentially in the future [1,2].

However, the increasing demand for cheese from large scale industrial and home-used has led to a vast commercial production of cheese. As a result, dairy farming expended worldwide and has contributed to animal cruelty and exploitation by the breeding industry in order to fulfil the demands [3]. Other than that, the whey from cheese has pollutant characteristics that can affect the environment [4]. Moreover, as consumers' vegetarian diet group increases, a vegetarian diet substitute is critical [5]. A better and sustainable product should be consumed to preserve the environment better [6]. Besides, the consumer with a common digestive disorder *of lactose intolerance* can consume coconut milk's cheese [3]. Indeed, most of the cheese product in Malaysia and Asia is largely imported, resulting in expensive retail prices [3].

The *Cocos Nucifera L.* is *monocotyledon* palm under *palmaceae* family [7] and widely available in Asian countries at a low price. Coconut milk is a thick white liquid extracted from the shredded pulp of coconut that contributed to the rich taste and opacity of the coconut milk. Coconut milk has an important element of coconut as the milk is an oil emulsion in water that is stable with proteins that can serve as thickening agents and emulsifiers [8]. The characteristic contributes to potential use as a substitute for making cheese. The previous studies had tested the combination of fresh milk and coconut to produce Warankashi traditional milk cheese-like product [9,10] and employed traditional cheese coagulation from soy milk and coconut milk in making cheese [11]. However, a paucity of research investigated the use of 100 per cent of coconut milk with other plant-based ingredients to make a 100 per cent vegan and non-lactose-based Cheddar Cheese-like substitute, with a short production time without a traditional coagulation process.

Therefore, this work is directed to study the characteristics and attributes of Cheddar Cheese-like produced using coconut milk from *Cocos Nucifera L* and plant-based ingredients.

2. Experimental

The extracted, sterilised, and homogenised coconut cream extract from *Cocos Nucifera L.* originated from Indonesia was used in this study. Agar powder (seaweed gelatine), tapioca (*Manihot Esculenta*) starch, salt (*sodium chloride*), water (H_2O), garlic powder and nutritional yeast (a strain of *Saccharomyces Cerevisiae*) were also used. 200 grams (g) of coconut cream, 1 ½ tablespoon (tbsp) of nutritional yeast, 1 tbsp of tapioca starch, 1 teaspoon (tsp) of salt, and ¼ tsp of garlic powder were mixed in a pot. Meanwhile, ½ cup of plain water and 2 tbsp of seaweed gelatine has been mixed in a different pot over medium heat and stir continuously until it transforms into a paste texture. Combine the coconut cream mixture with a seaweed gelatine paste in a pot and boil while continuously stirring using a hand whisk. Once boiling reduce the heat to simmer and constantly stir for 6 minutes until a smooth and thick texture is achieved. Immediately pour into the prepared silicone mould or moulds that were grease with vegetable oil. Let it cool with the lid off for 15 minutes at room temperature, then transfer to the refrigerator (0°C to 4°C) at least for 2 hours to firmly set.

The shelf life of Cheddar Cheese-like was then evaluated to determine how long it is safe to eat if stored in the refrigerator at 0°C to 4°C. As supported by a previous study, the cheese-like were placed in the refrigerator between 0°C to 4°C for 0 day, 3 days and 6 days [12]. The same samples were used for the colour test using Chroma Meter CR-410. The L*, a* and b* value was used in the colour test to signify the colour management. The L* value indicates the luminosity, the extent of lightness from black colour to white colour. The a* value indicates the green colour as negative and the red colour as positive. While b* value indicates the blue colour as negative and the yellow colour as positive [13]. Tapioca starch was added as one of the ingredients to mimic the melting purposes like traditional dairy cheese. Therefore, the meltability experiment was conducted to test whether the cheese-like melting or not.

The Cheddar Cheese-like were tested for acceptability by 30 random respondents using the Comparison Test between traditional dairy cheese and cheese-like as supported by the previous scholar. 30 respondents could lead to about 80 per cent power, which is suggested as a minimum for a study provided a medium to large effect size [14]. Five attributes were used for evaluation: appearance, aroma, taste, texture, and consistency. The assessment of respondents' preferences and perception of the cheese-like was similarly conducted using the Hedonic Scale. The sensory properties were presented on a seven-point scoring scale (1) Dislike (Extremely) (2) Dislike (Moderately) (3) Dislike (Lightly) (4) Neither like nor dislike (5) Like (Lightly) (6) Like (Moderately) (7) Like (Extremely).

3. Result and discussion

3.1. Shelf life study

The shelf life is a time length of a particular product that can be kept without having spoiled effects [12,15]. For this experiment, the cheese-like was refrigerated between 0°C to 4°C for 0 day, 3 days and 6 days, as shown in table 1.

Table 1. Results of the changes the Cheddar Cheese-like's shelf life in the refrigerator.

Sample	0 day	3 days	6 days
Sample 1	No changes	No changes	No changes
Sample 2	No changes	No changes	No changes
Sample 3	No changes	No changes	No changes

Based on data displayed in table 1, no changes observed in 0, 3, and 6 days when the cheese-like is stored in a refrigerator (4°C to 0°C) as supported by other studies [15,16]. The cheese-like was stored out of danger zone temperature and not affected by other microorganisms that can spoil the product, such as molds or bacteria [17]. The temperature is an important factor controlling cheese-like shelf life, chemical and microbiological consistency, and the inappropriate storage temperature can prolong the microorganisms' survival [15].

3.2. Colour measurement

Based on table 2, the L* value from sample 1 to sample 3 is between 91.30 until 93.93. This result was different from the previous study [18], as the L* value is between 70.9 to 75.6. The finding indicates that Cheddar Cheese-like's colour is more neutral and whiter compared to dairy cheese. The score of a* value from sample 1 to sample 3 is between 1.02 until 2.15. The finding is contradicted with the previous study [18], as a result of the study was 7.7 until 8.9. It shows that the cheese-like's has lacked a red colour than normal cheese. The b* value from sample 1 to sample 3 is between 15.18 until 15.49. The value also differs from the previous study [18], as the a* value in their study was 7.3 until 9.2. Thus, Cheddar Cheese-like's colour has lacked a yellow colour than ordinary cheese. In general, assessing food quality depends on many variables, including taste, colour, texture and nutritional value [13].

Table 2. Results of the changes Cheddar Cheese-like's colour.

	T^1	T^2	T^3
Sample 1 (0 day)	L=91.30	L=91.90	L=92.12
	a=2.15	a=2.01	a=2.00
	b=15.28	b=15.18	b=15.18
Sample 2 (3 days)	L=92.32	L=92.63	L=92.61
	a=1.07	a=1.04	a=1.02
	b=15.45	b=15.47	b=15.48
Sample 3 (6 days)	L=93.89	L=93.93	L=93.90
	a=1.83	a=1.84	a=1.84
	b=15.48	b=15.49	b=15.49

3.3. Cheese-like meltability evaluation

The tapioca starch was added for this experiment, and the result can be found in table 3. The tapioca starch was suggested by previous studies to mimic the effect of melting cheese [19–22]

IOP Conf. Series: Materials Science and Engineering 1176 (2021) 012044 doi:10.1088/1757-899X/1176/1/012044

Table 3. Results of Cheddar Cheese-like meltability evaluation

Changes	**Sample 1 (200°C)**	**Sample 2 (180°C)**
Samples before heating	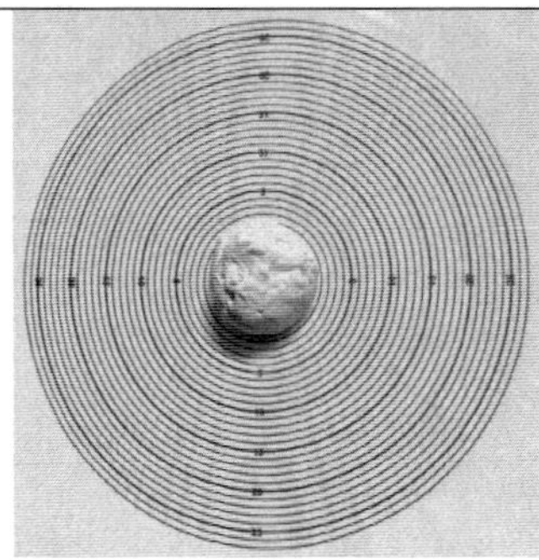	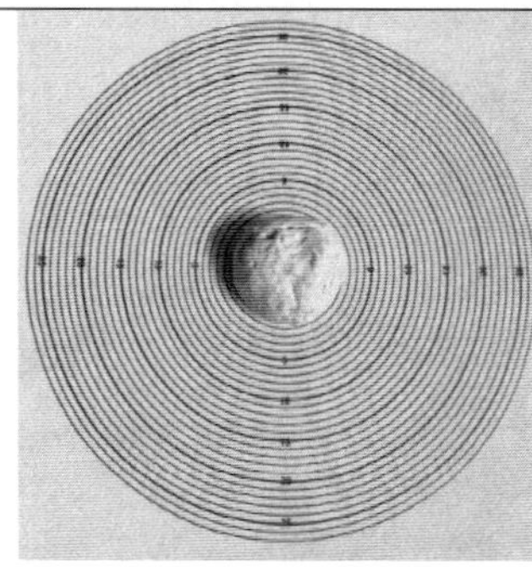
Diameter	1mm	1mm
Height	1mm	1mm
Samples after heating	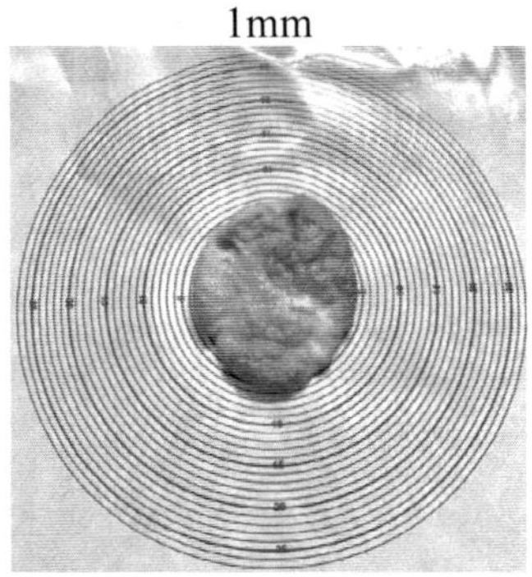	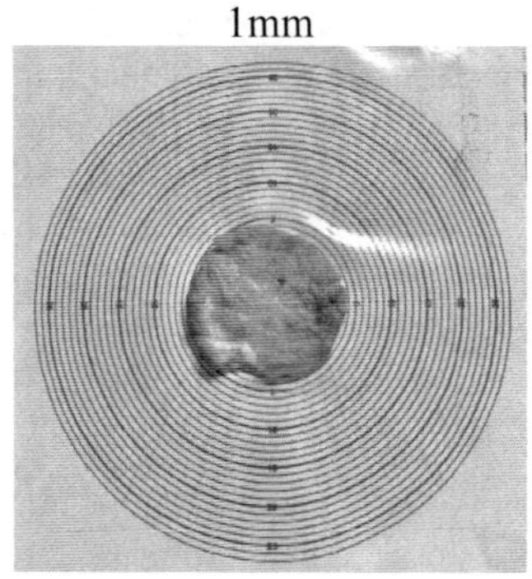
Diameter	4mm	3mm
Height	6mm	6mm

Cheese-like meltability was tested by heating two cheese-like samples. In a tray, sample 1 and 2 (weight = 20gm; diameter = 1mm, and height 1mm) were heated in an oven at 200°C and 180°C. After 5 minutes, the subjects were removed from the oven. The cheese-like was measured using spread diameter and height. As shown in table 3, sample 1 (diameter = 4mm and height 6mm) while sample 2 (diameter = 3mm and height 6mm).

3.4. Respondents' acceptability
To examine the consumers' acceptability of a product, the sensory evaluation was used through a scientific experiment involves humans' smell, sight, taste, and touch [23]. Therefore, the sensory evaluation was tested for this research using the comparison test between the dairy cheese and Cheddar Cheese-like made from coconut milk and the cheese-like preferences [23]. The result can be accessed from table 4.

3.5. Paired compared test
Based on the results in table 4, most of the respondents accept the appearance, aroma, taste, texture and consistency of the Cheddar Cheese-like made from coconut milk comparable to cheddar cheese made from dairy milk. So, the appearance, aroma, and taste of Cheddar Cheese-like are acceptable and have a better preference than regular cheddar from dairy milk. However, the regular cheddar from dairy milk has better texture preferences compared to Cheddar Cheese-like.

Table 4. Paired Compared Test between Cheddar Cheese-like and dairy cheese

Products Characteristic	Cheddar Cheese-like	Frequency (n=30)	Percentage 100%	Regular cheddar	Frequency (n=30)	Percentage 100%
Appearance	Acceptable	28	93.3	Acceptable	20	66.7
Aroma	Acceptable	26	86.7	Acceptable	20	66.7
Taste	Acceptable	28	93.3	Acceptable	21	70.0
Texture	Acceptable	22	73.3	Acceptable	24	80.0
Consistency	Acceptable	26	86.7	Acceptable	26	86.7

3.6. Consumers' preferences analysis using Hedonic Scale.

Based on table 5, most of the respondents lightly like the Cheddar Cheese-like, which accumulated to 46.7 per cent (n=14). The second highest was neither like nor dislike, which accumulated to 23.3 per cent (n=7). However, some respondents likely dislike the cheese-like in which accumulated 16.6 per cent (n=5). Some respondents chose moderately like in which accumulated to 10 per cent (n=3). Lastly, there was one respondent who answered moderately dislike in which accumulated to 3.3 per cent. To conclude, 56.7 per cent (n=17) of respondents like (lightly and moderately) the Cheddar cheese-like.

Table 5. The Cheddar Cheese-like by using hedonic scale.

		Frequency	Per cent
Valid	Dislike (Moderately)	1	3.3
	Dislike (Lightly)	5	16.6
	Neither like nor dislike	7	23.3
	Like (Lightly)	14	46.7
	Like (Moderately)	3	10.0
	Total	30	100.0

4. Conclusion

This study attempts to find a substitute for cheese from cow milk by experimenting with the character and attributes of Cheddar cheese-like based on coconut milk that is used as the primary raw material. Based on the analysis, the findings suggest that the experimented product can imitate the cheese characteristic despite lacking colour compared to the original product. The findings also indicate that most of the respondents who participate in the test favour the cheese-like from coconut milk and remarks that it is similar to milk-based cheese in terms of appearance, aroma, taste, texture and consistency. Therefore, the product produced from this experiment can be further made on a larger scale and substitute for the cheese from cow milk product in the market.

References

[1] OECD/FAO 2019 *OECD-FAO Agriculture Outlook 2019-2028* (Rome)

[2] Cabral C F S, Veiga L B E, Araújo M G and de Souza S L Q 2020 Enviromental life cycle assessment of goat cheese production in Brazil: A path towards sustainability *Food Sci Technol.* **129** 1–6.

[3] Ismail S N, Latip M S A, Ghazali A M, Alias M A M and Mohamad A M 2020 Klapper cheese-like spread *3rd International Innovation, Invention and Design Competition 2020 (Terengganu)* p 167–9.

[4] Palmieri N, Forleo M B and Salimei E 2017 Environmental impacts of a dairy cheese chain including whey feeding: An Italian case study *J. Clean Prod.* **140** 881–9.

[5] Latip M S A Newaz F T Noh I and Mohamad A M 2020 Individual green consideration model: A conceptual study *Int. J. Manag.* **11** 849–58

[6] Latip M S A, Newaz F T, Ramasamy R, Tumin S A and Noh I 2020 How do food safety knowledge and trust affect individual's green considerations during the COVID- 19 pandemic

in Malaysia? *Malaysian J. Consum. Fam. Econ.* **24** 261–85
[7] Chavan R S and Jana A 2018 Coconut milk and coconut oil: Their manufacture associated with protein functionality *J. Food Sci.* **83** 2019–27.
[8] Zhu X, Zhao Z, Wang L and Zhang L 2014 A new method to measure fat content in coconut milk based on Y-type optic fiber system *Optik* **125** 6172–8
[9] Okon E and Ojimelukwe P 2017 Potentials of coconut milk as a substitute for cow milk in cheese making *J. Adv. Microbiol.* **4** 1–9.
[10] Balogun M A, Kolawole F L, Joseph J K, Adebisi T T and Ogunleye O T 2016 Effect of fortification of fresh cow milk with coconut milk on the proximate composition and yield of warankashi, a traditional cheese *Croat. J. Food Sci. Technol.* **8** 10–4.
[11] Ayodeji A A, Ahure D, Efiong E E and Acham I O 2020 Production and quality evaluation of cheese from soy and coconut milk using selected coagulants *Eur. J. Nutr. Food Saf.* **12** 1–12.
[12] Jalilzadeh A and Tuncturk Y 2015 Extension shelf life of cheese: A review *International Journal of Dairy Science* **10** 44-60.
[13] Markovic I, Ilic J, Markovic D, Simonovic V and Kosanic N 2013 Color measurement of food products using CIE L * a * b * and RGB color space *J. Hyg. Eng. Des.* **4** 50–3.
[14] Van, Voorhis C R and Morgan B 2007 Understanding power and rules of thumb for determining sample sizes *Tutor Quant. Methods Psychol.* **3** 43–50.
[15] Perveen K, Alabdulkarim B and Arzoo S 2011 Effect of temperature on shelf life, chemical and microbial properties of cream cheese *African J. Biotechnol.* **10** 16929–36.
[16] Park J, Shin J, Bak D, Kim N, Lim K, Yang C et al. 2014 Determination of shelf life for butter and cheese products in actual and accelerated conditions *Korean J. Food Sci. Anim. Resour.* **34** 245–51
[17] Felfoul I, Attia H and Bornaz S 2017 Shelf life determination of fresh cheese subjected to different modified atmospheres packaging *J. Agric. Sci. Technol.* **19** 847–60.
[18] Wadhwani R and McMahon D J 2012 Color of low-fat cheese influences flavor perception and consumer liking *J. Dairy Sci.* **95** 2336–46.
[19] Klemaszewski J L, Fonteyn D, Bouron F and Lemonnier L 2016 *Cheese product with modified starches* United States of America: World Intellectual Property Organization; WO 2016/195814 A1.
[20] Sipahioglu O and Alvarez V B 1999 Structure, physico-chemical and sensory properties of feta cheese made with tapioca starch and lecithin as fat mimetics *Int. Dairy J.* **9** 783–9.
[21] Diamantino V R, Costa M S Taboga S R Vilamaior P S L, Franco C M L and Penna A L B 2019 Starch as a potential fat replacer for application in cheese: Behaviour of different starches in casein/starch mixtures and in the casein matrix *Int. Dairy J.* **89** 129–38.
[22] Fu W and Nakamura T 2018 Effects of starches on the mechanical properties and microstructure of processed cheeses with different types of casein network structures *Food Hydrocoll.* **79** 587–95.
[23] Sharif M K, Butt M S, Sharif H R and Nasir 2017 Sensory evaluation and consumer acceptability *Handbook of Food Science and Technology* editors 361–85.

Effects of Different Drying Methods on Antioxidant Properties of Malaysian Ginger

I Mustafa[1*], N L Chin[2], Y A Yusof[2] and M F Husin[1]

[1]Department of Chemical Engineering, University of Teknologi MARA Cawangan Terengganu, Kampus Bukit Besi, Terengganu, Malaysia

[2]Department of Process and Food Engineering, University of Putra Malaysia 43400 UPM Serdang, Selangor

*Corresponding author: iswaibahmustafa@uitm.edu.my

Abstract. Malaysian ginger is one the sought herbs that contains a lot of beneficial properties that could contribute to health. However, most usage of ginger from its fresh form which is prone to microorganisms' spoilage. Therefore, effects of shade, sun, oven, vacuum oven, and freeze drying on phytochemical contents, ferric reducing antioxidant power (FRAP), phosphomolybdenum assays, 2,2′azinobis (3-ethylbenzothiozoline-6-sulfonic acid) disodium salt ($ABTS^{\bullet+}$), 2,2-diphenyl-1-picryl-hydrazyl ($DPPH^{\bullet}$), hydroxyl ($OH^{\bullet}$) radical and metal chelating properties of *Halia Bara* were studied. Dried and fresh ginger crude extracts were extracted with ethanol. The freeze-dried extract had highest level of total phenolic of about 20.07 mg GAE/g dry extract as compared to fresh ginger extract at 10.52 mg GAE/g dry extract. For antioxidant activity of FRAP and phosphomolybdenum, sun-dried extract exhibited the highest values with increase of 3.95-fold and 4.29-fold from fresh ginger extract, respectively. In scavenging $ABTS^{\bullet+}$ radical, sun-dried extract also exhibited the highest values with increase of 2.07-fold from fresh ginger extract. Sun-dried extracts also had the lowest IC_{50} of 14.69 µg/ml. The ascorbic acid of ginger types was below 1.5 mg AA/g extract. Sun-dried ginger extract exhibited most significant antioxidant potential and as free radical scavengers.

1. Introduction

Antioxidants play an important role in humans system as defense mechanism by stabilizing or converting the free radicals to less reactive molecules before attacking the cells [1,2]. Antioxidants is defined as a substance which prevent oxidation of process [3], inhibit radical chain reactions, scavenge free radicals, and retard process of lipid peroxidation [4]. In an oxidative stressed conditions, the amount of antioxidant present may be insufficient to neutralize or balance the free radical generated [1]. Hence, there has been an increased interest in natural antioxidants from plants antioxidant rich dietary materials. Several plants and their active constituents are rich source of antioxidant and play a crucial role in prevention of degenerative diseases [5]. These natural antioxidants are also known as redox-active secondary metabolites of plants [3,6].

Ginger, *Zingiber officinale Roscoe* contains a large number of antioxidants and has the capability to reduce oxidation of lipid and inhibit pathogenesis diseases. Previous study reported that some phenolic substances and bioactive constituents present in ginger have shown a vital effect in anti-oxidative and anti-inflammatory properties and exert anti-mutagenic and anti-carcinogenic activities [3,5,7,8]. *Halia Bara* is popular and sought of ginger types in Malaysian and neighbouring countries as main seasoning in diet and also for medicinal reasons. Recent practice of using sun-dried ginger for suppressing illnesses due to free radicals in local community prompted studies on drying effects of ginger [9,10]. Sun drying is a drying source that may be preferred to other drying methods because it is abundant, unlimited, renewable, inexpensive, non-pollutant and environmentally friendly [11]. Malaysia weather benefits from a tropical climate with uniform temperature, high humidity and sunny has prompted this study.

In general, drying is a postharvest processing method meant to preserve and extend the shelf life of food products. Dried products would inhibit microbial growth and prevent changes in biochemical components. Although drying process is known to give adverse effects on physical properties, nutrients and also on the antioxidant activity [7,12], information on effect of different drying methods on various antioxidant properties and activities of Malaysian ginger are not available. Henceforth, phytochemical composition in terms of antioxidant properties and activities of ethanolic extracted dried and fresh Malaysian ginger types were studied.

2. Research Method

2.1 Drying and Extraction Procedures

Halia Bara was obtained in the farm area of Tendong, Pasir Mas, Kelantan and was subjected to different methods of drying [13]. The dried ginger then was extracted using maceration techniques [13].

2.2 Determination of phenolic content, FRAP, TAA and DPPH• inhibition

The phenolic content and antioxidant activities of ginger extracts (FRAP, scavenging of $ABTS^{\bullet+}$ and $DPPH^{\bullet}$ inhibition) were reported somewhere [13].

2.3 *Ascorbic acid content*

Ascorbic acid of ginger was determined following the method described by Klein and Perry [14]. 150 mg of crude extract was re-extracted with 10 ml (1% w/v) metaphosphoric acid for 45 min at room temperature and filtered using Whatman filter paper (No. 4). 1 ml of filtrate was mixed with 9 ml (0.005% w/v) of 2,6-dichlorophenolindophenol (DCPIP), and the absorbance was measured immediately at 515 nm against a blank. The content of ascorbic acid was calculated on the basis of the calibration curve of authentic L-ascorbic acid and the results were expressed in mg ascorbic acid (AA)/g dry extract of ginger.

2.4 Phosphomolybdenum assay

The phosphomolybdenum method is based on the reduction phosphate-molybdenum (VI) to phosphate-molybdenum (V) by the antioxidant at acidic pH. The phosphomolybdenum assay of ginger extract was evaluated by the method of Prieto et al. [15]. 100 µl of extract was added with 3.6 ml of reagent (0.6 M sulfuric acid, 28 mM sodium phosphate and 4 mM ammonium molybdate) in test tubes. The test tubes were capped and incubated at 95 °C in water bath for 30 min. The mixtures were cooled to room temperature and the absorbance of each solution was measured at 695 nm against the reagent blank. The results were reported mean values expressed in grams of ascorbic acid equivalents per 100-gram sample (AEAC).

2.5 Free radical scavenging activity on hydroxyl OH•

The scavenging activity of ginger extracts on hydroxyl ($OH^{\bullet}$) radical was measured following Klein's method [16] with some modifications. The extracts at concentrations from 0 to 100 µg/ml were added with 1 ml of iron-ethylenediamine- tetraacetic acid (Fe-EDTA) solution (0.13% w/v ferrous ammonium sulfate and 0.26% w/v EDTA), 0.5 ml of EDTA solution (0.018% w/v) and 1 ml of dimethyl sulfoxide (DMSO) (0.85% v/v in 0.1 M phosphate buffer, pH 7.4). The reaction was initiated by adding 0.5 ml of ascorbic acid (0.22% w/v) and was incubated at 88°C for 15 min in a water bath. After incubation, the reaction was terminated by the addition of 1 ml of ice-cold trichloroacetic acid (TCA) (17.5% w/v). 3 ml of Nash reagent (7.6g of ammonium acetate, 0.3 ml of glacial acetic acid and 0.2 ml of acetyl acetone were mixed and made up to 100 ml with distilled water) was added and measured at 412 nm against reagent blank. The reaction mixture without sample was used as control. The percentage hydroxyl radical scavenging activity was expressed as the inhibition of free radicals by the samples Equation (1).

$$\text{OH}^{\cdot}\text{ scavenging activity (\%)} = \frac{(\text{Control absorbance} - \text{Sample absorbance})}{\text{Control absorbance}} \times 100\% \qquad (1)$$

2.6 Determination of metal chelating potential of ginger extracts
The chelating of ferrous ions by ginger extract was estimated by method of Dinis et al. [17] with minor modification. The extract samples (100 μl) were added to a solution of 2mM ferrous chloride (0.05 mL). The reaction was initiated by the addition of 5 mM ferrozine (0.2 mL) and made up to 5 mL with distilled water. The mixture was shaken vigorously and left standing at room temperature for 10 min. The absorbance of the solution was then measured spectrophotometrically at 562 nm. The chelating activity of the extracts was evaluated using EDTA as standard. The results were expressed as mg EDTA equivalent/g dry extract.

2.7 Statistical Analysis
The data obtained were presented as means of 3 replicate determinations ±standard deviation (SD). The data were subjected to one- way analysis of variance (ANOVA) and the significance of the difference between means was determined by Tukey comparison test ($P < 0.05$) using statistical Minitab 16 software (Minitab Inc., State College, PA, USA).

3. Results and Discussion

3.1 Recovery percent, total phenolic and ascorbic acid contents
Table 1 shows the extract yields, total phenolic and ascorbic acid contents of fresh and dried ginger prepared by different drying methods. The extract yield of *Halia Bara* ranged from 2.57% to 9.37%. The maximum extract yield was obtained from sun drying at 9.37%. Extract yield from fresh extract was the lowest with 2.57%.

Phenolic extracts are considered as secondary metabolites components. They are directly responsible for antioxidant activities Freeze-dried extract of *Halia Bara* had highest level of total phenolic with 20.07 mg GAE/g extract. The phenolic values of freeze-dried extract was 1.9 fold significantly higher ($P < 0.05$) than fresh extracts. Vitamin C is a powerful water-soluble antioxidant found in many dietary plants, which plays an important role in the inhibition of free radicals [18]. Generally, ascorbic acid blocks 'free radicals' before they attack cellular membrane by acting as reducing agent by electron donation [7]. In this study, ascorbic acid content in both gingers extract were less than 1.5 mg AA/g extract, which is approximately less than 0.15g/100g (recalculated). Other studies claimed that ascorbic acid content in ginger is not available [7] or very low at 0.91 mmol/100g and 1.83 mmol/100g low in white and red ginger, respectively [19]. To the best of our finding, result shows *Halia Bara* contain less amount of ascorbic acid.

Table 1. Extract yield, total phenolic and ascorbic acid contents of *Halia Bara*

Sample	Extract yield (%)	Total phenolic (mg GAE/g dry extract)	Ascorbic acid (mg AA/g dry extract)
Fresh	2.57	10.526 ± 0.21^{d}	0.690 ± 0.01^{bc}
Shade-Dried	6.84	16.438 ± 0.17^{c}	0.579 ± 0.10^{c}
Sun-Dried	9.37	18.941 ± 0.29^{b}	0.813 ± 0.04^{b}
Oven-Dried	6.85	16.083 ± 0.52^{c}	1.137 ± 0.13^{a}
Vacuum Oven-Dried	6.56	19.533 ± 0.31^{ab}	0.719 ± 0.06^{bc}
Freeze-Dried	8.86	20.068 ± 0.52^{a}	0.863 ± 0.02^{b}

Values are mean (n=3) ± standard deviation.

Mean values followed by different superscripts in a column are significantly different ($P < 0.05$).

3.2 Reduction power by FRAP and phosphomolybdenum assays

Table 2 shows the FRAP and phosphomolybdenum assay prepared by different drying methods. Sun-dried extract had highest ferric reducing antioxidant activity with 4033.1 mmol Fe (II)/g dry extract. FRAP values of sun-dried extracts was 3.95-fold significantly higher ($P < 0.05$) than fresh extracts. Sun-dried also exhibited a higher phosphpmolybdenum reduction of 81.12 g AA/100 g dry extract. The phosphpmolybdenum reductive ability was significantly increased ($P < 0.05$) 4.29-fold than fresh extracts. The reducing capacity of ginger extracts suggesting that *Halia Bara* may serve as significant electron donors in antioxidant activity.

Table 2. FRAP and phosphomolybdenum activity of *Halia Bara*

Sample	**FRAP (mmol Fe(II) / g dry extract)**	**AEAC (g AA/100 g dry extract)**
Fresh	1021 ± 29.7^{d}	18.90 ± 0.56^{f}
Shade-Dried	3229 ± 25.2^{c}	74.27 ± 2.68^{b}
Sun-Dried	4033 ± 29.9^{a}	81.12 ± 0.58^{a}
Oven-Dried	3584 ± 61.1^{b}	59.34 ± 1.56^{e}
Vacuum-Oven-Dried	3174 ± 117.3^{c}	64.72 ± 1.51^{d}
Freeze-Dried	3219 ± 72.5^{c}	69.47 ± 1.26^{c}

Values are mean (n=3) ± standard deviation.
Mean values followed by different superscripts in a column are significantly different ($P < 0.05$).

3.3 Radical scavenging activities by $ABTS^{\bullet+}$, $DPPH^{\bullet}$ and $OH^{\bullet}$

Table 3 shows that sun-dried had highest scavenging $ABTS^{\bullet+}$ activity at 1712 mmol equivalent Trolox/g dry extract. The TEAC values of sun-dried extracts was 2.07-fold significantly higher ($P < 0.05$) than fresh extracts. The increase of TEAC values is reflective to higher polyphenols content of the dried extracts. It has been reported that an increase in antioxidant activity after drying process has been attributed to the release of bound phenolic compounds as cellular constituents would breakdown and new compounds with enhanced antioxidant properties were formed [19].

Scavenging the $DPPH^{\bullet}$ radical is another method to study the potential of radical scavengers. The antioxidant activity of plant extracts is reflected to phenolic components as their capacity to be donors of hydrogen atoms or electrons and to capture the free radicals [20]. Thus, $DPPH^{\bullet}$ radical will be reduced by antioxidant analyte to α, α-diphenyl- β-picrylhydrazine (yellow), with respect to the degree of reduction. A lower concentration of ginger extract to decrease the initial concentration of $DPPH^{\bullet}$ by 50% (IC_{50}) indicates higher antioxidant activity. The best free radical scavenging activity was exerted by sun-dried extract with IC_{50} 14.69 µg/ml. This was higher than the reference standard butylated hydroxyanisole (BHA) at 28.59 µg/ml. The fresh extracts ginger had high values of IC_{50} which possess lowest antioxidant activity. The increased DPPH activity of dried ginger extracts is related to an accumulation and increase of phenolic in dried substance [20,21].

$OH^{\bullet}$ radical is one of the reactive free radicals in living systems that cause DNA blockage and lead to strand breakage. This can contribute to carcinogenesis, cytotoxicity and mutagenesis in living systems [22]. *Halia Bara* extracts at different concentrations were tested for hydroxyl radical scavenging activity. The percentage of $OH^{\bullet}$ radical scavenging activity was not significant different for the different drying extracts (figure 1).

Table 3. **ABTS$^{•+}$ and DPPH$^{•}$ scavenging activities of *Halia Bara***

Sample	TEAC (mmol/g dry extract)	IC_{50} of DPPH (μg/ml)
Fresh	829 ± 44.0^{d}	65.82 ± 6.83^{a}
Shade-Dried	1569 ± 21.2^{ab}	27.19 ± 0.54^{b}
Sun-Dried	1712 ± 1.7^{a}	14.69 ± 2.34^{c}
Oven-Dried	1428 ± 51.8^{bc}	27.97 ± 1.92^{b}
Vacuum Oven-Dried	1419 ± 90.9^{bc}	28.08 ± 0.29^{b}
Freeze-Dried	1336 ± 72.8^{c}	28.59 ± 0.83^{b}
Standard		
BHA		28.59 ± 0.10

Values are mean (n=3) ± standard deviation.
Mean values followed by different superscripts in a column are significantly different ($P < 0.05$)

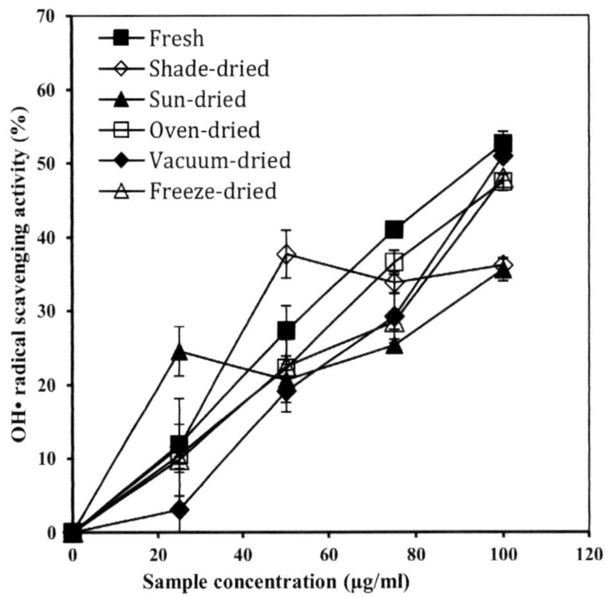

Figure 1. Hydroxyl radical scavenging activity of Halia Bara.
Values are mean ± SD (n=3).

3.4 Metal chelating ability

Iron is a redox active metal which tends to undergo redox cycling reactions such as Fenton reaction and generate many reactive oxygen species (ROS) which lead to lipid peroxidation, hydroperoxide decomposition, protein modification and other effects [18]. Chelation of iron by extract will avoid its participation in redox reaction and prevent subsequent oxidative stress. The binding of ferrous ion by fresh and dried ginger extracts were estimated by disrupting the Fe^{2+}-ferrozine complex which results the decrease of red colour intensity of the complex. Colour reduction would indicate that the extracts could chelate irons and possesses metal chelating activity. Table 4 shows that all sample extracts had potential chelating ability on Fe^{2+} but not unanimous in trend. Chelating agents are effective as secondary antioxidants as they capable to reduce redox potential, thereby stabilizing the oxidized form of metal ion [12]. Significant differences of metal chelating activities were identified in the table 4.

Table 4. Metal chelating activity (mg EDTA/g dry extract)

Sample	*Halia Bara*
Fresh	32.23 ± 7.50^{ab}
Shade-Dried	44.69 ± 0.27^{ab}
Sun-Dried	45.47 ± 9.49^{a}
Oven-Dried	51.39 ± 7.91^{a}
Vacuum oven-Dried	47.80 ± 3.06^{a}
Freeze-Dried	21.32 ± 15.29^{b}

Values are mean (n=3) ± standard deviation.
Mean values followed by different superscripts in a column are significantly different ($P < 0.05$).

3.5 Correlation analysis

In this study, a significant positive correlation ($P < 0.05$) was found between total phenolic content (TPC) and antioxidant properties from ginger extracts as tabulated in table 5. Phenolic content of *Halia Bara* were highly correlated with FRAP, phosphomolybdenum (PM), $ABTS^{\bullet+}$ and $DPPH^{\bullet}$. This strong relationship was due to hydroxyl group of phenolic compounds which enable them to participate in redox reactions in living mechanism. Reducing power of ginger extract was also positively correlated with radical scavenging activity. Correlation of antioxidant properties with ascorbic acid (AA) and metal chelating were low.

Table 5. Correlation (R^2) among phytochemical content, antioxidant and metal chelating activities of ginger extracts

	Phytochemical content		Reducing power		Radical scavenging		Metal chelating
Halia Bara							
TPC	1						
AA	0.137	1					
FRAP	**0.821**	0.362	1				
PM	**0.867**	0.085	**0.939**	1			
$ABTS^{\bullet+}$	**0.739**	0.127	**0.938**	**0.953**	1		
$DPPH^{\bullet}$	**0.801**	0.234	**0.978**	**0.961**	**0.954**	1	
Fe^{2+} chelating activity	0.052	0.164	0.356	0.240	0.437	0.332	1

4. Conclusion

Drying methods could enhance ginger's antioxidant capacity and as free radical scavengers. Drying process might breakdown the cellular constituents, thus promoting the release of bound phenolic compounds from food matrix. Sun dried gingers exhibited high total phenolic contents with 18.94 mg GAE/g dry extract and antioxidant activities through measured of FRAP (4033 mmol Fe(II)/g dry extract), $ABTS^{\bullet+}$ (1712 mmol/g dry extract), $DPPH^{\bullet}$ (14.69 µg/ml) and AEAC assay (81.12 g AA/100 g dry extract). Further work is on identification of composition and components of fresh and dried ginger extracts at molecular levels.

Acknowledgements

We would like to thank Universiti Teknologi MARA and the Ministry of Higher Education of Malaysia (UPM-MOHE) for providing scholarship to Iswaibah under Young Lecturer Scheme.

References

[1] Boligon A A, Machado M M and Athayde M L 2014 Medicinal chemistry Technical Evaluation of Antioxidant Activity *Med. Chem.* **4** 517–22.

[2] Saumya S M and P M B 2011 in Vitro Evaluation of Free Radical Scavenging Activities of Panax Ginseng and Lagerstroemia Speciosa : a Comparative Analysis *Int. J. Pharm. Pharm. Sci.* **3** 1–5.

[3] Ghasemzadeh A, Jaafar H Z E and Rahmat A 2010 Antioxidant activities, total phenolics and flavonoids content in two varieties of malaysia young ginger (Zingiber officinale Roscoe) *Molecules* **15** 4324–33.

[4] Yesiloglu Y, Aydin H and Kilic I 2013 In Vitro Antioxidant Activity of Various Extracts of Ginger (Zingiber officinale L.) Seed *Asian J. Chem.* **25** 3573–8.

[5] Rahmani A H, Al F M and Aly S M 2014 Active ingredients of ginger as potential candidates in the prevention and treatment of diseases via modulation of biological activities *Int. J. Physiol. Pathophysiol Pharmacol* **6** 125–36.

[6] Sowndhararajan K, Joseph J M, Arunachalam K and Manian S 2010 Evaluation of Merremia tridentata (L.) Hallier f. for in vitro antioxidant activity *Food Sci. Biotechnol.* **19** 663–9.

[7] Gümüşay O A, Borazan A A, Ercal N and Demirkol O 2015 Drying effects on the antioxidant properties of tomatoes and ginger. *Food Chem.* **173** 156–62.

[8] Mojani M and Ghasemzadeh A 2014 Assessment of bioactive compounds, nutritional composition and antioxidant activity of Malaysian young ginger (Zingiber officinale Roscoe). *Int. Food ...* **21** 1931–5.

[9] Jayashree E, Visvanathan R and John Zachariah T 2012 Quality of dry ginger (Zingiber officinale) by different drying methods *J. Food Sci. Technol.* **51** 3190–8.

[10] Prasad J, Prasad A and Vijay V K 2006 Studies on the drying characteristics of Zingiber officinale under open sun and solar biomass (hybrid) drying *Int. J. Green Energy* **3** 79–89.

[11] Chong C H, Law C L, Cloke M, Abdullah L C and Wan Daud W R 2009 Drying Models and Quality Analysis of Sun-Dried Ciku *Dry. Technol.* **27** 985–92.

[12] Sowndhararajan K, Siddhuraju P and Manian S 2011 Antioxidant and free radical scavenging capacity of the underutilized legume, Vigna vexillata (L.) A. Rich *J. Food Compos. Anal.* **24** 160–5.

[13] Mustafa I, Chin N L, Fakurazi S and Palanisamy A 2019 Comparison of phytochemicals, antioxidant and anti-inflammatory properties of sun-, oven- And freeze-dried ginger extracts *Foods* **8**.

[14] Klein B P and Perry A K 1982 Ascorbic Acid and Vitamin A Activity in Selected Vegetables from Different Geographical Areas of the United States *J. Food Sci.* **47** 941–5.

[15] Prieto P, Pineda M and Aguilar M 1999 Spectrophotometric quantitationof antioxidant capacity through the formation of a phosphomolybdinum complex : specific application to the determination of Vitamin E. *Anal. Biochem.* **269** 337–41.

[16] Klein S M, Cohen G and Cederbaum A I 1981 Production of formaldehyde during metabolism of dimethyl sulfoxide by hydroxy *Biochemistry* **20** 6006–12.

[17] Dinis T C P, Madeira V M C and Almeida L M 1994 Action of Phenolic Derivatives (Acetaminophen, Salicylate, and 5-Aminosalicylate) as Inhibitors of Membrane Lipid Peroxidation and as Peroxyl Radical Scavengers *Arch. Biochem. Biophys.* **315** 161–9.

[18] Liu H, Cao J and Jiang W 2015 Evaluation and comparison of vitamin C, phenolic compounds, antioxidant properties and metal chelating activity of pulp and peel from selected peach cultivars *LWT - Food Sci. Technol.* **63** 1042–8.

[19] Oboh G, Akinyemi A J and Ademiluyi A O 2012 Antioxidant and inhibitory effect of red ginger (Zingiber officinale var. Rubra) and white ginger (Zingiber officinale Roscoe) on Fe2+ induced lipid peroxidation in rat brain in vitro *Exp. Toxicol. Pathol.* **64** 31–6.

[20] Chumroenphat T, Khanprom I and Butkhup L 2011 Stability of Phytochemicals and Antioxidant Properties in Ginger (Zingiber officinale Roscoe) Rhizome with Different Drying Methods *J.*

Herbs. Spices Med. Plants **17** 361–74.
[21] Ho S-C and Su M-S 2016 Optimized heat treatment enhances the anti-inflammatory capacity of ginger *Int. J. Food Prop.* **19** 1884–98.
[22] Arunachalam K and Parimelazhagan T 2012 Evaluation of nutritional composition and antioxidant properties of underutilized Ficus talboti King fruit for nutraceuticals and food supplements .*Food Sci. Technol.* **51** 1260–8.

140121

ORIGINALITY REPORT

4%	4%	8%	5%
SIMILARITY INDEX	INTERNET SOURCES	PUBLICATIONS	STUDENT PAPERS

PRIMARY SOURCES

1	www.researchgate.net Internet Source	4%

Exclude quotes	On	Exclude matches	< 100 words
Exclude bibliography	On		

Lithology Identification, Shale Volume Estimation and Formation Wate Saturation of Fenchuganj Gas Field Using Wireline Log Data

M Rahman[1*], M R Mojid[1], I M Khan[2]

[1]Department of Petroleum Engineering, Universiti Teknologi PETRONAS, Bandar Seri Iskandar, Perak, 32610, Malaysia

[2]Department of Petroleum and Mining Engineering, Chittagong University of Engineering & Technology, Chittagong, 4349, Bangladesh

*Corresponding author: musfika_20000295@utp.edu.my

Abstract. This study has been undertaken with a view to interpreting and analyzing different types of logs, detecting hydrocarbon-bearing sand zone, identifying the lithology, estimating shale volume, and determining water saturation of the formation of the Fenchuganj gas field. Fenchuganj gas field is in the southern part of Bangladesh in the Sylhet division. It lies in the Surma basin and characterized as a water drive gas field. Gamma Ray (GR) log and Spontaneous Potential (SP) log analysis help to interpret the formation lithology. After interpreting lithology, the gamma-ray log and SP log were analyzed further, and the hydrocarbon-bearing zone was detected. From the analysis of different logs, for well 4, so many gas-bearing sand zones have been detected. The gas-bearing zones are found between (1585-1595) m, (1600-1610) m, (2409-2420) m, (2265-2270) m, (2295-2303) m, (2435-2450) m, (2465-2480) m, (3150-3155) m, (2955- 2960) m. These are the gas-bearing zones of fenchugonj well-4 determined from the composite log of this field. Water-bearing sand zones are between (1708-1714) m and the fully compacted shale zone is between (1920-1930) m. From the analysis of different logs, for well 5, so many gas-bearing sand zones have been detected. The gas-bearing zones are found between (2365-2385.5) m, (1820-1830) m, (1782- 1803) m, (305-308) m, (336.5-338.5) m, (607.5-609.5) m. These are the gas-bearing zone of fenchugonj well-5 determined from the composite log of the field. Shale volume has been estimated from the gamma-ray log and from resistivity logs by the true resistivity method. In well 4, the calculated value of shale volume by Gamma Ray method and true Resistivity methods were respectively 15% and 6%, in well 5 these values were 7.25% and 7.64%. Formation water resistivity was determined from the formula and taken as 0. 1ohm-m and mud filtrate resistivity 0.76 ohm-m. for well-4 and for well-5 it was taken as 0.7 ohm-m (from the provided composite log of well 4 and 5). Formation true resistivity R_t was estimated directly from the ILD log and true resistivity log. The average value of R_t for well-5 was 33.75 ohm-m and for well-4 the value is 34.09 ohm-m. From the RFOC log, flushed zone resistivity R_{xo} was calculated directly. Water saturation has been determined by three different techniques named Archie, Indonesia, and Simandoux method. For well 5, the value of average water saturation found from Archie, Indonesia, and Simandoux model was 8.5%,22%, and 24.84%, and for well-4. The value of average water saturation was found as 7.3%,18%, 20.51% respectively.

1. Introduction

In petroleum geology, reservoir is the most common element of petroleum system which can accumulate hydrocarbons (oil or gas). Reservoir rock is that place in which fluids can be stored and accumulated. The fluids might be water, oil, and gas. The most important geologic information is reservoir that is termed as a subsurface volume of rock. It has adequate porosity and permeability so that it can store hydrocarbon under presence of seals and traps. Migration of hydrocarbon form the source rock to porous and permeable rock occurred in conventional reservoir [1].

Integration of all available data to define the geometry, distribution of physical parameters, and flow properties of a petroleum reservoir is known as Reservoir Characterization. there are some physical properties of reservoir which is needed to characterize the reservoir. They are: Lithology, Porosity, Permeability, Compressibility, Fluid saturation, Capillary characteristics, Rock stress, Fluid rock interaction. Permeability is one of the flow properties [2]. In oil and gas industry, the most elementary methods for reservoir characterization well logging is used all over the world. To know about the petrophysical properties of the reservoir and detect the hydrocarbon bearing zone, evaluate the hydrocarbon volume well logging is used mostly. Well log data are routinely used for stratigraphic interpretation of the Earth's subsurface. By using wireline log data, one can calculate: Shale Volume (V_{sh}), Water Saturation (S_w), Porosity (Φ), and Permeability (k). Several steps are required to interpret wireline log data and analyzation should not be done randomly because it might cause error in result. Basically, petrophysical properties (shale volume, water saturation, permeability, etc.) will be used in reservoir characterization which are mostly based on geology. The steps will be determination of petrophysical properties (shale volume, water saturation, permeability, etc.) of reservoir rock, evaluation of their nature and volume, at last its essential to estimate their economic feasibility [3].

Reserves are known as those quantities of petroleum resources which are economically feasible to recover commercially from the known accumulations soon from given data forward using the commercial methods under the regulations of the government and regulatory agencies. Thus, reserves must satisfy four criteria: it must be discovered, recoverable, commercial, and remaining. Proved reserves, possible reserves and probable reserves are the types of reserve. Proven reserves are the number of natural resources that a business fairly hopes to obtain from a structure that is formed using engineering and geological information collected by exploratory excavation and seismic testing. It should fulfil at odds 90% or more likelihood of regeneration. Probable stocks are those unproven resources that suggest that analysing geological and engineering information is more probable than unrecoverable. The likelihood of regeneration should be at least 50% or more of the expected amount plus likely stocks. Possible reserves are those unproved reserves which analysis of geological and engineering data suggests are less likely to be recoverable than probable reserves. Probability of recovery should be at least 10% or more of the sum of the estimated proved plus probable plus possible reserves [4]. Different approaches of reserve estimations are: Volumetric Estimate, Material Balance Estimate, Decline Curve and Reservoir Simulation. After discovery and at the early stage of production most commonly used method is volumetric method. The parameters needed for volumetric methods are: Area, Thickness, Porosity, Water saturation, Formation volume factor and Recovery factor.

Research was taken out in the Niger Delta to characterize hydrocarbon reservoirs using well registers to assess potential hydrocarbon, delineate hydrocarbon and water-bearing areas and petrophysical characteristics of interesting hydrocarbon reservoirs. The research used data from four composite well records consisting of the records of Gamma Ray, Resistivity, Neutron, and Density. Results from the research found that nine out of twenty-two concern areas (mud fields) were delineated and linked across to identify prospective hydrocarbon reservoirs. The outcome also shows an improvement in porosity with a rise in permeability. The petrophysical parameter assessed showed that porosity varies from (18-31%), air saturation (14-44%), hydrocarbon saturation (56-86%), permeability (138-10662) [5]. A research was conducted in order to classify the reservoirs in well-4 Rashidpur, Bangladesh, by quantitatively analysing petrophysical parameters in the gas field. The study identified 20 permeable areas. Four were gas-bearing zones among them, while the remainder were hot. In these areas, the hydrocarbon form may be a gas that shows relatively high resistivity, low neutron, and low-density

IOP Conf. Series: Materials Science and Engineering 1176 (2021) 012046 doi:10.1088/1757-899X/1176/1/012046

record responses than in the water areas. The average pore content in those areas ranged from 18-27% to 17-40% for water saturation and 60-82% for hydrocarbon saturation. Hydrocarbon is moved into the tank because the hydrocarbon mobility index indicates less than 0.70. The volume of average bulk was between 0.04 and 0.13. The lithology is fine to very fine grain sandstone with irreducible water saturation and all the calculated values confirms this. Zone-4 was shown to be comparatively highly thick, porous, moist, and permeable to other regions. [6]. From time-to-time Petro Bangla includes consultants in the carrying out of specific jobs and undertake studies to establish various gas fields on its behalf. Petro Bangla typically engages consultants in seismic surveying and interpreting, drilling, completion, work on, precision survey, reserve assessment etc., on the side of exploration and development. Some of these consultancy companies have performed studies on behalf of Petro Bangla to estimate and update the gas and reserves in place. BAPEX carried out research on the gas sector of Fenchuganj [7]. The objectives of this research are to interpret and analyse different types of logs, identify the lithology of the formation, and estimate shale volume and water saturation of the formation using log data.

2. Location and exploration of the field

Fenchuganj gas field is in southern part of Bangladesh and in Sylhet division. It lies in Surma basin and characterized in water drive gas field. It is in Fenchuganj upazila and about 40km south-east of Sylhet town. It is bounded by Longitude E 91° 53' –92°and Latitude N 24° 30' –24° 37'and is 30 km long and 8 km wide. Fenchuganj gas field was discovered by Petro Bangla in 1988.This field is operated by Bangladesh Petroleum Exploration and Production Company Limited (BAPEX). The sequences of lithologies found in the region range from the age of Oligocene to Pliocene. It is an oval trough that is negative. It is an oval trough. In Lower Bokabil to Upper Bhuban formations in the late Miocene period, the gas bearing sands of the structure are found. Miocene sediments like alternating gray to dark gray clay and very fine to fine grained sandstone is found in the reservoir [8]. In Lower Bokabil to Upper Bhuban formations in the late Miocene period, the gas bearing sands of the structure are found. The seismic analysis reveals a major reverse fault of the Fenchuganj formation on the eastern flank. The line of the steep North Flank and the slightly south Flank is an asymmetrical antichange. The Fenchuganj sediments consist of a different proportion of shale and sandstone from Oligocene to the recent period. [9]. Fenchuganj gas fields' structural pattern in the whole region is smooth and uniform. The Fenchuganj structure's litho-stratigraphy is created based on good knowledge. The sequences of lithologies found in the region range from the age of Oligocene to Pliocene. This structure was a reverse fault with a trending axis of NNE-SSW. The number of dips on the eastern side ranges from 30°-35°; on the western side, the number varies from 20°-25° [9]. In Lower Bokabil to Upper Bhuban formations in the late Miocene period, the gas bearing sands of the structure are found. The seismic analysis reveals a major reverse fault of the Fenchuganj formation on the eastern flank. The line of the steep North Flank and the slightly south Flank is an asymmetrical antichange. The Fenchuganj sediments are composed of alternate shell and sandstones, which differ according to age. There has been no extensive research on the structural development of the region since this area has been discovered. The goal of this research is therefore to re-structure structural maps of the linkages between formations, to identify canal, fault and other structure, to delegate seismic disturbances and finally to interpret the results analysed to understand the structural development of the region [9].

In Bangladesh, most of the onshore part has been explored widely for natural gas, yet there is more possibility to explore the resource inland by remodelling and re-studying with modern techniques. Fenchuganj Gas Field is one of the producing gas fields located at north-eastern part of Bangladesh.

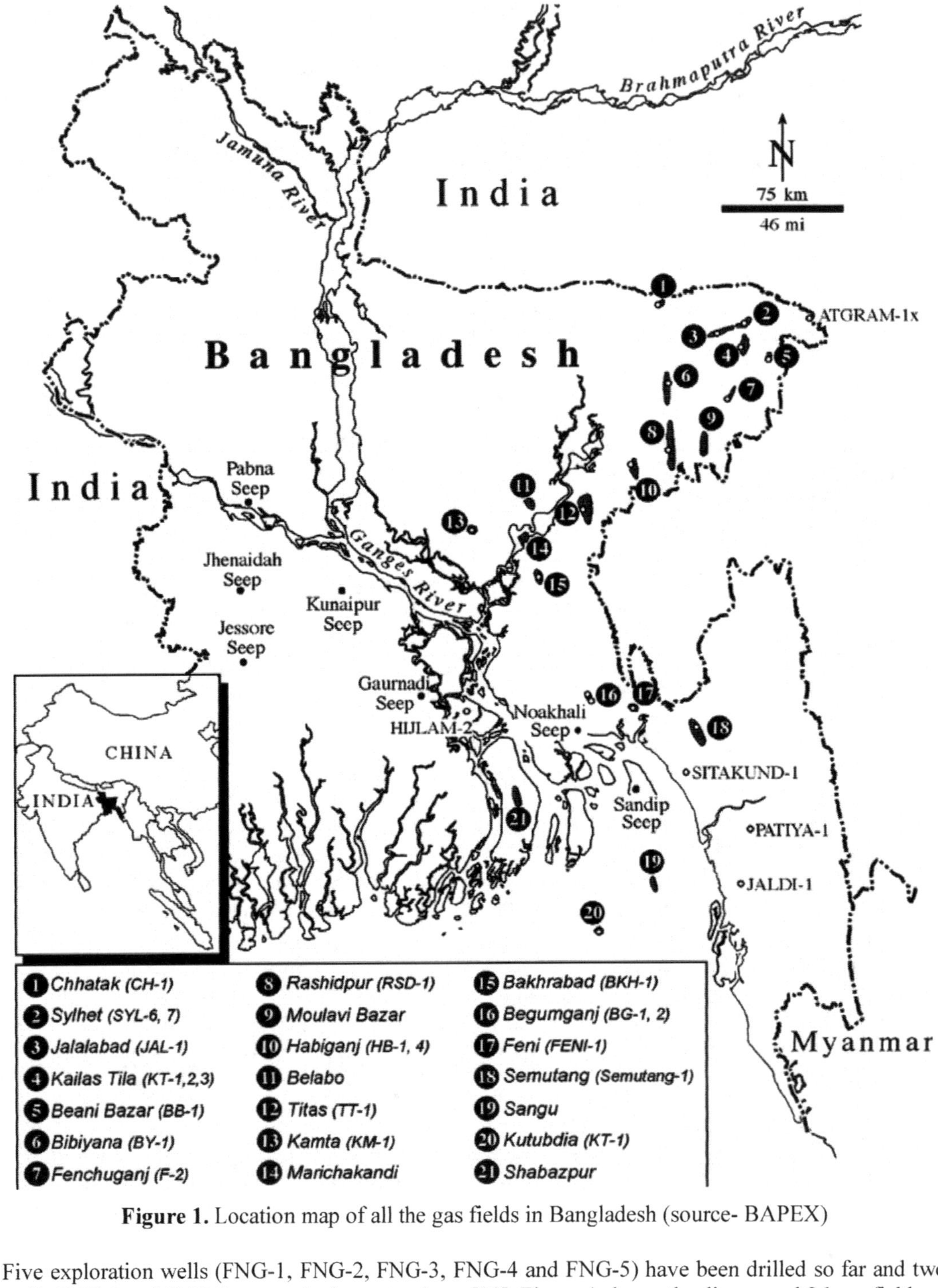

Figure 1. Location map of all the gas fields in Bangladesh (source- BAPEX)

Five exploration wells (FNG-1, FNG-2, FNG-3, FNG-4 and FNG-5) have been drilled so far and two wells FNG-2 and FNG-3 are producing natural gas [10]. Figure 1 shows the discovered 26 gas fields of the country including 2 offshore ones and searching for other prospects both on offshore and onshore area. The total initial gas reserve in the 26 gas fields has been estimated at 27.12 Tcf (Proved and Probable) and 20.08 TCF (recoverable) of natural gas enough to meet our demand of energy consumption till 2019. Fenchuganj is one of the most productive and prospective gas fields in Bangladesh and falls in gas block #14. Fenchuganj gas field was discovered in 1988. First well #FNG-1 was drilled in 1960 down to 2439 m and abandoned as a dry hole. Later, January 21st, 1985 drilling of the well #FNG-2 started and reached TD on November 4, 1986. After producing 24 BSCF gas, the well#2 was suspended due to excessive water and sand production and the well was recompleted at lower zone. But due to much amount of water and well head pressure down (1020 psi), well#2 shut on July 21, 2013. The well #FNG-3 started on February 6, 2004 and completed as production well on May 14, 2004. Drilling of the well #FNG-4 started on November 2010 and was completed in May 2011 [10].

3. Methodology

3.1. Lithology identification and shale volume estimation

There are several reasons why it is important to know about the lithology of a zone (i.e., sandstone, limestone, or dolomite). The best logs for lithological purposes are those that are most influenced by rock properties and least influenced by fluid properties. The most useful of the commonly available logs are: Gamma Ray (GR) log and Spontaneous Potential (SP) log.

Gamma Ray log can be used to calculate volume of shale in porous reservoirs. Shale is more radioactive than sand or carbonate. The volume of shale expressed as a decimal fraction or percentage is called V_{sh}. Calculation of the gamma ray index is the first step needed to determine the volume of shale from gamma ray log. The gamma ray log has several nonlinear empirical responses as well as linear response which have been expressed as equation (1) and (2). The nonlinear responses are based on geographic area or formation age.

Linear response ($V_{sh}= I_{GR}$) [11]:

$$V_{sh} = I_{GR} = \frac{GR_{\log} - GR_{\min}}{GR_{\max} - GR_{\min}} \tag{1}$$

Nonlinear response: Larionov (1969) for Tertiary rocks [11]:

$$V_{sh} = 0.083\left(2^{3.7 I_{GR}} - 1\right) \tag{2}$$

Where, V_{sh} = shale volume, I_{GR}= Gamma Ray index, GR_{log} = Log value of Gamma Ray, GR_{max} = Maximum value of Gamma Ray of the formation and GR_{min} = Minimum value of Gamma Ray of the formation.

Spontaneous potential log, commonly called the self-potential log or SP log, is a measurement taken by oil industry well loggers to characterize rock formation properties. The log works by measuring small electric potentials (measured in millivolts) between depths in the borehole and a grounded electrode at the surface. A log of the natural difference in electrical potential, in millivolts, between an electrode in the borehole and a fixed reference electrode on the surface. The magnitude of the deflection depends mainly on the salinity. The difference between drilling mud and formation water and the clay content of the permeable bed. The spontaneous potential (SP) log is used to detect permeable beds from the formation [12].

3.2. Determination of water saturation

There are several methods to determine the water saturation and among these Archie, Simandoux and Indonesia method are most used, and these have been expressed in equation (3), (4) and (5) respectively.

Archie Formula:

Water saturation (Sw) of a reservoir's uninvaded zone is calculated by Archie (1942) formula [11].

$$S_w = \left(\frac{a * R_w}{R_t * \phi^m} \right)^{1/n} \tag{3}$$

Simandoux Equation: [13]

$$S_w = \left[\left(\frac{0.4 R_w}{\phi_e^2} \right) \right] * \left[\sqrt{ \left(\frac{V_{sh}}{R_{sh}} \right)^2 + \frac{(5\phi_e^2)}{(R_w * R_t)} } - \left(\frac{V_{sh}}{R_{sh}} \right) \right] \tag{4}$$

Indonesia Method:
Water saturation for un-invaded zones [13],

$$S_w = \left(\frac{1}{R_t} \right) * \left[\left\{ \frac{V_{cl}^{(1-0.5V_{cl})}}{(R_{cl})^{0.5}} \right\} + \left\{ \frac{\phi_e^{0.5m}}{(aR_w)^{0.5}} \right\} \right] \tag{5}$$

(5)

Water saturation for flushed zones [13],

$$S_{xo} = \left(\frac{1}{R_{xo}} \right) * \left[\left\{ \frac{V_{cl}^{(1-0.5V_{cl})}}{(R_{cl})^{0.5}} \right\} + \left\{ \frac{\phi_e^{0.5m}}{(aR_{mf})^{0.5}} \right\} \right] \tag{6}$$

Where, S_w= Water saturation, R_w= Formation water resistivity, R_t =True resistivity of formation, Φ = Porosity, m= cementation factor, n= saturation exponent, V_{sh}= Shale volume, V_{cl}= Clay volume, Φe= Effective porosity, R_t = True resistivity of formation, R_{cl} = Clay zone resistivity, R_{xo} = Flushed zone resistivity, R_{mf} = Mud filtrate resistivity.

4. Results and discussions

At first the analyzation of available log data takes place carefully to interpret lithology of the formation. In the formation there exist some gas bearing sand zone, some water bearing sand zone and fully compact shale zone. These zones are found from the log data. After detecting hydrocarbon bearing zones, different types of reservoir properties like shale volume, porosity, formation water resistivity, flushed zone resistivity, formation true resistivity, water saturation etc. Gamma Ray log, SP log, ILD log, sonic log, neutron log, density log etc. logs are used to determine the different types of reservoir properties.

4.1. Detection of lithology and hydrocarbon bearing zones

Gamma ray log and SP log analysis helps to interpret the formation lithology. After interpreting lithology, the gamma ray log and sp log were analyzed further and the hydrocarbon bearing zone was detected. When the value of GR log is low, resistivity log is high, neutron log is low and density log is low at the same time the zone is detected as hydrocarbon bearing zone. The relation between neutron log and density log helps to determine the hydrocarbon bearing zone easily and perfectly. From the analysis of different logs, for well 4, so many gas bearing sand zones has been detected. The gas bearing zones are found between (1585-1595) m, (1600-1610) m, (2409-2420) m, (2265-2270) m, (2295-2303) m, (2435-2450) m, (2465-2480) m, (3150-3155) m, (2955- 2960) m. These are the gas bearing zones of fenchugonj well-4 determined from the composite log of this field. Water bearing sand zones are between (1708-1714) m and fully compacted shale zone is between (1920-1930) m. From the analysis of different logs, for well 5, so many gas bearing sand zones has been detected. The gas bearing zones are found between (2365-2385.5) m, (1820-1830) m, (1782- 1803) m, (305-308) m, (336.5-338.5) m ,(607.5-609.5)m. These are the gas bearing zone of fenchugonj well-5 determined from the composite log of the field.

4.2. Resistivity and shale volume calculation

Formation water resistivity was determined from formula and taken as 0.1ohm-m and mud filtrate resistivity 0.76ohm.m. for well-4 and for well-5 it was taken as 0.7 ohm-m (from the provided composite log of well 4 and 5). Formation true resistivity R_t was estimated directly from ILD log and true resistivity log. The average value of R_t for well-5 was 33.75 ohm-m and for well-4 the value is 34.09 ohm-m. from RFOC log, flushed zone resistivity R_{xo} is calculated directly. Shale volume was calculated by using Gamma ray log values and True resistivity method. For well 4, the average shale volume calculated using Gamma ray log was 15.47% and the average shale volume estimated using true resistivity method 6.22%. For well 5 the average shale volume calculated using Gamma ray log was 7.25% and the average shale volume estimated using true resistivity method was 7.64%. Table 1 and table 2 show the shale volume calculation of well 4 zone (1600-1610) m and well 5 zone (2365-2385.5) m, by using gamma ray and true resistivity methods. These shale volume results are used to find one of the most important reservoir properties known as porosity on which the determination of water saturation also dependent.

Table 1. Shale volume (Vsh) estimation of well 4 from GR and TR methods

Well 4	Vsh from Gamma Ray log					Vsh from True resistivity method			
Depth, m	Log values (API)			Calculated		Log values (ohm-m)			Calculated
	GRlog	GRmax	GRmin	Ish	Vsh (%)	Rt (max)	Rt	Rclay	Vsh (%)
1600	95	130.0	35.0	0.632	33.63	42	15.0	5	27.02
1601	78	130.0	35.0	0.453	18.20	42	30.0	5	7.87
1602	72	130.0	35.0	0.389	14.24	42	30.0	5	7.87
1603	66	130.0	35.0	0.326	10.87	42	30.0	5	7.87
1604	70	130.0	35.0	0.368	13.05	42	33.0	5	5.90
1605	65	130.0	35.0	0.316	10.36	42	42.0	5	0.00
1606	67	130.0	35.0	0.337	11.39	42	40.0	5	1.79
1607	68	130.0	35.0	0.347	11.93	42	38.0	5	2.99
1608	75	130.0	35.0	0.421	16.14	42	40.0	5	1.79
1609	72	130.0	35.0	0.389	14.24	42	40.0	5	1.79
1610	75	130.0	35.0	0.421	16.14	42	37.0	5	3.56
Average	73.00			0.40	15.47		34.09		6.22

Table 2. Shale volume (Vsh) estimation of well 5 from GR and TR methods

Well 5	Vsh from Gamma Ray log					Vsh from True resistivity method			
Depth (m)	Log values (API)			Calculated		Log values (ohm-m)			Calculated
	GRlog	GRmax	GRmin	Ish	Vsh (%)	Rt (max)	Rt	Rclay	Vsh (%)
2365	80	145.0	55.0	0.278	8.62	42	40.0	5	1.79
2366	70	145.0	55.0	0.167	4.43	42	33.0	5	5.90
2367	70	145.0	55.0	0.167	4.43	42	30.0	5	7.87
2368	65	145.0	55.0	0.111	2.74	42	30.0	5	7.87
2369	72	145.0	55.0	0.189	5.17	42	28.0	5	9.34
2370	66	145.0	55.0	0.122	3.06	42	27.0	5	10.14
2371	68	145.0	55.0	0.144	3.72	42	28.0	5	9.34
2372	65	145.0	55.0	0.111	2.74	42	28.0	5	9.34
2373	70	145.0	55.0	0.167	4.43	42	30.0	5	7.87
2374	67	145.0	55.0	0.133	3.38	42	33.0	5	5.90
2375	70	145.0	55.0	0.167	4.43	42	30.0	5	7.87
2376	68	145.0	55.0	0.144	3.72	42	40.0	5	1.79
2377	75	145.0	55.0	0.222	6.38	42	34.0	5	5.30
2378	82	145.0	55.0	0.300	9.62	42	27.0	5	10.14
2379	120	145.0	55.0	0.722	44.61	42	13.0	5	32.70
2380	102	145.0	55.0	0.522	23.38	42	30.0	5	7.87
2381	70	145.0	55.0	0.167	4.43	42	42.0	5	0.00
2382	63	145.0	55.0	0.089	2.13	42	42.0	5	0.00
2383	70	145.0	55.0	0.167	4.43	42	40.0	5	1.79
2384	67	145.0	55.0	0.133	3.38	42	35.0	5	4.71
2385.5	66	145.0	55.0	0.122	3.06	42	24.0	5	12.89
Average	73.62			0.21	7.25				7.64

4.3. Water Saturation estimation

Water saturation was calculated using three different methods because from the proper analyzation, the water saturation of the zone can be analyzed effectively. Methods are archie, Indonesia and simandoux to determine the water saturation. From figure 2, we can see that at 1600 m the degradation of all the graph takes place simultaneously that means the water accumulation takes place in low quantity from this depth. Archies ranging from (6-13) %, Indonesia ranging from (15-30) % and Simandoux ranging from (17-36) %.

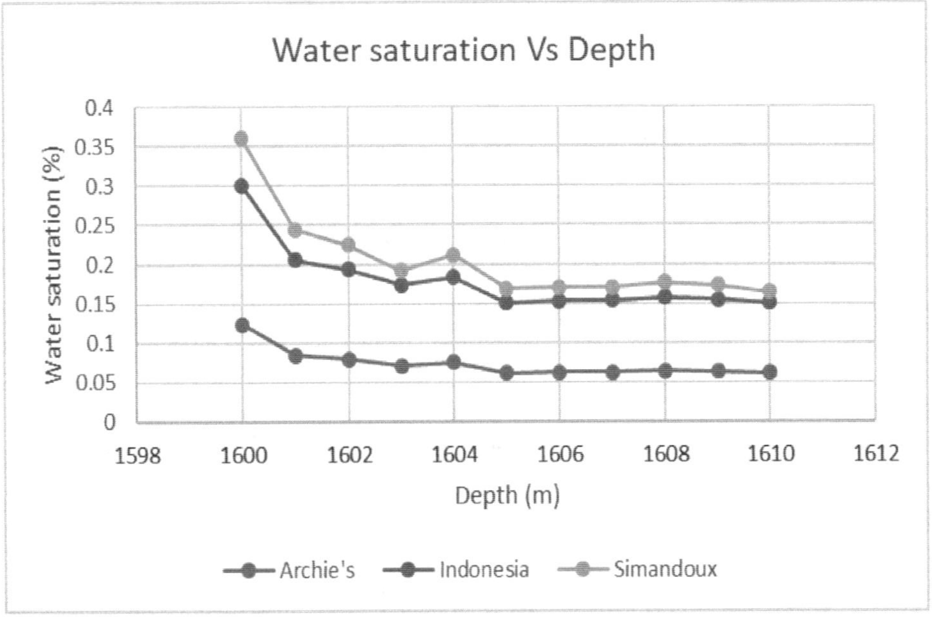

Figure 2. Depth vs water saturation calculated using Archie, Indonesia, Simandoux formula (well 4)

For well 5, from figure 3 the value of average water saturation found from archie, Indonesia and simandoux model were 8.5% ,22% and 24.84% and for well-4, The value of average water saturation was found as 7.3% ,18%, 20.51% respectively. Table 3 and table 4 show the water saturation estimation of well#4 zone (1600-1610) m and well#5 zone (2365-2385.5) m, by using Archie, Indonesia and Simandoux methods.

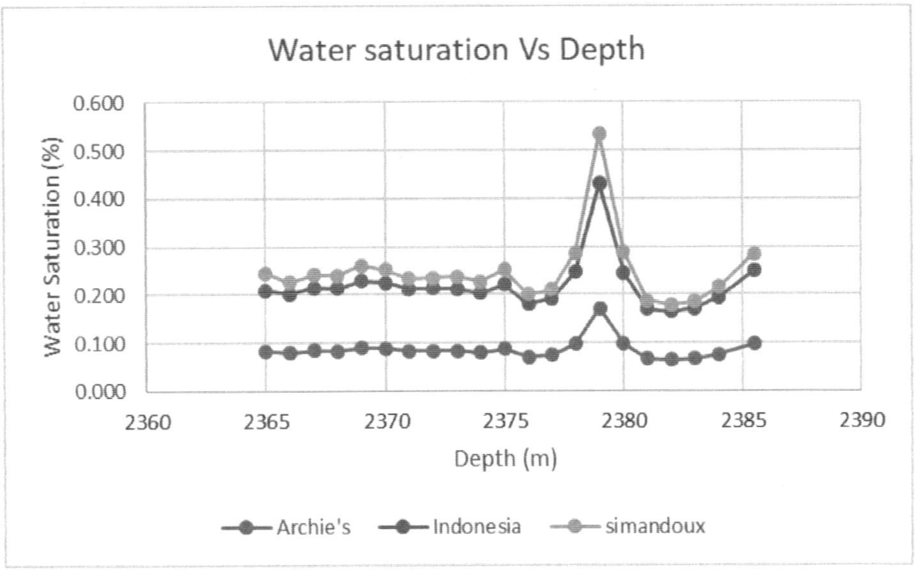

Figure 3. Depth vs water saturation calculated using Archie, Indonesia, Simandoux formula (well 5)

Table 3. Water saturation estimation by Archie, Indonesia and Simandoux method for well 4

Depth (m)	R_t (ohm-m)	Shale volume (%)	Porosity (%)	S_w (%) (Archie)	S_w (%) (Indonesia)	S_w (%) (Simandoux)
1600	15	33.63	20.22	0.124	0.300025883	0.36043735
1601	30	18.20	21.13	0.085	0.205753857	0.244073591
1602	30	14.24	23.01	0.079	0.193799707	0.224227987
1603	30	10.87	26.88	0.070	0.173607344	0.192023191
1604	33	13.05	23.19	0.075	0.183748748	0.212110295
1605	42	10.36	25.91	0.061	0.150602297	0.168337876
1606	40	11.39	26.20	0.062	0.153083395	0.170541937
1607	38	11.93	26.90	0.062	0.15415073	0.170431268
1608	40	16.14	25.26	0.064	0.157102533	0.17682069
1609	40	14.24	25.80	0.063	0.154795822	0.173198376
1610	37	16.14	28.28	0.061	0.150747927	0.164254594

Table 4. Water saturation estimation by Archie, Indonesia and Simandoux method for well 5

Depth (m)	**R_t (ohm-m)**	**Shale volume (%)**	**Porosity (%)**	**S_w (%) (Archie)**	**S_w (%) (Indonesia)**	**S_w (%) (Simandoux)**
2365	40	8.622715	18.21396	0.082146	0.20918706	0.245325537
2366	33	4.426625	21.82683	0.078677	0.20159023	0.225504018
2367	30	4.426625	21.38717	0.08382	0.21462849	0.241374679
2368	30	2.736589	21.52886	0.083395	0.21358546	0.23981585
2369	28	5.173009	20.52583	0.089552	0.22899739	0.260316396
2370	27	3.055612	21.58433	0.087732	0.22471174	0.252135623
2371	28	3.72159	22.8569	0.082434	0.21152072	0.233799071
2372	28	2.736589	22.7181	0.082821	0.21247525	0.23524243
2373	30	4.426625	21.7158	0.082842	0.21222704	0.237723136
2374	33	3.383857	21.69967	0.079032	0.2024615	0.226842693
2375	30	4.426625	20.55268	0.086429	0.22101947	0.25117183
2376	40	3.72159	22.38763	0.07008	0.17970494	0.199699825
2377	34	6.375457	23.0751	0.074262	0.1906082	0.210118348
2378	27	9.615189	19.02027	0.096705	0.24664374	0.285972127
2379	13	44.60505	14.67312	0.17019	0.42972172	0.532971986
2380	30	23.37634	17.84607	0.096356	0.24519046	0.288775666
2381	42	4.426625	23.466	0.065958	0.16937922	0.185922107
2382	42	2.125179	24.45081	0.063902	0.16430187	0.178467003
2383	40	4.426625	24.1401	0.066129	0.16996156	0.185196798
2384	35	3.383857	22.19251	0.075425	0.19335897	0.215374238
2385.5	24	3.055612	20.31366	0.097505	0.24924582	0.284158564

5. Conclusion

Determination of reservoir characteristics and assessment of reserves are very essential and essential to the decision making of a reservation. Estimation offers the trust that producing from a reservoir is economically feasible, whether it is capable of price or not. Estimation of a hydrocarbon reserve is a complex process because it involves integrating geologic and engineering data. In this study, hydrocarbon i.e., gas bearing zones have been detected using composite log interpretation and found between (1585-1595) m, (1600-1610) m, (2409-2420) m, (2265-2270) m, (2295-2303) m, (2435-2450) m, (2465-2480) m, (3150-3155) m, and (2955- 2960) m for well 4 and between (2365-2385.5) m, (1820-1830) m, (1782- 1803) m, (305-308) m, (336.5-338.5) m, and (607.5-609.5) m for well 5. In this research different methods are used for estimating some of the reservoir properties of Fenchugonj gas field well -4 and well-5. The water saturation was estimated by three different models as Archie's, Indonesia and Simandoux method, respectively. For well 4, the value of average water saturations found from Archie, Indonesia and Simandoux model were 7.3% ,18%, 20.51% respectively and for well 5 the value of average water saturations was found as 8.5% ,22% and 24.84% respectively. Results showed that the water saturation obtained from Archie's formulae was lower than the value of Indonesia and Simandoux model.

Acknowledgement

The authors would like to acknowledge Universiti Teknologi PETRONAS (UTP) for providing a conducive environment to do research and its Graduate Assistant (GA) scheme for financial support.

References

[1] Wiki A 2017 *Well Log Analysis for Reservoir Characterization*
[2] Dictionary C E 2014 *Collins english dictionary* Complete and Unabridged.
[3] Serra O 2008 *Well logging handbook.* Editions Ophrys.
[4] Mitchell J V 2004 *Petroleum reserves in question.* Royal Institute of International Affairs.
[5] Horsfall O I, Davies D H and Davies O A 2015 Hydrocarbon reservoir characterization using well log in Niger Delta Basin of Nigeria *International Journal of Applied and Natural Sciences* **4(5)** 55-64.
[6] Islam A R M T, Habib M A, Islam M T and Mita M R 2013 Interpretation of wireline log data for reservoir characterization of the Rashidpur Gas Field, Bengal Basin, Bangladesh *IOSR Journal of Applied Geology and Geophysics* **1(4)** 47-54.
[7] Quader A K M A and Gomes E 2002 *CPD Occasional Paper Series.*
[8] Chowdhury S and Chowdhury M 2018 Natural Gas Properties Analysis of Bangladesh: A Case Study of Fenchuganj Gas Field *IOSR Journal of Applied Geology and Geophysics* **66** 1-9.
[9] Deb P K, Howladar M F, Miah M I, Faruque M O, Islam M S, Mia M B and Quamruzzaman C 2014 Structural interpretation of Fenchuganj gas field *J. Eng. Geol. Hydrogeol* **2** 29-36.
[10] Mamud M L, Woobaidullah A S M, Baki M A, Chowdhury R C, Sazal M Z H, Islam M S and Khan M A 2016 Structural Modeling of Fenchugang Gas Field, Sylhet, Bangladesh *International Journal of Emerging Technology and Advanced Engineering* 6(3) 29-39.
[11] Asquith G B, Krygowski, D and Gibson C R 2004 *Basic well log analysis (Vol. 16)* Tulsa: American Association of Petroleum Geologists.
[12] Passey Q R, Yin H, Rendeiro C M and Fitz D E 2005 Overview of high-angle and horizontal well formation evaluation: issues, learnings, and future directions *SPWLA 46th Annual Logging Symp* (*New Orleans*).
[13] Gibson C R 1982 *Basic well log analysis for geologists.* American Association of Petroleum Geologists.

Eco-Friendly Hybrid Solar Cells using *Eugenia Claviflora* as New Alternatives Sensitizer to TiO_2/PEDOT:PSS

N H Kamarulzaman[1,4], H Salleh[2*], A N Dagang[1], M S M Ghazali[2], N Ishak[1], M Abdullah[4], S M Muda[4]

[1] School of Ocean Engineering, Universiti Malaysia Terengganu, 21030 Kuala Nerus, Terengganu, Malaysia.

[2] Centre of Fundamental and Liberal Education, Universiti Malaysia Terengganu, 21030 Kuala Nerus, Terengganu, Malaysia.

[3] School of Fundamental Science, Universiti Malaysia Terengganu, 21030 Kuala Nerus, Terengganu, Malaysia.

[4] Faculty of Applied Sciences, Universiti Teknologi MARA, 23200, Dungun, Terengganu, Malaysia.

*Corresponding author: hasiahsalleh1@gmail.com

Abstract. Studies on HSCs have been focussing on exploration of natural dyes obtained from *Eugenia claviflora* (EC). This work mainly aimed to enhance device efficiencies related to new dye sensitizer particularly *Eugenia claviflora*. These samples were then characterised based on morphology, thickness, optical absorption, electrical conductivity and solar cell's performance in term of efficiency, voltage and current using various techniques including FESEM, EDS, profilometer, UV-Vis, FTIR and autolab with irradiance of 100 mW/m^2. TiO_2/PEDOT:PSS/ EC obtained an open –circuit voltage (Voc) of 0.43 V with fill factor of 0.2 leading to an efficiency of 1.33%. The absorption of TiO_2/PEDOT:PSS/EC was located at 550 nm (between 450 nm to 755 nm) which exhibited an intense broad band in visible light. Hence, fabrication of HSCs was successful thus improving its' performance.

1. Introduction

Numerous researches focused on Hybrid Solar Cell (HSC) and Dye-Sensitized Solar Cell (DSSC) in photovoltaic systems. The performance of DSSC and HSC were still limited as compared to silicon solar cell. This possibly is enhanced by using proper selection of material, design and energy gap. Problems also can be filtered out by the development of dye (sensitizer) in the HSC. Rutheniums are usually used in DSSC and HSC research's as synthetic dye and are reported to be more efficient plus possessed higher efficiency when combined with nanoporous TiO_2 electrode [1]. Meanwhile, Rutheniums are undesirable to environment safety due to the chemical used and varieties of procedure to be done. It is time consuming thus production become very expensive. Due to this problem, natural dye has been discovered as the most promising material presents to next generation of solar cells.

As compared to such synthetic dye, dye extracted from natural resources are more desireable due to its abundant in resources, natural extraction i.e stems, flowers, fruits and etc. Additionally, the advantages also include less effect to environment, simple preparation technique applications', low production cost, easier handling process and low toxicity [2]. Interestingly, investigation of various dyes obtained is from waste material. To date, *Eugenia claviflora* was used as a new sensitizer, eventhough the efficiency is still limited due to degradation of dye. It can be overcome by selection of the dye itself plus the achitectural design of HSC. Stability problem of dye is because of degradation of dye in the presence of sunlight radiation. It has been also suggested that combination of dye with semiconductor material and conductive polymer will provide high conductivity and stability [3]. Recently, this understanding allows further development in designing thin film to improve performance of HSC. Hybrid solar cell is the combination of organic and inorganic material, and is one of the major concerns in solar energy technologies. Metal oxide semiconductor consist of TiO_2 [4], ZnO [5, 6], $CuSO_4$ [7] and FeO_3 [8] have been studied extensively in hybrid solar cells. Among them, TiO_2 is the most promising acceptor materials for hybrid solar cells, owing to its environment-friendly and easy availablibility. Currently, one of the most promising conductive polymer is Poly(3,4-ethylenedioxythiophene): polystyrenesulfonate (PEDOT: PSS) because of its higher light transmittance, low cost in production process, easy device processability with potential applications in photovoltaic [9].

This work mainly aimed to enhance device efficiencies related to new dye sensitizer particularly *Eugenia claviflora*. However, understanding solar cell mechanism and moderate energy conversion efficiencies evolved as major challenges encountered in application of thin film solar cells based HSCs. *Eugenia claviflora* was chosen for this research to investigate the ability of this waste as a natural dye towards electrical properties for the purpose of solar application. Dye extraction from *Eugenia claviflora* was a rich source of anthocyanin which is widely used as antioxidant properties. These pigments differ from other flavonoid due to its positive charge in the central ring structure that provides color intensity to plants and fruits. Hybrid solar cells are formed with inorganic materials (titanium dioxide) and conjugated polymer of PEDOT:PSS. To this date, there is not yet a research on hybrid solar cells using *Eugenia clavifora* as photosensitizer. Recently, this understanding allows further development in designing the thin film to improve performance of HSCs.

2. Experimental Details

2.1. Material's Preparation

First, ITO substrates (sheet resistance 7Ω/sq with dimension of 2 cm x 2 cm) were cleaned in detergent, acetone (R&M Chemical), isopropanol (R&M Chemical) and distilled water. And they were ultrasonic (JEIOTECH) cleaned to remove residual organics. Later, the ITO coated glass substrates were dried using dryer before being kept in a Petri dish. Natural dye, *Eugenia claviflora* (EC) was chosen based on colour observed which was purplish blackish (figure 1). Then, the skin of EC were cleaned with distilled water and dried at $60^{\circ C}$. After, they were crushed into fine powder and 10 g sample of powdered pulp were placed into 100 ml of ethanol (R&M Chemical). The mixture was then kept for a week at room temperature, filtered and extraction's solution was collected. The solution was then placed in an ultrasonic bath for 10 minutes around temperature of 30 $^{\circ C}$ leaving it to become apparently concentrated. The concentrated solution was eventually used as a dye to TiO_2 nanoparticles for the fabrication of solar cell [1, 3].

Figure 1. *Eugenia claviflora* fruits

2.2. Fabrication and Characterization of HSC

0.2M of TiO_2 colloid was prepared using TiO_2 powder (anatase, 25 nm by Sigma Aldrich Chemical) with ethanol (95%, HmbG Chemicals). This sample was then deposited using doctor blade technique onto the ITO glass at 45o angle. These substrates were afterward annealed for 30 mins at 450 ^{oC} on a hot plate. Sample was allowed to cool at room temperature and then transferred to glove box for subsequent procedures. PEDOT:PSS was spin coated on a prepared TiO_2 film at 2000 rpm for 10 s with a thickness of 28 nm approximately [13, 35]. The films were dried for 15 mins and then immersed into EC's solution for 5, 10 and 15 minutes respectively. Lastly, Aluminum (Al) electrode was deposited on top of the films. Physical vapour deposition (PVD) was used to control evaporation thickness. Figure 2 shows schematic diagram of hybrid solar cell.

Fourier-transform infrared (FTIR spectroscopy IRTracer-100, Shimadzu) was used to study the functional group. Surface morphology images of the films were characterized using FESEM. Absorption spectra and thickness of these films were measured via UV-Vis spectrometer (Model Lambda 25 with UV Winlab V2.85 software, Perkim Elmer) and Dektak 150 surface profiler, respectively. The electrical conductivity was measured with four point probes, under dark and light condition.

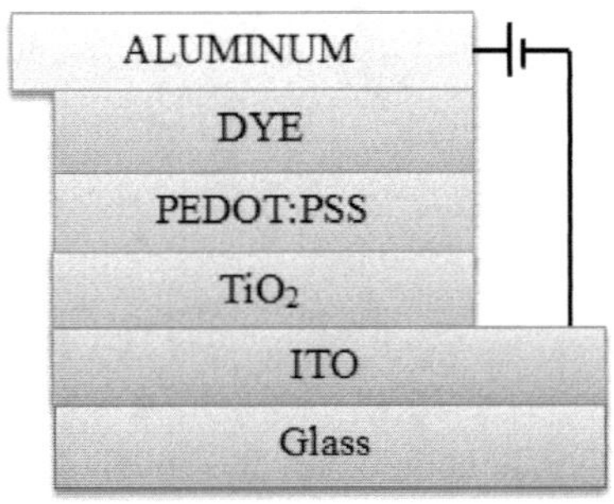

Figure 2. Schematic diagram of hybrid solar cell

3. Experimental Details

3.1. Absorption Spectra

Optical properties of *Eugenia claviflora* were investigated using UV-Vis spectrometer. Figure 3 presents the absorption spectra for EC dye extract, TiO_2/PEDOT and TiO_2/PEDOT:PSS/EC. Figure 3 clearly shows that *Eugenia claviflora* exhibits broad absorption band at wavelength from 450 nm to 600 nm with its peak allocated at 530 nm. The color observed by *Eugenia claviflora* was purple however; the pigment absorbs green color in the range of 500 nm to 560 nm. Cyanidin 3-glucoside was a pigment in anthocyanin found in *Eugenia claviflora*. In addition, the highest concentration of cyanidin found was in seeds and skin of the fruits [10]. The broadest absorption spectrum was *Eugenia claviflora* due to colour ranges that produce difference absorption characteristics [11]. Wu et al stated that wavelength of TiO_2 was recorded 380 nm in ultraviolet (UV) light spectrum [12]. TiO_2 showed high transparency in visible region and a sharp fall in UV region yet it does not absorb in visible region. Hence, absorption observed at 350 nm to 600 nm was due to PEDOT:PSS.

The absorption of anthocyanin onto the surface of TiO2/PEDOT:PSS was due to the presence of carbonyl and hydroxyl group on anthocyanins. Electrons from dye molecule were excited to transfer to conduction band of TiO_2. Thus, it will enhance the efficiency of hybrid solar cell. The absorption of EC dye was obtained to be red shifted at 550 nm upon chelation to TiO_2/PEDOT:PSS.

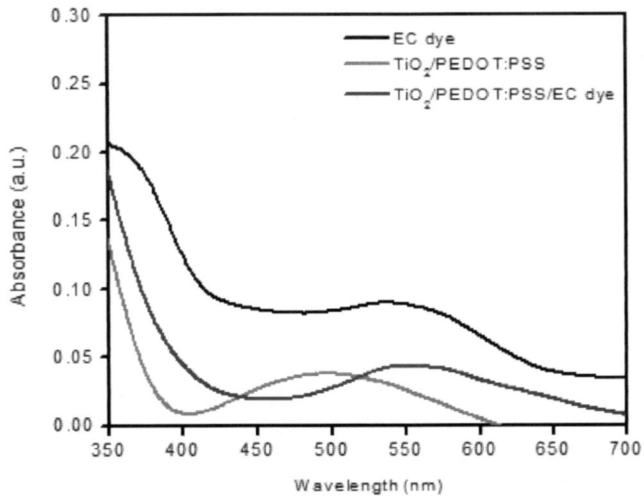

Figure 3. Absorption spectra of EC dye, $TiOTiO_2$/PEDOT:PSS without dye and TiO_2/PEDOT:PSS/EC dye

3.2. *Energy Gap of TiO_2, PEDOT:PSS and EC dye*

Figure 4 exhibits formalism showing sensitization mechanism. Energy gap of EC dye was calculated at 2.04 eV with anthocyanin pigments that are effective in electron transfer [3]. The lowest energy gap will promote electrons from dye molecule easily excite to another level. Absorption of the ultra-violet radiations results due to the excitation of the electrons to transfer from ground state to higher energy state.

The attachment of EC dye and TiO_2 has been suggested to possess an effect on HOMO and LUMO level of anthocyanin due to the decreasing of energy gap which results in the red shifted's absorption peak. The energy level of the *Eugenia claviflora* (-2.81 eV) was far closer to HOMO energy level of PEDOT:PSS (-3.5 eV) and conduction band of TiO_2 (-4.2 eV) . The closeness of the energy levels between the *Eugenia claviflora,* PEDOT:PSS and TiO_2 are beneficial for a better hole extraction to the ITO. An efficient hole extraction reduces charge recombination thus improving hole collection efficiency.

The lowest unoccupied molecular orbital (LUMO) energy levels of *Eugenia claviflora* calculated as 2.87eV. Energy of conduction band (CB) for TiO_2 and HOMO energy for PEDOT:PSS were obtained at -7.4eV and -5.1eV, respectively. The valence band (VB) and LUMO energy for TiO_2 and PEDOT:PSS were recorded at -4.2eV and -3.5 eV, respectively. When dye molecules (*Eugenia claviflora*) on TiO_2 absorb incident photons, electrons in the molecules become excited and promoted from photosensitizer (*Eugenia claviflora*) which is HOMO to LUMO [3, 13]. Then, positive charge (electron hole) remained in HOMO. Consequently, molecule oxidation process were performed by these resulting holes. Then, electrons were transfered from dye molecule (*Eugenia claviflora*) to the conductive polymer (PEDOT:PSS) and TiO_2 conductive band (CB). It can be concluded that *Eugenia claviflora* exhibits proper alignment of excited state's energy level between dye and conduction band edge. This means, decreased of energy gap were enhanced by the photo response of titania thin film and due to the excellent conjugated polymer of PEDOT:PSS.

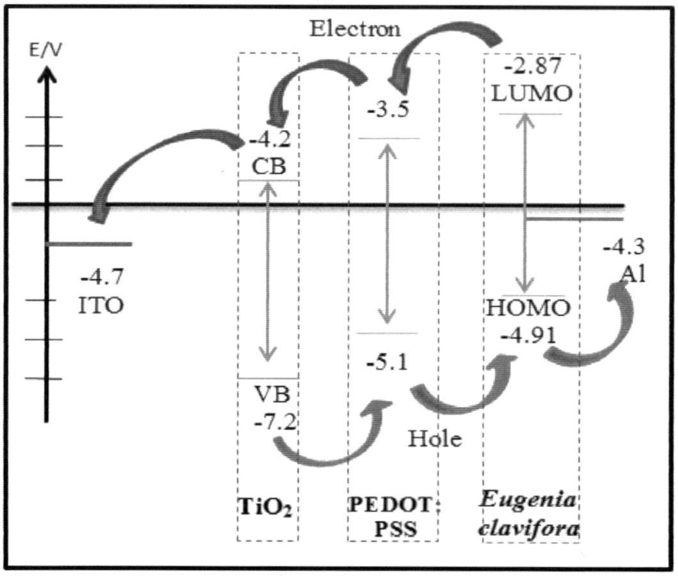

Figure 4. Formalism showing sensitization mechanism

3.3. Dyes Structure

Eugenia claviflora presents C-H aliphatic and 3335 cm^{-1} as shown in Figure 5. Observed strong bands at 1080 cm^{-1} correspond to the symmetric vibrational mode of C-O stretching mode. Structure of dye provides the significant impact to both dyes' efficiencies. However, the stability of anthocyanin was reduced due to the methoxyl contained in isoflavones. Meanwhile, hydroxyl group (-OH) exists in flavones increased anthocyanin's stability [14, 15]. As referring to FTIR analysis, Carbonyl (C=O) and -OH contained in anthocyanin able to stabilize excited state and develop maximum absorption by having lower energy [16]. Al-Alwani et. al also was agreed with the statement of all pigments in dye can absorb light photon and lead to production of excited electrons [17]. Consequently, the broadest region in -OH group was quite important to maintain the stability of all dyes and can be promoted to be applied in excitation of electron.

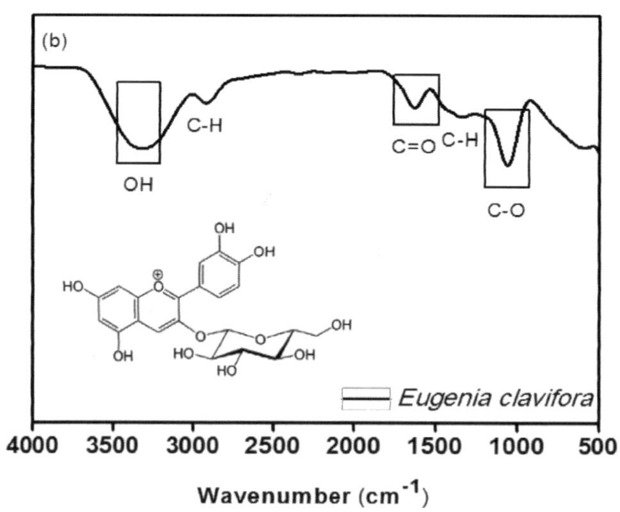

Figure 5. FTIR Spectra of Natural Dye Solution Extracted from *Eugenia claviflora*

3.3.1 Morphological Studies of EC Dye in Hybrid Thin Film. Particles' size are estimated to be in a range of 60 nm to 100 nm as observed in figure 6. It can be seen that the surface morphology differs a bit in TiO_2/PEDOT:PSS and TiO_2/PEDOT:PSS/EC. Moreover, uniform distribution of surface potential in TiO_2/PEDOT:PSS implies that the nanomaterial is coated uniformly by PEDOT:PSS in accordance with observation from FESEM images.

Thickness of TiO_2/PEDOT:PSS increased to 74 nm. Meanwhile, thickness of TiO_2/PEDOT:PSS/EC was recorded at 162.9 nm. This increase was due to the amount of *Eugenia claviflora* loading onto the surface of TiO_2/PEDOT:PSS. Therefore it can be confirmed that *Eugenia claviflora* was detected during fabrication of thin film. Table 1 shows sulphur and carbon elements were observed clearly in the EDS analysis. Therefore, both elements were detected in TiO_2/PEDOT:PSS. Other than that, mass percentage of carbon were also increase for TiO_2/PEDOT:PSS/EC compared to TiO_2/PEDOT:PSS which was depicted in table 1. Hence, a highly magnified FESEM image of TiO_2/PEDOT:PSS/EC suggests that it consist of PEDOT:PSS and *Eugenia claviflora* in the fabrication of thin film. This results were also supported by the thickness incrementation for TiO_2/PEDOT:PSS and TiO_2/PEDOT:PSS/EC.

Fabrication of all devices with each layer thickness was recorded below than 100 nm, leading to a shorter exciton diffusion length. This shorter exciton length was necessary in connection with strong light absorption and due to this requirement, the interface between electron accepting and hole accepting material received more volume heterojunction. Consequently, this allows the highest performance of power conversion efficiency to be achieved in this fabrication [18].

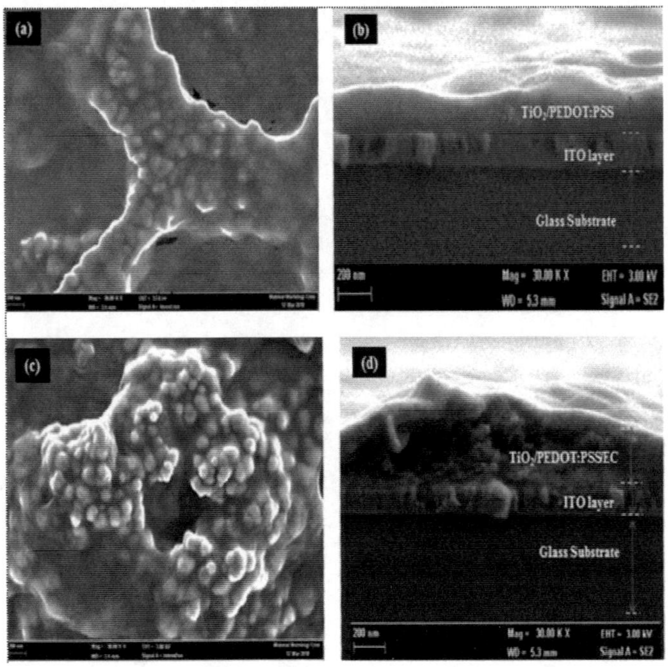

Figure 6. (a)Surface morphology of TiO_2/PEDOT:PSS without dye (b) cross section of TiO_2/PEDOT:PSS without dye (c) surface morphology of TiO_2/PEDOT:PSS with dye EC (d) cross section of TiO_2/PEDOT:PSS with dye EC

According to Kalaiselvi and Murtaza, high efficiency can be achieved if the thickness of active layers is less than 500 nm because of enough sunlight absorption by the sample. [18-20]. If immersion time is longer, thickness of the sample will increased and correspond to low efficiency performance. Sengupta et.al also agreed with the statement and stated that, with respect to the thickness of hybrid solar cells, the focus has been done to control the absorption of dye molecules within semiconductor, electronic charge transport as well as may contribute to enhance in power conversion efficiency of the device [21].

Table 1. EDS of TiO_2/PEDOT:PSS with and without dye EC

Element	Mass (%) TiO_2/PEDOT:PSS without dye	Mass (%) TiO_2/PEDOT:PSS/EC dye
In	34.0	37.2
O	22.9	19.7
C	12.1	14.3
Si	8.5	11.2
Ti	16.5	10.8
Sn	5.7	6.4
S	0.3	0.4

3.4. Electrical conductivity
Figure 7 illustrates the electrical conductivity under light condition for bilayer heterojunction thin film of TiO_2/PEDOT:PSS and TiO_2/PEDOT:PSS/EC with conductivity being calculated at 0.028 Scm^{-1} and 0.045 Scm^{-1} respectively. Electrical conductivity of the TiO_2/PEDOT:PSS/EC was the highest compared to TiO_2/PEDOT:PSS with increment of 60.7%. This is due to the absorption of light by the samples since anthocyanin in *Eugenia claviflora* absorb more light from the sunlight and convert it into electricity. Electrical conductivity was increased under light condition because the samples absorbed light energy and change it into electrical energy [22]. This light energy was absorbed in the visible region by anthocyanin pigment. Meanwhile, TiO_2 absorbed energy from sunlight in infrared region.

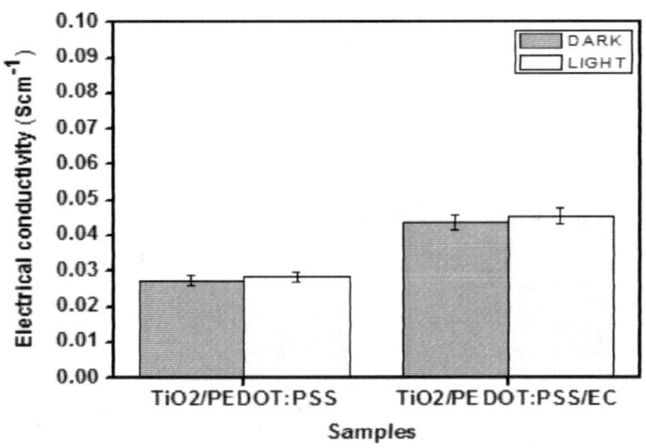

Figure 7. Electrical conductivity of TiO_2/PEDOT:PSS with and without dye EC at different condition

3.5. Performance of Hybrid Solar Cells
Performance of HSC was carried out and the performance of hybrid solar cell is generally evaluated by the current–voltage (I–V) measurements [28, 29].

3.5.1 Effect of Dye Loading Period on the Photovoltaic Performance. Table 2 represents power conversion efficiency (PCE) for TiO_2/PEDOT:PSS/EC at different dye loading period. Immersion time for natural dye plays an important role in determining the best efficiency. After deposition of TiO_2 and PEDOT:PSS, samples were immersed in natural dye solution for three consecutive time; i) five minutes, ii) ten minutes and, iii) fifteen minutes. Futhermore, film absorption of dye molecule depends on the dye loading period [23-25]. Thus, more electrons were excited to transfer and emit. Lower performance obtained at 5 minutes and 15 minutes compared to 10 minutes caused by reduction in dye concentration [26]. Thickness sample will increase with longer immersion time and may also worsen cell performance. Therefore, through this approaches 10 minutes of immersion time was proceed to next investigations.

Table 2. Efficiency of TiO_2/PEDOT:PSS with and without dye EC at different dye loading period

Samples	Dye loading period (mins)	PCE (%)
TiO_2/PEDOT:PSS	0	0.005±0.001
TiO_2/PEDOT:PSS/EC	5	0.57±0.25
	10	1.33±0.39
	15	0.21±0.07

Photovoltaic parameters includes short circuit current (I_{sc}) [27], open circuit voltage (V_{oc}), fill factor (FF) [28] and efficiency (PCE) were presented in Table 3. As shown in Table 3, TiO_2/PEDOT:PSS/EC obtained the best performance with I_{sc} of 2.5E-5 A, V_{oc} of 0.43V, FF of 0.2 and leading to an efficiency of 1.33%. It was reported that photoelectric conversion efficiency and fill factor are the most important parameters to describe solar cell performance [28]. The value of FF, V_{oc}, and I_{sc} were decreased equivalently to low PCE performance due to interfacial recombination losses [29]. Highest efficiency of TiO_2/PEDOT:PSS/EC cause by its better photon response in visible spectrum and reducing recombination of electron – hole throughout the device [30].

Table 3: Power conversion efficiency of TiO_2/PEDOT:PSS with and without dye EC

Samples	J_{sc} (mA/cm^2)	I_{sc} (μA)	V_{oc} (V)	FF	PCE (%)
TiO_2/PEDOT: PSS	0.058	0.125	0.4	0.17	0.005± 0.001
TiO_2/PEDOT: PSS/EC	0.12	0.025	0.43	0.20	1.33±0. 39

4. Conclusion

Low energy gap of TiO_2/PEDOT:PSS/*EC* was found as the best material with lower energy gap compared to other samples. In order to obtain greater understanding on performance of HSCs, efficiency measurement was performed. Among all devices, TiO_2/PEDOT:PSS/*Eugenia claviflora* demonstrated as the best device performance; with fill factor of 0.2 and an overall efficiency of 1.33%. The combination of TiO_2/PEDOT:PSS/*Eugenia claviflora* may reveals the best efficiency with increasing conductivity among all samples. Efficiency value is nearly dependent on the FF value which is directly related to device capability of light absorption.

Furthermore, performance of HSC also depends on a broader wavelength in absorption spectrum. Wavelength of TiO_2/PEDOT:PSS/*Eugenia claviflora* ranging from 400 nm to 750 nm exhibits higher efficiency due to higher broadband absorption value. Consequently, these devices are capable to generate more carriers. These carriers were important in excitement of electrons transferred from ground state to a higher state. Eventhough this device exhibits lower efficiencies than silicon solar cells, its efficiency is still encouraging and can be enhanced by further researches. Data and knowledge acquired possibly be used for other researches in HSC. Studies on new sensitizers are still vastly unexplored which left for future exploration.

Acknowledgement
The authors would like to thank Universiti Malaysia Terengganu (UMT), Fundamental Physics Laboratory UMT's members and Nano Research Team (UMT) for supporting in this research. We also acknowledge the financial support from UiTM and Ministry of Education Malaysia (SLAB/SLAI).

References

[1] Sengupta D, et al 2015 Visible light absorption and photo-sensitizing properties of spinach leaves and beetroot extracted natural dyes Spectrochim *Acta A. Mol. Biomol. Spectrosc* **148** p 85-92.
[2] Richhariya G et al 2017 Natural dyes for dye sensitized solar cell: A review. *Renewable and Sustainable Energy Reviews* **69** p 705-718.
[3] Jaafar S N H et al 2017 Natural dyes as TiO_2 sensitizers with membranes for photoelectrochemical water splitting: An overview *Renewable and Sustainable Energy Reviews* **78** p 698-709.
[4] Ravirajan P et al 2004 Hybrid nanocrystalline TiO_2 solar cells with a fluorene–thiophene copolymer as a sensitizer and hole conductor *J. Appl. Phys.* **95(3)** p 1473-1480.
[5] Beek W J et al 2005 Hybrid zinc oxide conjugated polymer bulk heterojunction solar cells *J. Phys. Chem. B* **109(19)** p 9505-9516.
[6] Oosterhout S D et al 2009 The effect of three-dimensional morphology on the efficiency of hybrid polymer solar cells *Nature materials* **8(10)** p 818.
[7] Hung L I et al 2010 Room-Temperature Formation of Hollow Cu2O Nanoparticles *Advanced Materials* **22(17)** p 1910-1914.
[8] Han Z et al 2010 Synthesis and photovoltaic property of poly (3-octylthiophene)/titanium dioxide/ferric oxide composite *Synth. Met.* **160(19-20)** p 2167-2174.
[9] Thomas J P et al 2014 High-efficiency hybrid solar cells by nanostructural modification in PEDOT: PSS with co-solvent addition *J. Mater. Chem. A Mater.* **2(7)** p 2383-2389.
[10] Marcin H, Ryszard K, Anna G and Henryk D 2008 Anthocyanins of Fruits and Vegetables-Their Occurrence, Analysis and Role in Human Nutrition *Vegetable Crops Research Bulletin* **68** p 5-22.
[11] Wongcharee K et al 2007 Dye-sensitized solar cell using natural dyes extracted from rosella and blue pea flowers *Solar Energy Materials and Solar Cells* **91(7)** p 566-571.
[12] Wu J et al 2013 An ultraviolet responsive hybrid solar cell based on titania/poly(3-hexylthiophene) *Sci. Rep.* **3** p 1283.
[13] Amirtharajan S et al 2015 Electrical investigation of TiO_2 thin films coated on glass and silicon substrates—effect of UV and visible light illumination *Appl. Nanosci.* **6(4)** 591-598.
[14] Marpaung A M et al 2015 Spectral Characteristics and Color Stability of Melastomataceae and Clitoria ternatea L. Extracts *Innovative ASEAN Food Research towards the World.*
[15] Vadwala Y and Kola N 2015 Natural dyes extracted from waste leaves of *terminalia catappa* locally known as tropical almond and its application on silk fabrics pretreated with eco friendly and noneco-friendly mordants *International Journal of Research* **5** p 125-137.
[16] Syafinar R, Nair G, Irwanto M, Fareq M and Irwan Y M 2015 Optical Characterization using Nature Based Dye Extracted from Hibiscus's Flower *Asian Research Publishing Network (ARPN)* **10** p 6336-6340.
[17] Al-Alwani M A M, Mohamad A B, Kadhum A A H and Ludin N A 2015 Effect of Solvents on the Extraction of Natural Pigments and Adsorption onto TiO_2 for Dye-Sensitized Solar Cell Applications *Spectrochim. Acta A. Mol. Biomol. Spectrosc.* **138** p 130-137.
[18] Gondek E et al 2012 Influence of TiO_2 nanoparticles on the photovoltaic efficiency of the ITO/PEDOT: PSS/fluorine copolymers/polythiophene: TiO_2/Al architecture *J. Mater. Sci.*: Materials in Electronics **23(11)** p 2057-2064.

[19] Kalaiselvi C R et al 2018 Importance of halide perovskites for next generation solar cells – A review *Materials Letters* **219** p 198-200.
[20] Murtaza G and Ahmad I 2011 First principle study of the structural and optoelectronic properties of cubic perovskites CsPbM3 (M= Cl, Br, I) *Physica B: Condens. Matter* **406(17)** p 3222-3229.
[21] Sengupta D et al 2016 Effects of doping, morphology and film-thickness of photo-anode materials for dye sensitized solar cell application – A review *Renewable and Sustainable Energy Reviews* **60** p 356-376.
[22] Salleh H et al 2014 Study the Electrical Properties and the Efficiency of Polythiophene with Dye and Chlorophyll as Bulk Hetero-Junction Organic Solar Cell *Advanced Materials Research* **895** 513-519.
[23] Ruhane T A et al 2017 Photo current enhancement of natural dye sensitized solar cell by optimizing dye extraction and its loading period *Optik - International Journal for Light and Electron Optics* **149** p 174-183.
[24] Hossain M K et al 2017 Effect of dye extracting solvents and sensitization time on photovoltaic performance of natural dye sensitized solar cells *Results in Physics* **7** p 1516-1523.
[25] Kim J-K et al 2010 The analysis of the change in the performance and impedance of dye-sensitized solar cell according to the dye-adsorption time *Curr. Appl. Phys.* **10(3)** p S418-S421.
[26] Chou T P et al 2007 Effects of dye loading conditions on the energy conversion efficiency of ZnO and TiO_2 dye-sensitized solar cell *J. Phys. Chem. C* **111(50)** p 18804-18811.
[27] Paredes-López O and Delgado-Vargas F 2003 *Natural colorants for food and nutraceutical uses* CRC Press LLC.
[28] Azrul L et al 2014 Tannins quantification in Terminalia catappa leaves extract and antihelmenthic potential evaluation *Journal of Natural Products* **7**.
[29] Hochbaum A I and Yang P 2019 Semiconductor nanowires for energy conversion *Chem. Reviews* **110(1)** p 527-546.
[30] Saleem H and Habib A 2016 Study of band gap reduction of TiO_2 thin films with variation in GO contents and use of TiO_2/Graphene composite in hybrid solar cell *J. Alloys Compd.* **679** p 177-183.

AUTHOR INDEX